ÉTUDES ON THEORETICAL PHYSICS

ÉTUDES ON THEORETICAL PHYSICS

Collection of Works Dedicated to 65th Anniversary of the Department of Theoretical Physics of Belarusian State University

Editors

L M Barkovsky
I D Feranchuk
Belarusian State University, Belarus

Yakov M Shnir
University of Oldenburg, Germany

World Scientific

NEW JERSEY • LONDON • SINGAPORE • BEIJING • SHANGHAI • HONG KONG • TAIPEI • CHENNAI

Published by

World Scientific Publishing Co. Pte. Ltd.

5 Toh Tuck Link, Singapore 596224

USA office: 27 Warren Street, Suite 401-402, Hackensack, NJ 07601

UK office: 57 Shelton Street, Covent Garden, London WC2H 9HE

British Library Cataloguing-in-Publication Data
A catalogue record for this book is available from the British Library.

ETUDES ON THEORETICAL PHYSICS
Collection of Works Dedicated to 65th Anniversary of the Department of Theoretical Physics of Belarusian State University

ISBN 981-256-081-5

Printed in Singapore.

FOREWORD

This special volume devoted to the 65th anniversary of the Department of Theoretical Physics of the Belarusian State University presents the papers written by its leading scientists, researchers of the department as well as its graduates. The department was established in 1938 and its first head was Professor F.I. Fedorov who in fact became the founder of the School of Theoretical Physics in Belarus. His fundamental works on crystallooptics and acoustics, field theory and mathematical physics have received worldwide recognition. He also established high standards for teaching theoretical physics as the basis of physics education at the Belarusian state University — one of the universities in the former USSR. To do justice to F.I. Fedorov's contribution to the various fields of theoretical physics, one of the papers is devoted to his life and scientific career.

All other heads of the department followed the best traditions established by F.I. Fedorov and at the same time introduced their individual priorities in the researching and educational directions. In 1962 Professor A.E. Levashev headed the department during this period. The investigations in the field of gravitation theory were developed more actively. A number of important results in the quantum theory of condensed matter were obtained by scientists and researchers of the department when Professor L.I. Komarov was appointed its head in 1973. Professor G.V. Shishkin (head from 1982) made a considerable contribution to the development of the studies connected with the analysis of the Dirac equation in the general relativity. Professor L.M. Barkovsky headed the department in 1983 and, at the present moment, fundamental and applied problems of the electromagnetic and acoustic studies on the wave interaction with complex media are actively carried out under his supervision.

The papers in this jubilee edition cover a wide variety of research topics of modern theoretical physics. They have been researched for many years by the leading professors of the department and well-known scientists who graduated from it and are now working in research centers all over the world. Each contribution to this volume represents a brief review of the methods and techniques, numerical simulations and results in the corresponding fields of theoretical physics. We hope that this collection of papers will be of interest to scientists, researchers, post-graduates and students because the discussion of the qualitative and physical peculiarities of different problems was the focus of each author.

On behalf of all the authors of this edition, we would like to express sincere gratitude to Le Van Hoang, Yuri Zeldin and Mikhail Shurim who had graduated from the Department of Theoretical Physics but had remained faithful to their "alma mater" by supporting this project.

Editors
L. M. Barkovsky
I. D. Feranchuk
Ya M. Shnir
Minsk 20.6.2004

CONTENTS

Part II: Mathematical Methods in Theoretical Physics

Part III: Condensed Matter and Complex Systems

F.I. FEDOROV'S CONTRIBUTION TO THE CREATION AND DEVELOPMENT OF THEORETICAL PHYSICS IN BELARUS

Tatsiana A. Alexeeva†, Leonid M. Barkovsky†, Andrey A. Bogush‡
†*Department of Theoretical Physics, Belarusian State University,*
4, Fr. Skarina Ave., Minsk, 220050, Belarus
E-mail: Alexeeva@bsu.by; Barkovsky@bsu.by
‡*B.I.Stepanov Institute of Physics, National Academy of Sciences of Belarus,*
68, Fr. Skarina Ave., Minsk, 220072, Belarus
E-mail: bogush@dragon.bas-net.by

The Department of Theoretical Physics of the Belarusian State University was founded by Academician Fedor Ivanovich Fedorov (1911–1994), a legendary figure of the Soviet physics community. Nowadays one can regard the year 1938 as an advent of the belarusian theoretical physics school which became well-known all over the world. F.I.Fedorov headed the Department for 25 years and worked there till the last days of his life. Fedorov had a wide range of scientific interests which essentially determined the research branches for the Department. The subjects of scientific research there comprise many modern problems of optics, acoustics, quantum physics, elementary-particle and high-energy physics as well as theory of gravitation, astrophysics and cosmology. Today there are a lot of scientific centers in Belarus, which carry out theoretical physics research. Highly qualified specialists work at these centers. They are graduates of the department of theoretical physics. Many of them were students of F.I.Fedorov.

F.I.Fedorov graduated from the Belarusian State University in 1931, and took up a post-graduate course at Institute of Physics of Leningrad State University (1933–1936) under the supervision of a well-known physicist-theoretician of high renown V.A.Fock. In 1936 F.I.Fedorov successfully defended his Ph. D. thesis devoted to electromagnetic interactions of elementary particles. In 2004 it is the 50th anniversary since he had successfully acquired his habilitation degree[1]. The title of the thesis was

"Invariant methods in optics of anisotropic media". It had substantially influenced the development of many branches of theoretical physics, especially crystal optics. "He entirely reorganized the area of knowledge by his proceedings, saying the thing that had been expected for fifty years". It was said by a well-known Soviet physicist and historian of science T.P.Kravets during the defence. In his thesis F.I.Fedorov developed and applied a novel, general and effective approach called a covariant method, which is applicable to a wide range of theoretical physics problems.

F.I.Fedorov realized the significance of having effective methods of theoretical research[2] during his post-graduate studies. He shared Dirac's credo that physical laws must have beautiful mathematical formulation. His covariant method had been widely used in his numerous journal publications and 6 large original monographs[3–7]. The new approach allowed to consider many physical phenomena of quantum field theory, gravity, optics, acoustics, crystallography on a unified basis.

The specific point of Fedorov's method is the use of invariance and covariance concepts, underlying the symmetry properties of the studied physical phenomena. It is based on specially developed and adapted algebraic and geometric techniques that allowed to formulate a physical theory without using any coordinates and bases of multidimensional spaces. Valuable physical information can be extracted from general equations by a methodical use of invariant matrix properties belonging to the equations. Hence, the basic relations of the theory are formulated in a general form. They turn out to have rather a compact form which is easy to investigate to investigate and is convenient for physical interpretation. Mathematical operations, transformations and calculations are very often greatly simplified and shortened in that formalism. In his papers, F.I.Fedorov devoted special attention to optics and acoustics, which have been a proving ground for upgrading of the mathematical methods of physics for a long time. However, the use of the covariant method in different areas of theoretical physics turned out to be extremely effective. It is partly represented in the present collection "Etudes on theoretical physics".

F.I.Fedorov was adapt at attracting young physicist and infusing them with his enthusiasm and experience. In this way, he played a major role in the development of theoretical physics in Belarus. In the 40s of the last century he initiated at the Department the research on electromagnetic theory of light in anisotropic and gyrotropic media, as well as gravity and theory of elementary particles. At that time (1949) the general construction method of Green function for relativistic wave equations of the first order describing elementary particles with an arbitrary spin, interacting with an external electromagnetic field was suggested[8]. Relativistic wave equations of the 1-st order in the matrix form for a particle with the spin equal to 2 were obtained and investigated for the first time[9]. Three fundamental Fedorov's ideas are especially important: the general method of projection operators in the theory of elementary particles (1958); three-dimensional complex vector parametrization of finite transformations of Lorentz group (1958s–1962s); the universal nonlinear matrix equations for some interacting (self-acting) fields (1968). They underlay specific applications of the covariant approach to the theory of elementary particles and fundamental interactions problems (see contributions to this collection). The theory of relativistic wave equations of the first order on the basis of multiple representations of homogeneous Lorentz groups was developed. It enabled to give natural description of internal electromagnetic structure of elementary particles.

In the 50s of the last century F.I.Fedorov was particulary interested in the theory of crystals. To describe propagation of electromagnetic waves in such a medium he developed a new fundamental approach which is based on the notion of the full complex vector of electric (or magnetic) field without decomposition of it into components instead of the classical methods of describing polarization which originated in XIX century. The description of polarization in such formalism acquires an unusual form. However, it is quite convenient and general because it allows to determine the wave polarization state directly from Maxwell's equations where field vectors appear.

F.I.Fedorov was the first to use a coordinate-free approach for considering electromagnetic waves propagation in media with arbitrary anisotropy[10,11]. In these fundamental papers he introduced a refraction vector which is convenient for solving boundary problems of electrodynamics and optics and analyzed of the inverse boundary problems of crystal optics. Fresnel equation and eigenwave polarization equation were obtained and used for absorbing, magnetic and gyrotropy crystals of all types of symmetry.

F.I.Fedorov wrote a series of fundamental papers on the covariant optics of absorbing crystals. It was shown that the "transfer principle", which is included in classical monographs, was incorrect[12]. It was found that in crystals of the lowest syngonies the "screw" axes (or "circular" axes), may exist, along which light with arbitrary polarization can propagate. Such axes were named "isotropic optical axes". The singularities of refraction and absorption surfaces and the regularity of light propagation along "circular" axes were investigated[13]. Finally, it was established that instead of three types of absorbing crystals, as stated in previous theories, 16 types of such crystals may exist. They differ by the number and character of optical axes and by other optical properties as well.

Meanwhile the theory of inhomogeneous waves, which appear in the total reflection from the interface of a transparent isotropic medium, was developed. It was shown that for general polarization of the incident wave the average flux of energy in the second medium is directed at a certain angle to the plane of incidence. The consequence of this fact was the prediction of a lateral displacement of the reflected beam, accompanied by its exit from the plane of incidence. This interesting optical phenomenon was confirmed experimentally about 10 years later and was called "Fedorov shift"[14]. Nowadays many works are devoted to the theoretical and experimental study of this phenomenon.

For the first time the general and rigorous theory for transparent media with arbitrary anisotropy of magnetic permeability was formulated on the basis of covariant methods[15,16].

The consistent description scheme of optical properties of active crystals was formulated[5]. The electrodynamic law of energy conservation was first used by Fedorov as a condition that restricts the type of optical activity tensors and characterizes the phenomenon of optical activity[17]. It resulted in a rather general phenomenological derivation of constitutive equations for the electrodynamics of gyrotropic media with both dielectric and magnetic anisotropy. Covariant expressions for tensors describing the gyrotropy in all crystal classes, where applicable, were found. It was established that crystals of planar classes of middle syngonies, that were not included previously to the class of optically active media, do possess this property, though they do not rotate the plane of polarization. It was shown that such crystals differ from non-active crystals in the sense that light reflected from them is elliptically polarized in spite of the absence of absorption. Thus, it was proved that the number of crystallographic classes in which optical activity is possible is 18, and not 15, as it was considered earlier[18,19]. In 1976 Academician F.I.Fedorov was awarded the State Award of the USSR in the field of optics of anisotropic media.

The number of variants describing properties of optically active crystals shows the complexity of the theory. In[20] different variants were analyzed and it was noted that all of them prove to be the approximations of general unified methods, characterized by redefinition of electromagnetic field vectors. A discussion with the Nobel Prize laureate, academician V.L.Ginzburg took place[21–23] on the form of material equations for electrodynamics of gyrotropic media. This discussion came to an end when the opponents admitted the correctness of the concept defended in[24–26]. Presently the aspects stated in Fedorov's works on optical activity have become generally accepted. They are partly included in well-known concluding collections of selected papers on these problems[26,27].

There was a need of an exact and general theory of optical properties of gyrotropic crystals, even when neglecting magnetic anisotropy. Earlier, all of the authors[18,19,28] restricted themselves to the simplest case, in which not only terms with the squared gyrotropic parameters were neglected, but also the products of parameters of gyrotropy and anisotropy of crystals. In[29,30] this question was analyzed and it was shown that in some cases the generally accepted theory lead to conclusions that were fundamentally wrong. So, in[29] it was shown that if the equations of electric and magnetic fields of plane waves in gyrotropic crystals were limited to terms of first order of gyrotropy, the electric field equation appears to be incompatible with the magnetic field equation. Moreover, opposite physical conclusions follow from them. For example, in the case of planar crystals of middle syngonies, it follows from the approximate (first-order) equation for the electric field that there is an ordinary wave in the crystal whose velocity does not depend on either the direction of propagation or the presence of gyrotropy. Meanwhile, from the approximate equation of the same order for the magnetic field it follows that there are no ordinary waves in crystals. Instead, the wave actually propagates with a velocity that depends weakly on either the direction or the gyrotropy. Thus an internal contradiction was found in the approximate theory if limited to first-order terms. In[29] it was shown that if one proceeds from equations for electric and magnetic fields taking into account second-order terms of gyrotropy, this contradiction disappears and the theory becomes completely self-consistent. The theory proved self-consistent even when people tried to return to the old theory of natural gyrotropy see Ref.[31]

In all vast practical applications of crystal optics the major attention is given to the behavior of a light beam on the crystal interface, i. e. the problem of reflection and refraction of light. The special significance that this problem has acquired in connection with the development of solid state laser systems and optical networks. A series of works (1956s–1971s) is devoted to a detailed treatment of this question for transparent and absorbing crystals and isotropic media. Fedorov obtained general expressions in compact covariant vector form. He also gave a principal solution to the problem of determining all optical parameters of a transparent or absorbing medium based on the observation of light reflected from an arbitrarily oriented interface. In spite of a well-known role of the surface layer, this method is now the only one with which the optical parameters of strongly absorbing media can be determined. Basic results obtained earlier are systimized and generalized[3].

In 1976 F.I.Fedorov published a book[6]. The geometrical reflection and refraction laws on transparent uniaxial and biaxial crystals interface with isotropic medium were generalized. The relations allowing to calculate the amplitude coefficients for any crystal cut and any incident plane were obtained. The conditions for inhomogeneous waves in crystal and in isotropic medium were considered.

The covariant method of the description of partially polarized light with the help of the beam tensor was generalized by F.I.Fedorov[12]. It was shown that invariants of this tensor were more convenient than Stokes parameters for total characteristic of partially polarized light. This method allowed to solve a series of complex problems related to the definition of the light polarization patterns by the dispersion, luminescence etc.[3,27].

Harmonic electromagnetic and acoustic plane waves incident on anisotropic medium surface can excite Voigt's waves and those of a more general form. The amplitudes of such waves contains component which are linear, quadratic and cubic functions of coordinates. Such electromagnetic waves are Maxwell's equations solutions. They are not at all some exotic solutions. Similar optical waves with linear amplitude coordinates dependencies were considered in Fedorov's papers and were called Fedorov-Petrov waves[32]. These ideas underwent further development in[33–35]. Well-known english theorist M.Berry showed a great interest in such waves[36].

At the beginning of the 1960s F.I.Fedorov carried out fundamental research connected with the elastic wave theory in crystals[4]. The existence of a new type of inhomogeneous waves in absorbing crystals was predicted, i.e. waves with linear dependence of amplitude on coordinates, so-called eigenwaves, as well as the approximate theory of quasilongitudinal elastic waves in crystals was formulated. He advanced an original method for the description of elastic waves in crystals, i.e. the method of comparison with isotropic medium, and applied it to the calculation of Debye temperatures of solid states. In the foreword to the english edition Prof. H.B.Huntington noted: "In many respects his approach will appears quite novel as his treatment introduces several concepts and parameters previously unfamiliar to the literature of the West." The works of F.I.Fedorov in physics of elastic waves in crystals was marked by the State Awards of the republic of Belarus in 1972.

With the 1960s in works by F.I.Fedorov and his followers the research of induced anisotropy and gyrotropy effects, generated by external electric, magnetic and acoustic fields was began. Such nonlinear effects are called as electrooptical, magnetooptical, acoustooptical effects. They found wide application for modulation of controlled deviation and laser emission focussing. Covariant theory of forced anisotropy and gyrotropy effects for arbitrary direction of light beam and external fields in crystals was formulated. The effective methods of the polyadic representation of physical higher-rank tensors were developed[5]; the external and internal symmetries were investigated.

To the 1980s F.I.Fedorov had found many followers working in optics and acoustics of complex media. Their works were recognized both in USSR and abroad. It is enough to mention the review[37], in which the development of optical activity theory is divided into three periods: pre-Fedorov, Fedorov and post-Fedorov. Particular recognition of belarusian scientists works in branches of optics success can be shown by the following unprecedented plagiarism.

In 1983 in the USA the book "Theory of Electromagnetic Waves. A Coordinate-Free Approach"[38] by Hollis C.Chen was published. The book contained a brief abstract where it is stated that "this new book, for the first time, treats the subject matter completely from a coordinate-free approach". However, this covariant coordinate-free approach to the theory of electromagnetic waves in a media with arbitrary anisotropy that had been first proposed, thoroughly grounded and developed by F.I.Fedorov more than 50 years before. The coincidence of book subjects by F.I.Fedorov[3,5] and H.C.Chen[38] is not a key moment. Detailed studying of Chen's book contents, has shown absolute identity of basic contents between his book and Fedorov's books. H.C.Chen copied more than 1000 formulas from Fedorov's books. He saved the order of information and statements, seldom adding the relations being trivial consequence of Fedorov's equations and even the figures are virtually identiced to Fedorov's figures.

The presence of references in text is generally accepted in monographs and text-books. However, despite all scientific traditions, it has not a single reference to any publication from supplied list. There was not a single reference to belarusian scientist's works in the bibliography.

Moreover, in 1978 at the request of H.C.Chen F.I.Fedorov had sent him one of the books[5].

Therefore, it is absolutely unbelievable that H.C.Chen is not familiar with Fedorov's papers, which had been published earlier and were well-known in scientific world.

In 1980s the information about Chen's papers like Fedorov's ones appeared. The question was of Chen's report "A coordinate-free approach to wave reflection from an anisotropic medium" on International URSI Symposium in Munich (1980). This report was published two times: in proceedings of symposium and in journal "Radio Science" (1981). A bit late he published two more papers in "Journal of Applied Physics" (1982) and "IEEE Trans. Microwave Theory and Techniques" (1983). It turned out, that earlier published Chen's papers on these problems are also duplicates of Fedorov's works. There are nothing in the Chen's report that would not be contained in Fedorov's thesis[1] or his monographs[3,5].

The question of plagiarism is still open. There are authors of works which known about cogent priority of belarusian physicist in crystal optics. Nevertheless they refer to Chen's book.

F.I.Fedorov with followers developed operator technique for calculation of propagation and reflection of electromagnetic waves in complex anisotropic circuits with non-parallel boundaries. This technique was applied for construction of cavity electro-optical modulators, the corner reflectors, Porro's prisms, interferometers in laser emission control systems[39]. A series of fundamental concepts of optics and acoustics were generalized[40–54]. These concepts describe the complex media dynamics without subwave decomposition. (see also contributions to this collection)

In the 1960s–1970s F.I.Fedorov fulfilled a series of research on general relativity. It was shown that the decomposition of antisymmetric tensors in Riemann space of general relativity to auto- (self-) and antidual components leads to the form of curvature tensor which underlied Petrov's classification of gravity fields. F.I.Fedorov suggested the general method of calibration of gravity field tetradic potential. This method based on analysis of minimum deviation of their components from values corresponding to a missing gravity field. F.I.Fedorov with followers investigated the problem of gauge approach in the theory of gravitational interactions, homogeneous isotropic cosmological models in gauge theory of gravity, the problem of classical fields dynamics with material sources in Riemann-Carthan's space etc.[55–69].

The fundamental contribution to the classical (quantum-relativistic, but without secondary quantization) field theory of elementary particles was also made in Belarus due to the covariant approach. The description and calculation of specific processes and their electromagnetic interaction was given. A simple working scheme of electromagnetic scattering amplitude calculation was developed. The solution of equations for self-acting fields found with the help of the causal Green functions in the framework of perturbation theory was used[70] in the development of foregoing Fedorov's paper[8]. At the same time obtained results are similar to those given by quantum electrodynamics, but they are easier and take less time to obtain. Similar calculation scheme for processes of weak interaction was developed[71].

It should be noted that it was a time-consuming process for the classical field theory of elementary particles to become recognized as a physical theory. The seeningly impenetrable barriers which for a long time had divided classical (initially-quantified) and quantum (secondary quantified) field theory in particle physics were finally eliminated in the frameworks of universally recognized gauge conception.

In the 1960s the investigations of nonlinear universal matrix equations for different of gauge interactions including gravity were begun by F.I.Fedorov with his followers. The description of all four types of fundamental interactions (gravitational, electromagnetic, weak (electroweak) and strong) using universal nonlinear Fedorov's equation was given. It was made on a united mathematical base. As a result, detailed analysis of theory versions was carried out. These equations have recently been developed into investigations of nonlinear dynamics of various nature complex systems[72,73].

Covariant description of spin was introduced on the basis of combining method of projective operators with method of vector parametrization of Lorentz group. A covariant method (Bogush-Fedorov method) was developed. General calculation of electromagnetic interaction matrix elements for longitudinally polarized Dirac's particles was proposed[74]. F.I.Fedorov radically improved this method. He disseminated it on the case of arbitrarily polarized particle interaction. He also completed the general calculation of quantum electrodynamics matrix elements. A new approach to description of spin particle states including their changing in scattering processes was elaborated. A series of important relativistic kinematics problems was solved (see[7]).

The ideas of vector parametrization and basic results obtained in it framework for rotation Lorentz's groups were disseminated and used in the theory of many others Lie groups, for example, complex Lorentz's group, higher orthogonal and unitary groups, nonuniform Galilei's group and Poincare's group[75]. (see also contributions to this collection)

Today we can state, that the progress in the field of theoretical physics in Belarus were realized only due to creative and organizational talent by unforgettable F.I.Fedorov.

The material of many monographs[42,43,65,70,71,76–80], a series of international conferences (see too[27,26,20,81]) and works included in the book show that Fedorov's methods are actual in theoretical physics. They are not only widely used but generalized and intensified. The scientific Fedorov's school is going on successfully.

Bibliography

1. F. I. Fedorov, *Invariant methods in optics of anisotropic media*, habilitation degree thesis. (State Optical Institute, Leningrad, December, 1954).
2. P. A. M. Dirac. *From a Life of Physics, Evening Lectures the International Center of Theoretical Physics in Triest* (Wiena, IAIA, 1969).
3. F. I. Fedorov, *Optics of Anisotropic Media* (Acad. Nauk BSSR, Minsk, 1958).; (URSS, Moscow, 2004).
4. F. I. Fedorov *Theory of Elastic Waves in Crystals* (Nauka, Moscow, 1965); (Plenum Press, New York, 1968).
5. F. I. Fedorov, *Theory of Gyrotropy* (Nauka i Teknika, Minsk, 1976).
6. F. I. Fedorov and V. V. Filippov, *Reflection and Refraction of Light by Transparent Crystals* (Nauka i Tekhnika, Minsk, 1976).
7. F. I. Fedorov, *Lorentz Group* (Nauka, Moscow, 1979); (URSS, Moscow, 2003).
8. F. I. Fedorov, *Dokl. Akad. Nauk SSSR* **65**, 813 (1949).
9. F. I. Fedorov, *Uchenye Zapiski BSU* **12**, 156 (1951).
10. B. I. Stepanov, *Introduction in Modern Optics. Basic Conception of Optical Science on the Frontier of XX Century.* (Nauka i tekhnika, Minsk, 1989).; *Sketches by History of Optical Science.* (Nauka i tekhnika, Minsk, 1978).
11. F. I. Fedorov, *Dokl. Akad. Nauk SSSR* **84**, 1171 (1952).
12. F. I. Fedorov, *Opt. Spektrosk.* **1**, 807 (1956).
13. F. I. Fedorov and A.M.Goncharenko, *Crystallographia* **3**, 587 (1958).
14. T. Tamir, *Optik* **36**, 209 (1972).
15. F. I. Fedorov and L. M. Tomilchik, *Opt. Spektrosk.* **5**, 462 (1958).
16. F. I. Fedorov and L. M. Tomilchik, *Opt. Spektrosk.* **5**, 601 (1958).
17. F. I. Fedorov, *Opt. Spectrosk.* **6**, 85 (1959).
18. F. Pockels, *Luhrbuch der Kristalloptik* (Teubner, Leipzig and Berlin, 1906).
19. G. Szivessy and I. W. Münster, *Kristalloptik.* in *Handbuch der Physik* **20**, (Springer, Berlin), 634 (1928).
20. I. Semchenko, S. Tretyakov and A. Serdyukov, *Prog. Electromagn. Res.* **12**, 335 (1996).
21. F. I. Fedorov, *Usp. Fiz. Nauk* **108**, 763 (1972).
22. V. L. Ginzburg, *Usp. Fiz. Nauk* **108**, 749 (1972).
23. V. M. Agranovich and V. L. Ginzburg, *Zh. Eksperim. i Teor. Fiz.* **63**, 838 (1972).

24. V. M. Agranovich and V. L. Ginzburg, *Crystalooptics Taking into Account Spatial Dispersion and Theory of Excitons* (Nauka, Moscow, 1979).
25. V. L. Ginzburg, *Theoretical Physics and Astrophysics.* (Nauka, Moscow, 1987)
26. *Selected Papers on Natural Optical Activity.* Ed. by A. Lakhtakia, and B. J. Thompson SPIE Milestone Ser. MS 15. (Bellingham, WA: SPIE Opt. Eng. Press, 1990).
27. L. M. Barkovsky and G. N. Borzdov, in *Advances in Complex Electromagnetic Materials*, Ed. A. Priou, (Kluwer, Dordrecht, the Netherlands, 1997), p. 3.
28. M. Born, *Optik.* (Springer, Berlin, 1933).
29. F. I. Fedorov, *Vestsi Akad. Nauk BSSR, Ser. Fiz. Mat. Nauk*, N5, 70 (1973).
30. F. I. Fedorov, *Vestsi Akad. Nauk BSSR, Ser. Fiz. Mat. Nauk*, N1, 48 (1974).
31. L. M. Barkovsky and G. N. Borzdov, *Opt. Spektrosk.* **95**, 114 (2003).
32. F. I. Fedorov and N. S. Petrov, *Opt. Spektrosk.* **14**, 256 (1963); *Opt. Spektrosk.* **15**, 792 (1963).
33. L. M. Barkovsky and G. N. Borzdov, A. V. Lavrinenko *Dokl. Akad. Nauk BSSR* **35**, 313 (1990)
34. G. N. Borzdov, *Cristallographia* **35**, 313 (1990)
35. G. N. Borzdov, *Pramana — Journal of Physics* **46**, 245 (1996)
36. M. Berry, *Current Science* **67**, 220 (1994)
37. J. Kenneth O'Loane, *Chem. Rev.* **80**, 41 (1980).
38. H. C. Chen, *Theory of Electromagnetic Waves. A Coordinate-free Approach* (New York: McGraw-Hill Book Company, 1983).
39. L. M. Barkovsky et. al. *J. Phys. D: Appl. Phys.* **29**, 289 (1996).
40. L. M. Barkovsky, G. N. Borzdov and A. V. Lavrinenko, *J. Phys. A: Math. Gen.* **20**, 1095 (1987).
41. L. M. Barkovsky, G. N. Borzdov, F. I. Fedorov, *J. Mod. Opt.* **37**, 85 (1990).
42. L. M. Barkovsky and A. N. Furs, *Polarized waves in complex media* (BSU, Minsk, 2002).
43. A. N. Furs and L. M. Barkovsky, *Operator Methods for Description of Optical Fields in Complex Media* (Belaruskaya Navuka, Minsk, 2003).
44. G. N. Borzdov, *in Electromagnetic Fields in Unconventional Materials and Structures.* Ed. by O. N. Singh, A. Lakhtakia. (New York, Wiley Intersience) 83 (2000).
45. A. N. Furs and L. M. Barkovsky *Journ. Tekhn. fiziki* **73**, N 4, 9 (2003).
46. G. N. Borzdov, *Phys. Rev. E.* **61**, 4462 (2000).
47. G. N. Borzdov, *Phys. Rev. E.* **65**, 066612(16) (2002).
48. A. N. Borzdov, *Electromagnetics* **19**, 501 (1999).
49. D. N. Chigrin, et al, *Optics & Photonic News* **10**, 33 (1999).
50. D. N. Chigrin and A. V. Lavrinenko, *Optics & Photonic News* **10**, 5 (1999).
51. A. N. Furs and L. M. Barkovsky, *J. Opt. A: Pure Appl. Opt.* **1**, 109 (1999).
52. L. M. Barkovsky and A. N. Furs, *J. Phys. A: Math. Gen.* **33**, 3241 (2000).
53. D. N. Chigrin, et al, *Appl. Phys. A.* **68**, 25 (1999).
54. D. N. Chigrin, et al, *J. of Light. Techn.* **17**, 2018 (1999).
55. A. V. Minkevich and F. I. Fedorov, in *Fundamental Vzaimodeistvia* (MGPI,

Minsk), 141 (1984).
56. A. V. Minkevich and F. I. Fedorov, *Acta Phys. Polon.* **B11**, 367 (1980).
57. A. V. Minkevich and F. I. Fedorov, V. I. Kudin, *Dokl. Akad. Nauk BSSR* **25**, 406 (1981).
58. A. V. Minkevich and F. I. Fedorov, *Izv. Akad. Nauk BSSR, Ser. Fiz. Mat. Nauk* **5**, 45 (1985).
59. A. V. Minkevich, Nguen Hoang and F. I. Fedorov, *Dokl. Akad. Nauk BSSR* **28**, 211 (1984).
60. A. V. Minkevich, Nguen Hoang and F. I. Fedorov, *Dokl. Akad. Nauk BSSR* **29**, 697 (1985).
61. F. I. Fedorov, V. I. Kudin and A. V. Minkevich, *Acta Phys. Polon.* **B15**, 107 (1984).
62. A. V. Minkevich, *Phys. Lett.* **A80**, 232 (1980).
63. A. V. Minkevich, ArXiv gr-qc/0312068; to be published in *Annals of European Academy of Sciences* (2004).
64. A. V. Minkevich, *Vestsi Akad. Nauk BSSR, Ser. Fiz. Mat. Nauk*, N 2, 87 (1980).
65. A.K.Gorbatsievich, *Quantum Mechanics in General Relativity Theory: Basic Principles and Elementary Particles.* (BSU, Minsk, 1985); (URSS, Moscow, 2003).
66. A. K. Gorbatsievich, *Acta Physica Polonica* **B17**, 111 (1986).
67. A. V. Minkevich and A. S. Garkun, *Class. Quantum Grav.* **17**, 3045 (2000).
68. A. V. Minkevich, *Cov. Meth. in Theor. Phes. Elem. Part. Phys. and Relat. Theory.* (IP AN BSSR, Minsk) (2001).
69. A. K. Gorbatsievich, *General Relativity and Gravitation* **33**, (2001).
70. A. A. Bogush and L. G. Moroz. *Introduction to the Theory of Classical Fields* (Nauka i tekhnika, Minsk, 1968); (URSS, Moscow, 2003).
71. A. A. Bogush, *Introduction to the Gauge Field Theory of Electroweak Interactions* (Ibid., 1987); (URSS, Moscow, 2003).
72. V. I. Kuvshinov and A. V. Kuzmin, *Phys. Let. A.* **316**, 391 (2003).
73. V. I. Kuvshinov, A. V. Kuzmin and R. G. Shulyakovsky *Phys. Rev. E* **67**, 015201(R) (2003).
74. A. A. Bogush and F. I. Fedorov, *Vecti Akad. Nauk BSSR, ser. fiz.-tekhn. nauk* N 2, 26 (1962).
75. A. A. Bogush and F. I. Fedorov, *Rep. Math. Phys.* **11**, N 1, 37 (1977).
76. A. A. Bogush, *Introduction to the field theory of elementary particles* (Nauka i tekhnika, Minsk, 1981); (URSS, Moscow, 2003).
77. V. I. Strazjev and L. M. Tomil'chik, *Electrodynamics with Magnetic Charge* (Nauka i tekhnika, Minsk, 1975).
78. V. I. Kuvshinov and V. I. Strazhev, *From Scientific Hypothesis to Physical Fact* (Nauka i tekhnika, Minsk, 1977).
79. V. G. Kiselev, Ya. M. Shnir and A. Ya. Tregubovich, *Introduction in Quantum Theory* (Gordon & Breach, New-York, 2000).
80. V. I. Strazhev. *Polarization Symmetry* (Nauka i obrazovanije, Minsk, 1996).
81. I. I. Gancheryonok and A. V. Lavrinenko, *Handbook of Thin Film materials* in 5 vol. Ed. H. S. Nalva. (Academic Press, New York), **2**, 597 (2002).

Part I

PARTICLES AND FIELDS

THE IMPROVED DEUTERON MODEL WITH THE HYPERBOLIC RELATIVE SPACE

Vladimir V. Kudryashov, Lev M. Tomilchik
B.I.Stepanov Institute of Physics, National Academy of Sciences of Belarus
68 F. Skaryna Ave., 220072 Minsk, Belarus
E-mail: kudryash@dragon.bas-net.by, lmt@dragon.bas-net.by

In the proposed deuteron model, relative motions of nucleons are considered in the space of constant negative curvature (hyperbolic space). The interaction is described by Yukawa-type potentials. The corresponding Schrödinger equation is solved analytically. Modern deuteron data are exactly reproduced.

1. Introduction

The geometric approach to description of the strongly coupled systems is known for a long time. Several authors indicated the possibility to describe hadrons as strongly curved universes (see, e.g.,[1]). Less drastic proposal means the use of some bag-like representation of the anti-de Sitter geometry[2]. The confining properties of the conformally flat static effective metric are revealed in[3,4]. The efficiency of such approach is demonstrated by phenomenological hadronization model developed in[5] to explain the data on charmed mesons decays. The model of non-Euclidean space with spherical geometry is successfully applied to describe the excitons confined in the semiconductor quantum dots[6]. While the relative space of the positive constant curvature has been chosen in the case of the quark-antiquark system[7,8] now we choose the relative space of the negative constant curvature in the case of the nucleon-nucleon system.

All realistic models of the nucleon-nucleon interaction contain different variants[9,10,11,12] of the quadratic momentum-dependent terms in addition to the linear momentum dependence of the spin-orbit term. In this work we investigate new version of the quadratic momentum dependence by means of the geometric approach in order to describe the bound state of the proton-neutron system (deuteron). The present consideration is the improvement of approach[13]. We assume that the space of the relative nucleon motions has the hyperbolic structure with the metric tensor containing an arbitrary function $y(r)$. The special quantization procedure provides the free quantum Hamiltonian $\hat{H}_0$ which is $SO(3,1)$ invariant operator quadratic in momentum. This operator is inserted into the Schrödinger equation instead of the conventional kinetic term $\hat{\mathbf{p}}^2/2m$. In our deuteron model the central, tensor and spin-orbit potentials are the linear combinations of the generalized Hulthen potentials in variable y and the squared ones. The Schrödinger equation is solved analytically for an arbitrary $y(r)$. The explicit form of the function $y(r)$ is determined by the requirement that our model potential is equal to the realistic Yukawa potential.

2. Non-Euclidean relative space

The approach developed in[14] is based on the partial geometrization of the motions in the nonrelativistic N-particle system. The configuration space R^{3N} can be separated in the following way $R^{3N} = R^3 \otimes R^{3N-3}_{rel}$ where R^3 and R^{3N-3}_{rel} are spaces associated with the center of mass motion and relative motion respectively.

The separation can be achieved by introducing the so-called rationalized Jacobi coordinates

$$\mathbf{R} = \sum_{i=1}^{N} \frac{m_i \mathbf{x}_i}{M}, \quad \mathbf{r}_j = \left(\frac{\mu_j}{m}\right)^{1/2} \left(\frac{\sum_{i=1}^{j} m_i \mathbf{x}_i}{\sum_{j=1}^{j} m_i} - \mathbf{x}_{j+1} \right), \quad 1 \leq j \leq N-1, \tag{1}$$

where

$$M = \sum_{i=1}^{N} m_i, \quad \mu_j = \frac{m_j \sum_{j=1}^{j} m_i}{\sum_{j=1}^{j+1} m_i}, \quad m = \left(\frac{\prod_{i=1}^{N} m_i}{M} \right)^{\frac{1}{N-1}}. \tag{2}$$

R^3 is the usual Euclidean space which corresponds to a standard description of the system motion as a whole. However relative motion is described in a nonstandard way. The relative space R^{3N-3}_{rel} is supposed to be in general not Euclidean one but it is the uniform space with the metric tensor

$$g(\mathbf{r}) = \left(\frac{dy}{dr}\right)^2 \frac{\mathbf{r}\cdot\mathbf{r}}{r^2} + \frac{\rho^2 \sinh^2(\rho^{-1} y)}{r^2} \left(I - \frac{\mathbf{r}\cdot\mathbf{r}}{r^2} \right) \tag{3}$$

where I is identity operator and $y(r)$ is some increasing function of r. We consider the case $\rho^2 > 0$ corresponding to the negative curvature (hyperbolic case). Different realizations of the relative space geometry correspond to various physical systems under consideration. The obtained metric is finally determined by one parameter ρ and the single characteristic function $y(r)$ possessing rather arbitrary form. The relative kinetic energy takes the quadratic form in momenta more general than the usual classical kinetic energy because of the modification of the metric.

The group quantization of such classical N-particle spinless system gives the Hamiltonian for a free relative motion:

$$\hat{H}_0 = \frac{1}{2m}\left(\hat{\mathbf{\Pi}}^2 - \frac{1}{\rho^2}\hat{\mathbf{L}}^2\right) - \frac{\hbar^2}{2m\rho^2}\left(\frac{3N-4}{2}\right)^2 \tag{4}$$

where

$$\hat{\mathbf{\Pi}} = (\det g(\mathbf{r}))^{1/4}\mathbf{\Pi}(\det g(\mathbf{r}))^{-1/4},$$

$$\mathbf{\Pi} = -i\hbar\left(\rho^{-1}r\coth(\rho^{-1}y)\left(I - \frac{\mathbf{r}\cdot\mathbf{r}}{r^2}\right) + \left(\frac{dy}{dr}\right)^{-1}\frac{\mathbf{r}\cdot\mathbf{r}}{r^2}\right)\nabla,$$

$$\hat{L}^2 = \frac{1}{2}\hat{L}_{kl}\hat{L}_{kl}, \quad \hat{L}_{kl} = -i\hbar\left(r_k\frac{\partial}{\partial r_l} - r_l\frac{\partial}{\partial r_k}\right), \quad k,l = 1,2,.....3N-3.$$

In the presence of a potential interaction, the radial Schrödinger equation can be represented in the form

$$\left(\frac{\hbar^2}{2m}\frac{d^2}{dy^2} - \frac{\hbar^2}{2m}\frac{l'(l'+1)}{\rho^2\sinh^2(\rho^{-1}y)} + E - V(y)\right)\phi(y) = 0 \tag{5}$$

where $l' = l + (3N-6)/2$. The function $\phi(y)$ is connected with the complete wave function $\Psi(\mathbf{r}) = R(r)Y_l^m(\mathbf{r}/r)$, where $Y_l^m(\mathbf{r}/r)$ are the spherical harmonics, via formula

$$R(r) = \left(r^{-(3N-4)}\rho^{-1}\frac{dy}{dr}\right)^{1/2}\phi(y).$$

In our further consideration of the Scrödinger equation we shall use the condition of the analytical solvability as a crucial one.

3. Two-fermion system

Let us consider the system of two particles with masses m_1 and m_2 and with the position vectors $\mathbf{r}_1$ and $\mathbf{r}_2$. The spin operators are represented by the Pauli matrices as follows $\hat{\mathbf{S}}_{1,2} = (\hbar/2)\vec{\sigma}_{1,2}$. In the framework of the quasi-relativistic description of the fermion-fermion interaction, the phenomenological potential energy $V(\mathbf{r})$ in the simplest case can be presented as a sum of central, tensor and spin-orbit terms

$$V(\mathbf{r}) = V_C(r) + V_T(r)\hat{T} + \hbar^{-2}V_{LS}(r)\hat{\mathbf{L}}\hat{\mathbf{S}} \tag{6}$$

where

$$\hat{T} = 3r^{-2}(\vec{\sigma}_1\mathbf{r})(\vec{\sigma}_2\mathbf{r}) - (\vec{\sigma}_1\vec{\sigma}_2), \quad \hat{\mathbf{S}} = \hat{\mathbf{S}}_1 + \hat{\mathbf{S}}_2, \quad \mathbf{r} = \mathbf{r}_1 - \mathbf{r}_2.$$

Taking into account the supposed non-Euclidean structure of the relative two-fermion space we have the following Schrödinger equation

$$(\hat{H}_0 + V(\mathbf{r}) - E)|\Psi >= 0 \tag{7}$$

where $\hat{H}_0$ is determined by (4) with $N = 2$.

The integrals of motion in this case are $\hat{\mathbf{J}} = \hat{\mathbf{L}} + \hat{\mathbf{S}}$ and $\hat{\mathbf{S}}^2$. The eigenvalues of $\hat{\mathbf{S}}^2$ are $\hbar^2 s(s+1)$ where $s = 0, 1$. We shall consider the physically interesting triplet state ($s = 1$). Let us introduce eigenvectors $|j, l>$ such that

$$\hat{\mathbf{J}}^2|j,l>= \hbar^2 j(j+1)|j,l>, \quad \hat{\mathbf{L}}^2|j,l>= \hbar^2 l(l+1)|j,l>.$$

We represent $|\Psi>$ as a linear combination

$$|\Psi>= r^{-1}\left(\frac{dy}{dr}\right)^{1/2}\sum_{l=j-1}^{j+1} u_l(y)|j,l>. \tag{8}$$

Substituting vector (8) into the Schrödinger equation (7) we obtain three equations

$$\left(\frac{\hbar^2}{2m}\frac{d^2}{dy^2} - \frac{\hbar^2}{2m\rho^2}\frac{(j+2)(j-1)}{\sinh^2(\rho^{-1}y)} + E - V_C\right)u_{j-1}(y) \tag{9}$$
$$= (j-1)\left(V_{ef} - \frac{2V_T}{(2j+1)}\right)u_{j-1}(y) + \frac{6[j(j+1)]^{1/2}}{(2j+1)}V_T u_{j+1}(y),$$

$$\left(\frac{\hbar^2}{2m}\frac{d^2}{dy^2} - \frac{\hbar^2}{2m\rho^2}\frac{(j+2)(j-1)}{\sinh^2(\rho^{-1}y)} + E - V_C\right)u_{j+1}(y) \tag{10}$$
$$= -(j+2)\left(V_{ef} + \frac{2V_T}{(2j+1)}\right)u_{j+1}(y) + \frac{6[j(j+1)]^{1/2}}{(2j+1)}V_T u_{j-1}(y),$$

$$\left(\frac{\hbar^2}{2m}\frac{d^2}{dy^2} - \frac{\hbar^2}{2m\rho^2}\frac{(j+2)(j-1)}{\sinh^2(\rho^{-1}y)} + E - V_C\right)u_j(y) \tag{11}$$
$$= (-V_{ef} + V_T)u_j(y)$$

for the functions $u_l(y)$, where $V_{ef} = V_{LS} - \hbar^2[m\rho^2 \sinh^2(\rho^{-1}y)]^{-1}$.

The presence of the nondiagonal tensor operator $\hat{T}$ leads to considerable difficulties in solving for the functions $u_{j\pm1}(y)$. In order to obtain the solvable problem we demand that the spin-orbit and tensor potential functions are connected by the following additional condition

$$V_{LS} = 2BV_T + \hbar^2[m\rho^2 \sinh^2(\rho^{-1}y)]^{-1} \tag{12}$$

where B is some numerical parameter. At the same time we conserve freedom in the choice of functions $V_C(y), V_T(y)$ and $y(r)$. Under condition (12) the system of radial equations (9)–(10) allows diagonalization and we can use the following representation

$$u_{j-1}^{\pm}(y) = \pm\frac{u_{\pm}(y)}{\sqrt{2}}\left(1 + x^2 \mp x\sqrt{x^2+1}\right)^{-1/2},$$
$$u_{j+1}^{\pm}(y) = \frac{u_{\pm}(y)}{\sqrt{2}}\left(1 + x^2 \pm x\sqrt{x^2+1}\right)^{-1/2}$$

where

$$x = \frac{(2j+1)^2B+3}{6[j(j+1)]^{1/2}}.$$

Then we obtain the separate equations

$$\left(\frac{\hbar^2}{2m}\frac{d^2}{dy^2} - \frac{\hbar^2}{2m\rho^2}\frac{(j+2)(j-1)}{\sinh^2(\rho^{-1}y)} + E - V_C - \epsilon_i V_T\right)u_i(y) = 0 \tag{13}$$

for functions $u_i(y)$ $(i = -, 0, +)$ where

$$\epsilon_{\pm} = -(1+3B) \pm [9 + 6B + (2j+1)^2B^2]^{1/2}, \quad \epsilon_0 = 2(1-B).$$

Thus in the framework of the supposed simplified model, the problem of description of triplet two-fermion bound state is reduced to the solution of three equations (13) of the same type for radial wave functions $u_i(y)$,

4. Exactly solvable deuteron model

The deuteron is the bound system of proton and neutron with the quantum numbers $j = 1, l = 0, 2$. In this case it is convenient to define new parameter ω with the help of equation $B = \frac{1}{3}(2\sqrt{2}\cot 2\omega - 1)$, instead of parameter B. Now we have relations $\epsilon_{\pm} = \pm 2\sqrt{2}(\tan\omega)^{\pm1}$ and formulas for the effective potentials in (13):

$$V_{\pm}(y) = \left(V_C(y) \pm 2\sqrt{2}(\tan\omega)^{\pm1}V_T(y)\right). \tag{14}$$

We choose the potentials $V_{\pm}(y)$ as the Hulthen potentials

$$V_{+}(y) = -\frac{4}{\rho^2}\frac{\hbar^2}{2m}\left((1+k)(e^{2y/\rho}-1)^{-1}\right), \tag{15}$$

$$V_{-}(y) = \frac{12}{\rho^2}\frac{\hbar^2}{2m}\left(2(e^{2y/\rho}-1)^{-2}-(1+k)(e^{2y/\rho}-1)^{-1}\right) \tag{16}$$

in variable y.

To describe the state with the positive parity we consider only two equations

$$\left(\frac{\hbar^2}{2m}\frac{d^2}{dy^2}+E-V_{\pm}(y)\right)u_{\pm}(y)=0. \tag{17}$$

Then we obtain the analytical expressions

$$u_{+}(y) = \left[\frac{2\Gamma(k+3)}{\rho\Gamma(k)\Gamma(3)}\right]^{1/2} e^{-ky/\rho}\left(1-e^{-2y/\rho}\right), \tag{18}$$

$$u_{-}(y) = \left[\frac{2\Gamma(k+7)}{\rho\Gamma(k)\Gamma(7)}\right]^{1/2} e^{-ky/\rho}\left(1-e^{-2y/\rho}\right)^3 \tag{19}$$

for the normalized eigenfunctions and eigenvalue

$$E = -\frac{\hbar^2}{2m}\frac{k^2}{\rho^2}. \tag{20}$$

We demand the generalized Hulthen potential to be equal to the realistic Yukawa potential:

$$2(e^{2y(r)/\rho}-1)^{-1} = \coth\left(\frac{y(r)}{\rho}\right)-1 = \frac{\nu\rho}{r}\exp\left(-2\frac{\beta r}{\rho}\right). \tag{21}$$

As a consequence we find the explicit form of the function $y(r)$:

$$y(r) = \frac{\rho}{2}\ln\left(1+\frac{2r}{\nu\rho}e^{2\beta r/\rho}\right). \tag{22}$$

The deuteron state $|\Psi_d>$ is described by the linear combination

$$|\Psi_d> = \cos\delta|\Psi^+> + \sin\delta|\Psi^->$$

with a mixture angle δ, where

$$|\Psi^+> = r^{-1}\left(\frac{dy}{dr}\right)^{1/2} u_{+}(y)\left(\cos\omega|1,0> + \sin\omega|1,2>\right),$$

$$|\Psi^-> = r^{-1}\left(\frac{dy}{dr}\right)^{1/2} u_{-}(y)\left(-\sin\omega|1,0> + \cos\omega|1,2>\right).$$

This state is usually represented in the form

$$|\Psi_d >= \frac{1}{r}\left(u(r)|1,0> + w(r)|1,2>\right)$$

with the help of the radial wave functions $u(r)$ (S-wave) and $w(r)$ (D-wave).

Finally, we obtain the explicit formulas for the deuteron wave functions

$$u(r) = \left(\frac{dy}{dr}\right)^{1/2} \left(\cos\omega \cos\delta u_+(y) - \sin\omega \sin\delta u_-(y)\right), \tag{23}$$

$$w(r) = \left(\frac{dy}{dr}\right)^{1/2} \left(\sin\omega \cos\delta u_+(y) + \cos\omega \sin\delta u_-(y)\right). \tag{24}$$

The functions $u(r)$ and $w(r)$ satisfy conventional normalization requirement

$$\int_0^\infty (u^2(r) + w^2(r))dr = 1. \tag{25}$$

We calculate the following quantities: the binding energy $E_B = -E$, the asymptotic D/S ratio $\eta = \lim_{r\to\infty} \frac{w(r)}{u(r)}$, the quadrupole moment

$$Q_d = \frac{1}{20}\int_0^\infty (2\sqrt{2}uw - w^2)r^2 dr,$$

the deuteron radius

$$r_d = \frac{1}{2}\left(\int_0^\infty (u^2 + w^2)r^2 dr\right)^{1/2}$$

and the D-state probability

$$P_D = \int_0^\infty w^2 dr$$

which determines the magnetic moment μ_d via formula

$$\mu_d = \mu_p + \mu_n - \frac{3}{2}\left(\mu_p + \mu_n - \frac{1}{2}\right) P_D$$

where μ_p and μ_n are the magnetic moments of proton and neutron.

If the values of the model parameters are chosen as follows

$$\rho = 12.0323 fm, \quad \beta = 0.9997, \quad \nu = 0.779535,$$
$$k = 2.78676, \quad \omega = 0.269709, \quad \delta = -0.0909187,$$

then we obtain the deuteron properties

$$E_B = 2.224575 MeV, \quad \eta = 0.0256, \quad Q_d = 0.286 fm^2,$$
$$r_d = 1.975 fm, \quad \mu_d = 0.857415 \quad (P_D = 0.0393)$$

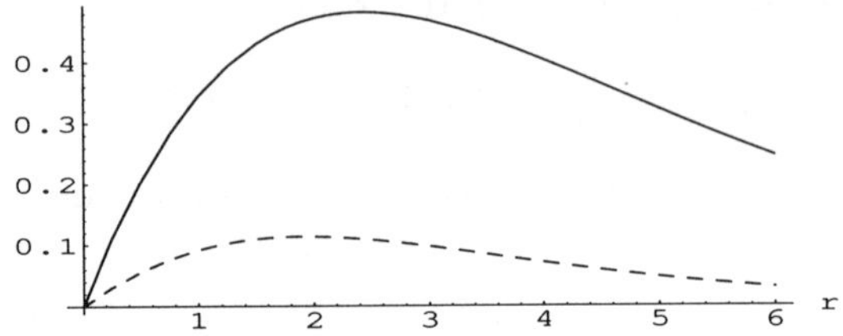

Figure 1. Dependence of functions u (solid line) and w (dashed line) on $r(fm)$

which reproduce the modern experimental data[15] exactly.

Figure 1 shows our model deuteron wave functions. Note that these simple wave functions are similar to the multi-parameter functions which are obtained with the help of very complicated realistic nucleon-nucleon potentials[16]. More detailed verification of our wave functions by calculation of the electromagnetic form factors is now in progress.

Bibliography

1. A. Salam and J. Strathdee, *Phys. Rev.* **D18**, 4596 (1978).
2. E. van Beveren and C. Dullemond, *Phys. Rev.* **D30**, 1103 (1984).
3. A. K. Gorbatsevich and L. M. Tomilchik, in *Problems of HEP and field theory* (Nauka, Moscow, 1987) p.378.
4. M. Gasperini, *Phys. Lett.* **B195**, 453 (1987); E. R. Chaaniello *at al*, *Phys. Lett.* **A132**, 82 (1988).
5. I. I. Basdevant, I. Bediaga and E. Predazzi, *Nucl. Phys.* **B294**, 1054 (1987).
6. V. V. Gritsev and Yu. A. Kurochkin, *Phys. Rev.* **B64**, 035308 (2003).
7. V. V. Kudryashov and L. M. Tomilchik, *Dokl. Akad. Nauk BSSR* **34** (N9), 794 (1989).
8. A. A. Izmest'ev, *Yad. Fiz.* **52**, 1697 (1990).
9. M. Lacomb *at al*, *Phys. Rev.* **C21**, 861, (1980).
10. J. J. de Swart, W. A. van der Sanden and W. Derks, *Nucl. Phys.* **A416**, 199 (1984).
11. R. Machledit, K. Holinde and Ch. Elster, *Phys. Rep.* **149**, 1 (1987).
12. R. B. Wiringa, A. Arriaga and V. R. Pandharipanda, *Phys. Rev.* **C68**, 054006 (2003).
13. V. V. Kudryashov, V. I. Murzov and L. M Tomilchik, in *Topological phases in quantum theory* (World Scientific, Singapore, 1989) p. 393.
14. V. V. Kudryashov, V. I. Murzov and L. M. Tomilchik, in *Problems of HEP and field theory* (Nauka, Moscow, 1988) p. 485.
15. D. R. Entem and R. Machleidt, *Phys. Lett.* **B524**, 93 (2002).
16. F. Gross, *Eur. Phys. J.* **A17**, 407 (2003).

NON-SINGULAR COSMOLOGY AND GAUGE THEORIES OF GRAVITATION

Albert V. Minkevich

Department of Theoretical Physics, Belarusian State University, 4, Fr. Skorina Ave., 220050, Minsk, Belarus
Department of Physics and Computer Methods, Warmia and Mazury University in Olsztyn, Poland
E-mail: MinkAV@bsu.by; awm@matman.uwm.edu.pl

The resolution of the problem of cosmological singularity in the framework of gauge theories of gravitation is discussed. Generalized cosmological Friedmann equations for homogeneous isotropic models filled by interacting scalar fields and usual gravitating matter are deduced. It is shown that the generic feature of cosmological models of flat, open and closed type is their regular bouncing character.

1. Introduction

As it is well known the problem of cosmological singularity (PCS) is one of the most principal problems of general relativity theory (GR) and relativistic cosmology. There were many attempts to resolve PCS in the frame of GR as well as various generalizations of Einstein's theory of gravitation. A number of regular cosmological solutions was obtained in the frame of metric theories of gravitation and also other theories, in the frame of which gravitation is described by using more general geometry than the Riemannian one (see[1–3] and Refs given herein). In connection with this note that the resolution of PCS means not only obtaining regular cosmological solutions, but also excluding singular solutions of cosmological equations, as result generic feature of cosmological solutions has to be their regular character. Moreover, gravitation theory and cosmological equations have to satisfy the correspondence principle with Newton's theory of gravitation and GR in the case of usual gravitating systems with sufficiently small energy densities and weak gravitational fields excluding nonphysical solutions. The greatest part of existent attempts to resolve the PCS does not satisfy indicated conditions. Let's take for instance some examples. Although GR permits to build regular closed inflationary models, the greatest part of cosmological inflationary solutions of GR is singular[4,5]. Metric theories of gravitation based on gravitational Lagrangians including terms quadratic in the curvature tensor lead to cosmological equations with high derivatives, and although these theories permit to obtain regular cosmological solutions with Friedmann and de Sitter asymptotics[7], however, they possess nonphysical solutions also. At last time single regular cosmological solutions were found in the frame of superstring theory (brane cosmology) by using specific suppositions (see for example[8–11]). To build cyclic model of the Universe specific negative scalar field potential was introduced in Refs[12,13].

As it was shown in a number of our papers[14–24, 1–3], the gauge theories of gravitation (GTG) – the Poincare GTG, metric-affine GTG – possess important regularizing properties and permit to resolve the PCS. Satisfying the correspondence principle with GR in the case of usual gravitating systems with rather small energy densities and pressures, the GTG lead to essentially different consequences in the case of gravitating systems at extreme conditions with extremely high energy densities and pressures, namely the GTG lead to conclusion on possible existence of limiting energy density for gravitating systems, nearby of which gravitation has the character of repulsion, but not attraction[14]. The investigation of homogeneous isotropic cosmological models in GTG shows, that by certain restrictions on equation of state of gravitating matter at extreme conditions and indefinite parameter of cosmological equations of GTG generic feature of cosmological models including inflationary models is their regular bouncing character. Such conclusions were obtained, in particular, in our recent papers[2,3] for inflationary cosmological models filled by noninteracting scalar fields and gravitating matter with linear equation of state.

Present paper is devoted to further study of regular cosmology in the frame of GTG. In Section 2 cosmological equations for homogeneous isotropic models filled by interacting scalar fields and gravitating matter with equation of state in general form are deduced. In Section 3 general mathematical properties of deduced equations and their solutions are analyzed. As illustration of discussed theory in Section 4 particular solution for regular inflationary cosmological model is given.

2. Generalized cosmological Friedmann equations in GTG

Homogeneous isotropic models in GTG are described by the following generalized cosmological Friedmann equations (GCFE)

$$\frac{k}{R^2} + \left\{ \frac{d}{dt} \ln \left[R\sqrt{|1-\beta(\rho-3p)|} \right] \right\}^2 = \frac{8\pi}{3M_p^2} \frac{\rho - \frac{\beta}{4}(\rho-3p)^2}{1-\beta(\rho-3p)}, \quad (1)$$

$$\frac{\left[\dot{R} + R\left(\ln \sqrt{|1-\beta(\rho-3p)|} \right) \right]}{R} = -\frac{4\pi}{3M_p^2} \frac{\rho + 3p + \frac{\beta}{2}(\rho-3p)^2}{1-\beta(\rho-3p)}, \quad (2)$$

where $R(t)$ is the scale factor of Robertson-Walker metrics, $k = +1, 0, -1$ for closed, flat, open models respectively, β is indefinite parameter with inverse dimension of energy density, M_p is Planckian mass, a dot denotes differentiation with respect to time[a]. (The system of units with $\hbar = c = 1$ is used). At first the GCFE were deduced in Poincare GTG[14], and later it was shown that Eqs.(1)–(2) take place also in metric-affine GTG[25,26]. In the frame of considered theory the conservation law in usual form takes place

$$\dot{\rho} + 3H(\rho + p) = 0, \quad (3)$$

where $H = \frac{\dot{R}}{R}$ is the Hubble parameter. Besides cosmological equations (1)–(2) gravitational equations of GTG lead to the following relation for torsion function S and nonmetricity function Q

$$S - \frac{1}{4}Q = -\frac{1}{4}\frac{d}{dt}\ln|1-\beta(\rho-3p)|. \quad (4)$$

In Poincare GTG $Q = 0$ and Eq. (4) determines the torsion function. In metric-affine GTG there are three kinds of models[26]: in the Riemann-Cartan space-time ($Q = 0$), in the Weyl space-time ($S = 0$), in the Weyl-Cartan space-time ($S \neq 0$, $Q \neq 0$, the function S is proportional to the function Q). The value of $|\beta|^{-1}$ determines the scale of extremely high energy densities. The GCFE (1)–(2) coincide practically with Friedmann cosmological equations of GR if the energy density is small $|\beta(\rho-3p)| \ll 1$ ($p \neq \frac{1}{3}\rho$). The difference between GR and GTG can be essential at extremely high energy densities $|\beta(\rho-3p)| \gtrsim 1$. Ultrarelativistic matter ($p = \frac{1}{3}\rho$) and gravitating vacuum ($p = -\rho$) with constant energy density are exceptional systems because Eqs. (1)–(2) are identical to Friedmann cosmological equations of GR in these cases independently on values of energy density and $S = Q = 0$. The behaviour of solutions of Eqs. (1)–(2)

---

[a]Parameter β is defined as $\beta = -\frac{1}{3}(16\pi)^2 f M_p^{-4}$, where f is linear combination of coefficients at terms of gravitational Lagrangian quadratic in the curvature tensor.

depends essentially on equation of state of gravitating matter at extreme conditions and on sign of parameter β. In the case of models filled by gravitating matter without scalar fields the GCFE (1)–(2) lead to regular in metrics cosmological solutions by the following restrictions: 1) $\beta > 0$ and at extreme conditions $p < \frac{1}{3}\rho$, 2) $\beta < 0$ and $p > \frac{1}{3}\rho$[1]. The investigation of models including scalar fields on the base of Eqs (1)–(2) shows that the choice $\beta < 0$ permits to exclude the divergence of time derivative of scalar fields[2]. In connection with this we put below that parameter β is negative and in the frame of our classical description $|\beta|^{-1} < 1 \cdot M_p^4$.

By using GCFE (1)–(2) we will study homogeneous isotropic models filled by interacting scalar field ϕ minimally coupled with gravitation and gravitating matter with equation of state in general form $p_m = p_m(\rho_m)$[b]. Then the energy density ρ and pressure p take the form

$$\rho = \frac{1}{2}\dot{\phi}^2 + V + \rho_m, \qquad p = \frac{1}{2}\dot{\phi}^2 - V + p_m, \tag{5}$$

where scalar field potential $V = V(\phi, \rho_m)$ includes the interaction between scalar field and gravitating matter. In the most important particular case of radiation (ultrarelativistic matter) the expressions of $V(\phi, \rho_m)$ can be obtained by taking into account temperature corrections for given scalar field potentials[27] and the following relation for energy density $\rho_m \sim T^4$ (T is absolute temperature). By using the scalar field equation in homogeneous isotropic space

$$\ddot{\phi} + 3H\dot{\phi} = -\frac{\partial V}{\partial \phi} \tag{6}$$

we obtain from Eqs. (3), (5), (6) the conservation law for gravitating matter

$$\dot{\rho}_m \left(1 + \frac{\partial V}{\partial \rho_m}\right) + 3H\left(\rho_m + p_m\right) = 0. \tag{7}$$

By virtue of Eqs.(5)–(7) the GCFE (1)–(2) can be transformed to the following form

$$\left\{ H \left[Z - 3\beta\dot{\phi}^2 - \frac{3\beta}{2}\frac{\rho_m + p_m}{1 + \frac{\partial V}{\partial \rho_m}} \left(\frac{d}{d\rho_m}\left(3p_m - 4V\right) - 1 \right) \right] - 3\beta \frac{\partial V}{\partial \phi}\dot{\phi} \right\}^2$$
$$+ \frac{k}{R^2} Z^2 = \frac{8\pi}{3M_p^2} \left[\rho_m + \frac{1}{2}\dot{\phi}^2 + V - \frac{1}{4}\beta\left(4V - \dot{\phi}^2 + \rho_m - 3p_m\right)^2 \right] Z \tag{8}$$

[b]The generalization for the case with several scalar fields can be made directly.

$$\dot{H}\left\{Z-3\beta\left[\dot{\phi}^2+\frac{1}{2}\frac{\rho_m+p_m}{1+\frac{\partial V}{\partial\rho_m}}\left(\frac{d}{d\rho_m}(3p_m-4V)-1\right)\right]\right\}Z$$

$$+H^2\left\{\left[Z+15\beta\dot{\phi}^2-\frac{3\beta}{2}\frac{\rho_m+p_m}{1+\frac{\partial V}{\partial\rho_m}}\left(\frac{d}{d\rho_m}(3p_m-4V)-1\right)\right.\right.$$

$$+\frac{9\beta}{2}\frac{\rho_m+p_m}{\left(1+\frac{\partial V}{\partial\rho_m}\right)^2}\left(\left(1+\frac{dp_m}{d\rho_m}\right)\left(\frac{d}{d\rho_m}(3p_m-4V)-1-3\frac{\rho_m+p_m}{1+\frac{\partial V}{\partial\rho_m}}\frac{\partial^2V}{\partial\rho_m^2}\right)\right.$$

$$\left.\left.+3(\rho_m+p_m)\frac{d^2p_m}{d\rho_m^2}\right)\right]Z-\frac{9}{2}\beta^2\left[\frac{\rho_m+p_m}{1+\frac{\partial V}{\partial\rho_m}}\left(\frac{d}{d\rho_m}(3p_m-4V)-1\right)\right.$$

$$\left.\left.+2\dot{\phi}^2\right]^2\right\}-12\beta H\dot{\phi}\left\{\frac{3\beta}{2}\frac{\partial V}{\partial\phi}\left[\frac{\rho_m+p_m}{1+\frac{\partial V}{\partial\rho_m}}\left(\frac{d}{d\rho_m}(3p_m-4V)-1\right)+2\dot{\phi}^2\right]\right.$$

$$\left.-\left[\frac{\partial V}{\partial\phi}+\frac{9}{8}\frac{\rho_m+p_m}{\left(1+\frac{\partial V}{\partial\rho_m}\right)^2}\frac{\partial^2V}{\partial\phi\partial\rho_m}\left(1+\frac{1}{3}\frac{d}{d\rho_m}(p_m+2V)\right)\right]Z\right\}$$

$$-3\beta\left[\frac{\partial^2V}{\partial\phi^2}\dot{\phi}^2-\left(\frac{\partial V}{\partial\phi}\right)^2\right]Z-18\beta^2\left(\frac{\partial V}{\partial\phi}\right)^2\dot{\phi}^2$$

$$=\frac{8\pi}{3M_p^2}\left[V-\dot{\phi}^2-\frac{1}{2}(\rho_m+3p_m)-\frac{1}{4}\beta\left(4V-\dot{\phi}^2+\rho_m-3p_m\right)^2\right]Z, \quad (9)$$

where $Z=1-\beta\left(4V-\dot{\phi}^2+\rho_m-3p_m\right)$. Relation (4) takes the form

$$S-\frac{1}{4}Q=\frac{3\beta}{2Z}\left\{H\left[\dot{\phi}^2+\frac{1}{2}\frac{\rho_m+p_m}{1+\frac{\partial V}{\partial\rho_m}}\left(\frac{d}{d\rho_m}(3p_m-4V)-1\right)\right]+\frac{\partial V}{\partial\phi}\dot{\phi}\right\}. \quad (10)$$

Unlike GR the cosmological equation (8) leads to essential restrictions on admissible values of scalar field and gravitating matter. Imposing $\beta<0$, we obtain from Eq. (8) in the case $k=0,+1$

$$Z\geq 0, \qquad \text{or} \qquad \dot{\phi}^2\leq 4V+|\beta|^{-1}+\rho_m-3p_m. \quad (11)$$

The condition (11) is valid also for open models discussed below. This means that relation (11) is fulfilled for all cosmological models independently on their type ($k=0,+1,-1$). Now let us introduce the 3-dimensional space P with axes $(\phi,\dot{\phi},\rho_m)$. The domain of admissible values of scalar field ϕ, time derivative $\dot{\phi}$ and energy density ρ_m in space P determined by (11) is limited by bound L defined as

$$Z = 0 \qquad \text{or} \qquad \dot{\phi} = \pm\left(4V + |\beta|^{-1} + \rho_m - 3p_m\right)^{\frac{1}{2}}. \tag{12}$$

From Eq. (8) the Hubble parameter on the bound L is equal to

$$H_L = -\frac{\dfrac{\partial V}{\partial \phi}\,\dot{\phi}}{\dot{\phi}^2 + \dfrac{1}{2}\,\dfrac{\rho_m + p_m}{1 + \frac{\partial V}{\partial \rho_m}}\left(\dfrac{d}{d\rho_m}\left(3p_m - 4V\right) - 1\right)}. \tag{13}$$

The right-hand part of Eq. (10) with the Hubble parameter determined by (13) is equal to $\frac{1}{2}H$, this means that the torsion (nonmetricity) in this case will be regular, if the Hubble parameter is regular.

3. Solutions properties of GCFE

Let us consider the most important general properties of cosmological solutions of GCFE (8)–(9). At first note, by given initial conditions for variables $(\phi,\ \dot{\phi}, \rho_m)$ and value of R there are two different solutions corresponding to two values of the Hubble parameter following from Eq. (8):

$$H_\pm = \frac{3\beta\,\dfrac{\partial V}{\partial \phi}\,\dot{\phi} \pm \sqrt{D}}{Z - 3\beta\left[\dot{\phi}^2 + \dfrac{1}{2}\,\dfrac{\rho_m + p_m}{1 + \frac{\partial V}{\partial \rho_m}}\left(\dfrac{d}{d\rho_m}\left(3p_m - 4V\right) - 1\right)\right]}, \tag{14}$$

where

$$D = \frac{8\pi}{3M_p^2}\left[\rho_m + \frac{1}{2}\dot{\phi}^2 + V - \frac{1}{4}\beta\left(4V - \dot{\phi}^2 + \rho_m - 3p_m\right)^2\right] Z - \frac{k}{R^2} Z^2. \tag{15}$$

Unlike GR, the values of H_+ and H_- in GTG are sign-variable and, hence, both solutions corresponding to H_+ and H_- can describe the expansion as well as the compression in dependence on their sign. Below we will call solutions of GCFE corresponding to H_+ and H_- as H_+-solutions and H_--solutions respectively. Any cosmological solution contains both H_-- and H_+-solution. In points of bound L we have $D = 0$, $H_+ = H_-$ and the Hubble parameter is determined by (13). The GCFE (8)–(9) are satisfied on the bound L, corresponding solutions of GCFE – L–solutions – are their particular solutions. The scalar field ϕ and energy density ρ_m for L-solutions can be found by integration of Eqs. (6)–(7), where the Hubble parameter is defined by (13), and by using the constrain relation $Z = 0$.

L-solutions can be regular as well singular in dependence on equation of state of gravitating matter, scalar field potential V and initial conditions[3]. Trajectories of particular solutions situated on the bound L have with $H_\pm$-solutions common points, H_--solutions reach the bound L and H_+-solutions originate from them, the surface $Z = 0$ containing trajectories of particular solutions is envelope in space P for cosmological solutions. By using the following formula obtained for $H_\pm$-solutions

$$\dot{Z} = 3\beta \frac{-2\,\dfrac{\partial V}{\partial \phi}\,\dot{\phi}\,Z \mp \sqrt{D}\left[\dfrac{\rho_m + p_m}{1 + \frac{\partial V}{\partial \rho_m}}\left(\dfrac{d}{d\rho_m}\left(3p_m - 4V\right) - 1\right) + 2\dot{\phi}^2\right]}{Z - 3\beta\left[\dot{\phi}^2 + \dfrac{1}{2}\,\dfrac{\rho_m + p_m}{1 + \frac{\partial V}{\partial \rho_m}}\left(\dfrac{d}{d\rho_m}\left(3p_m - 4V\right) - 1\right)\right]} \tag{16}$$

it is easy to show that

$$\lim_{Z\to 0} \dot{H}_\pm = \dot{H}_L - \frac{2\pi}{3\beta M_p^2}\,\frac{\rho_m + p_m + \dot{\phi}^2}{\dot{\phi}^2 + \frac{1}{2}\,\dfrac{\rho_m + p_m}{1 + \frac{\partial V}{\partial \rho_m}}\left(\frac{d}{d\rho_m}\left(3p_m - 4V\right) - 1\right)}. \tag{17}$$

From (17) follows that in points of bound L the derivatives $\dot{H}_+$ and $\dot{H}_-$ are equal and their values do not depend on the model type, as result we have the smooth transition from H_--solution to H_+-solution on bound L, and corresponding cosmological solutions for all types models are regular in metrics, Hubble parameter and its time derivative. At the same time according to (17) the value of the time derivative $\dot{H}_L$ for L-solutions is not equal to $lim_{Z\to 0}\dot{H}_\pm$ and smooth transition from H_--solutions to L-solutions and from L-solutions to H_+-solutions without jump of the derivative $\dot{H}$ is impossible. Note, that according to Eq. (16) the functions S and Q have the following asymptotics for H_+- and H_--solutions at $Z \to 0$:

$$\lim_{Z\to 0}(S - \frac{1}{4}Q) = -\frac{1}{4}\lim_{Z\to 0}\frac{\dot{Z}}{Z} \sim \mp Z^{-\frac{1}{2}} \tag{18}$$

Unlike flat and open models, for which $H_+ = H_-$ only in points of bound L and regular inflationary models include H_+– and H_-–solutions reaching bound L, in the case of closed models the regular transition from H_--solution to H_+-solution is possible without reaching the bound L. It is because by certain value of R according to (14)–(15) we have $H_+ = H_-$ in the case $Z \neq 0$. Such models are regular also in torsion and/or nonmetricity. Regular inflationary solution of such type was considered in Ref.[6].

In order to study the behaviour of cosmological models at the beginning of cosmological expansion, let us analyze extreme points for the scale factor

$R(t)$: $R_0 = R(0)$, $H_0 = H(0) = 0$. (This means that in the case of H_+-solutions $H_{+0} = 0$ and in the case of H_--solutions $H_{-0} = 0$). Denoting values of quantities at $t = 0$ by means of index "0", we obtain from (8)–(9):

$$\frac{k}{R_0^2}\,Z_0^2 + 9\beta^2\left(\frac{\partial V}{\partial\phi}\right)_0^2\dot{\phi}_0^2 = \frac{8\pi}{3M_p^2}\left[\rho_{m0} + \frac{1}{2}\dot{\phi}_0^2 + V_0 - \frac{1}{4}\beta\left(4V_0 - \dot{\phi}_0^2 + \rho_{m0} - 3p_{m0}\right)^2\right] Z_0, \quad (19)$$

$$\dot{H}_0 = \left\{\frac{8\pi}{3M_p^2}\left[V_0 - \dot{\phi}_0^2 - \frac{1}{2}(\rho_{m0} + 3p_{m0}) - \frac{1}{4}\beta\left(4V_0 - \dot{\phi}_0^2 + \rho_{m0} - 3p_{m0}\right)^2\right] + 3\beta\left[\left(\frac{\partial^2 V}{\partial\phi^2}\right)_0\dot{\phi}_0^2 - \left(\frac{\partial V}{\partial\phi}\right)_0^2\right] + 18\beta^2\left(\frac{\partial V}{\partial\phi}\right)_0^2\dot{\phi}_0^2\,Z_0^{-1}\right\}$$
$$\times\left\{Z_0 - 3\beta\left[\dot{\phi}_0^2 + \frac{1}{2}\,\frac{\rho_{m0} + p_{m0}}{1 + \left(\frac{\partial V}{\partial\rho_m}\right)_0}\left(\frac{d}{d\rho_m}\,(3p_m - 4V)_0 - 1\right)\right]\right\}^{-1}, \quad (20)$$

where $Z_0 = 1 - \beta\left(4V_0 - \dot{\phi}_0^2 + \rho_{m0} - 3p_{m0}\right)$. A bounce point is described by Eq. (19), if the value of $\dot{H}_0$ is positive. By using Eq.(19) we can rewrite the expression of $\dot{H}_0$ in the form

$$\dot{H}_0 = \left\{\frac{8\pi}{M_p^2}\left[V_0 + \frac{1}{2}(\rho_{m0} - p_{m0}) - \frac{1}{4}\beta\left(4V_0 - \dot{\phi}_0^2 + \rho_{m0} - 3p_{m0}\right)^2\right] + 3\beta\left[\left(\frac{\partial^2 V}{\partial\phi^2}\right)_0\dot{\phi}_0^2 - \left(\frac{\partial V}{\partial\phi}\right)_0^2\right] - \frac{2k}{R_0^2}\,Z_0^2\right\}$$
$$\times\left\{Z_0 - 3\beta\left[\dot{\phi}_0^2 + \frac{1}{2}\,\frac{\rho_{m0} + p_{m0}}{1 + \left(\frac{\partial V}{\partial\rho_m}\right)_0}\left(\frac{d}{d\rho_m}\,(3p_m - 4V)_0 - 1\right)\right]\right\}^{-1}. \quad (21)$$

We see from (21) unlike GR the presence of gravitating matter (with $p_m \le \rho_m$) does not prevent from the bounce realization[c] Eq. (19) determines in space P extremum surfaces depending on the value of β and in the case of closed and open models also parametrically on the

[c]In GR a bounce is possible only in closed models if the following condition $V_0 - \dot{\phi}_0^2 - \frac{1}{2}(\rho_{m0} + 3p_{mo}) > 0$ takes place.

scale factor R_0. In the case of various scalar field potentials applying in inflationary cosmology the value of $\dot{H}_{+0}$ or $\dot{H}_{-0}$ is positive on the greatest part of extremum surfaces, which can be called "bounce surfaces"[d]. By giving concrete form of potential V and choosing values of R_0, ϕ_0, $\dot{\phi}_0$ and ρ_{m0} at a bounce, we can obtain numerically particular bouncing solutions of GCFE for various values of parameter β.

The analysis of GCFE shows, that properties of cosmological solutions depend essentially on parameter β, i.e. on the scale of extremely high energy densities. From physical point of view interesting results can be obtained, if the value of $|\beta|^{-1}$ is much less than the Planckian energy density[6], i.e. in the case of large in module values of parameter β (by imposing $M_p = 1$). In order to investigate cosmological solutions at the beginning of cosmological expansion in this case, let us consider the GCFE by supposing that

$$\left|\beta\left(4V - \dot{\phi}^2 + \rho_m - 3p_m\right)\right| \gg 1,$$
$$\rho_m + \frac{1}{2}\dot{\phi}^2 + V \ll |\beta|\left(4V - \dot{\phi}^2 + \rho_m - 3p_m\right)^2. \tag{22}$$

Note that the second condition (22) does not exclude that ultrarelativistic matter energy density can dominate at a bounce. We obtain:

$$\begin{aligned}
&\frac{k}{R^2}\left(4V - \dot{\phi}^2 + \rho_m - 3p_m\right)^2 + \\
&\left\{H\left[4V + 2\dot{\phi}^2 + \rho_m - 3p_m + \frac{3}{2}\frac{\rho_m + p_m}{1 + \frac{\partial V}{\partial \rho_m}}\left(\frac{d}{d\rho_m}(3p_m - 4V) - 1\right)\right]\right. \\
&\qquad\left. + 3\frac{\partial V}{\partial \phi}\dot{\phi}\right\}^2 = \frac{2\pi}{3M_p^2}\left(4V - \dot{\phi}^2 + \rho_m - 3p_m\right)^3
\end{aligned} \tag{23}$$

$$\begin{aligned}
&\dot{H}\left[4V + 2\dot{\phi}^2 + \rho_m - 3p_m - \frac{3}{2}\frac{\rho_m + p_m}{1 + \frac{\partial V}{\partial \rho_m}}\left(\frac{d}{d\rho_m}(3p_m - 4V) - 1\right)\right] \\
&\quad \times \left(4V - \dot{\phi}^2 + \rho_m - 3p_m\right) + H^2\left\{\left[4V - 16\dot{\phi}^2 + \rho_m - 3p_m\right.\right. \\
&\quad + \frac{3}{2}\frac{\rho_m + p_m}{1 + \frac{\partial V}{\partial \rho_m}}\left(\frac{d}{d\rho_m}(3p_m - 4V) - 1\right) - \frac{9}{2}\frac{\rho_m + p_m}{\left(1 + \frac{\partial V}{\partial \rho_m}\right)^2}
\end{aligned}$$

[d] If $|\beta|^{-1} \ll M_p^4$ the derivative $\dot{H}_0$ is negative in the neighbourhood of origin of coordinates in space P that leads to appearance of oscillating solutions of GCFE[28].

$$\times\left(\left(1+\frac{dp_m}{d\rho_m}\right)\left(\frac{d}{d\rho_m}(3p_m-4V)-1-3\frac{\rho_m+p_m}{1+\frac{\partial V}{\partial\rho_m}}\frac{\partial^2 V}{\partial\rho_m^2}\right)\right.$$
$$\left.+3(\rho_m+p_m)\frac{d^2p_m}{d\rho_m^2}\right)\Bigg]\left(4V-\dot\phi^2+\rho_m-3p_m\right)$$
$$\left.-\frac{9}{2}\left[\frac{\rho_m+p_m}{1+\frac{\partial V}{\partial\rho_m}}\left(\frac{d}{d\rho_m}(3p_m-4V)-1\right)+2\dot\phi^2\right]^2\right\}$$
$$-12H\dot\phi\left\{\left[\frac{\partial V}{\partial\phi}+\frac{9}{8}\frac{\rho_m+p_m}{\left(1+\frac{\partial V}{\partial\rho_m}\right)^2}\frac{\partial^2 V}{\partial\phi\partial\rho_m}\left(1+\frac{1}{3}\frac{d}{d\rho_m}(p_m+2V)\right)\right]\right.$$
$$\times\left(4V-\dot\phi^2+\rho_m-3p_m\right)+\frac{3}{2}\frac{\partial V}{\partial\phi}\left[\frac{\rho_m+p_m}{1+\frac{\partial V}{\partial\rho_m}}\left(\frac{d}{d\rho_m}(3p_m-4V)-1\right)\right.$$
$$\left.\left.+2\dot\phi^2\right]\right\}+3\left[\frac{\partial^2 V}{\partial\phi^2}\dot\phi^2-\left(\frac{\partial V}{\partial\phi}\right)^2\right]\left(4V-\dot\phi^2+\rho_m-3p_m\right)$$
$$-18\left(\frac{\partial V}{\partial\phi}\right)^2\dot\phi^2=\frac{2\pi}{3M_p^2}\left(4V-\dot\phi^2+\rho_m-3p_m\right)^3, \quad (24)$$

Equations (23)–(24) do not contain the parameter β. According to Eq. (23) the Hubble parameter in considered approximation is equal to

$$H_\pm=\left[4V+2\dot\phi^2+\rho_m-3p_m+\frac{3}{2}\frac{\rho_m+p_m}{1+\frac{\partial V}{\partial\rho_m}}\left(\frac{d}{d\rho_m}(3p_m-4V)-1\right)\right]^{-1}$$
$$\times\left[-3\frac{\partial V}{\partial\phi}\dot\phi\pm\left|4V-\dot\phi^2+\rho_m-3p_m\right|\sqrt{\frac{2\pi}{3M_p^2}\left(4V-\dot\phi^2+\rho_m-3p_m\right)-\frac{k}{R^2}}\right] \quad (25)$$

and extreme points of the scale factor are determined by the following condition

$$\frac{k}{R_0^2}+9\left[\frac{\left(\frac{\partial V}{\partial\phi}\right)_0\dot\phi_0}{4V_0-\dot\phi_0^2+\rho_{m0}-3p_{m0}}\right]^2=\frac{2\pi}{3M_p^2}\left(4V_0-\dot\phi_0^2+\rho_{m0}-3p_{m0}\right). \quad (26)$$

From Eq. (24) the time derivative of the Hubble parameter at extreme points is

$$\dot{H}_0 = \left\{ \frac{2\pi}{3M_p^2} \left(4V_0 - \dot{\phi}_0^2 + \rho_{m0} - 3p_{m0}\right)^2 \right.$$
$$\left. -3\left[\left(\frac{\partial^2 V}{\partial \phi^2}\right)_0 \dot{\phi}_0^2 - \left(\frac{\partial V}{\partial \phi}\right)_0^2\right] + \frac{18\left(\frac{\partial V}{\partial \phi}\right)_0^2 \dot{\phi}^2}{4V_0 - \dot{\phi}_0^2 + \rho_{m0} - 3p_{m0}} \right\}$$
$$\times \left[4V_0 + 2\dot{\phi}_0^2 + \rho_{m0} - 3p_{m0} - \frac{3}{2}\frac{\rho_{m0} + p_{m0}}{1 + \left(\frac{\partial V}{\partial \rho_m}\right)_0} \left(\frac{d}{d\rho_m}\left(3p_m - 4V\right)_0 - 1\right)\right]^{-1} \tag{27}$$

Obviously Eqs. (26)–(27) correspond to (19)–(20) in considered approximation.

The analysis given in this Section confirms conclusions obtained in our previous papers[2,3]. The existence of limiting bound L and bounce surface in space P in the case of models including scalar fields and gravitating matter ensures regular character of cosmological solutions in metrics, Hubble parameter and its time derivative; corresponding restriction on equation of state of gravitating matter and potential V at extreme conditions is: $\left(1 + \frac{\partial V}{\partial \rho_m}\right)^{-1} \left(\frac{d}{d\rho_m}\left(3p_m - 4V\right) - 1\right) > 0$ (see (14)). In absence of scalar fields the bound L is reduced to a point, particular solution of GCFE is stationary ($H{=}0$), and the condition $Z = 0$ determines a bounce point and the value of limiting energy density simultaneously if at extreme conditions $p_m > \frac{1}{3}\rho_m$[1].

4. Regular inflationary cosmological models

As illustration of discussed theory regular cosmological models in the simplest particular case will be considered in this Section. We will consider models including noninteracting scalar field with potential $V(\phi)$ and ultrarelativistic matter ($p_m = \frac{1}{3}\rho_m$)[2]. In this case the bound L in space P is reduced to two cylindric surfaces $\dot{\phi} = \pm\left(4V + |\beta|^{-1}\right)^{\frac{1}{2}}$. Bounce surface is reduced also to cylindric surfaces in the case under consideration, when the scale of extremely high energy densities is much smaller than the Planckian energy density (see (26)). In connection with this we will consider instead of space P the plane of variables $(\phi,\ \dot{\phi})$ and intersections of the bound L and bounce surface with this plane. We have in this plane two bound $L_\pm$–curves and in the case of flat models two bounce curves B_1 and B_2

determined by equation[e]

$$4V_0 - \dot{\phi}_0^2 = 3\left(\frac{M_p^2}{2\pi}\,V_0'^2\,\dot{\phi}_0^2\right)^{\frac{1}{3}}.$$

Each of two curves $B_{1,2}$ contains two parts corresponding to vanishing of H_+ or H_- and denoting by (B_{1+}, B_{2+}) and (B_{1-}, B_{2-}) respectively. If V' is positive (negative) in quadrants 1 and 4 (2 and 3) on the plane (ϕ, $\dot{\phi}$), the bounce will take place in points of bounce curves B_{1+} and B_{2+} (B_{1-} and B_{2-}) in quadrants 1 and 3 (2 and 4) for H_+-solutions (H_--solutions) (see Fig. 1). To analyze flat bouncing models we have to take into account that

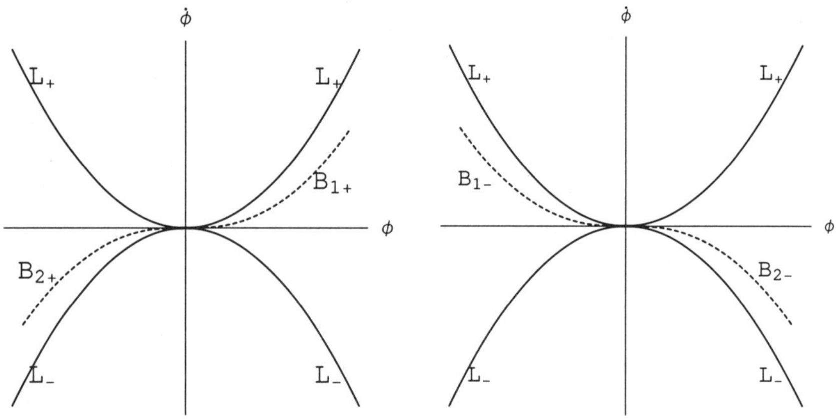

Figure 1. Bound $L_\pm$-curves and bounce curves for flat models in the case of potential $V = \frac{1}{4}\lambda\phi^4$.

besides regions lying between curves $L_\pm$ and corresponding bounce curves the sign of values H_+ and H_- for applying potentials is normal: $H_+ > 0$, $H_- < 0$. The Hubble parameter H_+ is negative in regions between curves (L_+ and B_{1+}), (L_- and B_{2+}), and the value of H_- is positive in regions between curves (L_+ and B_{1-}), (L_- and B_{2-}). As it was noted above any cosmological solution has to contain both H_-- and H_+-solution. The regular transition from H_--solution to H_+-solution takes place in points of $L_\pm$ where $H_+ = H_-$. In the case of open and closed models Eq.(26) determines 1-parametric family of bounce curves with parameter R_0. Bounce curves of closed models are situated in region between two bounce curves B_1 and B_2

[e]The neighbourhood of origin of coordinates is not considered in this approximation, the behavior of bounce curves near origin of coordinates was examined in Ref. [28], where scalar fields superdense gravitating systems were discussed.

of flat models, and in the case of open models bounce curves are situated in two regions between the curves: L_+ and B_1, L_- and B_2. In general case, when approximation (22) is not valid, bounce surface in space P of cosmological models including scalar field and ultrarelativistic matter determined by Eq.(19) in space P depends on parameter β and it is not more cylindric surface. The situation concerning cosmological solutions of Eqs. (8)–(9) does not change.

Note that the analysis of inflationary solutions reaching the bound L by numerical integration of Eqs. (9) and (6) is difficult, because the coefficient at $\dot{H}$ in Eq. (9) tends to zero at $Z \to 0$.

Below particular bouncing cosmological inflationary solution for flat model by using scalar field potential in the form $V = \frac{1}{4}\lambda\phi^4$ $(\lambda = 10^{-12})$ is given. The solution was obtained by numerical integration of Eqs. (6), (9) and by choosing in accordance with Eq.(19) (or (26)) the following initial conditions at a bounce: $\phi_0 = \sqrt{2} \cdot 10^{3/2} M_p$, $\dot{\phi}_0 = 1,998 \cdot 10^{-3} M_p^2$ $(\beta = -10^{18} M_p^{-4})$; initial value of R_0 can be arbitrary. As was noted above, the radiation energy density does not have influence on the dynamics of cosmological model, if the value of ρ_m satisfies the condition (22). A bouncing solution includes: quasi-de-Sitter stage of compression, the stage of transition from compression to expansion, quasi-de-Sitter inflationary stage, stage after inflation. The dynamics of the Hubble parameter and scalar field is presented for different stages of obtained bouncing solution in Figs. 2–4 (by choosing $M_p = 1$). The transition stage from compression

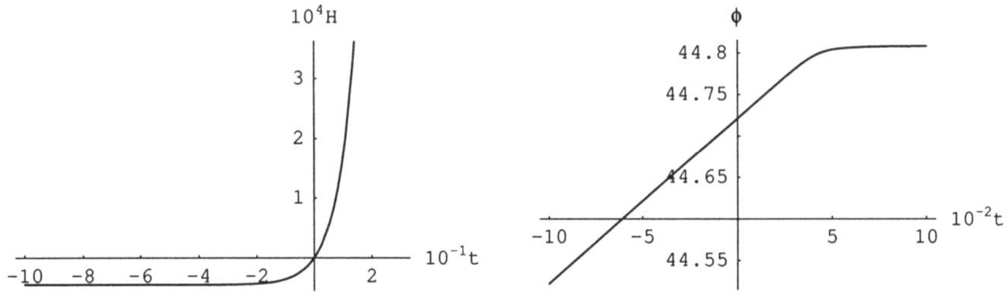

Figure 2. The stage of transition from compression to expansion.

to expansion (Fig. 2) is essentially asymmetric with respect to the point $t = 0$ because of $\dot{\phi}_0 \neq 0$. In course of transition stage the Hubble parameter changes from maximum in module negative value at the end of compression stage to maximum positive value at the beginning of expansion stage. The scalar field changes linearly at most part of transition stage, the derivative $\dot{\phi}$ grows at first being positive to maximum value $\dot{\phi} \sim \dot{\phi}_0$ and then the

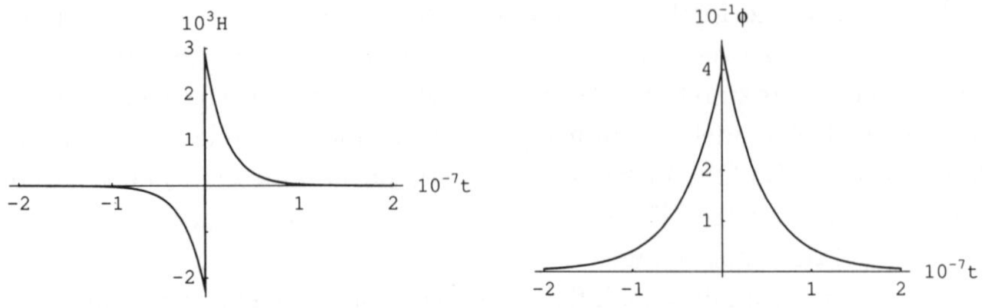

Figure 3. Quasi-de-Sitter stage of compression and inflationary stage.

value of $\dot{\phi}$ decreases and becomes negative. Quasi-de-Sitter inflationary stage and quasi-de-Sitter compression stage are presented in Fig. 3. The amplitude and frequency of oscillating scalar field after inflation (Fig. 4) are different than that of GR, this means that approximation of small energy densities $\left|\beta\left(4V - \dot{\phi}^2 - 2\rho_1\right)\right| \ll 1$ at the beginning of this stage is not valid; however, the approximation (23)–(24) is not valid also because of dependence on parameter β of oscillations characteristics[6]. The behaviour of the Hubble parameter after inflation is also noneinsteinian, at first the Hubble parameter oscillates near the value $H = 0$, and later the Hubble parameter becomes positive and decreases with the time like in GR. Before quasi-de-Sitter compression stage there are also oscillations of the Hubble parameter and scalar field not presented in Figs. 2–4. Ultrarelativistic matter, which could dominate at a bounce has negligibly small energy densities at quasi-de Sitter stages. At the same time the gravitating matter could be at compression stage in more realistic bouncing models, and scalar fields could appear only at certain stage of cosmological compression. As it follows from our consideration regular character of such inflationary cosmological models has to be ensured by cosmological equations of GTG.

The interaction between scalar fields and radiation leads to quantitative corrections of considered inflationary cosmological models. It is necessary to note that corrections for scalar field potentials quadratic in the temperature $T^2 \sim \rho_m^{1/2}$ can play the important role for more late stages of cosmological evolution, when energy densities are sufficiently small and consequences of GTG and GR coincide. In accordance with Eq. (7) these terms change essentially the connection between scale factor R and radiation energy density. By certain restrictions on parameters of potentials V indicated terms can be greater than radiation energy density and scalar field energy density (without terms of interaction) at late stages of cosmological expansion leading to acceleration of cosmological expansion.

Note that bouncing character have solutions not only in classical region, where scalar field potential, kinetic energy density of scalar field and energy density of gravitating matter do not exceed the Planckian energy density, but also in regions, where classical restrictions are not fulfilled and according to accepted opinion quantum gravitational effects can be essential.

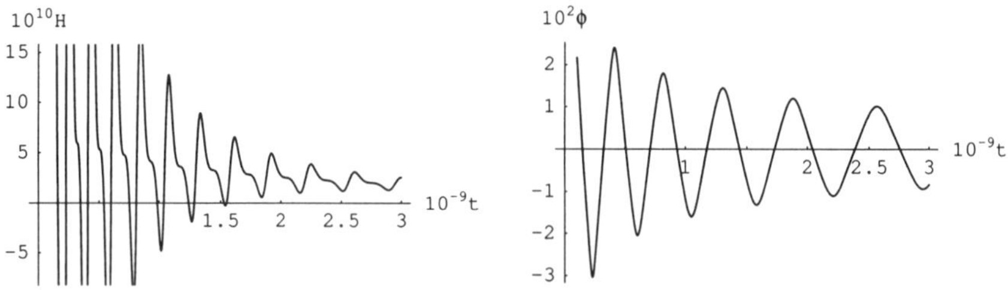

Figure 4. The stage after inflation.

5. Conclusion

As it is shown in our paper, GTG permit to build non-singular cosmology and at first of all regular inflationary cosmology, if gravitating matter and scalar fields satisfy certain restrictions at extreme conditions and indefinite parameter β is negative. All cosmological solutions for flat, open and closed models are regular in metrics, Hubble parameter, its time derivative and, hence, not limited in the time. The presence of scalar fields leading to appearance of inflation in cosmological models changes essentially the structure of GCFE, as result a family of closed models regular also in torsion and/or nonmetricity appears. To build realistic cosmological models we have to know properties of matter filling the evolving Universe, and at first of all the change of equation of state of gravitating matter and properties of scalar fields by evolution of the Universe.

Acknowledgements

I am grateful to Alexander Garkun and Andrey Minkevich for help by preparation of this paper. I am thankful to my wife Eleonora for her permanent support.

Bibliography

1. A. V. Minkevich, *Acta Phys. Polon.* **B29**, 949 (1998).

2. A. V. Minkevich, *Nonlinear Phenomena in Complex Systems,* **6**, N4, 824 (2003); gr-qc/0307026.
3. A. V. Minkevich, gr-qc/0312068; to be published in *Annals of European Acad. Sci.* (2004).
4. A. A. Starobinsky. *Pis'ma v Astron. Zhurn.* **4**, 155 (1978).
5. V. A. Belinsky, L. P. Grishchuk, Ya. B. Zeldovich,I. M. Khalatnikov, *Zhurn. Exp. Teor. Fiz.* **89**, 346 (1985).
6. A. V. Minkevich, gr-qc/0303022.
7. R. Brandenberger, V. Mukhanov, A. Sornborger, *Phys. Rev.,* **D48**, N 9, 1629 (1993).
8. Varun Sahni, gr-qc/0208047.
9. M. Gasperini, G. Veneziano, *Phys. Rep.,* **373**, 1 (2003); *hep-th/0207130.*
10. F. Quevedo, Lectures on String/Brane Cosmology, hep-th/0210292.
11. A. A. Coley and S. Hernik, gr-qc/0303003.
12. P. J. Steinhardt and N. Turok, *Phys. Rev.,* **D65**, 126003 (2002).
13. N. Turok and P. J. Steinhardt, hep-th/0403020.
14. A. V. Minkevich, *Vestsi Akad. Nauk BSSR.* Ser. fiz.-mat., no. 2, 87 (1980); *Phys. Lett.* **A80**, 232 (1980).
15. A. V. Minkevich, *Dokl. Akad. Nauk BSSR.* **29**, 998 (1985).
16. A. V. Minkevich, *Dokl. Akad. Nauk BSSR.* **30**, 311 (1986).
17. A. V. Minkevich, N. H. Chuong, F. I. Fedorov, *Dokl. Akad. Nauk BSSR.* **29**, 697 (1985); *Class. Quantum Grav.* **5**, 515 (1988).
18. A. V. Minkevich, N. H. Chuong, N. V. Hoang, In: *Gravitation and Electromagnetism*, Minsk, BSU, p. 103 (1987).
19. A. V. Minkevich, *Vestsi Akad. Nauk Belarus.* Ser. fiz.-mat., no. 5, 123 (1995); in *Proc. of the Third Alexander Friedmann Intern. Seminar on Gravitation and Cosmology,* St.-Petersburg, July 4–12, 1995, p. 12–22.
20. A. V. Minkevich, N. H. Chuong, *Vestsi Akad Nauk BSSR.* Ser. fiz.-mat., no. 5, 63 (1989).
21. A. V. Minkevich, I. M. Nemenman, *Class. Quantum Grav.* **12**, 1259 (1995).
22. A. V. Minkevich, A. S. Garkun, *Gravitation & Cosmology.* **5**, 115 (1999).
23. A. V. Minkevich, *Int. J. Mod. Phys..* **A17**, 4442 (2002).
24. A. V. Minkevich, *Phys. Lett.* **95A**, 422 (1983).
25. A. V. Minkevich, *Dokl. Akad. Nauk Belarus.* **37**, 33 (1993).
26. A. V. Minkevich and A. S. Garkun, *Class. Quantum Grav.* **17**, 3045 (2000).
27. A. D. Linde, *Physics of Elementary Particles and Inflationary Cosmology.* (Nauka, Moscow, 1990).
28. A. V. Minkevich, A. S. Garkun, Yu. G. Vasilevski, *Nonlinear Phenomena in Complex Systems,* **7**, N1, 78 (2004); gr-qc/0310060.

THE ELECTRODYNAMICS OF MEDIA UNDER TIME-REVERSAL-VIOLATION CONDITION (CP-VIOLATION) AND POSSIBILITY OF T-ODD (CP-ODD) INTERACTION TESTING BEYOND THE STANDARD MODEL

Vladimir G. Baryshevsky

Nuclear Problems Research Institute
11 Bobruiskaya Str., Minsk 220080 Belarus
E-mail: bar@@inp.minsk.by

T-noninvariance (CP-violation) results in the existence of experimentally unstudied new physical phenomena: time-reversal-violating generation of static electric and magnetic fields and CP-odd optical gyrotropy of a medium (photon polarization plane rotation in an external electric field and birefringence depending on mutial orientation of electric and magnetic fields). The above penomenon enables to improve the sensitivity of measurement of the electrical dipole moment of an electron (nucleon, nucleus) and the constants, describing T-odd (CP-odd) interactions.

1. The phenomenon of time-reversal violating generation of static magnetic and electric fields is a basis of a new method for measurement of the electron EDM and T-odd P-odd constants of electron interaction with a nucleon beyond Standard Model

Violation of the time reversal symmetry remains one of the greatest unsolved problems in the elementary particle physics. This fact stimulates plenty of experimental attempts to observe T-odd phenomena.

Particularly, violation of the time reversal invariance of nature laws provides for elementary particles (nuclei, atoms) the possibility to possess an additional quantum characteristic - the electric dipole moment (EDM) $\vec{d}$, which could exist along with other characteristics such as electric charge, magnetic dipole moment, electric and magnetic polarisabilities. Plenty of experiments set the limits for the EDM for different particles, atoms and nuclei[1-6]. The attained experimental level gives, for example, for the electron EDM $d_e < 1.6 \cdot 10^{-27}\ e \cdot cm^4$. The similar evaluations are provided for nuclei. Meanwhile, being combined with calculations the experimental limits for d can be interpreted in terms of fundamental parity and time-invariance-violating (P-,T-odd) constants. These limits tightly constrain competing theories of CP-violation.

So, it is very important to consider new possibilities for measuring constants describing T and CP-odd interactions[1,7–10].

Nowadays there is an appreciable progress in development of methods for measurement of ultra weak magnetic and electric fields. Therefore, new experiments for measurement of the electric dipole moment(EDM) d of electron[1] are being prepared.

F.L.Shapiro's idea[5] to measure the electron EDM by applying a strong electric field to a substance that has an unpaired electron spin is being used for the EDM search[1,6].

The interaction W_E of the electron electric dipole moment $\vec{d}$ with an electric field $\vec{E}$ depends on their orientation:

$$W_E = -\vec{d}\vec{E}, \tag{1}$$

where $\vec{d} = d\frac{\vec{J}}{J}$, $\vec{J}$ is the atom spin, d is the EDM. Spins of electrons

Figure 1. Splitting of levels in an electric field

(atoms) at a low temperature appear to be polarized due to (1) similar to the polarization (magnetization) of electrons by a magnetic field in paramagnetic substances due to the interaction W_B of an electron (atom) magnetic moment $\vec{\mu}$ with a magnetic field $\vec{B}$

$$W_B = -\vec{\mu}\vec{B}. \tag{2}$$

Spins of electrons (atoms) polarized by an electric field induce the magnetic field $\vec{B}_E$ (Fig. 2 (a)) and change in the magnetic flux Φ at the surface of a flat sheet of material[1]:

$$\Delta\Phi = 4\pi\chi A d E^*/\mu_a, \tag{3}$$

$$B_E = \frac{\Delta\Phi}{A} = 4\pi\chi\frac{d}{\mu_a}E^*, \tag{4}$$

where χ is the magnetic susceptibility, $\chi \approx \frac{\rho \mu_a{}^2}{3k_BT}$, ρ is the number density of spins of interest, k_B is Boltzmann's constant and T is the sample temperature. In the cases where simple Langevin paramagnetism is applicable, E^* is the effective electric field at the location of the spins, $\mu_a = g\sqrt{J(J+1)}\mu_B$, where μ_B is the Bohr magneton, μ_a is the atomic or ionic magnetic moment, g is the Lande factor and A is the sample area.

If an external magnetic field acts on either a para- or a ferromagnetic material, the spins in the substance become polarized due to substance magnetization. Therefore, the electric dipole moments appear polarized, too. This results in the induction of an electric field $\vec{E}_B$ (Fig. 2 (b)) (see Ref. D. DeMille in[1]):

$$E_B = 4\pi\rho d P(B), \tag{5}$$

where P represents the degree that the spins are polarized in the sample. According to the analysis[1], modern methods for measurement of B_E and

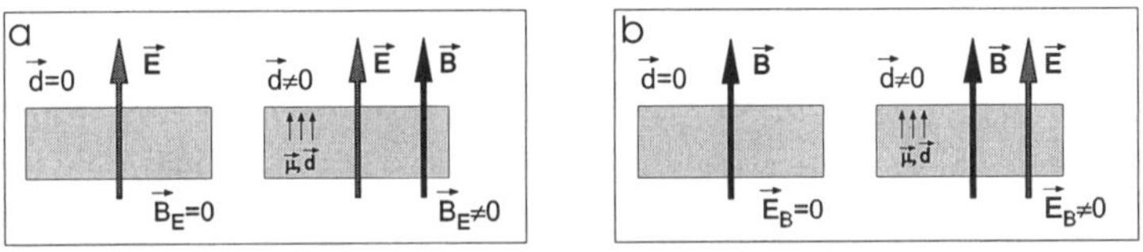

Figure 2. Spins of electrons (atoms) polarized by an electric field induce the magnetic field $\vec{B}_E$ and change in the magnetic flux Φ at the surface of a flat sheet of material (a) and an external magnetic field induces the electric field $\vec{E}_B$ (b).

E_B provide sensitivity for electric dipole moment measurement about 10^{-32} $e\ cm$ and in some cases even 10^{-35} $e\ cm$.

It is important to pay attention to another mechanism of time-reversal violating generation of magnetic and electric fields, which have been discussed by Baryshevsky in[7,8]. According to the idea of[7,8], an induced magnetic moment $\vec{\mu}(\vec{E})$ (and, as a result, a magnetic field (Fig. 3 (a)) of a particle appears due to the action of a field $\vec{E}$ under conditions of violation of P- and T-invariance (and similarly, an induced electric dipole moment $\vec{d}_B$ (an electric field Fig. 3 (b)) of a particle appears due to the action of a field $\vec{B}$). This new effect does not depend on temperature. An effect magnitude is determined by a P-odd T-odd tensor polarizability β_{ik}^T of a particle (atom, molecule, nucleus, neutron, electron and so on). For an atom (molecule), β_{ik}^T arises due to P- and T-odd interaction of electrons with a nucleus. For the stationary state of an atom (molecule) $|N_0\rangle$ the tensor β_{ik}^T

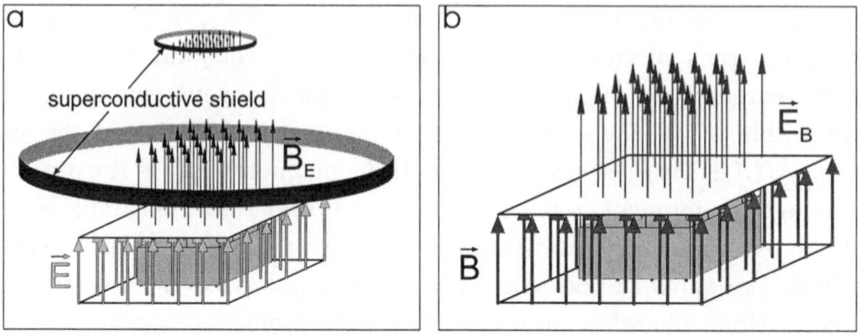

Figure 3. In a substance placed into an electric field the magnetic field is induced (a) and similarly, in a substance placed into a magnetic field the electric field is induced (b).

is as follows:

$$\beta_{ik}^{T}=\sum_{F}\frac{\langle N_0|\widehat{d}_i|F\rangle\,\langle F|\widehat{\mu}_k|N_0\rangle+\langle N_0|\widehat{\mu}_i|F\rangle\,\langle F|\widehat{d}_k|N_0\rangle}{E_F-E_{N_0}},\qquad(6)$$

where $|F\rangle$ is the wave function of a stationary state of the atom, considering T-odd interaction V_w^T, E_F and E_{N_0} are the energies of the atom (molecule) stationary states, $\widehat{\vec{d}}$ and $\widehat{\vec{\mu}}$ are the operators of the electric dipole moment and magnetic moment, respectively and $i,k=1,2,3$ correspond to the axes x,y,z.

Let us place an atom (molecule) into an electric field $\vec{E}$. The induced magnetic dipole moment $\vec{\mu}(\vec{E})$ appears in thiscase[7,8] :

$$\mu_i(\vec{E})=\beta_{ik}^{T}E_k,\qquad(7)$$

The tensor β_{ik}^T (like any tensor of rank two) can be expanded into scalar, symmetric and antisymmetric parts.

The antisymmetric part of the tensor β_{ik}^T is proportional to $e_{ikl}J_l$, where e_{ikl} is the totally antisymmetric tensor of rank three. The symmetric part of the tensor β_{ik}^T is proportional to the tensor of quadrupolarization $Q_{ik}=\frac{3}{2J(2J-1)}[J_iJ_l+J_kJ_l-\frac{2}{3}J(J+1)\delta_{ik}]$. As a result

$$\beta_{ik}^{T}=\beta_s^T\delta_{ik}+\beta_v^Te_{ikl}J_l+\beta_t^TQ_{ik},\qquad(8)$$

where $\beta_s^T,\beta_v^T,\beta_t^T$ are the scalar, vector and tensor P-, T-odd polarizabilities of the particle, respectively. For a substance with the nonpolarized spins, $Sp\ \rho(J)\vec{J}=0$ and $Sp\ \rho(J)Q_{ik}=0$ (here $\rho(J)$ is the atom (molecule) spin density matrix). As a result, for such a substance, β_{ik}^T appears to be a scalar $\beta_{ik}^T=\delta_{ik}\beta_s^T$.

Placement of a nonpolarized atom (molecule, nucleus) into an electric field induces the magnetic dipole moment $\vec{\mu}_E$:

$$\vec{\mu}(\vec{E}) = \beta_s^T \vec{E}, \tag{9}$$

where $\beta_s^T = \sum_F \frac{\langle N_0|\widehat{d}_z|F\rangle \langle F|\widehat{\mu}_z|N_0\rangle + \langle N_0|\widehat{\mu}_z|F\rangle \langle F|\widehat{d}_z|N_0\rangle}{E_F - E_{N_0}}$, $\widehat{d}_z$ and $\widehat{\mu}_z$ are the z components of the operators of the electric dipole moment and magnetic moment, respectively, axis z is parallel to the electric field $\vec{E}$. It should be emphasized that for strong fields (when the distance between atom (molecule) levels is comparable with the energy of interaction with an electric $\vec{E}$ (magnetic $\vec{B}$) field) β_s^T depends on $\vec{E}$ ($\vec{B}$).

Weak interaction is much weaker than strong and electromagnetic interactions. Therefore, to find the wave function $|F\rangle$, the perturbation theory can be applied:

$$|F\rangle = |f\rangle + \sum_n \frac{\langle n|V_w^T|f\rangle}{E_f - E_n}|n\rangle = |f\rangle + \sum_n \eta_{nf}^T|n\rangle, \tag{10}$$

where $|f\rangle$ is the wave function of an atom in the absence of weak interactions and the mixing ratio is $\eta_{nf}^T = \frac{\langle n|V_w^T|f\rangle}{E_f - E_n}$. It should be mentioned that for theoretical analysis of β_s^T in a substance it is necessary to find the wave function of an excited state of an atom in the substance, which is difficult to do.

It follows from (9) that in a substance placed into an electric field the magnetic field is induced[7,8] (Fig. 3 (a)):

$$\vec{B}_E^{ind} = 4\pi\rho\beta_s^T \vec{E}^*. \tag{11}$$

Vice versa, if an atom (molecule, nucleus) is placed into a magnetic field, the induced electric dipole moment $\vec{d}(B)$ (Fig. 3 (b)) appears[7,8],

$$d_i(B) = \chi_{ik}^T B_k, \tag{12}$$

where the tensor polarizability χ_{ik}^T is $\chi_{ik}^T = \beta_{ki}^T$. The dipole moment $\vec{d}(B)$ leads to the induction of an electric field in the substance: $E_i^{ind}(B) = 4\pi\rho\beta_{ki}^T\vec{B}_k^*$,where $\vec{B}^*$ is the local magnetic field, acting on the considered particle in the substance. If an atom is found in a point with the cubic symmetry (or in a liquid), then $\langle Q_{ik}\rangle = Sp\rho(J)Q_{ik} = 0$ and $\left\langle \overrightarrow{J} \right\rangle = Sp\rho(J)\overrightarrow{J} \parallel \overrightarrow{B}^*$. As a result, the terms including β_v^T and β_t^T turn to zero ($e_{ikl}\langle J_l\rangle B_k^* = e_{ikl}B_l^*B_k^* = 0$).

As a consequence, in this case

$$E_{ind}(B) = 4\pi\rho\beta_s^T \vec{B}^*, \tag{13}$$

Hence, analyzing the results of the experiment proposed in[1], one should consider that the appearance of the induced magnetic and electric fields is caused by:

1. A magnetic field induced due to interaction of the electric dipole moment of an atom with an external electric field[1,5] (see (3),(4))
2. A magnetic field induced due to mechanism[7,8] (see (11))
3. A magnetic field appearing as a result of polarization (magnetization) of atom magnetic moments by the local induced magnetic field $\vec{B}_{loc}$ due to interaction W of the magnetic dipole moment of an atom with this field

$$W = -\vec{\mu}_a \vec{B}_{loc}. \tag{14}$$

The local field $\vec{B}_{loc}$ is the sum of two contributions:

$$\vec{B}_{loc} = \vec{B}_{E\ loc} + \vec{B}^{ind}_{loc}, \tag{15}$$

where the field $\vec{B}_{E\ loc}$ is the local magnetic field acting on the atom from the polarized (by mechanism[7,8], see (3), (4)) magnetic moments of the other atoms of the sample. This field depends on temperature and its contribution could be neglected for those temperature values, which provide $\chi \ll 1$. But for temperature $T < 1K$ the susceptibility $\chi \sim \frac{1}{T}$ becomes comparable with 1 and higher, and the energy of interaction of two magnetic dipoles for neighbour atoms occurs of order of $k_B T$ and greater. Thus, in this case, the collective effects, well-known in the theory of phase transition in magnetism, should be taken into account while considering magnetization by an electric field.

The field $\vec{B}^{ind}_{loc} = \vec{B}^{ind}_{1\ loc} + \vec{B}^{ind}_{2\ loc}$ does not depend on temperature.

The field $\vec{B}^{ind}_{1\ loc} = \chi^T_{1\ loc} \vec{E}^*$ is the local magnetic field produced in the point of the considered atom location by the magnetic moments of atoms of the substance (except for the considered atom) induced by the aid of mechanism[7,8] (see (6), (8)); $\chi^T_{1\ loc} \sim \rho \beta^T_s$ is the local P-,T-odd susceptibility of the substance. In general case $\chi^T_{1\ loc}$ is a tensor, but if atom surrounding posseses cubic symmetry, then the principle contribution to this tensor is made by its scalar part ($\chi^T_{1\ loc}$ depends on the substance density and sample shape: for sphere $\chi^T_{1\ loc} = \frac{8\pi}{3} \rho \beta^T_s$; for cylinder $\chi^T_{1\ loc} = 4\pi \rho \beta^T_s$).

The field $\vec{B}^{ind}_{2\ loc}$ is the self-induced magnetic field of the considered atom. The magnetic moment (T-odd current) of the atom induced by an electric field acting on the atom due to mechanism[1,5] causes appearance of the magnetic field inside the atom:

$$\vec{H}^T_E(\vec{r}) = rot\ \vec{A}^T_E(\vec{r}) \tag{16}$$

with the vector potential $\vec{A}^T_E(\vec{r}) = \frac{1}{c}\int \frac{j^T_E(\vec{r}')}{|\vec{r}-\vec{r}')|} d^3r'$, $j^T_E(\vec{r}')$ is the T-odd part of the transition current density operator for an atom (molecule) placed in an electric field[7,8] (it is calculated by the use of wavefunctions (10)). The magnetic interaction Hamiltonian of an atom with the field $\vec{A}^T_E$ can be expressed as[9]:

$$W_{2\ loc} = -\frac{1}{2c}\int (\vec{j}(\vec{r})\vec{A}^T(\vec{r}) + \vec{A}^T(\vec{r})\vec{j}(\vec{r}))d^3r, \tag{17}$$

where $\vec{j}(\vec{r})$ is the atom transition current density operator calculated with the atom wavefunction without consideration of P-,T-odd interactions, $\vec{j}(\vec{r}) = c\ rot\vec{\mu}(\vec{r})$ and $\vec{\mu}(\vec{r})$ is the operator of the atom magnetic moment density. If the atom is found in a point with the cubic symmetry (or in a liquid) then we may omit contributions from atom multipoles.

In this case the above expression can be written as:

$$W_{2\ loc} = -\vec{\mu}_a \vec{B}_{ind} = -\chi^T_{at}\vec{\mu}_a \vec{E}^*_{loc}, \tag{18}$$

where $\vec{\mu}_a = \mu_a \frac{\vec{J}}{J}$ and χ^T_{at} is the T-odd atom susceptibility, which does not depend on the substance density and sample shape, $\chi^T_{at} \sim \beta^T_s \frac{1}{a^3}$ (here a is the typical radius of distribution density of the magnetic moment induced by the electric field in the atom[7,8]). As a result one obtains:

$$\vec{B}^{ind}_{loc} = (\chi^T_{1\ loc} + \chi^T_{at})\vec{E}^*_{loc} = \chi^T_{loc(subst)}\vec{E}^*_{loc}, \tag{19}$$

$$\chi^T_{loc(subst)} = \chi^T_{1\ loc} + \chi^T_{at}. \tag{20}$$

The interaction (2) of the magnetic moment of an atom with the induced magnetic field causes the appearance of the magnetic field due to different population of magnetic levels of the atom in the field $\vec{B}_{loc}$ in thermal equilibrium

$$\vec{B}'_{ind} = 4\pi\chi\vec{B}_{loc} \approx 4\pi\chi\vec{B}^{ind}_{loc} = 4\pi\frac{\rho\mu_a^2}{3k_BT}\chi^T_{loc(subst)}\vec{E}^*, \tag{21}$$

the field $\vec{B}_{loc} = \vec{B}_{E\ loc} + \vec{B}^{ind}_{loc}$, but $\vec{B}_{E\ loc}$ contribution could be neglected for those temperature values, which provide $\chi \ll 1$, and it is omitted here.

Therefore, the flux $\Delta\Phi$, which is going to be measured in the experiment proposed in[1] should be written as:

$$\Delta\Phi = AB_E = 4\pi A(\chi\frac{d}{\mu_a} + \rho\beta_s^T + \chi\chi_{loc(subst)}^T)E^* =$$
$$= 4\pi A[\chi(\frac{d+\mu_a\chi_{at}^T}{\mu_a} + \chi_{1\ loc}^T) + \rho\beta_s^T]E^*, \quad (22)$$

$$\vec{B}_E = 4\pi[\chi(\frac{d+\mu_a\chi_{at}^T}{\mu_a} + \chi_{1\ loc}^T) + \rho\beta_s^T]E^*, \quad (23)$$

where $\chi = \frac{\rho\mu_a^2}{3k_BT}$. It should be noted that the quantity $d + \mu_a\chi_{at}^T$ is

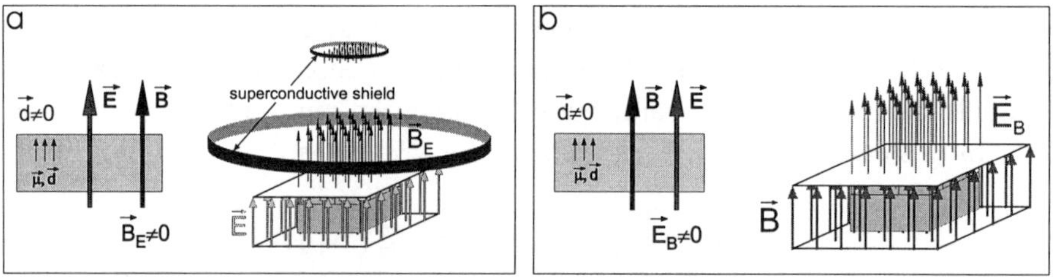

Figure 4. The magnetic field that is observed in an experiment is the sum of magnetic fields produced by the polarized spins of particles and induced by the external electric field ($\sim \beta^T$) (a) and, visa versa, the electric field measured in the experiment is also the sum of two contributions (24) (b).

the electric dipole moment of an atom available for measurement in conventional EDM experiments studying atom (molecule) spin precession in an external electric field. It is well known that the atom EDM is arisen from several mechanisms: the contribution proportional to d_e, the contribution due to T-,P-odd interaction of atom electrons with nuclear nucleons (dependent and independent on nuclear spin)[12] (d also contains contribution $\sim \beta_s^T \frac{\mu_a}{a^3}$ induced according to (12) by the magnetic field $\sim \frac{\mu_a}{a^3}$, which is produced inside the paramagnetic atom by its electrons).

According to (18,23) there is one more addition to the EDM $\sim \mu_a\chi_{at}^T$. Note that the contribution to the atom EDM proportional to the nucleus spin I is equal to zero if the sample temperature is high ($\langle I \rangle = 0$ at high temperatures).

Let us consider now the experiment to detect the electric dipole moment of an electron by means of measurement of the electric field[1] (see (5)). In this case we also should take into consideration the effect[7,8] of the EDM induction by the magnetic field (12). Thus, the electric field measured in the experiment[1] is as follows (see (5), (13), Fig. 4 (b)):

$$E_B = 4\pi\rho(d_a P(B) + \beta_s^T B^*), \tag{24}$$

d_a is the atom EDM containing contribution $\sim \beta_s^T \frac{\mu_a}{a^3}$ induced according to (12) by the magnetic field $\sim \frac{\mu_a}{a^3}$, which is produced inside the paramagnetic atom by its electrons.

So, measurement of $\Delta\Phi$ and E_B provides knowledge about the atom EDM and β_s^T. To distinguish these contributions one should consider the fact that χ and $P(B)$ depend on temperature, while β_s^T does not. Therefore, studying B_E and E_B dependence on temperature allows one to evaluate contributions from the EDM and β_s^T to the measured effect.

It should be particularly emphasized that $\Delta\Phi$ and E_B differs from zero even when the electron EDM d_e is equal to zero.

It should be also emphasized that the polarizability β_s^T differs from zero even for atoms with the zero spin, for which EDM is absent. In this case (as well as for high temperatures when the average atom spin $\left\langle \vec{J} \right\rangle = 0$) only the effect[7,8], described by the terms (23), (24), proportional to β_s^T, contributes to the induced electric and magnetic fields. If the substance consists of several types of atoms, then their contribution to the induced field is expressed as a sum of contributions from different atoms:

$$\vec{B}_E = 4\pi\rho \sum_n c_n \beta_{ns}^T \vec{E}_n^*, \vec{E}_B = 4\pi\rho \sum_n c_n \beta_{ns}^T \vec{B}_n^*, \tag{25}$$

where c_n is the concentration of atoms of the type n, $\vec{E}_n^*$ and $\vec{B}_n^*$ are the local fields acting on atoms of the type n.

Now let us consider what information about constants of T-,P-odd interaction of an electron with a nucleus can be obtained from studying the effect[7,8] of time-reversal violating generation of fields $\vec{E}_B$ and $\vec{B}_E$ (describing by (24), (25)).

According to[1] we can expect a magnetic induction measurement sensitivity about 3×10^{-15} $G/\sqrt{Hz}$. In ten days of averaging the sensitivity is $\sim 10^{-18}$ G. This leads to the sensitivity for d_e measurement of about 10^{-32} e cm[1].

Such sensitivity of magnetic induction measurement provides for polarizability β_s^T measurement the sensitivity $\beta_s^T = \frac{B_E}{4\pi\rho E^*} \sim 10^{-43}$ cm^3 (it is supposed that $\rho \approx 2 \div 3 \cdot 10^{22}$ and $E^* \sim 10$ $\frac{kV}{cm}$, for example, for liquid and solid Xe $\rho \approx 2 \cdot 10^{22}$).

Let us consider now the possibilities given by the experiment studying the electric field, which is induced by a magnetic field[7,8] for measurement of the polarizability β_s^T. Analysis[1] shows that existing methods of the electric field measurement allow to measure electric fields $E \sim 10^{-13} \div 10^{-14} \frac{V}{cm} \sim 3 \cdot 10^{-16} \div 3 \cdot 10^{-17}$ CGSE in ten days operation. Therefore, for $\beta_s^T = \frac{E_B}{4\pi\rho B}$ we can get the estimation $\beta_s^T \approx 10^{-43} \div 10^{-45}$ cm^3 ($\rho \approx 2 \cdot 10^{22}$, $B \approx 10^4 \div 5 \cdot 10^4$ Gauss).

The obtained evaluation for β_s^T ($\beta_s^T \sim 10^{-43} \div 10^{-45}$ cm^3) and the expressions (6), (10) allow us to evaluate the mixing ratio η_{nf}^T. Recall that conventional T-,P-even polarizability of an atom can be expressed similarly (6) with replacement of the matrix element μ by the matrix element d (and $\eta_{nf}^T = 0$). Therefore, we can estimate $\beta_s^T \sim \beta_s \alpha \eta_T$, where β_s is the conventional T-,P-even polarizability of the atom, $\alpha = \frac{1}{137}$ is the fine structure constant and η_T is some average value for the coefficient of mixing of opposite parity levels η_{nf}^T. The estimation for η_T follows from the above: $\eta_T \sim \frac{\beta_s^T}{\beta_s \alpha} \approx 10^{-17} \div 10^{-19}$ (the atom polarizability is of the order $\beta_s \sim 10^{-24}$ cm^3, for example, for Xe according to[13] $\beta_s = 2.7 \cdot 10^{-24}$ cm^3).

Two types of T-,P-odd interactions contribute to the constant of mixing of opposite parity levels η_{nf}^T: the interaction of the electron EDM with the coulomb field of the nucleus and the T-,P-odd interaction of the electron with the nucleus nucleons.

The spin-independent part of the T-,P-odd interaction of an electron with nucleons is described by two constants[12]: k_{1p} describes interaction with protons and k_{1n} describes interaction with neutrons. For example, calculation adduced in[12] for Cs provides $\eta_T = 3.7 \cdot 10^{-11}(0.41k_{1p} + 0.59k_{1n})$, where the sum $(0.41k_{1p} + 0.59k_{1n}) < 5 \cdot 10^{-4}$. Therefore, for Cs $\eta_T \lesssim 10^{-14}$. The same limits for the sum k_{1p} and k_{1n} were obtained from the EDM measurements for ^{129}Xe[12] $(0.4k_{1p} + 0.6k_{1n}) \lesssim 10^{-4}$.

Let us note that in the experiment[2] planned to measure the EDM of Cs atoms trapped in optical lattices it is expected to obtain the sensitivity for Cs EDM measurement of about $d_{Cs} \approx 3 \cdot 10^{-25}$ $e \cdot cm$. This value of the atom EDM provides to improve estimation for η_T and to get limits $\eta_T \lesssim 10^{-16}$ and the sum of k_{1p}, $k_{1n} \lesssim 5 \cdot 10^{-6}$.

As it was shown above the experiments studying β_s^T provide for mixing coefficient the value $\eta_T \lesssim 10^{-17} \div 10^{-19}$. Therefore, these experiments give hope to reduce (three orders) the estimations for the sum k_{1p}, k_{1n} (it is expected to be $\lesssim 10^{-7} \div 10^{-9}$). This value for k_{1p}, k_{1n} is significantly lower than the limitation that could be obtained from the results of measurements of the atom dipole moment those have been done earlier (see, for example,[12]) and from the proposed experiment[2].

Let us consider now what limits for the electron EDM d_e can be obtained from study of the effect[7,8]. To estimate contribution from d_e to polarizability $\beta_s^T \sim \beta_s \alpha \eta_T$ let us use connection of mixing coefficient η_T^e (caused by the electron EDM d_e) with the atom EDM d_A induced by the electron EDM d_e: $d_A \sim ea\eta_T^e \sim Rd_e$[12], R is the atomic EDM enhanced factor. Therefore, $\eta_T^e \sim \frac{Rd_e}{ea}$. Using the mentioned estimations for β_s^T and η_T^e we obtain $d_e \sim \frac{a\beta_s^T}{\beta_s \alpha R}\ e \cdot cm$ i.e. $d_e \sim 10^{-27} \div 10^{-30}\ e \cdot cm$ ($\beta_s^T \approx 10^{-43} \div 10^{-45}\ cm^3$, $R \approx 10^2 \div 5 \cdot 10^2$). Let us remind that the upper limit $d_e \lesssim 1.6 \cdot 10^{-27}\ e \cdot cm$ follows from the experiments with Tl[4].

Thus, experimental study of the effect[7,8] provides to get more (three or four orders) strict limits for constants those describe T-,P-odd interactions of an electron with nucleons and for d_e. Such improvement of estimations constrains theories beyond the Standard Model. To study the effect[7,8] one could use different atoms (molecules) and substances (for example, ferroelectric crystals providing very high electric fields for heavy atoms) and this study does not require target cooling to ultralow temperatures (1 K and lower). Spins of atoms (molecules) can be nonpolarized or atoms (molecules) can be spinless.

2. About possibility to measure an electric dipole moment (EDM) of nuclei in the range $10^{-27} \div 10^{-32}\ e \cdot cm$ in experiments for search of time-reversal violating generation of magnetic and electric fields.

Very sensitive solid state based electron EDM experiments are preparing now[1]. They will provide for the electron EDM measurement the sesitivity of about $10^{-32}\ e \cdot cm$ or even better ($10^{-35}\ e \cdot cm$)[1].

The idea of the experiment is based on the measurement of a magnetic field, which appears since the electron spin (and, therefore, the magnetic moment) is oriented along an external electric field E due to interaction of the electron EDM with the field E[5,6].

According to[1] measurement of the electron EDM at the sensitivity rate of about $10^{-32}\ e \cdot cm$ requires substance cooling up to $T \sim 10^{-2}$ K, while for $10^{-35}\ e \cdot cm$ the temperature $T \sim 10\mu K$ is necessary.

But it was shown in[9] that for temperature values $T \approx 10^{-1} \div 10^{-2}$ the magnetic susceptibility of matter χ becomes compartible with 1 and higher. The energy of interaction of two electron magnetic dipoles for neighbour atoms occurs of order of $k_B T$ and greater (k_B is Boltzmann's constant). Thus, in this case, the collective effects, well-known in the theory of phase transitions in magnetism, should be taken into account while considering magnetization by an electric field. Particularly, spontaneous magnetization of a system can appear. Emergent great magnetic field is caused by fluctuations of a magnetic field and weak residual magnetic fields rather than external electric fields. In such conditions an electric field would not affect distinctly on the value of the measured magnetic field. Therefore, measurement of the electron EDM in such conditions becomes difficult.

In the present section it is shown that the above described problems make considerably attractive carrying out the experiment to search the EDM of nuclei by measurement of the magnetic (electric) field that appears at polarization of nuclear spins due to interaction of the nuclear EDM with an external electric (magnetic) field. The magnetic moment of a nucleus is small. Therefore, considering the degree of polarization to be the same for electrons and nuclei we obtain the value of this magnetic field for nuclei of two or three orders lower then for electrons. But the phase transition of nuclear spins in an ordered state appears at temperatures $10^4 \div 10^5$ times lower comparing with electrons. According to[11] the nuclear dipole ordering appears at $T_N \sim 10^{-6}$ K. Let us consider the parameter $\varkappa = \frac{d_N E}{kT}$, which describes the degree of nuclei spin polarization due to interaction of the nuclear EDM d_N with an electric field E in the temperature range above the temperature of phase transition. The temperature T_N is $10^4 \div 10^5$ times lower than the similar temperature for electrons T_e ($T_e \sim 10^{-1} \div 10^{-2}$ K). Hence, for nuclei $\varkappa$ is $10^4 \div 10^5$ times greater than for electrons.

As a result, the magnetic field produced by nuclei cooled to the temperature T_N is even stronger than that produced by electrons cooled to the temperature T_e. As a consequence, the nuclear EDM can be measured with sensitivity about 10^{-32} $e \cdot cm$.

3. Nuclei in an electric field at low temperatures

Thus, let us consider a substance, placed into an electric field.

Interaction W_E of the EDM $\vec{d}_N$ of a nucleus with an electric field $\vec{E}$ is similar to interaction of magnetic moment of a nucleus with a magnetic field and can be expressed as:

$$W_E = -\vec{d}_N \vec{E} \tag{26}$$

where $\vec{d}_N = d_N \frac{\vec{J}}{J}$, $\vec{J}$ is the nucleus spin.

If the nucleus environment in the substance does not possess cubic symmetry, then the energy of interaction of nuclear quadrupole moment with a nonuniform electric field field produced nucleus environment should be added to (26). If an external magnetic field also presents, then the energy of interaction of nuclear magnetic moment with this field should be added to (26), too.

But here we will omit these contributions, assuming that there is no an external magnetic field and an elementary cell of the substance is cubic.

The interaction of nucleus with the field $\overrightarrow{E}$ (26) makes spins of nuclei at low temperature polarized similar to the polarization (magnetization) of nuclei by a magnetic field due to the interaction W_B of a nucleus magnetic moment μ_N with a magnetic field $\overrightarrow{B}$:

$$W_B = -\overrightarrow{\mu}_N \overrightarrow{B} \tag{27}$$

Spins of nuclei polarized by an electic field induce the magnetic field $\overrightarrow{B}_E$ and change in the magnetic flux Φ at the surface of a flat sheet of material (see Fig. 2 (a)):

$$\Delta\Phi_N = 4\pi\chi_N A \frac{d_N}{\mu_N} E^* \tag{28}$$

$$B_E = \frac{\Delta\Phi_N}{A} = 4\pi\chi_N \frac{d_N}{\mu_N} E^*, \tag{29}$$

where A is the sample area, χ_N is the magnetic susceptibility of nuclei subsystem.

For the temperature range above the temperature of nuclear magnetic ordering the susceptibility χ_N is described by Langevin's formula as follows:

$$\chi_N \approx \frac{\rho\mu_N^2}{3k_B T}, \tag{30}$$

ρ is the number of nuclear spins in cm^3, k_B is Boltzmann's constant, $\mu_N = g_N\sqrt{J(J+1)}\mu_{NB}$, μ_{NB} is the nucleus Bohr magneton, g_N is the gyromagnetic ratio, E^* is the effective electric field at the location of the nuclear spin. Reasons providing interaction of an electric dipole moment with an electric field in spite of electrostatic shielding are specified in[1,12].

Let us compare (29) for $\Delta\Phi_N$ with the similar expression in[1] for the magnetic flux $\Delta\Phi_e$ produced by electrons of atoms:

$$\Delta\Phi_e = 4\pi\chi_e A \frac{d_e}{\mu_a} E^*, \tag{31}$$

where χ_e is the magnetic susceptibility of a material caused by paramagnetic atoms, μ_a is the magnetic moment of the electron shell of the atom.

As it has been already mentioned, the magnitude of the susceptibility χ grows with temperature T tending to the temperature of magnetic ordering and appears equal to 1 at some temperature value T_1. At this temperature (note that for nuclei $T_{1N} \sim 10^{-6}$ K, while for atom spins $T_1 \sim 10^{-1} \div 10^{-2}$ K) the magnetic flux produced by nuclei is ($T_{1N} \sim 10^{-6}$ K)

$$\Delta\Phi_N(T_{1N}) = 4\pi A \frac{d_N E^*{}_N}{\mu_N}, \tag{32}$$

the magnetic flux produced by atoms ($T_{1e} \sim 10^{-1} \div 10^{-2}$ K)

$$\Delta\Phi_e(T_{1e}) = 4\pi A \frac{d_e E^*{}_a}{\mu_a}. \tag{33}$$

Let us consider temperature range close to the temperature of nuclear magnetic ordering. Suppose the nuclear EDM is of the same order as the EDM of an electron (atom) $d_N E^*{}_N \sim d_e E^*{}_a$.

Then, from (32), (33) we obtain the magnetic flux $\Delta\Phi_N(T_{1N}) = \frac{\mu_a}{\mu_N}\Delta\Phi_e(T_{1e})$ i.e. the magnetic flux produced by nuclei is $\frac{\mu_a}{\mu_N}$ times greater than the magnetic flux produced by atom spins!

According to estimations[1] the sensitivity for the atom EDM measurement $\sim 10^{-30}$ $e\cdot$ cm in 10 days of averaging (and even 10^{-32} $e \cdot cm$) can be obtained for the material temperature $T_a = 10^{-2}$ K.

Cooling of nuclear system to the temperature $T_N \approx 10^{-5}$ K provides the magnetic flux $\Delta\Phi_N = \frac{\mu_N}{\mu_a}\frac{T_a}{T_N}\Delta\Phi_e$. Due to $\frac{\mu_a}{\mu_N}\frac{T_a}{T_N} \approx 1$ the sensitivity for measurement of the nuclear EDM appears the same as for the electron (atom) EDM i.e. $d_N \approx 10^{-30} \div 10^{-32}$ $e\cdot$ cm.

Thus, the method for measurement of the nuclear EDM by means of measurement of the induced magnetic field gives hope to obtain considerably substantial limit for the nuclear EDM. Let us note that in ferroelectric materials there are very intense internal electric fields $E*$ that are considerably more intense than external fields. This provides to advance to d_N values less than 10^{-32} $e\cdot$ cm.

4. Nuclei in a magnetic field at low temperatures

If an external magnetic field acts on a material, the nuclei spins become polarized due to nucleus magnetization. Therefore, the nucleus electric dipole moments appears polarized, too. This results in the induction of an electric field $\vec{E}_B$ (see Fig. 2 (b)) (similar to the electron case[1]):

$$E_B = 4\pi\rho d_N P(B), \tag{34}$$

where P represents the degree that the spins of nuclei are polarized in the sample.

There are methods providing $P(B) \sim 1$ for nuclei at low temperatures. According to the analysis[1] the technique for electric fields measurement provides a sensitivity $\sim 10^{-30}$ $e \cdot cm$ in 10 days of operation. The same sensitivity can be obtained for the nuclear EDM due to high polarization degree available for nuclear spins at low temperatures.

An electric field induced by a magnetic field in vacuum can be increased by the use of electrostatic shielding or by selection of target shape (for example, a target having an edge). This gives hope for further improvement of the possibile limits for the EDM measurement ($d < 10^{-30}$ $e\cdot$ cm) in such experiments.

5. Time-reversal violating generation of static magnetic and electric fields

It is important to pay attention that carrying out the proposed experiment to measure the nuclear EDM one should consider additional effect of time-reversal violating generation of static magnetic and electric fields caused by the T-odd polarizability of atoms, molecules and other particles[7,8,9,10] (see sections 1–2).

Therefore, the magnetic field $\vec{B}^*$, which is going to be measured in the proposed experiment, should be written as:

$$\vec{B}^* = \vec{B}_E^N + \vec{B}^{ind}(E) + \vec{B}_E^A, \tag{35}$$

where $\vec{B}_E^A$ is the magnetic field produced by atom spins, $\vec{B}^{ind}(E) = 4\pi\rho\beta_{ik}^T\vec{E}_k^*$.

Let us consider now the experiment to detect the electric dipole moment of the nucleus by means of measurement of the electric field (see (30)). In this case we also should take into consideration the effect of electric field induction by the magnetic field and the electric field E_B^A produced by the polarized EDMs of atoms.

Thus, the electric field measured in the proposed experiment is as follows (see (5), (13)):

$$\vec{E}_B = \vec{E}_B^N + \vec{E}^{ind}(B) + \vec{E}_B^A. \tag{36}$$

So, measurement of $\vec{B}_E$ and $\vec{E}_B$ provides knowledge about the nuclear EDM, the electron EDM and β_s^T. To distinguish these contributions one should consider the fact that studying B_E and E_B dependence on temperature allows one to evaluate different contributions from[9].

It should be emphasized that the atom EDM does not contribute to the discussed phenomena if the substance is diamagnetic (atom spin is equal to zero or spin atoms are nonpolarized)[9].

If the substance consists of several types of atoms (nuclei), then their contribution to the induced field is expressed as a sum of contributions from different atoms:

$$\vec{B}_E = \sum_n \vec{E}^*_n, \vec{E}_B = \sum_n \vec{B}^*_n, \tag{37}$$

6. Time-reversal-violating optical gyrotropy

The time reversal violating weak interactions yield to the appearance of diverse interesting optical phenomena. In the present section the optical anisotropy of medium caused by T-odd interactions of electrons and nuclei in atoms and molecules is investigated. The experiments are discussing for observation of the polarization plane rotation phenomena caused by the pseudo-Zeeman splitting of the atom (molecule) levels by an external electric field $\overrightarrow{E}$ due to interaction $W = -\overrightarrow{d_a} \cdot \overrightarrow{E}$ of the atom (molecule) EDM $\overrightarrow{d_a}$ with an electric field[18–22] (this effect is similar to the magneto-optic Macaluso-Corbino effect[23]).

It should be noted that the mentioned experiments use the possible existence of such intrinsic quantum characteristic of an atom (molecule) as a static EDM. According to[24–26] together with the EDM there is one more characteristic of atom (molecule) describing its response to the external field effect - the T- and P-odd polarizability of atom (molecule) β^T. This polarizability differs from zero even if the EDM of an electron is equal to zero and pseudo-Zeeman splitting of atom (molecule) levels is absent.

Both β^T and the EDM yield to appearance of several new optical phenomena. Let us mention two of them:

a). the T-odd photon birefringence effect[7] (i.e effect when plane polarized photons are converted to circular polarized ones and vice versa, this effect is similar to magneto-optic birefringence Cotton-Mouton effect[27])

b). the photon polarization plane rotation and circular dichroism in an optically homogeneous isotropic medium exposed to an electric field caused by the Stark mixing of atom (molecule) levels[8,44]. This T-odd phenomenon is a kinematic analog of the well known T-even phenomenon of Faraday effect of the photon polarization plane rotation in the medium exposed to a magnetic field due to Van-Vleck mechanism. Similarly to the well known P-odd T-even effect of light polarization plane rotation for which the intrinsic spin spiral of atom is responsible[12], this effect is caused by the atom magnetization appearing under external electric field action. Moreover, according to[8], the magnetization of atom appearing under action of static electric field causes the appearance of induced magnetic field $\overrightarrow{H_{ind}}(\vec{E})$. The energy of interaction of atom magnetic moment $\vec{\mu_a}$ with this field is $W_H = -\vec{\mu_a}\cdot \overrightarrow{H_{ind}}(\vec{E})$. Therefore, the total splitting of atom levels is determined by energy $W_T = -\vec{d_a}\cdot\vec{E} - \vec{\mu_a}\cdot\overrightarrow{H_{ind}}(\vec{E})$. As a result, the effect of polarization plane rotation (birefringence effect) deal with the energy levels splitting is caused not only by $\vec{d_a}$ interaction with electric field but by $\overrightarrow{H_{ind}}(\vec{E})$ interaction with $\vec{\mu_a}$, too. It is easy to see, that even for $\vec{d_a} = 0$ the energy of splitting differs from zero and the T-odd effect of polarization plane rotation and birefringence exist.

Let us consider the propagation of electromagnetic wave through the homogeneous isotropic medium placed to the stationary electric $\vec{E}$ and magnetic $\vec{H}$ fields. Tensor of dielectric permittivity of medium ε_{ik} is a certain function of these fields

$$\varepsilon_{ik} = \varepsilon_{ik}(\omega, \vec{k}, \vec{E}, \vec{H}),$$

where ω is the frequency and $\vec{k}$ is the wave vector of the photon.

Tensor ε_{ik} can be presented as the sum

$$\varepsilon_{ik} = \varepsilon_{ik}^{even}(\omega, \vec{k}, \vec{E}, \vec{H}) + \varepsilon_{ik}^{odd}(\omega, \vec{k}, \vec{E}, \vec{H}),$$

where $\varepsilon_{ik}^{even}(\omega, \vec{k}, \vec{E}, \vec{H})$ is the tensor of dielectric permittivity of medium in the absence of P,T-odd interactions and $\varepsilon_{ik}^{odd}(\omega, \vec{k}, \vec{E}, \vec{H})$ is the term caused by the T,P-odd weak interactions. Let us expand ε_{ik} into irreducible parts:

$$\varepsilon_{ik} = \varepsilon_0\delta_{ik} + \varepsilon_{ik}^{s} + \varepsilon_{ik}^{a}, \tag{38}$$

where δ_{ik} is the Kronecker symbol, indices $i, k = 1, 2, 3$ correspond (x,y,z), $\varepsilon_0 = \frac{1}{3}\sum_i \varepsilon_{ii}$ is the scalar, $\varepsilon_{ik}^{s} = \frac{1}{2}(\varepsilon_{ik} + \varepsilon_{ki}) - \varepsilon_0\delta_{ik}$ is the symmetric tensor (with trace equal to zero), $\varepsilon_{ik}^{a} = \frac{1}{2}(\varepsilon_{ik} - \varepsilon_{ki})$ is the antisymmetric tensor of rank two.

An antisymmetric tensor of rank two can be represented[7] as

$$\varepsilon_{ik}^{a} = ie_{ikl}g_l,$$

where $\vec{g}$ is the axled vector dual to the antisymmetric tensor ε_{ik}^{a}, e_{ikl} is the fully antisymmetrical unit tensor of rank three.

Tensors $\varepsilon_{ik}^{even}(\omega, \vec{k}, \vec{E}, \vec{H})$ and $\varepsilon_{ik}^{odd}(\omega, \vec{k}, \vec{E}, \vec{H})$ can be written by way of (38), too.

Tensor $\varepsilon_{ik}^{even}(\omega, \vec{k}, \vec{E}, \vec{H})$ describes optical anysotropy of the media in external fields for P,T-even world. Particularly, the symmetrical part of ε_{ik}^{even} is responsible for birefringence effects caused by the external fields $\vec{E}$ and $\vec{H}$ (i.e. effects of Kerr and Cotton-Mouton) and antisymmetrical part describes Faraday and Macaluso-Corbino effects.

Let us consider $\varepsilon_{ik}^{odd}(\omega, \vec{k}, \vec{E}, \vec{H})$. Weak interactions are lower than electromagnetic ones. Then, only linear terms in the expansion of ε_{ik}^{odd} over weak interaction constant should be taken into consideration. Therefore, ε_{ik}^{odd} depends on P,T-odd scalar and tensor combinations of vectors $\vec{E}, \vec{H}$ and $\vec{k}$ linearly:

$$\begin{aligned}\varepsilon_{ik}^{odd} &= \varepsilon_0^{odd}\delta_{ik} + \varepsilon_{ik}^{(s)\ odd} + ie_{ikl}g_l^{odd} = \\ &= [\chi_{sEH}^T(\vec{n_E}\,\vec{n_H}) + \chi_{sE}^T(\vec{n_\gamma}\,\vec{n_E}) + \chi_{sH}^P(\vec{n_\gamma}\,\vec{n_H})]\delta_{ik} + \\ &\quad + \chi_t^T[\frac{1}{2}(n_{E\,i}\,n_{H\,k} + n_{H\,i}\,n_{E\,k}) - \frac{1}{3}(\vec{n_E}\,\vec{n_H})] + \\ &\quad + i\chi_{sEH}^P e_{ikl}n_{EHl} + i\chi_s^P e_{ikl}n_{\gamma l} + i\chi_E^T e_{ikl}n_{El}\end{aligned} \tag{39}$$

where $\vec{n}_\gamma = \frac{\vec{k}}{|\vec{k}|}, \vec{n}_E = \frac{\vec{E}}{|\vec{E}|}, \vec{n}_H = \frac{\vec{H}}{|\vec{H}|}$ and $\vec{n}_{EH} = \frac{[\vec{E}\times\vec{H}]}{|[\vec{E}\times\vec{H}]|}$ are the unit vectors, repeated indices imply summation. Term, containing δ_{ik}, describes contribution of P,T-odd weak interactions in the scalar part of dielectric polarisability, where χ_{sEH}^T determines contribution of T,P-odd interactions, χ_{sE}^T is responsible for T-odd, P-even interactions (this contribution was first considered in[29]), χ_{sH}^P describes P-odd, T-even interactions. The second term in (39), which is proportional to χ_t^T, is responsible for T,P-odd birefringence effect (predicted in[7]). Terms proportional to tensor e_{ikl} describes light polarisation plane rotation around $\vec{n}_{EH}$ (P-odd, T-even), around electric field direction $\vec{n}_E$ (P, T-odd) and around $\vec{n}_\gamma$ (P-odd, T-even). The latter is well known and actively studied phenomenon of polarisation plane rotation and dichroism caused by P-odd T-even interactions[12,32,33]. If spins of atoms (molecules) are certainly oriented then additional contributions to ε_{ik}^{odd} can appear[25].

Detailed discussion see in[7,8].

Several words about birefringence effect that is described by the symmetrical part of T-odd polarisability tensor. Two effects contribute both in polarisation plane rotation and birefringence phenomena. They are:

1. splitting of atom (molecule) levels
2. mixing of ground state and opposite parity states in external fields.

Birefringence effect appears when photon moves orthogonally to the electric and magnetic fields. The magnitude of effect is proportional to the correlation $(\overrightarrow{E}\overrightarrow{H})$. Thus, one can distinguish the T-odd birefringence effect against a background of T-even one by changing the direction of $\overrightarrow{E}$ with respect to direction of $\overrightarrow{H}$.

The possibility to observe the time-reversal violating optical phenomena was considered in[7,8]. It was shown there that use of "photon traps" (see Fig. 5) provides to measure d_e with the sensitivity of about 10^{-33} e·cm

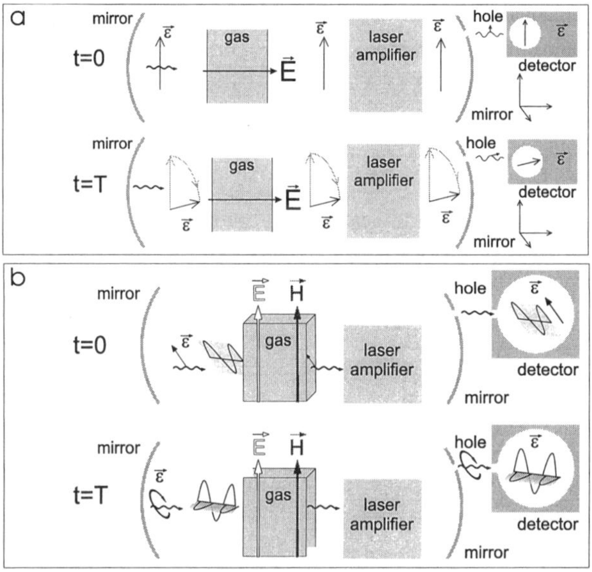

Figure 5. Rotation of polarisation plane of photon (a) is caused by P-odd T-even interactions in "photon trap". Owing to birefringence effect in "photon trap" linear polarisation is converted into circular one and vice versa (b)

Analysis of the tensor of dielectric permittivity of medium ϵ_{ik} in presence of P- and T-odd interactions results in conclusion of existence of several interesting T-noninvariant optical gyrotropy phenomena. Despite they are small,the use of photon "trap" allow to expect to make possible the experimental observation that let us obtain the important data on the value of T-odd weak interactions[7,8].

Bibliography

1. S. K. Lamoreaux, LANL e-print arXive: nucl-ex/0109014v4 (2002).
2. Cheng Chin, Veronique Leiber, Vladan Vuletic, Andrew J. Kerman, and Steven Chu Phys.Rev. A **63**, 033401 (2001).

3. M. V. Romalis, W.C. Griffith, and E.N. Fortson, LANL e-print arXive: hep-ex/0012001v1 (2000).
4. B. C. Regan, Eugene D. Commins, Cristian J. Schmidt and David DeMille, *Phys.Rev.Lett* **88**, n.071805-1 (2002).
5. F. L. Shapiro, *Sov. Phys. Usp.* **11**, 345 (1968).
6. B. V. Vasil'ev and E.V. Kolycheva, *Sov. Phys. JETP* **47**, 243 (1978) .
7. Baryshevsky V. G. LANL e-print archive hep-ph/9912438.
8. Baryshevsky V. G. LANL e-print archive hep-ph/9912270
9. V. G. Baryshevsky, LANL e-print arXive: hep-ph/0307291 (2003).
10. V. G. Baryshevsky and D. N. Matsukevich, *Phys.Rev.* **A66**, 062110 (2002).
11. A. Abraham, M. Goldman, *Nuclear magnetism: order and disorder*, Clarendon Press, Oxford 1982.
12. Khriplovich I. B. *Parity Nonconservation in Atomic Phenomena*, (London: Gordon and Breach, 1991).
13. V. A. Dzuba, V.V.Flambaum, J.Ginges, M.G.Kozlov, *Phys.Rev.* **A66**, 012111 (2002).
14. V. F. Dmitriev and R. A. Senkov, *Phys. Rev. Lett.* **91**, n21 (2003).
15. I.B.Khriplovich, R.V.Korkin, *Nucl. Phys.* **A665**, 365 (2000).
16. Lamoreaux S. K. Nucl. *Instrum. Methods* **A284**, 43 (1989).
17. Forte M. J. *Journ. Of Phys. G.: Nucl. Part. Phys.* **G9**, 745 (1983).
18. Yashchuk V., Budker D., Zolotorev M. *Preprint LBNL-42228*, August 1998
19. Barkov L. M., Melik-Pashayev D., Zolotorev M. *Opt. Commun.* **70(6)**, 467 (1989).
20. Barkov L. M. Zolotorev M. S., Melik-Pashayev D. *JETP Letters* **48(3)**, 144 (1988).
21. Sushkov O. P., Flambaum, V. V., *Zh. Eksp. Teor. Fiz.* **75(4)**, 1208 (1978).
22. Baranova N.B., Bogdanov Y.V., Zeldovich B.Ya. Uspehi Phiz. Nauk 123, 349 (1977).
23. Macaluso D., Corbino O. M. *Nuovo Cimento* **9**, 384 (1899).
24. Baryshevsky V.G. *Phys. Lett* **A177**, 38 (1993).
25. Baryshevsky V.G., Baryshevsky D.V. *Journ. Of Phys. B: At. Mol. Opt. Phys.* **27**, 4421 (1994).
26. Baryshevsky V.G. *J.High Energy Phys.* **04**, 018 (1998).
27. Landau L., Lifshitz E. Electrodynamics in Continuous Media , 1982, Moscow Science.
28. Baryshevsky V.G. *Phys. Lett* **A260**, 24 (1999).
29. Kozlov M.G., Porsev S.G. *Phys. Lett.* **A142**, 233 (1989).
30. Landau L., Lifshitz E. Quantum mechanics, 1989, Moscow Science.
31. Berestetskii V., Lifshitz E., Pitaevskii L. Quantum electrodynamics, 1989, Moscow Science.
32. Meekhof D.H., Vetter P.A., Majumber P.K. et.al. *Phys. Rev.* **A52**, 1895 (1995).
33. Edwards N.H., Phipp S.J., Baird G. et. al. *Phys. Rev. Lett.* **74**, 2654 (1995).
34. Cameron R., Cantatore G., Melissinos A.C. et. al. *Phys. Rev.* **D47**, 3707 (1993).

ON ONE-ELECTRON ATOM IN THE NEIGHBORHOOD OF BLACK HOLES

Alexander K. Gorbatsievich
Department of Physics, Belarusian State University
4, Fr. Skorina Ave., 220050, Minsk, Belarus
E-mail: gorbatsievich@bsu.by

It is argued that an external gravitational field may substantially influence atomic spectra. A hydrogen atom falling freely in Schwarzschild's and in Kerr space-time is investigated as an example. The hydrogen atom is considered at the level of traditional quantum mechanics. Starting from a two-component Hamilton operator describing an atomic electron in the presence of an external gravitational field we evaluate explicit expressions for the perturbations (i.e. splitting) of the energy levels. In the Kerr space-time we restrict our considerations to geodesics at which $\theta = \pi/2$ and to "conical spiral" trajectories. It is shown that the perturbations can reach observable values for a sufficiently high initial velocity at infinity.

1. Introduction

This paper is devoted to the following question: Does the gravitational field influence atomic spectra which give rise to the observable effects? This question is of particular importance in astrophysics, since most of the information on celestial bodies is available from spectral investigations[1–3]. Almost all observations, so far, are based on the assumption that the external gravitational field has no influence on the atomic spectra, apart from an overall redshift. A theoretical estimate[4] reveals that, even on the surface of a neutron star with $M \sim 0.4M_{\odot}$, the characteristic energies of the line splitting of a hydrogen atom are of the order of magnitude of $3.5 \cdot 10^{-12}\,\mathrm{eV}$. Recent investigation by Parker[5,6] of the spectrum of a hydrogen-like atom in an arbitrary curved space-time confirmed this result. In general, space-time must be highly curved (characteristic curvature radius about $10^{-3}\,\mathrm{cm}$) if gravitational corrections are to become of the order of magnitude of the Lamb shift ($4.4 \cdot 10^{-6}$ eV). Calculations similar to those by Parker have been performed by the author[8,9]. Further references on the subject can be found in the papers by Audretsch and Schäfer[10], which deal with an extensive study of the hydrogen-like atom in the expanding universe. However, we will show in the following that there are situations in which an external gravitational field may lead to corrections of the order of magnitude of the fine structure, even if the field is much weaker than that mentioned above (e.g., the field outside a Schwarzschild or Kerr Black Hole). The following classical considerations illustrate the idea. The equation of geodesic deviation implies that, besides the Coulomb force (caused by the charge of the nucleus), an additional "tidal force" (see, e.g.[11]) acts on the electron, which is proportional to $R^i{}_{jkl}u^j u^k \xi^l$ (ξ^k is the distance vector between electron and nucleus, u^i is the four-velocity of the nucleus). Because ξ^k is very small ($\xi^k \sim 10^{-8}$ cm), the "tidal force" can essentially contribute to the electron, energy only if either $R^i{}_{jkl}$ or u^i take large values. This appears to be the case in the vicinity of singularities (this case was discussed in the paper of Parker[5]) or under ultrarelativistic velocities of the atom (when $v \to c$ some components of u^i can take infinitely large values). The latter possibility is of much more interest for an external observer.

In this paper we consider the freely falling hydrogen atom in the gravitational field of a rotating black hole, described by the Kerr metric. First we introduce the explicit expression for the two-component, Hamilton operator, which is derived from the generally relativistic Dirac equation, then we derive the equation of motion of an atom in an external gravitational field, then we summarize some properties of geodesies, which will be useful in the discussion of the Hamilton operator of an hydrogen atom in the Kerr space-time.

Because of the negligible mass ratio of electron mass m and nucleus mass μ the motion of a classical mass point will be a good approximation for the motion of the nucleus $x^i = \xi^i(\tau)$, where τ is the proper time ($d\tau^2 = -\frac{1}{c^2}\, ds^2$). The motion of the electron is described by quantum mechanics, assuming only one-electron atom for simplification. Moreover, the assumed identity of atom and nucleus masses permits the consideration of the atomic motion as a nuclear one.

2. Explicit Expression for the Hamilton Operator

Contrary to the situation in quantum field theory only a few papers deal with quantum mechanics in general relativity. Most of them start from the generally covariant Dirac equation and rewrite it as a Schrödinger-type equation[a]:

$$i\hbar\frac{\partial\psi}{\partial\tau} = \hat{H}\psi\,, \tag{1}$$

where τ is a parameter which represents the time (often proper time) and the operator $\hat{H}$ is interpreted as the energy operator, assuming that both time-evolution and energy are given by $\hat{H}$. Such a procedure leads, in general, to a non-hermitean operator and other difficulties, which do not allow a simple physical interpretation of equation (1) (cf. e.g.[1–5,10]). A detailed discussion of these problems by the author (see[12–15] and references cited there) showed that it is always possible to treat the generally covariant Dirac equation as a special representation of the usual quantum evolution equation in a certain Hilbert space, and with a certain Hamilton operator, if one uses non-orthogonal basis vectors. A correct procedure was given to evaluate the Hamilton operator (whose expectation value in a quasistationary state is the energy corresponding to this state) for an electron of an (arbitrary moving) atom in an external gravitational and electromagnetic field. The atom is considered in the comoving "frame of the single observer" (see e.g.[16]). This reference frame is determined by the motion of a single mass point. The world line of this mass point ("single observer") is named basis. Using the world line $\xi^i(\tau)$ of the nucleus as basis, we obtain a convenient comoving reference frame for the atom. Along the basic line we establish an orthonormal vierbein $h_{(m)}{}^i$, defined by

$$h_{(4)}{}^i = u^i \equiv \frac{\mathrm{d}\xi^i(\tau)}{\mathrm{d}\tau}\,, \quad h_{(i)}{}^k h_{(j)k} = \eta_{(i)(j)}\,, \tag{2}$$

with $\eta_{(i)(j)} = \mathrm{diag}(1,1,1,-1)$ being the Minkowski tensor. The introduced vierbein is determined with exception of three-dimensional rotations. The three-dimensional physical space is given by a geodesic spacelike hypersurface f (related to τ), which lies orthogonal to the basic world line. In order to arithmetize the hypersurface f, at each point $P \in f$ we fix a set of three scalars $X^{(\alpha)} = \sigma_P h^{(\alpha)}{}_i k^i$, where σ_P is the value of the canonic parameter σ at P, defined along a spacelike geodesic in f and going through the point P, k^i is the tangent unit vector to that geodesic, defined at the point on the basic line ($\sigma = 0$). For a nonrotating frame (i.e. the vectors $h_{(n)}{}^i$ are displaced along the basic line $\xi(\tau)$ according to the

[a]Latin indices run from 1 to 4, Greek ones from 1 to 3. The signature of space-time is (+, +,+,-). We will use (except of some starting equations and final results) units $\hbar = c = 1$.

Fermi-Walker transport) the quantities $\{X^{(\alpha)}, c\tau\}$ correspond to the Fermi normal coordinates[8]. Analogous quantities $x^{\hat{\alpha}} = X^{(\alpha)}$, $x^{\hat{4}} = c\tau$ we treat as rotating Fermi coordinates. In these coordinates the metric tensor $g_{\hat{i}\hat{j}}$ becomes

$$g_{\hat{i}\hat{j}} = \eta_{(i)(j)} + \varepsilon_{(i)(j)}\,, \tag{3}$$

with

$$\varepsilon_{(\alpha)(\beta)} = -\tfrac{1}{3}\, R_{(\alpha)(\mu)(\beta)(\nu)} X^{(\mu)} X^{(\nu)} + O(\varrho^3)\,, \tag{4a}$$

$$\varepsilon_{(\alpha)(4)} = \frac{1}{c}\, e_{(\alpha)(\varkappa)(\tau)} X^{(\tau)} \omega^{(\varkappa)} + \Theta_{(\alpha)}\,, \tag{4b}$$

$$\varepsilon_{(4)(4)} = -2\Big(\Theta + \frac{1}{c^2}\, W_{(\alpha)} X^{(\alpha)}\Big)\,; \tag{4c}$$

$$\Theta_{(\alpha)} = \tfrac{2}{3}\, R_{(\alpha)(\mu)(\nu)(4)} X^{(\mu)} X^{(\nu)} + O(\varrho^3)\,, \tag{5a}$$

$$\Theta = \tfrac{1}{2}\left(R_{(4)(\mu)(4)(\nu)} - \tfrac{1}{2}\, R_{(4)(\mu)(4)(\nu)(\varkappa)} X^{(\varkappa)}\right) X^{(\mu)} X^{(\nu)} + O(\varrho^4)\,. \tag{5b}$$

Here, we used the following notations: $W^{(\alpha)} = h^{(\alpha)}{}_i \dfrac{\mathrm{D}u^i}{\mathrm{D}\tau}$ and $\omega^{(\alpha)} = \tfrac{1}{2}\, e^{(\alpha)(\varkappa)(\tau)} h_{(\tau)i} \dfrac{\mathrm{D}h_{(\varkappa)}{}^i}{\mathrm{D}\tau}$ are the acceleration and the angular velocity of the reference frame respectively, $e^{(\alpha)(\varkappa)(\tau)}$ is the three-dimensional Levi-Civita symbol, $R^{(i)}{}_{(\mu)(\nu)(k)}$ and $R^{(i)}{}_{(\mu)(\nu)(k)(l)} = h^{(i)}{}_a h_{(\mu)}{}^b h_{(\nu)}{}^c h_{(k)}{}^p h_{(l)}{}^r R^a{}_{bcp;r}$ are the vierbein-components of the Riemann curvature tensor and of its covariant derivative respectively. Because $\varrho \equiv \sqrt{X^{(\alpha)} X_{(\alpha)}}$ is in order of the atomic radii, the condition $\varepsilon_{(i)(j)} \ll 1$ is valid even for extremely strong (from a macroscopic viewpoint) gravitational fields, and without much loss of accuracy one can keep only linear terms in the $\varepsilon_{(i)(j)} \ll 1$.

The covariant Dirac equation can be interpreted as a special representation of the traditional quantum-mechanical law of motion in Hilbert space, which allows one to build quantum mechanics in arbitrary reference frames and under consideration of external gravitational fields[3] (for details see the book[9]). In particular, one can show that in quasirelativistic approximation the electron can be always described by a two-component Pauli-type equation

$$i\hbar\,\frac{\partial \Phi}{\partial \tau} = \hat{H}_{\Phi}\Phi \tag{6}$$

(with

$$\int_f \Phi^+ \Phi \mathrm{d}^3 X = 1\,, \tag{7}$$

and $\mathrm{d}^3 X = \mathrm{d}X^{(1)}\mathrm{d}X^{(2)}\mathrm{d}X^{(3)}$) in the reference frame of a single observer. The complete expressions for the two-component Hamiltonian $\hat{H}_\Phi$ one can find in the monograph[15] (see also[5]). A discussion of the complete operator $\hat{H}_\Phi$ shows that the influence of external gravitational field will always be more significant than effects caused by the non-Coulomb participants of the nuclear potentials[b]. Therefore, in a good approximation the Hamiltonian is given by

$$\hat{H}_\Phi = \hat{H}_0 + \hat{H}_1\,, \tag{8}$$

with

$$\hat{H} = \frac{P^2_{(\alpha)}}{2m} + \frac{qe}{\varrho} + \hat{H}_F \tag{9}$$

— the Hamiltonian of a non-disturbed electron which is bound by a nucleus (q and e are charge of the nucleus and of the electron respectively; $\hat{H}_F$ being the operator that gives the fine-structure splitting) and $\hat{H}_1$ describing the interaction with the external gravitational field

$$\begin{aligned}\hat{H}_1 \cong mc^2\Theta + \frac{c}{2}\left(e^{(\sigma)(\alpha)(\tau)}\Theta_{(\sigma),(\alpha)}S_{(\tau)} - \left\{\Theta^{((\alpha))}\,,\,P_{(\alpha)}\right\}\right)\\ - \omega^{(\alpha)}\left(L_{(\alpha)} + S_{(\alpha)}\right) + mW_{(\alpha)}X^{(\alpha)} + O(1/c^0)\,.\end{aligned} \tag{10}$$

In the last equation we introduced the following notations $P_{(\alpha)} = \dfrac{\hbar}{i}\dfrac{\partial}{\partial X^{(\alpha)}}$, $L_{(\alpha)} = e_{(\alpha)(\varkappa)}{}^{(\tau)}X^{(\varkappa)}P_{(\tau)}$, $S_{(\alpha)} = \dfrac{\hbar}{2}\sigma_{(\alpha)}$, where $\sigma_{(\alpha)}$ are the standard Pauli matrices.

3. Equation of Motion of a One-Electron Atom in an External Gravitational Field

From the expression (10) for the Hamilton operator $\hat{H}_\Phi$ describing nuclear electron follows that this operator essentially depends on the law of motion of atom as the whole. However, as it has been already mentioned above, influences of gravitational fields on atomic spectra will become critical only

[b]The electromagnetic potentials of the nucleus are obtained from Eqs. (14) by changing $e \to q$, $\zeta^{(\varkappa)} \rightleftarrows X^{(\varkappa)}$, and then $\zeta^{(\varkappa)} \to 0$.

in extremely strong gravitational fields, or in not very strong gravitational fields, e.g. outer range of macroscopic black holes, but for ultrarelativistic motion of atoms[8,12]. Apparently, in both cases the motion of an atom cannot be considered a priori as a geodesic one. Subsequently we shall assume an atom with quasistationary quantum states.

One can obtain the equation of motion for the atom (equation for $\xi^k(\tau)$) from the following suggested model.

The nucleus, in the classical mass-point approximation with intrinsic spin, is moving in an external gravitational field and interacting with the electromagnetic field of the electron. The electron which for his part is moving in the external gravitational field and in the nuclear electromagnetic field, is considered in a quasistationary bound quantum state. It was shown by different authors[17,7] that the expectation values of the position and spin operators for a Dirac electron fulfil Mathisson–Papapetrou equations. Applying these results (valid for any Dirac particles) to the nucleus, we describe its motion in our model by the system of Mathisson–Papapetrou equations[18]. To the first equation of this system we have to add a "Lorentz-force" term ($\frac{q}{c}\,F^{ij}u_j$) which describes in first approximation the interaction with mean electromagnetic field of the electron. This leads us to[c]

$$\mu\frac{\mathrm{D}u^i}{\mathrm{D}\tau} = \frac{1}{2c}\,R^i_{\ bcd}u^b\varepsilon^{cdpq}\bar{S}_p u_q + \frac{q}{c}F^{ij}u_j\,, \tag{11a}$$

$$\frac{\mathrm{D}\bar{S}^i}{\mathrm{D}\tau} = \frac{1}{c^2}\,u^i\bar{S}_n\frac{\mathrm{D}u^n}{\mathrm{D}\tau}\,, \quad u^n\bar{S}_n = 0\,, \tag{11b}$$

where q, $u^i = \dfrac{\mathrm{d}\xi^i}{\mathrm{d}\tau}$, and $\bar{S}^i$ denote the charge, the 4-velocity and the classical spin of the nucleus respectively. We shall regard the expectation value of the nuclear spin operator as classical spin. ε^{ijkl} denotes the Levi-Civita pseudotensor with $e^{1234} = (-g)^{-1/2}$ and $g = \det(g_{ij})$, R^{ijkl} is the Riemann curvature tensor and F^{ij} is the tensor of the electromagnetic field of the electron. Here, we have only to calculate the tensor F^{ij} to obtain an equation for $\xi^i(\tau)$.

In order to simplify calculations of the tensor F^{ij} we shall deal with a special, comoving reference frame of single observer, which was described in section 2. For further calculations we write down the system (11) in the

---

[c]We neglect $-\dfrac{1}{c^3}\,\varepsilon^{arq}{}_p\bar{S}_r u_q\dfrac{\mathrm{D}^2u^p}{\mathrm{D}\tau^2}$, and also terms which describe interaction between the field F^{ij} and the nuclear magnetic momentum (these terms can be found in[19]). Numerical calculations show that they are smaller than all other terms in (11).

rotating Fermi coordinates. From

$$u^{\hat{i}} = (0,0,0,c)\,, \quad g_{\hat{i}\hat{j}}(0) = \eta_{(i)(j)}\,, \quad \text{and} \quad \left(\frac{\partial x^i}{\partial x^{\hat{j}}}\right)_{X^{(\alpha)}=0} = h_{(j)}{}^{i} \tag{12}$$

it follows that

$$\mu W_{(\nu)} = -ce^{(\mu)(\varkappa)(\tau)} R_{(4)(\varkappa)(\nu)(\mu)} \bar{S}_{(\tau)} + qF_{(\nu)(4)}\,, \tag{13a}$$

$$\frac{\mathrm{d}}{\mathrm{d}\tau}\,\bar{S}_{(\alpha)} = e_{(\alpha)(\varkappa)(\nu)}\omega^{(\nu)}\bar{S}^{(\varkappa)}\,, \quad \bar{S}_{(4)} = 0\,. \tag{13b}$$

In equations (13) all quantities are taken along the basic line ($X^{(\alpha)} = 0$).

We shall calculate the vierbein-components of the mean electromagnetic field at the point $X^{(\alpha)} = 0$, generated by the atomic electron in the following way: first, solving Maxwell equations we determine the field of a stationary point charge e, which is located at the point $X^{(\alpha)}$. After that, we calculate the expectation value of that field, using the assumption that the electron is in a quasistationary state, and neglecting both the interaction of the electron magnetic field with the nuclear magnetic momentum and the effects of the retardation. It is clear that those retardation effects will become small enough because of the small velocity of the electron within the atom ($v/c \ll 1$). By means of the Maxwell equations we find that in a certain neighbourhood of the nucleus the electromagnetic field will be quasistationary (for the same limitations as made above to the space-time metric and to the law of motion for the atom, which guarantee the existence of quasistationary states). At the point with the coordinates $\zeta^{(\alpha)}$ the electromagnetic field is given by

$$\begin{aligned}\varphi(\zeta^{(\alpha)}) = \frac{e}{R}\Big[1 &+ \frac{1}{2c^2}\,W_{(\varkappa)}R^{(\varkappa)} + \tfrac{1}{6}\,R_{(4)(\varkappa)(4)(\tau)}\zeta^{(\varkappa)}\Big(2\zeta^{(\tau)} - X^{(\tau)}\Big)\\ &+ \frac{1}{R^2}\,\Big[R_{(\varkappa)(\tau)(\nu)(\mu)}\zeta^{(\varkappa)}\zeta^{(\nu)}X^{(\tau)}X^{(\mu)} + O(\varrho^3)\Big]\,,\end{aligned} \tag{14a}$$

$$\begin{aligned}A_{(\alpha)}(\zeta^{(\nu)}) = \frac{e}{R}\Big[&-\frac{1}{c}\,e_{(\alpha)(\tau)(\nu)}X^{(\tau)}\Omega^{(\nu)} - \tfrac{1}{6}\,R_{(4)(\tau)(\nu)(\alpha)}\zeta^{(\tau)}\zeta^{(\nu)}\\ &+\tfrac{1}{2}\,R_{(4)(\alpha)(\tau)(\nu)}X^{(\tau)}\zeta^{(\nu)} + \tfrac{1}{2}\,R_{(4)(\nu)(\tau)(\alpha)}X^{(\nu)}X^{(\alpha)} + O(\varrho^3)\Big]\end{aligned} \tag{14b}$$

with $\varphi = -A_{\hat{4}}$, $A_{(\alpha)} = A^{(\alpha)} = A_{\hat{\alpha}}$ the scalar and vector potentials and $R^{(\alpha)} = \zeta^{(\alpha)} - X^{(\alpha)}$, $R \equiv \sqrt{R^{(\alpha)}R_{(\alpha)}}$. To deduce Eqs. (14) we assume that the gravitational field is a vacuum one (i.e. $R_{ij} = 0$) within the atom, a restriction which is always valid since the curvature appearing here is that of the background geometry, and not that generated by atom itself or by an external electromagnetic field. From Eqs. (14) we obtain the following expansions for the electromagnetic field tensor

$$F_{(\alpha)(4)} = \left\langle \hat{F}_{(\alpha)(4)} \right\rangle , \tag{15}$$

where

$$\begin{aligned} \hat{F}_{(\alpha)(4)} &\cong - \left(\frac{\partial \varphi}{\partial \zeta^{(\alpha)}} \right)_{\zeta^{(\nu)}=0} \\ &\cong - \frac{e X_{(\alpha)}}{\varrho^3} - \frac{e}{\varrho} \left(\frac{1}{3} R_{(\alpha)(4)(4)(\tau)} X^{(\tau)} + \frac{1}{2c^2} W_{(\alpha)} \right) . \end{aligned} \tag{16}$$

For a non-disturbed atom we have

$$\left\langle \hat{F}_{(\alpha)(4)} \right\rangle_0 = - \frac{e}{2c^2} W_{(\alpha)} \left\langle \frac{1}{\varrho} \right\rangle_0 . \tag{17}$$

The notation $\langle \cdots \rangle_0$ means that the expectation values are calculated using eigenvectors of $\hat{H}_0$. If one neglects "deformation" of the atom by gravitational field, $F_{(\alpha)(4)}$ will affect only renormalization of the mass of the nucleus $\mu \to \mu + \Delta\mu$ with $\Delta\mu = \dfrac{eq}{2c^2 a}$ and $a = \langle \rho \rangle_0$ being the radius of the atom. Because of $\dfrac{\Delta\mu}{\ll} 1$, we shall not consider this renormalization. This means that we have to calculate only the expectation values with the eigenvectors of the full Hamiltonian $\hat{H}$ to determine $F_{(\alpha)(4)}$. Using perturbation theory one can show easily that

$$F_{(\alpha)(4)} = \left\langle - \frac{e X_{(\alpha)}}{\varrho^3} \right\rangle + \text{terms quadratic in } \varepsilon_{(i)(j)} . \tag{18}$$

Direct calculation of $\left\langle - \dfrac{e X_{(\alpha)}}{\varrho^3} \right\rangle$ is connected with several mathematical difficulties. Therefore we act as follows.

If $\hat{A}$ is any Hermitean operator, the operator

$$\mathring{A} \equiv \left(\frac{\partial \hat{A}}{\partial \tau} \right)_{\text{expl}} + \frac{i}{\hbar} \left[\hat{H} \, , \, \hat{A} \right] \tag{19}$$

will describe the observable "evolution of the observable $\hat{A}$", since

$$\frac{\mathrm{d}\hat{A}}{\mathrm{d}\tau} = \left\langle \mathring{A} \right\rangle . \tag{20}$$

If, besides, $\hat{A}$ does not depend on time τ explicitly, the expectation value of $\mathring{A}$ in a stationary state (for which the state vector is an eigenvector of the Hamiltonian) will be identically equal to zero. In our case the Hamiltonian

$\hat{H}_1$ (10) generally depends explicitly on proper time τ along the basic line, because Θ, $\Theta_{(\alpha)}$ and both $\omega^{(\alpha)}$ and $W^{(\alpha)}$ depend on time. However, generally, the time dependence of the operator $\hat{H}$ will be sufficiently weak (except of gravitational shortwave radiation). Indeed, even if the motion of the nucleus is ultrarelativistic, the atom covers only a short distance during a characteristic atomic time ($\sim 10^{-5}$ cm). In these dimensions, gravitational fields are actually homogeneous. Direct calculation shows that atomic states will be quasistationary and that the following equations for the operators of the 3-velocity and 3-acceleration

$$\left\langle \overset{\circ}{X}{}^{(\alpha)} \right\rangle \cong 0\,, \quad \left\langle \overset{\circ\circ}{X}{}^{(\alpha)} \right\rangle \cong 0 \tag{21}$$

will be a good approximation with high precision. Using explicit time-independence of the operators $X^{(\alpha)}$ and $\hat{P}_{(\alpha)}$ and the fact that the "gravitational potentials" Θ and $\Theta_{(\alpha)}$ vary slowly with time, by means of the above equations one finds

$$\begin{aligned}\left\langle -\frac{eqX_{(\nu)}}{\varrho^3} \right\rangle &\cong cR_{(\nu)(\varkappa)(4)(\mu)}e^{(\mu)(\varkappa)(\tau)}\left\langle L_{(\tau)} + S_{(\tau)} \right\rangle \\ &\quad - mW_{(\nu)} - mc^2\left\langle \Theta_{,(\nu)} \right\rangle .\end{aligned} \tag{22}$$

Note that we made use of the equality

$$R_{(4)(\mu)(\varkappa)(\nu)}\left\langle \left\{ X^{(\mu)}\,,\, \hat{P}^{(\nu)} \right\} \right\rangle \cong m\left\langle \overset{\circ}{\Theta}_{(\varkappa)} \right\rangle \cong 0\,. \tag{23}$$

By means of Eqs. (5) and (22), equation (13) leads (up to terms quadratic in $\varepsilon_{(i)(j)}$) to

$$\begin{aligned}(\mu + m)W_{(\nu)} = &- cR_{(4)(\varkappa)(\nu)(\mu)}\left[\bar{S}_{(\tau)} + \left\langle S_{(\tau)} + L_{(\tau)} \right\rangle_0\right]e^{(\mu)(\varkappa)(\tau)} \\ &+ \frac{mc^2}{4}\left[R_{(4)(\mu)(4)(\nu)(\tau)} + R_{(4)(\tau)(4)(\mu)(\nu)}\right. \\ &\left. + R_{(4)(\nu)(4)(\tau)(\mu)}\right]\left\langle X^{(\mu)}X^{(\nu)} \right\rangle_0 ,\end{aligned} \tag{24}$$

where neglecting of quadratic terms in $\varepsilon_{(i)(j)}$ allowed us to change from $\langle \cdots \rangle$ to $\langle \cdots \rangle_0$. From estimations made above, it follows that the second term on the r.h.s. of Eq. (24) becomes actually small and negligible with regard to the atomic dimensions, if we assume that full spin of the atom (nuclear spin + electron spin + orbital angular momentum of the electron) is non-zero. The calculations of the expectation values of the operator $\overset{\circ}{L}_{(\alpha)} + \overset{\circ}{S}_{(\alpha)}$ describing variation of the electron angular momentum, by means of

Eq. (13b), lead us to

$$\frac{\mathrm{d}}{\mathrm{d}\tau}\left(\bar{S}_{(\alpha)} + \langle L_{(\alpha)} + S_{(\alpha)} \rangle\right) \cong$$
$$\cong e_{(\alpha)}{}^{(\nu)(\varkappa)}\omega_{(\varkappa)}\left(\bar{S}_{(\nu)} + \langle L_{(\nu)} + S_{(\nu)} \rangle_0\right). \tag{25}$$

If we write down the both last equations (24) and (25) in the covariant form, we obtain

$$(\mu + m)\frac{\mathrm{D}u^i}{\mathrm{D}\tau} = \frac{1}{2c}\, R^i_{\ bcd}u^b\varepsilon^{cdpq}\mathbb{S}_p u_q\,, \tag{26a}$$

$$\frac{\mathrm{D}\mathbb{S}^i}{\mathrm{D}\tau} = \frac{1}{c^2}\, u^i\bar{S}_n\frac{\mathrm{D}u^n}{\mathrm{D}\tau}\,, \quad u^n\mathbb{S}_n = 0\,, \tag{26b}$$

where $\mathbb{S}_{(\alpha)} = \bar{S}_{(\alpha)} + \langle L_{(\alpha)} + S_{(\alpha)} \rangle$. In this way we demonstrated that the motion of an atom is given by Mathisson–Papapetrou equations in a good approximation if we put as a classical angular momentum the expectation value of the operator of the full angular momentum (including the nuclear and electron spins and the orbital angular momentum of the electron). We note that the consideration of external electromagnetic field ${}^{(0)}F^{ij}$ yields additional terms

$$\frac{1}{c}\,(e + q)\,{}^{(0)}F^{ij}u_j + \frac{1}{c}\,F^i_{\ m;n}\varepsilon^{mnsp}M_s u_p \tag{27}$$

in Eq. (26a), where M_i is the 4-vector of the atoms full magnetic moment. Particularly in a comoving frame of reference one gets

$$M^{(4)} = 0\,, \quad M_{(\alpha)} = \frac{e}{2mc}\,\langle L_{(\alpha)} + 2S_{(\alpha)} \rangle_0 + \frac{qg}{2\mu c}\,\bar{S}_{(\alpha)}\,. \tag{28}$$

4. Hydrogen Atom in the Kerr Space-Time

Let us consider a freely falling Hydrogen Atom in the gravitational field of a rotating Black Hole. As it has been shown above, the two-component Hamilton Operator $\hat{H}_\Phi$ for the one-electron atom, which is freely falling in the external gravitational field, has in the comoving reference frame of single observer the form (8). In order to analyse the quantum mechanics of the hydrogen atom in the Kerr space-time we put the operator $\hat{H}_1$ in the form

$$\hat{H}_1 = \hat{H}_{RZ} + \hat{H}_W + \hat{H}_{GS}\,. \tag{29}$$

The operator $\hat{H}_W = mW_{(\alpha)}X^{(\alpha)}$ is due to the acceleration $W^{(\alpha)}$ of the atom. As it was shown in the section 3 the equation of motion for the atom in external gravitational field is in a good approximation the Papapetrou equation (26). Therefore it follows that $W^i \cong \dfrac{1}{2c\mu} R^i{}_{bcd} u^b \varepsilon^{cdpq} \mathbb{S}_p u_q$. It can be shown that this is in the fact very small ($W^i \approx 0$) and for a freely falling atom the motion is indeed geodesic in a very good approximation, so that in this case we have $\hat{H}_W \approx 0$. The operator $\hat{H}_{Rz}$

$$\hat{H}_{Rz} = -\omega^{(\alpha)} J_{(\alpha)} \tag{30}$$

describes the influence of the rotation of the frame on the atomic electron. $J_{(\alpha)} = L_{(\alpha)} + \frac{\hbar}{2}\,\sigma(\alpha)$ is the total momentum operator. The operator $\hat{H}_{Rz}$ has some analogy to the operator, which causes the (magnetic) Zeeman effect. Note that a rotation of the frame is in general necessary to guarantee the existence of quasistationary states of the electron. The operator $\hat{H}_{Rz}$ leads to a shift of the energy:

$$\Delta E_{RZ} = -\omega(m_l + m_s)\,, \tag{31}$$

where m_l and m_s are the quantum numbers of the projections of the operators of the orbital momentum and spin of the electron to $\omega^{(\alpha)}$, ω is the module of $\omega^{(\alpha)}$ ($\omega = \sqrt{\omega^{(\alpha)}\omega_{(\alpha)}}$). The shift ΔE_{RZ} yields a splitting of spectral lines into three components (in full analogy to the so-called "normal" Zeeman effect).

The operator

$$\hat{H}_{GS} \cong \tfrac{1}{2}\, mc^2 R_{(4)(\alpha)f v(\nu)} X^{(\alpha)} X^{(\nu)} \tag{32}$$

describes the direct influence of the external gravitational field to the atomic electron. The operator $\hat{H}_{GS}$ is on one hand a generalization of the classical tidal forces, which are represented by the tensor $R_{ikjl}u^i u^j$ in the equation for the geodesic deviation, and on the other hand a generalization of the usual electric Stark effect to the spin-two gravitational field, i,e. the latter depends quadratically on the "position operator" $X^{(\alpha)}$.

A similar formulation for the Hamilton operator of the electron in a gravitational field was given by L. Parker, although he assumed nonrotating frames ($\omega^{(\alpha)}$) and geodesic motion of the atom only. In contrast, the power of the formalism, described in[12,15] lies in the rigorous solution of the problems with the non-hermitean Hamilton operator and the definition of the scalar product of the wave function of the electron, while in the papers[5,3] the generally covariant Dirac equation is used only in a formal sense as a purely "mathematical equation".

We note that it is easy to include external electromagnetic fields into the formalism, so that it is possible e.g. to study the hydrogen atom in the field of a magnetic massive dipole (see[13]).

We will now summarize some features of certain geodesic trajectories in the Kerrs space-time, which will play a central role in the following discussion.

4.1. *Geodesics with $\vartheta = const$ in the Kerr space-time*

Let us first give a brief review of some notations for the Kerr metric. We consider the Kerr metric in Boyer-Lindquist coordinates $\{r, \theta, \varphi, t\}$

$$\begin{aligned} \mathrm{d}s^2 = \varrho^2\Big(\frac{\mathrm{d}r^2}{\Delta} + \mathrm{d}\theta^2\Big) + \Big[r^2 + a^2 + \frac{2\mathfrak{M}r}{\varrho^2}\, a^2 \sin^2\theta\Big] \sin^2\theta \mathrm{d}\varphi^2 \\ - \frac{4\mathfrak{M}r}{\varrho^2}\, a \sin^2\theta \mathrm{d}\varphi \mathrm{d}(ct) - \Big(1 - \frac{2\mathfrak{M}r}{\varrho^2}\Big)\, \mathrm{d}(ct)^2 , \end{aligned} \tag{33}$$

where

$$\text{a)}\quad \varrho^2 = r^2 + a^2\cos^2\theta\,, \qquad \text{b)}\quad \Delta = r^2 - 2\mathfrak{M}r + a^2\,, \tag{34}$$

$\mathfrak{M}$ and $a \equiv \dfrac{\mathfrak{S}}{\mathfrak{M}}$ are the math in geometrical units and the angular momentum per units mass of the Kerr Black Hole respectively. One determines the first integrals of the geodesic equation easily from the Hamilton–Jacobi equation (cf.[21]). In Boyer–Lindquist coordinates we get

$$u^1 = \pm\frac{1}{\varrho^2}\sqrt{P^2 - \Delta(K + r^2)}\,, \tag{35a}$$

$$u^2 = \pm\frac{1}{\varrho^2}\sqrt{K - (a^2\cos^2\varphi + D^2)}\,, \tag{35b}$$

$$u^3 = \frac{1}{\varrho^2}\Big(\frac{aP}{\Delta} - \frac{D}{\sin\theta}\Big)\,, \tag{35c}$$

$$u^4 = \pm\frac{c}{\varrho^2}\Big[\frac{(r^2 + a^2)P}{\Delta} - a\sin\theta D\Big]\,, \tag{35d}$$

with

$$P = P(r) \equiv \frac{1}{\mu}\left[\mathcal{E}(r^2 + a^2) - a\mathcal{L}_z\right], \tag{36a}$$

$$D = D(\theta) \equiv \frac{1}{\mu}\Big[a\sin\theta\,\mathcal{E} - \frac{\mathcal{L}_z}{\sin\theta}\Big]\,. \tag{36b}$$

Here μ is the mass of the point mass (in our case the nucleus), $\mathcal{E}$ and $\mathcal{L}_z$ are the two well-known integrals of motion, corresponding to stationarity and axisymmetry of the Kerr solution of the Einstein equations. The quantity K is an integral of motion, which arises from the separation of variables in the Hamilton–Jacobi equation. Furthermore we restrict our considerations to geodesics on the "cone" θ = const. This includes two important and interesting subcases:

(1) the motion in the equatorial plane $(\theta = \frac{\pi}{2})$, and
(2) the motion along conical spirals $(\theta \neq \frac{\pi}{2})$,

and simplifies the calculations sufficiently.

The assumption θ = const has as a consequence the following condition (see[22]):

$$\frac{\mathrm{D}u^\theta}{\mathrm{D}\tau} = -\frac{\cot\theta}{\varrho^6}\left[\varrho^2 a^2 \sin^2\theta - 2a\sin\theta DP + \left(r^2 + a^2 + a^2\sin\theta\right)D\right] = 0\,. \tag{37}$$

For an atom moving in the equatorial plane Eq. (37) is fulfilled identically. On conical spirals, it can be rewritten as follows

$$\mathcal{E}^2 = \mu^2 + \left(\frac{\mathcal{L}_z}{a\sin^2\theta}\right)^2. \tag{38}$$

As we consider only infinite motion of the atom, we introduce the velocity at infinity v_∞ and a generalized impact parameter ℓ_0 by

$$\frac{\mathcal{E}}{\mu} = \gamma \equiv \frac{1}{\sqrt{1 - v_\infty^2}}\,, \quad \mathcal{L}_z = \mu\gamma v_\infty \ell_0 \sin\theta\,. \tag{39}$$

Then condition (38) reads

$$\ell_0 = \pm a \sin\theta \tag{40}$$

so that ℓ_0 is (up to the sign) determined by a and θ. According to the sign of $\mathcal{L}_z$ (we assume $a > 0$), we have to distinguish two principally different cases of motion:

(1) $\mathcal{L}_z < 0$: retrograde motion, and
(2) $\mathcal{L}_z > 0$: prograde motion.

It is favourable to express the quantity D (Eq. (36)) in terms of v_∞ and θ, which yields[7]

$$D_\pm = a\sin\theta\left(\frac{1 \mp v_\infty}{1 \pm v_\infty}\right)^{1/2} = \text{const}\,, \tag{41}$$

where upper signs belong to prograde motion. From Eq. (41) it follows immediately that the motion on conical spirals is essentially asymmetric, according to the sign of $\mathcal{L}_z$. This asymmetry is marked out by the fact that it is complete in the sense that the ranges for $D_\pm(v_\infty)$ do not intersect for different signs of $\mathcal{L}_z$:

$$0 < D_+ < a \sin\theta < D_- < \infty\,. \tag{42}$$

An important conclusion from this fact is discussed later. Note that the conical spiral trajectories are stable for small variations $\delta\mathcal{E}$ and $\delta\mathcal{L}_z$ [13].

Along the atoms trajectory we choose the comoving (cf. Eq. (2)) orthonormal vierbein $h_{(n)}{}^{i}$) in that way that it reflects maximally the symmetry of both the trajectory and the given motion, e.g. in our case $h_{(2)}{}^{i}$ will be parallel to the vector $\dfrac{\partial}{\partial\theta}$ and $h_{(3)}$ has only one spacelike coordinate component (along $\dfrac{\partial}{\partial\varphi}$). These conditions define the vierbein $h_{(n)}{}^{i}$ completely and allow to calculate explicit expressions for $\omega^{(\alpha)}$ and $R_{(i)(j)(k)(l)}$. We will not write down these explicit expressions and mention only that no singularities occur outside the horizon.

4.2. *Hamilton operator in the Kerr field. Observable effects*

In what follows, we give explicit expressions for the shift and the splitting of the energy levels. It will be shown that large effects occur only for ultrarelativistically moving atoms, i.e $\gamma \ll 1$, where γ is the so-called Lorentz factor, introduced by Eq. (39). This allows us to express the results in a power series of γ and take into account only the highest order of the Lorentz factor. Hereby special attention is paid to those effects which are (at least in principle) accessible to modern experimental technique. It can be shown that although the Hamilton operator explicitly depends on τ, practically for all of the interesting trajectories (including those regarded in the present paper) the gravitational field does not disturb the existence of quasistationary energy levels, i.e. the following inequality holds:

$$\frac{1}{\omega_{fi}}\,\frac{\mathrm{d}}{\mathrm{d}\tau}\,\left|\left\langle i \left| \hat{H}(\tau) \right| f \right\rangle\right| \ll \hbar\omega_{fi} \tag{43}$$

Here i and f indicate the initial and the final states correspondingly. Therefore the operators $\hat{H}_{GS}$ and $\hat{H}_{RZ}$ are regarded as corrections in the framework of Schrödinger's perturbation theory. We assume that the perturbations caused by these operators are greater than the fine-structure splitting, in order to simplify calculations. This consideration is not of principle importance. Explicit expressions for the energy splitting in the case of perturbations smaller than the fine-structure splitting (but greater than the hyperfine-splitting) are also easily obtained.

4.2.1. *Motion in the equatorial plane*

For motion in the equatorial plane the angular velocity has only one non-zero component:

$$\omega^{(1)} = \omega^{(2)} = 0\,, \tag{44a}$$

$$\omega = \omega^{(3)} = -\frac{1}{\Lambda^2 \varrho^4}\left[\left(r^2 - 3\mathfrak{M}r + 2a^2\right)DP - a\left(P^2 + \Delta D^2\right)\right], \tag{44b}$$

where $\Lambda = \dfrac{1}{\varrho}\sqrt{P^2 - \Delta D^2}$ has no singularities outside the horizon. The discussion of Eq. (44) and the explicit expression of the vierbein components $R_{(4)(\alpha)(4)(\varkappa)}$ (most favourably calculated in the Newman–Penrose formalism) shows, that sufficiently large effects occur only for

$$\left|\frac{D}{r}\right| \gg 1\,, \quad \text{i.e.} \quad |a - v_\infty \ell_0| \gg 1 \quad \text{and} \quad \ell_0 \neq a\,. \tag{45}$$

This means that important effects occur if the atom is nearby the black hole and is moving ultrarelativistically. In the ultrarelativistic limit we get:

$$\hat{H}_{GS} = \frac{3}{2}\frac{\mathfrak{M}}{r^5}\left(\ell_0 - a\right)^2\left[\left(X^{(2)}\right)^2 - \left(X^{(3)}\right)^2\right]mc^2\gamma^2\,, \tag{46}$$

$$\omega = \frac{1}{r^2}\left[\frac{(r^2 - a\ell + a^2)(\ell - a)(r^2 - 3\mathfrak{M}r + 2a^2)}{(r^2 - a\ell + a^2)^2 - \Delta(\ell - a)^2} + \frac{a(r^2 - a\ell + a^2)^2 - a\Delta(\ell - a)^2}{(r^2 - a\ell + a^2)^2 - \Delta(\ell - a)^2}\right]. \tag{47}$$

his yields

$$\frac{\left\langle \hat{H}_{GS}\right\rangle}{\left\langle \hat{H}_{RZ}\right\rangle} = \frac{3}{2\alpha}\,\frac{\mathfrak{M}}{r}\,\frac{\ell_0 - a}{r}\,\frac{R_B}{r}\,n^4\gamma^2 = \delta\,, \tag{48}$$

there $\alpha \approx 1/137$ is Sommerfeld's fine-structure constant and $R_B \approx 5.29 \cdot 10^{-9}$ cm, the Bohr radius of the hydrogen atom. Eq. (48) allows us to consider two limiting cases, for which we can give explicit results in a simple form:

1) Dominance of gravitation: $\delta > 1$. If we suppose that $\mathfrak{M}$, a and r have the same order of magnitude, this case is equivalent to

$$r \ll n^4 \gamma^2 R_B \,. \tag{49}$$

The explicit expressions for the energy level shift for levels $n = 1$ and $n = 2$ are

$$\Delta E_{GS}(n=1) = 0 \tag{50a}$$

$$\Delta E_{GS}(n=2) = \frac{18\mathfrak{M}}{r}\left(\frac{a-\ell}{r}\right)^2\left(\frac{R_B}{r}\right)^2 mc^2\,\gamma^2 m_l \,. \tag{50b}$$

The expressions for higher n are much more complicated (cf. the results of Parker[5,3]).

2) Dominance of rotational effects: $\delta < 1$. This case holds if

$$r \gg n^4 \gamma^2 R_B \tag{51}$$

and the shifts of the energy levels are given by

$$\Delta E_{RZ} = \hbar\omega(m_l + m_s)\,, \tag{52}$$

with the angular velocity given by Eq. (47). Note that the first order contribution to ΔE_{RZ} does not depend on the Lorentz factor. The Schwarzschild limit of Eq. (47) leads to:

$$(\Delta E)^{\rm Sch}_{RZ} = \frac{\hbar c \ell_0\left(1 - \dfrac{3\mathfrak{M}}{r}\right)}{r^2\left[1 - \left(1 - \dfrac{2\mathfrak{M}}{r}\right)\left(\dfrac{\ell_0}{r}\right)^2\right]} \,. \tag{53}$$

This expression vanishes for radial trajectories ($\ell_0 = 0$) and on the photon sphere ($r = 3\mathfrak{M}$). The latter is obviously connected with our use of the ultrarelativistic approximation ($\gamma \gg 1$). The results (52) and (53) agree with[23] where the atom in the field of Schwarzschild and Kerr black holes was considered.

4.2.2. *Motion on conical spirals*

As in the case of motion in the equatorial plane, large effects occur only if condition (45) holds. This together with Eq. (42) suggests what we have to suppose to get sufficiently large perturbations:

(1) The atom is sufficiently nearby the Kerr black hole.
(2) The atom moves ultrarelativistically: $\gamma \gg 1$.

(3) The motion of the atom is retrograde, i.e. $\mathcal{L}_z < 0$

Note that according to Eq. (42) large effects, even for ultrarelativistically moving atoms, do not occur for prograde motion. This is caused by the (complete) asymmetry, discussed in Sect. 4.1.

Using the explicit expressions for $\omega^{(\alpha)}$ and the vierbein components of the curvature tensor, we get:

$$\frac{\left\langle \hat{H}_{GS} \right\rangle}{\left\langle \hat{H}_{RZ} \right\rangle} = \frac{3}{\alpha} \frac{\mathfrak{M}}{r} \frac{R_B}{r} \tan\theta n^4 \ll 1\,, \tag{54}$$

so that we can omit the operator $\hat{H}_{GS}$ when considering atoms falling along conical spirals. Schrödinger's perturbation theory yields (in the ultrarelativistic approximation) the following expression for the shift of the energy levels:

$$\Delta E_{RZ} = \hbar c\Big(1 + \frac{8\mathfrak{M}r}{\varrho^2}\frac{a^2\sin^2\theta}{\varrho^2}\Big)\frac{a^2\sin 2\theta}{\Delta\varrho}\,\gamma^2(m_l + m_s)\,. \tag{55}$$

Note that the quantity ΔE_{RZ} does not depend on the parameters characterizing the atom and does not vanish on the ergosphere:

$$r_{\text{erg}} = \mathfrak{M} + \sqrt{\mathfrak{M}^2 - a^2\cos^2\theta} \tag{56a}$$

$$(\Delta E)^{\text{erg}}_{RZ} = \hbar c\gamma^2 \cot\theta\sqrt{\frac{2}{\mathfrak{M}r_{\text{erg}}}}\Big(1 + \frac{2a^2\sin^2\theta}{\mathfrak{M}r_{\text{erg}}}\Big) \tag{56b}$$

5. Conclusion

To give a more obvious representation of our results, let us make some numerical estimations. We will use $\mathfrak{M}$ as a free parameter, characterizing the scale of the system, and set for definitness $r = 3M$, $a = 1/2\mathfrak{M}$. For motion in the equatorial plane we set $\ell_0 = 3\mathfrak{M}$ and consider conical spirals with θ.

Thus we obtain the following values for the energy shift ΔE: For conical (see fig. 1) spirals the rotational Zeeman effect gives a splitting of order $\frac{M_\odot}{\mathfrak{M}}\,\gamma^2 \cdot 10^{-11}\,\text{eV}$, where $M_\odot$is the mass of the Sun. That means that γ has to be of order $10^5\frac{M_\odot}{\mathfrak{M}}$ to obtain effects of the order of the fine structure splitting. For motion in the equatorial plane (see fig. 2 we have $3\left(\frac{M_\odot}{\mathfrak{M}}\right)^2\gamma^2 \cdot 10^{-22}\,\text{eV}$ for high γ and $2\,\frac{M_\odot}{\mathfrak{M}} \cdot 10^{-11}\,\text{eV}$ for low γ.

Level splitting (eV)	$Z = \frac{\nu_0}{\nu} - 1$	Observable effects
m_j		
$n = 1$: $1/2$, $-1/2$		$3 \cdot 10^3 \sqrt{\frac{M}{M_\odot}} < \gamma < 4 \cdot 10^5 \sqrt{\frac{M}{M_\odot}}$
$n = 2$: $3/2$, $1/2$, $-1/2$, $-3/2$	$\sim 10^2 \gamma$	$4 \cdot 10^5 \sqrt{\frac{M}{M_\odot}} < Z < 5 \cdot 10^7 \sqrt{\frac{M}{M_\odot}}$

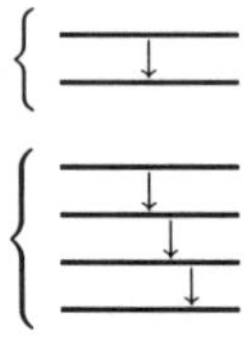

$$\Delta E \sim 7 \cdot 10^{-11} \sqrt{\frac{M_\odot}{M}}\, \gamma^2$$

$$(\Delta E)_{\mathrm{erg}} \sim 7 \cdot 10^{-10} \sqrt{\frac{M_\odot}{M}}\, \gamma^2$$

Figure 1. Energy levels ($n = 1$ and $n = 2$) splitting of freely falling in the Kerr space-time hydrogen atom. Motion on conical spirals. The numerical estimations correspond to the following parameters: $\theta = \text{const} = \frac{\pi}{4}$, $a = \frac{\mathfrak{M}}{2}$, $r = 2\mathfrak{M}$. The red shift was calculated using the radial isotropic geodesics only. As observable effects we denote the effects of the order of the fine structure splitting or greater: $\alpha^2 E_0 < E_0$, where E_0 is ionization energy of hydrogen atom.

It seems to be possible, that near black holes or compact objects there are such atoms which have very large Lorentz factors so that the effects considered in this paper (especially those for conical spirals) really could be observed by astronomers. Actually, we should not forget that in addition to the splitting of energy levels and spectral lines, we have a systematic red-shift of the spectrum (in the ultrarelativistic approximation of order $(1 \text{ to } 10) \cdot \gamma$), possibly up into the radio range. The obtained results let us hope contrary to the results of earlier papers; which dealt with non-moving atoms[1] that our results are not only of theoretical interest. Of course the established model is only a rough first approximation of more realistic ones, which should include an analysis of the mechanism of emission of photons (using QED), as well as ensembles of hydrogen atoms.

Level splitting (eV)	$Z = \frac{\nu_0}{\nu} - 1$	Observable effects
	$\gamma \ll 3 \cdot 10^7 \frac{M_\odot}{M}$	
m_j $n = 1$: $1/2$, $-1/2$ $n = 2$: $3/2$, $1/2$, $-1/2$, $-3/2$ $\Delta E \sim 3 \cdot 10^{-12} \frac{M_\odot}{M}$ $(\Delta E)_{\text{Sch}} \sim 2 \cdot 10^{-11} \sqrt{\frac{M_\odot}{M}}$	$\sim 10\gamma$	Only for mini BH: $2 \cdot 10^{-13} < \frac{M}{M_\odot} < 4 \cdot 10^{-9}$
	$\gamma \gg 3 \cdot 10^7 \frac{M_\odot}{M}$	
m_l $n = 1$: 0 $n = 2$: 1, 0, -1 $\Delta E \sim 3 \cdot 10^{-22} \left(\frac{M_\odot}{M}\right)^{-2} \gamma^2$ $(\Delta E)_{\text{Sch}} \sim 8 \cdot 10^{-22} \left(\frac{M_\odot}{M}\right)^{-2} \gamma^2$	$\sim 10\gamma$	$10^9 \frac{M}{M_\odot} < \gamma < 2 \cdot 10^{11} \frac{M}{M_\odot}$ $2 \cdot 10^{10} \frac{M}{M_\odot} < Z < 2 \cdot 10^{12} \frac{M}{M_\odot}$

Figure 2. Energy levels ($n = 1$ and $n = 2$) splitting of freely falling in the Kerr space-time hydrogen atom. Motion in the equatorial plane $\theta = \frac{\pi}{2}$. The numerical estimations correspond to the following parameters: $a = \frac{\mathfrak{M}}{2}$, $r = 2\mathfrak{M}$, $\ell_0 = \frac{3\mathfrak{M}}{2}$.

Bibliography

1. F. Pinto, *Phys. Rev. Lett.*, **70**, 3839 (1993).
2. F. Pinto, *Gen. Rel. Grav.*, **27**, 9 (1995).
3. L. Parker, D. Vollick, *Phys.Rev.*, **D56**, 2113-2117 (1997).
4. E. Fischbach, B. S. Freemann and Wey-Kwei Cheng, *Phys. Rev.* **D23**, 2157 (1981).
5. L. Parker, *Phys. Rev. Lett.* **44**, 1559 (1980); *Phys. Rev.* **D22**, 1922 (1980).
6. L. Parker, *Phys.Rev. Lett.*, **44**, 1559 (1980).
7. J. Audretsch, *J. Phys. A: Math. Gen*, **14**, 411 (1981).
8. A. Gorbatsievich, To Quantum Mechanics in Non-lnertial Reference Frame by Consideration of External Gravitational Fields, Dep. in V1NITI, No 1180-79 (1979).
9. A. Gorbatsievich, Dissertation, Jena 1980.
10. J. Audretsch, G. Schafer, *GRG Journal* **9**, 243, 489 (1978).
11. C. W. Misner, K. S. Thorne, J. A. Wheeler, *Gravitation* (W. H. Freeman and Company, San Francisco 1973).
12. A. K. Gorbatsievich, *Acta Phys. Pol.* **B16**, 21 (1985).
13. A. Priebe, A. K. Gorbatsievich, *Exp. Tech. Phys.* **34**, 77 (1986).
14. A. K. Gorbatsievich, A. Priebe, in *Problems of the Theory of Elementary Particles and Gravitation, Moscow* **16**, 98 (1985).
15. A. K. Gorbatsievich, *Quantum Mechanics in General Relativity* (University press, Minsk 1985; URSS, Moscow, 2003).
16. V. N. Mitskievich, A. P. Efremov, A. I. Nesterov, *Dynamics of Fields in General Relativity* (Moscow 1985).
17. A. K. Gorbatsievich, *Izv. Akad. Nauk BSSR*, No 2, 62 (1978)
18. M. Mathisson, *Acta Phys. Pol.* **6**, 163 (1937); A. Papapetrou, *Proc. Roy. Soc. (London)* **A209**, 242 (1951).
19. W. G. Dixon, *Nuovo Cimento* **34**, 317 (1964).
20. J. L. Synge, *Relativity: The General Theory* (North-Holland Publ. Co., Amsterdam 1960).
21. B. Carter, *Phys. Rev.* **174**, 1559 (1968).
22. O. P. Krivenko, K. A. Pyragas, I. T. *Zhuk, Astrophys. Space Set.* **40**, 39 (1976).
23. A. K. Gorbatsievich, *Acta Phys. Pol.* **B20**, 901 (1989).

PARTICLE PHYSICS AND QUANTUM CHAOS

Viatcheslav I. Kuvshinov, Andrei V. Kuzmin
Institute of Physics, NAS of Belarus
68 Fr. Skorina Ave., 220072, Minsk, Belarus
E-mail:kuvshino@dragon.bas-net.by

The relationship between chaos theory and particle physics is discussed. It is demonstrated that the mathematical apparatus used in gauge field theories of particle physics can be successfully applied to solve some problems connected with the stability and chaoticity of quantum mechanical systems and vice versa. Namely, we investigate the stability of holonomic quantum computations with respect to unavoidable errors of classical nature in assigning control parameters. We also demonstrate the approach to chaos assisted tunneling regime investigations based on chaotic instanton solutions. On the other hand, reverse influence of chaos theory on particle physics is also briefly discussed. We mainly focus on the problem of chaos in quantum field theory. This work does not give a complete review of all the results obtained in this direction. We discuss only some results on this problem.

1. Introduction

The deterministic chaos phenomenon plays a key role in nonlinear systems having various physical nature[1]. Particulary, chaotic motion is a common behavior regime of classical hamiltonian systems, whereas the case of integrable hamiltonian systems is rather exceptional[2,3]. Originally, the chaos was understood as the ability of dynamical system to behave randomly without effect of any random external forces and impacts. Afterwards, this notion was used to denote wider range of phenomena including manifestations of classical chaos in quantum system properties, for instance see[4–6].

There are few main approaches to the problem of chaos in quantum mechanics. The main progress is achieved under the semiclassical approximation. In the modern literature quantum chaos is often understood as studying statistical properties of the energy level spacing distributions of the systems demonstrating fully chaotic or mixed classical behavior[7,4,5]. Significant progress in this field is connected with the universal spectral fluctuations classes discovery, see[6,8]. Such understanding of the quantum chaos phenomenon is quite narrow from our point of view and it does not cover all results obtained in this direction. Particulary, some alternative methods for quantum chaos investigation stay out of consideration[9–11]. As well attempts to introduce the notion of chaos consistent with one accepted in classical physics, but for the quantum systems having no classical analogues or considered beyond the semiclassical approximation are also stay in shadow[12,13].

Classical phase space structure manifestations in quantum processes and quantum mechanical system properties are actively studied[4,5,14]. One of the research directions in this field is investigation of chaos assisted tunneling[15–18]. Particulary, it was shown that periodic in-time perturbation leading to the mixed behavior of the system in the classical limit results in the large enhancement of the quantum mechanical tunneling probability between KAM-tori[15]. Investigations in this direction are mainly performed by means of numerical techniques based on Floquet theory, see Ref.[19]. As well consideration is often restricted to the systems having one degree of freedom and affected by the time periodic perturbation[16,20–22]. At the same time the analytical methods for the classical chaos influence on quantum mechanical tunneling based on the path integral formalism are being developed[23–27]. In this work we are mainly focused on the closely related approach based on the instanton technique[28–31]. It is necessary to note that a lot of experiments on chaos assisted tunneling observation have been proposed and some of them have been implemented in practice[32–38].

Holonomic approach to quantum computations was primarily proposed in the Ref.[39]. Optical holonomic quantum computer was proposed to realize this idea in a non-linear Kerr medium with the degenerate states of laser beams to be interpreted as qubits and logical gates to be realized by employing the existing devices of quantum optics[40]. Another implementation of holonomic quantum computer based on trapped ions in optical cavity was discussed recently from the viewpoint of its resilience to control parameter errors (undesirable fluctuations of the laser beams amplitudes)[41]. Fidelity primarily proposed in the Ref.[42] is widely used as a measure of the stability of quantum systems. Fidelity close to unity means stability of quantum system with respect to small perturbations of the Hamiltonian (or more generally evolution operator)[43]. Particulary it defines the resilience of quantum computations. Another approaches to determine the stability of quantum systems were also proposed (for instance see the Ref.[44]). It was demonstrated in particular for cavity model of holonomic quantum computer that this realization of logical gates provides fidelity and a success rate close to unity[45]. We derived the general expression for the fidelity of any holonomic quantum computation. For this purpose the non-abelian Stokes theorem primarily utilized in connection with the confinement problem of quarks and gluons[46] was used. It is worth to note that fidelity in particular for holonomic quantum computer is mathematically similar to the Wilson loop defined in the theory of strong interactions of elementary particles (quantum chromodynamics), see the Ref.[46]. Namely, the "area law" for Wilson loop means confinement of quarks and gluons, as well the similar behavior of fidelity demonstrates the instability of quantum system. This corresponds to the connection between quantum chaos and confinement in the sense of the Ref.[44].

Chaos theory methods are also applied both in classical[47–55] and quantum[44,56–60] gauge field theories. Particulary, the investigation of deterministic chaos is of importance in connection with unsolved hadronization and quark confinement problems of quantum chromodinamics, see[46,61,44,58].

The dynamics of model spatially homogeneous systems of abelian and nonabelian classical gauge fields interacting with Higgs field were studied from the viewpoint of chaos theory[52–54,62]. Much attention is paid to the investigation of the chaotic behavior of gauge fields on a lattice both in classical and in quantum cases[59,63–65]. In the references mentioned above the question about the Higgs field influence on the clasical fields dynamics was arisen[49,50,52]. The conclusion the interaction with classical Higgs field to lead to the regularization of Yang-Mills fields chaotic behavior and appearance of the oreder-chaos transition when the density of the energy of field system rises was made[49,62,54].

In the massless electrodynamics on the example of spatially homogeneous fields it was shown that quantum corrections increase the energy threshold to be overcome in order the phase space region of chaotic motion to appear thus leading to regularization of the field dynamics at small densities of energy[66]. It is interesting to study the influence of the quantum corrections on the dynamics of nonabelian spatially homogeneous gauge fields interacting with Higgs field that was partially done in Refs.[62,67,68].

2. Generalized Toda criterion

Toda criterion of local instability for classical mechanical systems was at first formulated in[69]. Local instability of classical conservative Hamiltonian system with finite number of freedoms and finite (at finite energy) available phase space volume leads to mixing, destruction of the first integrals of motion and all that one calls chaos[2,3]. Agreement between Toda criterion and criterion of classical chaos based on KAM theory and conception of nonlinear resonance was checked on the particular example in[54].

Consider classical Hamiltonian system with any finite number of degrees of freedom

$$H = \frac{1}{2}\vec{p}^2 + V(\vec{q}) \quad , \quad \vec{p} = (p_{1,\dots,}p_N)\ ;\ \vec{q} = (q_{1,\dots,}q_N), \quad N > 1. \tag{1}$$

Behavior of the classical system is locally unstable if distance between two neighboring trajectories grows exponentially with time in some region of the phase space.

Consider small region Ω of the phase space near the point $(\vec{q}_0, \vec{p}_0)$. Suppose that there are two classical trajectories $(\vec{q}^{(1)}(t), \vec{p}^{(1)}(t))$ and $(\vec{q}^{(2)}(t), \vec{p}^{(2)}(t))$ in Ω. Then deviations of trajectories are

$$\delta\vec{q}(t) = \vec{q}^{(1)}(t) - \vec{q}^{(2)}(t) \quad , \quad \delta\vec{p}(t) = \vec{p}^{(1)}(t) - \vec{p}^{(2)}(t). \tag{2}$$

Evolution of these deviations in Ω is governed by the following linearized Hamilton equations

$$\frac{d}{dt}\begin{pmatrix}\delta\vec{q}\\ \delta\vec{p}\end{pmatrix} = G\begin{pmatrix}\delta\vec{q}\\ \delta\vec{p}\end{pmatrix} \quad , \quad G \equiv \begin{pmatrix} 0 & I \\ -\Sigma & 0\end{pmatrix}. \tag{3}$$

Here I is the $N\times N$ identity matrix, G is a stability matrix and matrix Σ is

$$\Sigma \equiv \left(\frac{\partial^2 V}{\partial q_i \partial q_j}\,|_{\vec{q}_0}\right). \tag{4}$$

Matrix Σ and stability matrix G are functions of the point $\vec{q}_0$ of configuration space of the system. Solution of the equations (3) valid in Ω has the following form

$$\begin{pmatrix}\delta\vec{q}(t)\\ \delta\vec{p}(t)\end{pmatrix} = \sum_{i=1}^{2N} C_i \exp\{\lambda_i t\}\begin{pmatrix}\delta\vec{q}(0)\\ \delta\vec{p}(0)\end{pmatrix}. \tag{5}$$

Here $\lambda_i = \lambda_i(\vec{q}_0)$ are eigenvalues of the stability matrix G. And $\{C_i\}$ is a full set of projectors. From (5) it is seen:

a) If there is i such as $\mathrm{Re}\,\lambda_i > 0$ then the distance between neighboring trajectories grows exponentially with time and motion is locally unstable.

According Liouville's theorem stretching of phase space flow in one direction ($\mathrm{Re}\,\lambda_i > 0$) is accompanied by its compression in other direction (directions) in order to keep phase space volume constant. That means the existence of $\mathrm{Re}\,\lambda_j < 0$. Thus for local instability of motion we can demand existence of $\mathrm{Re}\,\lambda_k \neq 0$.

b) If for any $i = \overline{1, 2N}$ $\mathrm{Re}\,\lambda_i = 0$ then there is no local instability and the motion is regular.

It is easy to see that $G^2 = \mathrm{diag}(-\Sigma, -\Sigma)$.Therefore if $(-\xi_i)$, $i = \overline{1, N}$ are eigenvalues of the matrix $(-\Sigma)$ then

$$(-\xi_i) = \lambda_i^2 \,,\ i = \overline{1, N} \quad , \quad \lambda_i^2 = \lambda_{i+N}^2 \,,\ i = \overline{1, N}. \tag{6}$$

Thus without loss of generality we can imply that $\mathrm{Re}\,\lambda_i \geq 0$. Notice that

$$\xi_i = -\lambda_i^2 = (\mathrm{Im}\,\lambda_i)^2 - (\mathrm{Re}\,\lambda_i)^2 - 2i\,\mathrm{Im}\,\lambda_i\,\mathrm{Re}\,\lambda_i \,,\ i = \overline{1, N}. \tag{7}$$

Since matrix Σ is real and symmetric its eigenvalues $\{\xi_i\}$, $i = \overline{1, N}$ are real. Therefore (7) leads to the expression

$$\mathrm{Im}\,\lambda_i\,\mathrm{Re}\,\lambda_i = 0 \qquad \forall i = \overline{1, N}. \tag{8}$$

Thus any eigenvalue of the stability matrix G is real or pure imaginary or equals zero. Therefore the generalized Toda criterion for classical Hamiltonian systems with any finite number of freedoms can be formulated as follows[44]:

a) If $\xi_i \geq 0$, $\forall i = \overline{1, N}$ then behavior of the system is regular near the point $\vec{q}_0$.

b) If $\exists i = \overline{1, N} : \xi_i < 0$ then behavior of the system is locally unstable near the point $\vec{q}_0$.

If one of these conditions holds in some region of the configuration space then the motion is stable or chaotic respectively in this region. These results for systems with two degrees of freedom coincide with ones obtained in[62].

3. Towards chaos criterion in quantum field theory

Now we give some qualitative arguments which bring us to formulation of chaos criterion in quantum mechanics and quantum field theory. From statistical mechanics and ergodic theory it is known that chaos in classical systems is a consequence of the property of mixing[2]. Mixing means rapid (exponential) decrease of correlation function with time[3]. In other words, if correlation function exponentially decreases than the corresponding motion is chaotic, if it oscillates or is constant then the motion is regular[1]. We expand criterion of this type for quantum field systems. All stated bellow remains valid for quantum mechanics, since mathematical description via path integrals is the same.

For field systems the analogue of classical correlation function is two-point connected Green function

$$G_{ik}(x,y) = -\frac{\delta^2 W[\vec{J}]}{\delta J_i(x)\delta J_k(y)} \mid_{\vec{J}=0} . \tag{9}$$

Here $W[\vec{J}]$ is generating functional of connected Green functions, $\vec{J}$ are the sources of the fields, x, y are 4-vectors of space-time coordinates.

Thus we formulate chaos criterion for quantum field theory and quantum mechanics in the following form:

a) If two-point Green function (9) exponentially or stronger goes to zero when $|x-y| \to \infty$ (4 - vector $x-y$ is inside the light cone $(x^0 - y^0) > 0$ then the system is chaotic.

b) If it oscillates or remains constant in this limit then we have regular behavior of quantum system.

To check the agreement between generalized Toda criterion and formulated quantum chaos criterion in semi-classical limit we calculate two-point Green function in semi-classical approximation of quantum mechanics[44]. The following expression is obtained under the assumption that characteristic time for quantum dynamics of the system is much shorter then classical one:

$$G_i(t_1,t_2) = \frac{i}{2}\,\mathrm{Re}\left(\frac{e^{-\lambda_i(t_1-t_2)}}{\lambda_i}\right), \quad t_1 > t_2. \tag{10}$$

From the expression (10) it is seen

a) If classical motion is locally unstable (chaotic) then according Toda criterion there is real eigenvalue λ_i. Therefore Green function (10) exponentially goes to zero for some i when $(t_1 - t_2) \to +\infty$. Opposite is also true. If Green function (10) exponentially goes to zero under the condition $(t_1 - t_2) \to +\infty$ for some i, then there exists real eigenvalue of the stability matrix and thus classical motion is locally unstable.

b) If all eigenvalues of the stability matrix G are pure imaginary, that corresponds classically stable motion, then in the limit $(t_1 - t_2) \to +\infty$ Green function (10) oscillates as a sine. Opposite is also true. If for any i Green functions oscillate in the limit $(t_1 - t_2) \to +\infty$ then $\{\lambda_i\}$ are pure imaginary for any i and classical motion is stable and regular.

Thus we have demonstrated that proposed quantum chaos criterion coincides with Toda criterion in the semi-classical limit (corresponding principle).

One of possible applications of proposed chaos criterion in field theory is an investigation of the stability of classical solutions with respect to small perturbations of initial conditions. Of course, this does not directly imply chaos, but advances us to it. To study the stability of certain classical solution of field equations one has to calculate (for instance, using one loop approximation) two-point Green function in the vicinity of considered classical solution.

To demonstrate this, consider real scalar φ^4-field

$$L = \frac{1}{2}\left(\partial_\mu \varphi\right)^2 - \frac{1}{2}m^2\varphi^2 - \frac{\lambda}{4!}\varphi^4. \tag{11}$$

Here $\lambda > 0$ is a coupling constant, m^2 is some parameter which can be larger or less then zero. In both cases $\varphi = 0$ is a solution of field equations. Asymptotic of two-point Green function calculated in the vicinity of the classical solution $\varphi = 0$ in the zero order of perturbation theory is

$$G(x,y)\underset{\rho\to\infty}{\frown}\rho^{-\frac{1}{2}}e^{im\sqrt{\rho}}. \tag{12}$$

Here $\rho = (x-y)^2$ and we accept that 4-vector $x-y$ is inside the light cone $(x^0 - y^0) > 0$, in other words $\rho > 0$. We can study the stability of considered solution with respect to small perturbations. Expression (12) shows that we have two different cases

a) Green function oscillates and slowly (non-exponentially) goes to zero when $\rho \to \infty$. According proposed chaos criterion considered solution is stable. Indeed, from (12) it follows that parameter m is real in this case. Therefore $\varphi = 0$ is a stable vacuum state.

b) Green function exponentially goes to zero in the limit $\rho \to \infty$. From proposed chaos criterion it follows that $\varphi = 0$ is an unstable solution. That is true since from (12) one can see that parameter m has to be pure imaginary. It is known that in this case state $\varphi = 0$ becomes unstable, two new stable vacuums are appeared and we obtain spontaneous symmetry breakdown[70].

4. Stability of Yang-Mills-Higgs models

We demonstrate that the "switching on" of quantum fluctuations of vector gauge fields leads to ordering at low densities of energy, order-to-chaos transition with the rise of density of energy of gauge fields occurs (compare with[62]). This phenomenon escapes one's notion under classical consideration and thus we make the step to understanding the role of chaos in GYMTs. Also we note that if the ratio of the coupling constants of

Yang-Mills and Higgs fields is large than some critical value then quantum corrections do not affect the chaotic dynamics of gauge and Higgs fields.

Consider $SU(2)\bigotimes U(1)$ gauge field theory with real massless scalar field ρ with the Lagrangian

$$\begin{aligned} L = -\tfrac{1}{4}G^a_{\mu\nu}G^{a\mu\nu} - \tfrac{1}{4}H_{\mu\nu}H^{\mu\nu} + \tfrac{1}{8}g^2\rho^2\left(W_1^2 + W_2^2 + \tfrac{Z^2}{\cos^2\theta_w}\right) \\ +\tfrac{1}{2}\partial_\mu\rho\partial^\mu\rho - \tfrac{1}{4!}\lambda\rho^4. \end{aligned} \tag{13}$$

Here λ denotes a self-coupling constant of scalar field, g - self-coupling constant of non-abelian gauge fields, θ_w is Weinberg angle, A_μ corresponds electro-magnetic field, W^1_μ, W^2_μ describe W-bosons and Z_μ - neutral Z-boson.

We used the following denotation

$$G^a_{\mu\nu} = \partial_\mu W^a_\nu - \partial_\nu W^a_\mu + g\varepsilon^{abc}W^b_\mu W^c_\nu, \quad a = 1,2,3; \tag{14}$$

$$H_{\mu\nu} = \partial_\mu W^0_\nu - \partial_\nu W^0_\mu. \tag{15}$$

To study the dynamics of classical gauge fields from the viewpoint of chaos we use spatially homogeneous solutions[71]. Their dynamics in the classical vacuum of scalar field $\rho = 0$ is chaotic at any densities of energy[72].

Situation qualitatively changes if we take into account quantum fluctuations of vector fields. It is caused by the known fact that the state $<\rho>= 0$ is not a vacuum state in this case. Here $<\rho>$ denotes vacuum quantum expectation value of scalar field related with its classical vacuum value ρ as follows $<\rho>= \rho + quantum \quad corrections$. To find a true vacuum of scalar field in this case we use the method of the effective potential[73].

One loop effective potential generated by the Lagrangian (13) has the form:

$$\begin{aligned} U(<\rho>) = \frac{1}{4!}\lambda <\rho>^4 + \frac{3g^4<\rho>^4}{128\pi^2}\left(-\frac{1}{2} + \ln\frac{g^2<\rho>^2}{2\mu^2}\right) \\ + \frac{3g^4<\rho>^4}{256\pi^2\cos^4\theta_w}\left(-\frac{1}{2} + \ln\frac{g^2<\rho>^2}{2\mu^2\cos^2\theta_w}\right). \end{aligned} \tag{16}$$

Where μ^2 is a renormalization constant. Here we took into account contributions of all Feynman diagrams with one loop of any (W_1, W_2, Z or A) gauge field and external lines of Higgs field. This potential leads to spontaneous symmetry breaking and non-zero vacuum expectation value of scalar $\rho-$field appears. Classical vacuum of scalar field is $\rho = 0$, but it is not so in quantum case, because of Coleman-Weinberg effect[73].

In the case of the Lagrangian (13) it is known that vacuum state of the scalar field becomes degenerated and $<\rho>\neq 0$ if we take into account quantum properties of gauge fields.

It is easy to calculate that the squared vacuum expectation value equals

$$<\rho>_v^2 = \frac{2\mu^2}{g^2} exp\left[\frac{\left(18g^4 \ln\cos\theta_w - 32\pi^2\lambda\cos^4\theta_w\right)}{9g^4\left(1+2\cos^4\theta_w\right)}\right]. \qquad (17)$$

Its existence qualitatively changes chaotic behavior of spatially homogeneous solutions in the quantum vacuum of scalar field compared to pure classical consideration. Order-to-chaos transition occurs with the rise of density of energy of non-abelian gauge fields (see also[62]).

Starting from the Lagrangian (13) we build the model field system including Yang-Mills and Higgs fields. We consider spatially homogeneous fields. We will investigate fields of the following form:

$$W_i^a = e_i^a q^a(\tau), \quad a = 1,2; \quad i = 1,2,3; \quad \left(\vec{e^a}\right)^2 = 1, \quad \vec{e^1}\vec{e^2} = \cos\xi. \quad (18)$$

Here e_i^a are constant vectors. Other *classical* gauge fields for simplicity are put to be equal to zero, but their quantum fluctuations are included and give contribution to the potential (16). In order to simplify the problem we put $\xi = 0$ which means that the fields of W^+ and W^- gauge bosons have the same linear polarization. Therefore the Hamiltonian of the model field system has the form

$$H = \frac{1}{2}(p_1^2 + p_2^2 + p^2) + \frac{1}{8}g^2 <\rho>^2 (q_1^2 + q_2^2) + U(<\rho>). \qquad (19)$$

Here p_1 and p_2 are momenta of gauge fields and p is a momentum of Higgs field. Now we make the substitution $q_1 = r\cos\varphi$ and $q_2 = r\sin\varphi$. It is easy to see that φ is a cyclical variable and therefore its conjugate momentum $p_\varphi = r^2\dot{\varphi}$ is a constant of motion. Hamiltonian (19) can be written is the form

$$H = \frac{1}{2}(p_r^2 + p^2) + \frac{1}{8}g^2 <\rho>^2 r^2 + U(<\rho>). \qquad (20)$$

Here p_r is a conjugate momrntum of r. We neglected by the term p_φ^2/r^2 compared to the term proportional to r^2 considering further classical Yang-Mills fields with high intensity. Using well known technique based on Toda criterion of local instability[62,44] one can obtain that there is an order to chaos transition with the rise of the density of energy of the system. Critical density of energy (minus vacuum density of energy which is non-zero) is given by the following relation

$$E_c = \frac{3\mu^4}{32\pi^2}\exp\left(2\alpha_w - \frac{2\lambda}{g^4}\beta_w\right)\left(1 + \frac{1}{2\cos^4\theta_w}\right)\left(1 - 7e^{-2}\right). \qquad (21)$$

Here the following denotations are used

$$\alpha_w = \frac{2\ln\cos\theta_w}{1+2\cos^4\theta_w}, \tag{22}$$

$$\beta_w = \frac{32\pi^2\cos^4\theta_w}{9(1+2\cos^4\theta_w)}. \tag{23}$$

Here θ_w denotes Weinberg angle. Equation (21) show that E_c is exponentially suppressed if the following condition is satisfied:

$$\frac{\lambda}{g^4} > \frac{2}{\beta_w}. \tag{24}$$

Thus if the ratio of Higgs field coupling constant and the forth order of the gauge fields coupling constant exceeds the critical value then the order-chaos transition becomes practically unobservable. Vice versa, if the ratio λ/g^4 is small then quantum corrections to the Higgs field potential do influence on the model system dynamics and lead to its regularization at the small densities of energy.

5. Stability of holonomic quantum computations

In holonomic quantum computer non-abelian geometric phases (holonomies) are used for implementation of unitary transformations (quantum gates) in the subspace C^N spanned on eigenvectors corresponding the degenerate eigenvalue of parametric isospectral family of Hamiltonians $F = \{H(\lambda) = U(\lambda)H_0U(\lambda)^+\}_{\lambda\in M}$[39]. The λ's are the control parameters and M represents the space of the control parameters. The subspace C^N is called quantum code (N is the dimension of the degenerate computational subspace). Quantum gates are realized when control parameters are adiabatically driven along the loops in the control manifold M. The unitary operator mapping the initial state vector into the final one has the form $\bigoplus_{l=1}^{R} e^{i\phi_l}\Gamma_\gamma(A_\mu^l)$, where l enumerates the energy levels of the system, ϕ_l is the dynamical phase, R is the number of different energy levels of the system under consideration and the holonomy associated with the loop $\gamma \in M$ is given by[74]:

$$\Gamma_\gamma(A_\mu) = \hat{P}\exp i\int_\gamma A_\mu d\lambda_\mu. \tag{25}$$

Here $\hat{P}$ denotes the path ordering operator, A_μ is the matrix valued adiabatic connection given by the expression:

$$(A_\mu)_{mn} = \int d^3x\psi_m^*(x,\lambda)\left(-i\frac{\partial}{\partial\lambda_\mu}\right)\psi_n(x,\lambda), \tag{26}$$

where integration goes over the spatial coordinates x and $\psi_n(x,\lambda)$, $n = \overline{1,N}$ are basis functions of the corresponding eigenspace C^N. Dynamical phase will be omitted bellow due to the suitable choice of the zero energy level[39]. We shall consider the single subspace (no energy level crossings are assumed).

Fidelity for holonomic quantum computer in the general case can be written as follows:

$$f = tr\left(\rho\Gamma_{\gamma'}^{-1}\Gamma_{\gamma_0}\right) = tr\left(\rho\hat{P}\exp\{i\int_{\delta\gamma} A_\mu d\lambda_\mu\}\right) \tag{27}$$

where γ_0 is the adiabatic loop implementing the desirable quantum gate, γ' is the actual loop with some deviations from γ_0 due to practically unavoidable errors in assignment of control parameters, ρ denotes the density matrix of the quantum state which stability is studied. The loop $\delta\gamma$ is defined by the relation $\delta\gamma = \gamma'^{-1}\cdot\gamma_0$. The ·-operation means that to obtain the contour $\delta\gamma$ we primarily go over the loop γ_0 in the straight direction and after that travel along the loop γ' in the opposite direction. For the definition and properties of this operation see Ref.[39]. The contour $\delta\gamma$ is assumed to be closed i.e. to form a loop.

Using the non-abelian Stokes theorem[46] is it easy to obtain the following expression for the fidelity (27):

$$f = tr\rho\hat{P}\exp\left(i\int_{\delta S}\left[\hat{P}e^{i\int_z^{x_0}A_\mu d\lambda_\mu}\right]F_{\chi\varrho}\left[\hat{P}e^{i\int_{x_0}^{z}A_\nu d\lambda_\nu}\right]d\sigma_{\chi\varrho}(z)\right). \tag{28}$$

Here $F_{\chi\varrho} = \partial_\chi A_\varrho - \partial_\varrho A_\chi - i[A_\chi, A_\varrho]_-$ is the curvature tensor in the space of the control parameters, $\partial_\chi \equiv \partial/\partial\lambda_\chi$, $[,]_-$ denotes the commutator, δS is the surface spanned on the loop $\delta\gamma$ and $d\sigma_{\chi\varrho}$ is the projection of the infinitesimally small surface element of δS on the coordinate plane $\chi\varrho$ of the control parameter space. In the formulae (28) and everywhere bellow the summation over the indices of the curvature tensor $F_{\chi\varrho}$ is performed under the condition $\chi > \varrho$. Current point z is located on the loop $\delta\gamma$ and point x_0 is the arbitrary one. In the Ref.[46] it was demonstrated that when the non-abelian Stokes theorem is applied the result does not depend on the particular form of δS and the point x_0 is arbitrary if there are no monopole-like and string-like topological structures. This is the case assumed here. The general case deserves further investigation.

The general expression (28) is not convenient for the practical use. To simplify it we consider more relevant limit of small errors. Namely, we assume that the errors in the assignment of control parameters $\delta\lambda_\mu$ are small and satisfy the restrictions $|\delta\lambda_\mu| < ||A_\mu||^{-1}$, $|\delta\lambda_\chi\delta\lambda_\rho| < ||F_{\chi\rho}||^{-1}$

$(\chi > \rho)$. The connection and curvature tensor are calculated in some point λ_0 defining the position of the small loop $\delta\gamma$. The norm is defined as follows $||B|| = sup\{\sqrt{<\psi|B^+B|\psi>} \mid |\psi>\in C^N, \quad <\psi|\psi>=1\}$. Here B denotes both the connection matrix components A_μ and the curvature tensor matrix components $F_{\chi\rho}$ $(\chi > \rho)$. Then the following Taylor expansion in the point $\lambda_0 \in M$ can be obtained for the fidelity (28):

$$f = 1 + itr\rho F_{\chi\varrho}\delta\lambda_\chi\delta\lambda_\varrho - tr\rho[A_\mu, F_{\chi\varrho}]_-\delta\lambda_\mu\delta\lambda_\chi\delta\lambda_\varrho + (itr\rho A_\mu F_{\chi\varrho}A_\nu - \\ -\frac{i}{2}tr\rho[F_{\chi\varrho}, A_\mu A_\nu]_+ - \frac{1}{2}tr\rho F_{\chi\varrho}F_{\mu\nu}\Big)\delta\lambda_\chi\delta\lambda_\varrho\delta\lambda_\mu\delta\lambda_\nu + O(\delta\lambda^5). \quad (29)$$

Here the point λ_0 defines the position of the small loop $\delta\gamma$ in the space of parameters M and $[,]_+$ denotes the anti-commutator. Under the stated restriction on the control parameter errors higher order terms in Eq.(29) give less significant contribution to the fidelity then the terms having the lower order on $\delta\lambda$. It is seen that the Taylor expansion for fidelity does not have the linear on $\delta\lambda$ term. The cancellation of the first order terms was noticed for the particular implementation of holonomic quantum computer on trapped ions[41]. We have demonstrated that this conclusion does not depend on particular realization of holonomic quantum computation. The reason is that the difference between two vector parallel transported along two different infinitesimally small paths from one point to another is proportional to the area enclosed by these contours and it does not depend on their lengths.

Equality of the fidelity to the unity means that the computations are stable. Namely, the result of computations is not changed, when the quantum gate is perturbed. Expression (29) shows that fidelity slightly deviates from unity for small errors in assignment of control parameters. For zero errors fidelity equals unity as it should be. The rate of fidelity deviation from unity extracted from the second item of the expansion (29) is determined by the curvature tensor:

$$\left|\frac{\delta f}{\delta S_{\mu\nu}}\right| = |tr\rho F_{\mu\nu}| \quad (30)$$

where $\delta S_{\mu\nu} = \delta\lambda_\mu\delta\lambda_\nu$ is the "error area" in the plane $\mu\nu$. We see that the less the curvature tensor leads the less the deviation of the fidelity from unity and therefore calculations are more stable. This result agrees with one of the Ref.[41], where for the particular realization of holonomic quantum computer on trapped ions it was demonstrated the one-qubit holonomic gate to be resilient to control errors for large values of the squeezing parameter (exponentially small values of the curvature tensor components). From the physical point of view this result can be understood as follows. The curvature tensor (as well as the area enclosed) determines the difference between two vector parallel transported along two different infinitesimally small paths from one point to another. Therefore the less the curvature means the less the difference between state vectors parallel transported along loops γ_0 and γ'. Thus the computation is more stable. The condition for robust holonomic quantum computation is $F_{\mu\nu} = 0$ for points belonging to the loop γ_0.

6. Chaotic instanton solutions

We consider one-dimensional quantum system with periodic in space potential affected by small periodic in time perturbation. We use methods created to describe chaos in classical Hamiltonian systems to investigate essentially quantum phenomenon of tunneling. It is achieved in the framework of instanton technique, where solutions of Euclidian equations of motion (instantons) play dominating role, by the use of standard methods from the viewpoint of chaos. For the systems with periodic in time perturbation energy is no more an exact integral of motion and the language of quasi-energies is more adequate[19]. For some estimations energy as an adiabatic invariant can also be used[3]. We study properties of chaotic instanton solutions and calculate the form of the spectrum and the lower bound for the width of the ground quasi-energy zone.

Hamiltonian of the system under consideration is taken in the form[28]:

$$\tilde{H} = \frac{1}{2}\tilde{p}^2 + \omega_0^2 \cos x - \epsilon x \sum_{n=-\infty}^{+\infty} \delta(t - n\tilde{T}), \tag{31}$$

$\tilde{T}$ is the real time period of perturbation, ϵ describes the strength of perturbation. The mass of the particle equals unit. Considered cosine potential corresponds nonlinear oscillator. Phase space of nonlinear oscillator has the topology of cylinder (points $(q, \tilde{p})$ and $(q + 2\pi, \tilde{p})$ are identified). Thus its (quasi-)energy spectrum is discrete. Chaos assisted tunneling between two major resonance islands (inside single potential well) for driven nonlinear oscillator has been studied numerically in Ref.[21]. In this work we study the system with x varying from $-\infty$ to $+\infty$. This results in band structure of the (quasi-)energy spectrum.

There are papers devoted chaos assisted tunneling where some analytical predictions for billiard systems basing on the path integral formulation of quantum mechanics are made[24]. Distinguishing feature of our work is analytical predictions for the system with smooth potential exploring for this purpose instanton technique adopted from quantum field theory[75].

For applying instanton technique we consider solutions of classical equations of motion in *imaginary* (Euclidian) time. Hamiltonian (3) has the same form (translated on π) in Euclidian time as in real one.

Euclidian Hamiltonian of the system is $H = H_0 + V$, where

$$H_0 = \frac{1}{2}p^2 - \omega_0^2 \cos x, \tag{32}$$

and

$$V = \alpha T x \sum_{n=-\infty}^{+\infty} \delta(\tau - nT). \tag{33}$$

Here H_0 is nonperturbed Euclidian Hamiltonian of the system and V is the Euclidian potential of the perturbation. We also introduced coupling constant $\alpha \ll 1$ instead of $\epsilon \equiv \alpha T$ in order to simplify formulas.

Nonperturbed instanton solution describes the motion on the separatrix of the Hamiltonian (32). It is well known that this separatrix is destroyed by any periodic perturbation and on its place stochastic layer is appeared[3]. Perturbed instanton solutions correspond to the motion in vicinity of the separatrix inside the layer. Therefore instead of one instanton solution connecting neighbor maxima of nonperturbed Euclidian potential (classical vacuum states in real time potential) we obtain a manyfold of instanton solutions of Euclidian equations placed inside the stochastic layer.

The width of stochastic layer is[28]:

$$\Delta H = |H_s - H_b| = \frac{8\pi\alpha\nu}{\omega_0} e^{-\pi\nu/2\omega_0}, \tag{34}$$

here H_b is the energy value on the bound of stochastic layer.

To calculate correlator for perturbed instanton solutions we use standard technique[3]. The correlator is

$$\begin{aligned} R(\tau, \tau_0) &= \frac{1}{2\pi} \int_0^{2\pi} d\psi_0 \exp\{i(\psi(\tau) - \psi_0)\} \\ &\sim \exp\left(-\frac{\tau - \tau_0}{\tau_R}\right), \end{aligned} \tag{35}$$

here $\psi_0 \equiv \psi(\tau_0)$ and the time of correlations decay is $\tau_R = 2\pi/(\omega \ln K_0)$. Exponential decrease of correlator shows that dynamics of the instanton solutions inside the stochastic layer ($K_0 > 1$) possesses the property of mixing (chaos).

Note that perturbed one-instanton solution due to stochastic layer connects not only neighbor vacua of real time potential but also two *arbitrary* chosen vacua. Remind that to describe tunneling between non-neighbor vacua of nonperturbed system one has to take into account contribution of multi-instanton configurations.

Let us consider the tunneling between neighbor vacua (from $x \approx -\pi$ to $x \approx \pi$ for distinctness) in presence of perturbation (33). In Euclidian time this tunneling process for the nonperturbed system is described by the solution of Euclidian equations of motion with asymptotes $x = -\pi, \quad p = 0$ at $\tau = -\infty$ and $x = \pi, \quad p = 0$ at $\tau = +\infty$. There is only one solution satisfying these conditions for nonperturbed system (32) (one-instanton solution)

$$x_0^{inst}(\tau - \tau_0) = -\pi + 4 \arctan e^{\omega_0(\tau - \tau_0)}. \tag{36}$$

Its Euclidian action is $S^{inst} = 8\omega_0$. The instanton's position is denoted by τ_0. Due to Euclidian equations of motion and anti-symmetry of x_0^{inst} when time is inverted with respect to the point τ_0 perturbation (33) does not change the Euclidian action of the one-instanton solution (36) in the first order on the coupling constant $S_{pert}^{inst} = S^{inst} + O(\alpha^2)$. The only manifestation of the perturbation in this approximation is the appearance of a number of the additional solutions of Euclidian equations of motion with energies close to the energy of nonperturbed one-instanton solution and placed inside the stochastic layer.

Let us consider firstly nonperturbed system at arbitrary energy $-\omega_0^2 + \varepsilon$, $0 < \varepsilon < 2\omega_0^2$. One half of truncated instanton action can be easily calculated

$$S[x^{inst}(\tau, \varepsilon)] = \int_{-a(\varepsilon)}^{a(\varepsilon)} \sqrt{2(\omega_0^2 \cos x - (-\omega_0^2 + \varepsilon))}\, dx =$$

$$= 4\sqrt{4\omega_0^2 - 2\varepsilon} E\left(a(\varepsilon), \frac{1}{1 - \frac{\varepsilon}{2\omega_0^2}}\right), \tag{37}$$

where $\pm a(\varepsilon) = \pm \arcsin \sqrt{1 - \frac{\varepsilon}{2\omega_0^2}}$ are turning points, function E is the elliptic integral of the second kind.

Then tunneling amplitude in *perturbed* system can be found by integration over energy of the tunneling amplitude in *nonperturbed* system

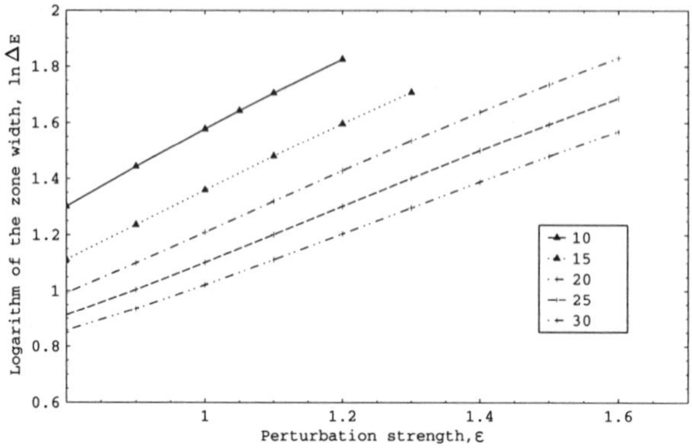

Figure 1. The dependence of the ground quasienergy zone width logarithm $\ln \Delta E$ on the perturbation strength ε at various perturbation frequency ν values.

with the action (37)

$$A = \int_0^{\Delta H} d\varepsilon \int_{x(\tau)\approx -\pi}^{x(\tau)\approx \pi} Dx \exp\left[-S[x^{inst}(\tau,\varepsilon)]\right], \tag{38}$$

where $\Delta H = 2|H_s - H_b|$ is the stochastic layer width. The contribution of the chaotic instanton solutions is taken into account by means of integration over ε. Expression (38) shows that the probability of tunneling (square of the absolute value of the tunneling amplitude) grows while chaotic region spreads (ΔH increases).

The result is obtained in the first order on coupling constant α and does not take into account possible structure of stochastic layer is valid if the layer is narrow and is in agreement with results of numerical[15] and real[35] experiments for similar problems. We have also correspondence in (38) with the nonperturbed case. Namely, if $\alpha = 0$ then $\Delta H = 0$ and the single solution describing the motion on the separatrix (nonperturbed one-instanton solution) contributes to the tunneling amplitude.

Formula (38) can be made more transparent if we use the approximate form of action (37) at $\varepsilon < 2\omega_0^2$

$$S[x^{inst}(\tau,\varepsilon)] \approx 8\omega_0 - \frac{\pi\varepsilon}{\omega_0}. \tag{39}$$

Then in Gauss approximation we obtain the following expression for the tunneling amplitude

$$\begin{aligned} A &= \int\limits_0^{\Delta H} d\varepsilon \int\limits_{-\infty}^{+\infty} dc_0 \sqrt{S[x^{inst}(\tau,\varepsilon)]} \exp\left(-S[x^{inst}(\tau,\varepsilon)]\right) \\ &\approx e^{-S^{inst}} \sqrt{S^{inst}} \Gamma F = \sqrt{8\omega_0} \Gamma e^{-8\omega_0} e^{\frac{\pi \Delta H}{\omega_0}}. \end{aligned} \tag{40}$$

where integration over c_0 gives the contribution of zero modes, Γ is a time of the tunneling. Formula (40) up to the factor i has the same form in real Minkovski time.

Expression (40) can be interpreted in the following way. Factor $F = \exp\left(\frac{\pi \Delta H}{\omega_0}\right) > 1$ in (40) differs perturbed and nonperturbed amplitudes and includes the contribution of the layer. One can think about F as a number of instanton solutions inside the stochastic layer.

We can find the minimal number of instanton solutions inside the stochastic layer. One can not distinguish instanton solutions within energy interval $\Delta E \sim 1/\Delta\tau$ (Heisenberg uncertainty relation). Here $\Delta\tau \sim \omega_0^{-1}$ denotes the time interval of observation. Thus the energy interval between neighbor instantons is $\Delta\varepsilon \sim \omega_0$. Therefore parameter F can be found as

$$F \sim 1 + \frac{\Delta H}{\Delta\varepsilon} \approx e^{\frac{\Delta H}{\omega_0}}, \tag{41}$$

where unit takes into account the nonperturbed separatrix.

The form of the spectrum of the lower quasi-energy zone is obtained using the amplitude (40) by means of standard technique (multi-instanton contributions are taken into account):

$$E_\theta \approx \frac{1}{2}\omega_0 - 2e^{-S^{inst}} \sqrt{S^{inst}} F \cos\theta, \tag{42}$$

here continuous variable θ parametrizes levels of the ground quasi-energy zone. Zone width

$$\Delta E \approx 4e^{-S^{inst}} \sqrt{S^{inst}} F \tag{43}$$

differs from the case of nonperturbed case by factor F reflecting the influence of perturbation. It follows from expressions (34), (41), (43) that perturbation (33) leads to exponential widening of the ground quasienergy zone. This conclusion was checked numerically by means of independent technique based on Floquet theory (see Ref.[19]). The dependence of the ground quasienergy zone width logarithm $\ln \Delta E$ on the perturbation strength ε is shown in Fig. 1. It is clearly seen that dependence is linear with good accuracy. It confirms analytically made conclusion that quasienergy zone width exponentially depends on the perturbation strength. Figure 2

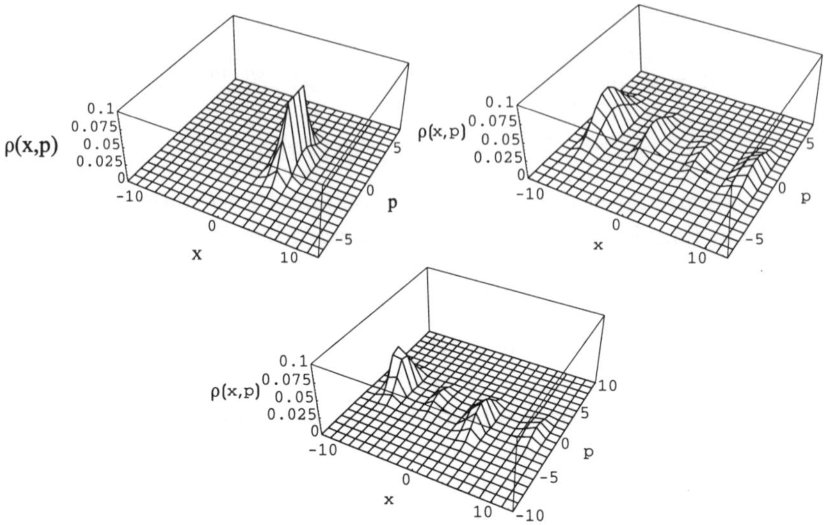

Figure 2. The first plot from the top of the page shows Husimi distribution of the initial gaussian wave packet. The second plot shows Husimi distributions for this packet after a lapse of some time in absence of the perturbation. The third plot shows Husimi distributions for this packet after a lapse of the time approximately thirteen times shorter then in the previous case but when the perturbation acts.

shows the results of numerical simulations of the gaussian wave packet dynamics. The first plot from the top of the page shows Husimi distribution of the initial gaussian wave packet. The second plot shows Husimi distributions for this packet after a lapse of some time in absence of the perturbation. The third plot shows Husimi distributions for this packet after a lapse of the time approximately thirteen times shorter then in the previous case but when the perturbation acts. It is seen that Hisumi distributions look similar, therefore at least for the parameter values chosen in this simulation the perturbation resulting in appearance of chaotic instanton solutions leads to the enhancement of the tunneling rate. This result is not trivial since the parameters are chosen in such a way that avoided crossing between relevant quasi-energy levels do not occur and thus three-level model is not applicable.

7. Conclusion

In this paper we discussed generalized Toda criterion for the Hamiltonian systems with any finite number of degrees of freedom. Basing on the formal analogy between statistical mechanics and quantum field theory chaos criterion for quantum mechanical and quantum field systems was proposed. We demonstrated that it corresponds to generalized Toda criterion in semi-classical limit of quantum mechanics. For real scalar φ^4-field we analyzed the stability of vacuum state and showed that spontaneous symmetry breakdown and degeneration of vacuum state can be regarded as signatures of quantum chaos.

As well we demonstrated that quantum fluctuations of non-abelian gauge fields leading to effective potential of Higgs field can regularize chaotic dynamics of Yang-Mills fields at low densities of energy. We showed also that if Higgs field is considered as a dynamical variable then under the conditions stated above quantum corrections do not affect classical dynamics of gauge and Higgs fields.

Exploring the similarity between the mathematical apparatus used in non-abelian gauge field theories and one used for description of holonomic quantum computation we obtained the general expression for fidelity of holonomic quantum computer. Its Taylor expansion in the small control errors limit is derived and the simple formulae for the rate of the fidelity deviation from the unity is obtained. Our general results are in agreement with ones obtained for the particular realization of holonomic quantum computer on trapped ions. The formal mathematical analogy between the fidelity and Wilson loop was noted. The question about the relationship between the confinement of the strongly interacting quarks and gluons and the stability of the corresponding systems is under the investigation.

We applied theory of *classical* chaos for investigation of the chaos assisted *tunneling* in terms of path integral formalism in imaginary time and instanton technique. We found the parameter of local instability and the width of the stochastic layer. Exponential decrease of the correlator for any perturbed instanton solution was also demonstrated, that means it to possess the property of mixing. Then properties of the stochastic layer and classical chaotic solutions in Euclidean space (chaotic instantons) were used for the calculation tunneling amplitude, ground quasi-energy zone spectrum in the presence of the perturbation and the zone width.

General tendency for chaos assisted tunneling regime (in average – if we abstract from fluctuations) is the increase of tunneling amplitude (probability) as the strength of perturbation increases[35,15]. It is confirmed here both analytically and numerically. The reason is the growth of the width of the chaotic layer and therefore the increase of the number of paths for particle to travel from one regular region to another. For small energies in Gauss approximation tunneling amplitude is increased by the factor $F > 1$. The life-time of the particle in the certain vacuum of the system decreases. It is connected with the widening of the (quasi-)energy zone (43).

We would like to emphasize that obtained results are not consequences of the particular choice of the nonperturbed Hamiltonian (32) or perturbation (33). Qualitatively they are valid for more general class of one-dimensional nonperturbed Hamiltonians with quadratic dependence on momentum and spatially periodic potential with single well in each period, as well as time dependence of homogeneous perturbation can be realized by any time periodic function. The reason is the universality of the separatrix destruction mechanism in these potentials affected by time-periodic perturbation[3].

8. Acknowledgements

We would like to thank Prof. Marko Robnik for our long-term useful contacts and discussions on the problems related with classical and quantum chaos. As well the fruitful discussions with Prof. J. Casati and Prof. T. Prosen are gratefully acknowledged. Part of the results presented in this paper was obtained in co-authorship with Dr. R.G. Shulyakovsky. His contribution was essential. Also we would like to thank as former students Theoretical Physics Department of the Belarusian State University Physics Faculty. One of us (V.K.) especially is grateful to his teacher and founder of the theoretical physics in Belarus academician F.I.Fedorov.

Bibliography

1. H. G. Schuster, Detrministic Chaos: An Introduction.– Physik-Verlag Weinheim, 1984.
2. A. J. Lichtenberg and M. A. Lieberman, Regular and Stochastic Motion.- Springer Publishing, 1983.
3. G. M. Zaslavskii and R. Z. Sagdeev, Introduction in nonlinear physics.- Moskow, 1988 (in Russian).
4. M. C. Gutzwiller, *Chaos in classical and quantum mechanics* (Springer - Verlag, New York, 1990).
5. O. Bohigas, S. Tomsovic and D. Ullmo, *Phys. Rep.* **223**, 43 (1993).
6. M. Robnik, *Open Sys. and Information Dyn.* **4**, 211 (1997).
7. O. Bohigas, P. Leboeuf and M. J. Sanchez, *Foundations of Physics* **31**, 489 (2001).
8. M. Robnik, *Nonl. Phenom. in Complex Syst.* **1**, 1 (1998).
9. G. P. Berman and G. M. Zaslavsky *Physica* **111A**, 17 (1982).
10. G. P. Berman and G. M. Zaslavsky, *Physica* **97A**, 367 (1979).

11. A. K. Pattanayak and W. C. Schieve, *Phys. Rev. A* **46**, 1821 (1992).
12. V. E. Bunakov, I. B. Ivanov and R. B. Panin, LANL E-Print: quant-ph/0105082v1.
13. F. Cametti, G. Jona-Lasinio, C. Presilla and F. Toninelli, LANL E-print: quant-ph/9910065v1.
14. B. Gremaud, LANL E-print: nlin.CD/0107031v1.
15. W. A. Lin and L. E. Ballentine *Phys. Rev. Lett.* **65**, 2927 (1990).
16. R. Utermann and T. Dittrich, Hängii P., *Phys. Rev. E* **49**, 273 (1994).
17. Coherent and incoherent chaotic tunneling near singlet-doublet crossings / Kohler S., Utermann R., Hänggi P., Dittrich T. *Phys. Rev. E* - 1998. - vol.58, 6. - P.7219-7230.
18. S. C. Creagh and S.-Y. Lee, *Annals of Phys.* **295**, 194 (2002).
19. J. H. Shirley, *Phys. Rev.* **138**, B979 (1965).
20. L. Bonci, A. Farusi, P. Grigolini and R. Roncaglia, *Phys. Rev. E* **58**, 5689 (1998).
21. M. Latka, P. Grigolini and B. J. West, *Phys. Rev. A* **50**, 1071 (1994).
22. F. Grossmann, T. Dittrich, P. Jung and P. Hängii *Phys. Rev. Lett.* **67**, 516 (1991).
23. E. Doron and S. D. Frischat, *Phys. Rev. Lett.* **75**, 3661 (1995).
24. S. D. Frischat and E. Doron, *Phys. Rev. E* **57**, 1421 (1998).
25. A. Shudo and K. S. Ikeda, *Physica D* **115**, 234 (1998).
26. A. Shudo and K. S. Ikeda, *Phys. Rev. Lett.* **76**, 4151 (1996).
27. F. Grossmann, *Phys. Rev. Lett.* **85**, 903 (2000).
28. V. I. Kuvshinov, A. V. Kuzmin and R. G. Shulyakovsky, *Phys. Rev. E* **67**, 015201(R) (2003).
29. V. I. Kuvshinov, A. V. Kuzmin and R. G. Shulyakovsky, *Acta Phys. Pol. B* **33**, 1721 (2002).
30. H. Jirari, H. Kröger, X. Q. Luo, K. J. M. Moriarty and S. G. Rubin, LANL E-print: quant-ph/9910116v3.
31. K.-I. Aoki, A. Horikoshi, M. Taniguchi and H. Terao, LANL E-print: hep-th/9812050v1.
32. Tunneling in complex systems // Ed. Tomsovic S. - World Scientific, Singapore, 1998.
33. D. A. Steck, W. H. Oskay and M. G. Raizen, *Phys. Rev. Lett.* **88**, 120406 (2002).
34. S. Keshavamurthy, LANL E-print: nlin.CD/0210051v1.
35. A. Mouchet, C. Miniatura, R. Kaiser, B. Gremaud and D. Delande, LANL E-print: nlin.CD/0012013v1.
36. X. R. Wang and Q. Niu, *Phys. Rev. B* **59**, R12755 (1999).
37. D. S. Saraga and T. S. Monteiro, *Phys. Rev. Lett.* **81**, 5796 (1998).
38. J. Zakrzewski, D. Delande and A. Buchleitner, *Phys. Rev. E* **57**, 1458 (1998).
39. P. Zanardi and M. Rasetti, *Phys.Lett.A* **264**, 94 (1999).
40. J. Pachos and S. Chountasis, *Phys.Rev. A* **62**, 052318 (2000).
41. J. Pachos, *Phys.Rev. A* **66**, 042318 (2002).
42. A. Peres, *Phys.Rev. A* **30**, 1610 (1984).

43. T. Prosen and M. Žnidarič, *J.Phys. A: Math. Gen.* **35**, 1455 (2002).
44. V. I. Kuvshinov and A. V. Kuzmin, *Phys. Lett. A* **296**, 82 (2002).
45. J. Pachos and H. Walther, *Phys.Rev.Lett.* **89**, 187903 (2002).
46. Yu. A. Simonov, *Usp. Fiz. Nauk* **166**, 337 (1996).
47. T. S. Biro, S. G. Matinyan and B. Muller, *Chaos and gauge field theory* (World Scientific, Singapore, 1994).
48. I. Ya. Aref'eva, A. S. Koshelev and P. B. Medvedev, *Mod. Phys. Lett. A* **13**, 2481 (1998).
49. T. Kawabe and S. Ohta, *Phys. Rev. D* **44**, 1274 (1991).
50. T. Kawabe, *Phys. Lett. B* **274**, 399 (1992).
51. T. Kawabe, *J. Phys. A: Math. Gen.* **26**, L1131 (1993).
52. T. Kawabe, *Phys. Lett. B* **343**, 254 (1995).
53. V. I. Kuvshinov and A. V. Kuzmin, *Nonl. Phenom. in Complex Syst.* **2**, 100 (1999).
54. V. I. Kuvshinov and A. V. Kuzmin, *Nonl. Phenom. in Complex Syst.* **4**, 64 (2001).
55. S. A. Gogilidze, A. M. Khvedelidze, D. M. Mladenov and H.-P. Pavel, *Phys. Rev. D* **57**, 7488 (1998).
56. Z. Cao and R. C. Hwa, *Phys. Rev. Lett.* **75**, 1268 (1995).
57. Z. Cao and R. C. Hwa, *Phys. Rev. D* **53**, 6608 (1996).
58. V. I. Kuvshinov and A. V. Kuzmin, *Prog. Theor. Phys. Suppl.* **150**, 126 (2003).
59. H. Markum, R. Pullirsch, K. Rabitsch, T. Wettig, *Nucl. Phys. Proc. Suppl.* **63**, 832 (1998).
60. H. Markum and R. Pullirsch, LANL E-print: nlin.CD/0303057v1.
61. H. B. Nielsen, H. H. Rugh and S. E. Rugh, LANL E-print: hep-th/9611128.
62. L. Salasnic, LANL E-print: nucl-th/9707035v1.
63. L. Casetti, R. Gatto and M. Pettini, *J. Phys. A: Math. Gen.* **32**, 3055 (1999).
64. T. S. Biro, N. Hörmann, H. Markum and R. Pullirsch, *Nucl. Phys. Proc. Suppl.* **86**, 403 (2000).
65. H. B. Nielsen, H. H. Rugh and S. E. Rugh, LANL E-print: chao-dyn/9605013.
66. S. G. Matinyan and B. Müller, *Phys. Rev. Lett.* **78**, 2515 (1997).
67. V. I. Kuvshinov and A. V. Kuzmin, *Nonl. Phenom. in Complex Syst.* **3**, 299 (2000).
68. V. I. Kuvshinov and A. V. Kuzmin, *J. Nonl. Math. Phys.* **9**, 382 (2002).
69. M. Toda, *Phys. Lett. A* **48**, 335 (1974).
70. P. W. Higgs, *Phys. Rev. Lett.* **13**, 508 (1964).
71. G. Z. Baseyan, S. G. Matinyan and G. K. Savvidy, *Pis'ma JETP* **29**, 641 (1979) (in Russian).
72. G. K. Savvidy, *Phys. Lett. B* **130**, 303 (1983).
73. S. Coleman and E. Weinberg *Phys. Rev. D.* **7**, 1888 (1973).
74. J. Pachos, P. Zanardi and M. Rasetti, *Phys. Rev. A* **61**, 010305 (2000).
75. A. Belavin, A. Polyakov, A. Schwarz, and Yu. Tyupkin, *Phys.Lett. B* **59**, 85 (1975).

POLARIZATION SYMMETRY OF AN ELECTROMAGNETIC FIELD

Vasilij I. Strazhev, Pjotr L. Shckol'nicov
Belarusian State University,
4 Fr. Skorina Ave., 220050, Minsk, Belarus
E-mail: strazhev@user.unibel.by

1. Introduction

Free-source Maxwell's equations are invariant not only under the transformation of a conformal group (see Ref.[1]) but under the transformation of dual rotations[2–4]:

$$\boldsymbol{E}' = \boldsymbol{E} cos\theta + \boldsymbol{H} sin\theta, \quad \boldsymbol{H}' = -\boldsymbol{E} sin\theta + \boldsymbol{H} cos\theta. \tag{1}$$

It is established[1,5,6], that the equations of an electromagnetic field possess a symmetry described also by $U(2) \times U(2)$ group too. A one-parametrical subgroup of it is the group of transformations (3). From the physical point of view the existence of this additional symmetry is determined by the fact that the free-source Maxwell equations "have not a sensation" of changing the electromagnetic wave's polarization. It is convinient to speak about the polarization symmetry of Maxwell's equations compared to $U(2)$ — the subgroup of additional invariance, which, in fact, exhausts all physical meaning.

The group of polarization symmetry enables to consider such well-known methods describing the polarization of electromagnetic waves as Jones matrix method[10], Stokes parameters[11], light beam tensor[12], polarization density matrix[13,14] on the basis of the theory of groups. (The proposed approach differs from the approaches used in[1,5,6].

An example we consider a plane monochromatic wave

$$\boldsymbol{e}(x) = \boldsymbol{e}(p)\, exp\,(ipx) \tag{2}$$

where $\boldsymbol{e}$ is a complex intensity vector of electromagnetic field. The light beam tensor witch gives a full describing of polarization property of the wave has form

$$\phi = \boldsymbol{e}(p) \otimes \boldsymbol{e}^*(p) \tag{3}$$

where the symbol $\otimes$ means the dyadic product and ϕ is a matrix (tensor) with the components $\phi_{jk} = e_j(p) \otimes e_k^*(p)$. The vector $\boldsymbol{e}(p)$ is a partial solution of the free-source Maxwell's equations in a momentum representation

$$\begin{aligned} p_0\boldsymbol{E}(p) &= [\boldsymbol{p}\boldsymbol{H}(p)], \quad \boldsymbol{p}\boldsymbol{E}(p) = 0, \\ p_0\boldsymbol{H}(p) &= -[\boldsymbol{p}\boldsymbol{E}(p)], \quad \boldsymbol{p}\boldsymbol{H}(p) = 0. \end{aligned} \tag{4}$$

Let's introduce the following (infinitesimal) transformations of the intencities $\boldsymbol{E}$, $\boldsymbol{H}$:

$$\begin{aligned} \delta\boldsymbol{E}(p) &= \alpha\phi(p)\boldsymbol{E}(p), \\ \delta\boldsymbol{H}(p) &= \alpha\breve{\phi}(p)\boldsymbol{H}(p), \end{aligned} \tag{5}$$

where

$$\phi(p) = \boldsymbol{e}(p) \otimes \boldsymbol{e}^*(p), \quad \breve{\phi}(p) = \boldsymbol{h}(p) \otimes \boldsymbol{h}^*(p), \tag{6}$$

and $\boldsymbol{e}(p)$, $\boldsymbol{h}(p)$ are any solutions of the equations (6), α is an infinitesimal parameter. It is obviously that the equations (6) are invariance on the transformations (7) witch do not touch a momentum but change a polarization of the wave. It is possible to determine a finite transformations making a representation of the group $U(2)$ of polarization symmetry of Maxwell's equations. In the coordinate representation the integro-differential operator is compared with the expression $\phi(p)$.

2. Polarization symmetry of Maxwell's equations

2.1. *About the additional invariance*

The Maxwell's equations for a free-sources electromagnetic field are

$$\begin{aligned} \partial_\nu F_{\mu\nu} &= 0, \\ F_{\mu\nu} &= \partial_\mu A_\nu - \partial_\nu A_\mu, \quad \mu, \nu = 1 \div 4. \end{aligned} \tag{7}$$

They have the followig matrix formulation[15,16]

$$(\beta_\mu \partial_\mu + \bar{P})\Psi(x) = 0, \tag{8}$$

where β_μ are the matrices of dimension 10×10 satisfying to the algebra of Daffine-Kemmer matrices and

$$\beta^\nu = \varepsilon^{\mu[\mu\nu]} + \varepsilon^{[\mu\nu]\mu}, \quad \varepsilon^{\mu[\mu\nu]} = -\varepsilon^{\mu[\nu\mu]} \tag{9}$$

The quantityes $\varepsilon^{A\,B}$ are the elements of a complete matrix algebra statisfying to the relations[17,18]

$$(\varepsilon^{A\,B})_{C\,D} = \delta^A_C \delta^B_D. \tag{10}$$

(Indexes $[\mu\nu]$) accept a row of values: ([2 3], [3 1], [1 2], [1 4], [2 4], [3 4])

The function of field $\Psi(x)$ is a column and it is connected with the potential A_μ and tensor of electromagnetic field $F_{\mu\nu}$ as follows

$$\Psi = (A_\mu,\ F_{\mu\nu}). \tag{11}$$

The matrix $\bar{P}$ in (8) is a projective matrix ($\bar{P}^2 = \bar{P}$) and it is determined by expression

$$\bar{P} = I_{10} - \varepsilon^{\mu\mu} = \begin{pmatrix} O_4 & 0 \\ 0 & I_6 \end{pmatrix}, \tag{12}$$

where I_{10} is the unit matrix with dimension 10×10, I_6 has dimension 6×6 O_4 is null matrix of 4×4 and others elements are equal to zero.

Using a Fourier decomposition

$$\Psi(x) = (2\pi)^{-\frac{3}{2}} \int \Psi(p) e^{ipx} d^4p \tag{13}$$

we can turn into a momentum representation

$$(i\hat{p} + \bar{P}) \cdot \Psi(p) = 0, \quad \hat{p} \equiv p_\mu \beta_\mu. \tag{14}$$

In the system of reference in witch $p_1 = p_2 = 0$ the equation (14) is written in the form

$$(i\hat{p}' + \bar{P}) \cdot \Phi(p') = 0, \tag{15}$$

where $\Phi(p') = T\Psi(p)$ and the matrix $T = T(\Lambda)$ corresponds to the superposition of inreduceble representations $(\frac{1}{2},\ \frac{1}{2}) \oplus (0,\ 1) \oplus (0,\ 1)$ of the main Lorenz group and $\Lambda_{\mu\nu} p_\nu = p'_\mu$, $p'_\mu = (0,\ 0,\ p'_3,\ p'_4)$. There is a set of matrixes $Q = \{Q_i\}$, witch commutate with the matrixes β_3, β_4. They have a form[7,19]

$$Q = \begin{pmatrix} a & & & & & \\ & I_2 & & & & \\ & & a' & & & \\ & & & I_1 & & \\ & & & & a & \\ & & & & & I_1 \end{pmatrix}, \quad a' = \frac{\tilde{a}^{-1}}{det\ a}, \tag{16}$$

where a is an arbitrary complex matrix with the dimension 2×2 and $\tilde{a}^{-1}$ is a inverse matrix and others elements of Q are equal to zero. For the unitari transformations when

$$a^{+} = a^{-1}, \quad Q^{+} = Q^{-1} \tag{17}$$

the matrix a may be represented as follows

$$a = e^{i\varphi}(\lambda_0 + i\lambda_k\sigma_k), \quad \lambda_0^2 + \lambda_k^2 = 1, \tag{18}$$

where $\lambda_0^* = \lambda_0$, $\lambda_k^* = \lambda_k$ and σ_k are Pauli matrices

$$\sigma_1 = \begin{pmatrix} 0 & 1 \\ 1 & 0 \end{pmatrix}, \quad \sigma_2 = \begin{pmatrix} 0 & -i \\ i & 0 \end{pmatrix}, \quad \sigma_3 = \begin{pmatrix} 1 & 0 \\ 0 & -1 \end{pmatrix}. \tag{19}$$

In this case matrices Q realize a fundamental representation of the $U(2)$ group. The transformations of wave function

$$\Psi'(p') = T^{-1}QT\Psi(p) = Q^L\Psi(p) \tag{20}$$

keeps the equations (14) in invariance because

$$[Q^L, \hat{p}]_- = 0. \tag{21}$$

It takes into account in the providing of (21) that

$$T^{-1}\beta_\mu T = \Lambda_{\mu\nu}\beta_\nu. \tag{22}$$

2.2. *Polarization symmetry*

It is possible to study only a transformations of the potential $\boldsymbol{A}$ because the matrix Q has a quasi-diagonal structure. The vector

$$\boldsymbol{A} = \begin{pmatrix} A_1 \\ A_2 \\ A_3 \end{pmatrix} \tag{23}$$

transforms by the infinitesimal transformations of the polarization symmetry group in the following way

$$\delta\boldsymbol{A}'(p') = Q_\delta(p')\boldsymbol{A}'(p') = \frac{i}{2}(\lambda_0(p') + \lambda_k(p')\sigma_k)\boldsymbol{A}'(p'), \tag{24}$$

where the matrix Q_δ has the same structure as the Q in (18) bat λ_0 and λ_k are an infinitysimal parameters.

Matrices $T(\boldsymbol{A}'(p')) = T(\boldsymbol{A}(p))$ give a associated (vectors) representation of the group of rotations $SO(3)$ witch is realized by real orthogonal set of matrices N with dimensional 3×3 such as

$$\tilde{N}N = 1, \; det\, N = 1. \tag{25}$$

In the case under consideration we interest of matrices that transform a vector $\boldsymbol{p} = (p_1, p_2, p_3)$ to the vector $\boldsymbol{p}' = (0, 0, p_3')$:

$$N\boldsymbol{p} = \boldsymbol{p}'. \tag{26}$$

The matrix N can be determined as

$$Np = \begin{pmatrix} e_1^{(1)} & e_2^{(1)} & e_3^{(1)} \\ e_1^{(2)} & e_2^{(2)} & e_3^{(2)} \\ n_1 & n_2 & n_3 \end{pmatrix}. \tag{27}$$

Then from (25)–(26) it follows that

$$\begin{aligned} (\boldsymbol{e}^{(1)})^2 = (\boldsymbol{e}^{(2)})^2 = \boldsymbol{n}^2 = 1, \\ \boldsymbol{e}^{(1)}\boldsymbol{e}^{(2)} = \boldsymbol{e}^{(1)}\boldsymbol{p} = \boldsymbol{e}^{(2)}\boldsymbol{p} = 0, \end{aligned} \tag{28}$$

in other words a set of real 3-vectors $\boldsymbol{e}^{(1)}(p)$, $\boldsymbol{e}^{(2)}(p)$, $\boldsymbol{n} = \frac{\boldsymbol{p}}{|\boldsymbol{p}|}$ forms an orthonormal triad of vectors:

$$[\boldsymbol{e}^{(1)}\,\boldsymbol{e}^{(2)}] = \boldsymbol{n}. \tag{29}$$

The next restrictions on vectors $\boldsymbol{e}^{(1)}(p)$, $\boldsymbol{e}^{(2)}(p)$ erase when we take into account a conditions of reality of electromagnetic field, that means for the potential $\boldsymbol{A}(p)$

$$\boldsymbol{A}^*(p) = \boldsymbol{A}(-p). \tag{30}$$

The consequence of (20) is that the transformations of potential $\boldsymbol{A}(p)$ are determined by quadratic combinations of vectors $\boldsymbol{e}^{(1)}(p)$, $\boldsymbol{e}^{(2)}(p)$ having a definite property of transformations on the replacement $p \to -p$ according to the condition (30).

For example

$$\boldsymbol{e}^{(1)}(-p) = \boldsymbol{e}^{(1)}(p). \tag{31}$$

As it follows from (30)

$$\boldsymbol{e}^{(2)}(-p) = -\boldsymbol{e}^{(2)}(p). \tag{32}$$

According to (20) and (24) we have in initial system of reference (in momentum representation)

$$\delta\boldsymbol{A}(p) = N^{-1}(p)\frac{i}{2}(\lambda_0(p') + \lambda_k(p')\sigma_k)N(p)\boldsymbol{A}(p),$$

such as

$$\begin{aligned} \boldsymbol{A}(p) = \frac{i}{2}[\lambda_0(p')(\boldsymbol{e}^{(1)} \otimes \boldsymbol{e}^{(1)} + \boldsymbol{e}^{(2)} \otimes \boldsymbol{e}^{(2)}) + \lambda_1(p')(\boldsymbol{e}^{(1)} \otimes \boldsymbol{e}^{(2)} \\ + \boldsymbol{e}^{(2)} \otimes \boldsymbol{e}^{(1)}) + \lambda_2(p')(\boldsymbol{e}^{(1)} \otimes \boldsymbol{e}^{(2)} - \boldsymbol{e}^{(2)} \otimes \boldsymbol{e}^{(1)}) \\ + \lambda_3(p')(\boldsymbol{e}^{(1)} \otimes \boldsymbol{e}^{(1)} - \boldsymbol{e}^{(2)} \otimes \boldsymbol{e}^{(2)})]\boldsymbol{A}(p), \end{aligned} \tag{33}$$

where the symbol $\otimes$ is a dyadic product. To the satisfaction of condition (30) it is necessary to realize of the following relations:

$$\lambda_0(p') = -\lambda_0(-p'), \quad \lambda_1(p') = \lambda_1(-p'),$$
$$\lambda_2(p') = -\lambda_2(-p'), \quad \lambda_3(p') = -\lambda_3(-p').$$

Introducing a notions

$$\lambda_0 = \frac{p_0}{|\,\boldsymbol{p}\,|}\epsilon_0, \ \lambda_1 = \epsilon_1, \ \lambda_2 = \frac{p_0}{|\,\boldsymbol{p}\,|}\epsilon_2, \ \lambda_3 = \frac{p_0}{|\,\boldsymbol{p}\,|}\epsilon_3,$$

(where ϵ_μ do not depend on momentum) we can write the transformations of polarization symmetry for a vector's potential of electromagnetic field in form

$$\delta\boldsymbol{A}(p) = i(\Pi_0(p)\epsilon_0 + \Pi_a(p)\epsilon_a)\boldsymbol{A}(p), \tag{34}$$

where

$$\begin{aligned} \Pi_0(p) &= \frac{p_0}{2\,|\,\boldsymbol{p}\,|}\boldsymbol{e}^{(\alpha)} \otimes \boldsymbol{e}^{(\alpha)}, \\ \Pi_1(p) &= \frac{1}{2}(\sigma_1)_{\alpha\beta}\boldsymbol{e}^{(\alpha)} \otimes \boldsymbol{e}^{(\beta)}, \\ \Pi_2(p) &= \frac{p_0}{2\,|\,\boldsymbol{p}\,|}(\sigma_2)_{\alpha\beta}\boldsymbol{e}^{(\alpha)} \otimes \boldsymbol{e}^{(\beta)}, \\ \Pi_3(p) &= \frac{p_0}{2\,|\,\boldsymbol{p}\,|}(\sigma_3)_{\alpha\beta}\boldsymbol{e}^{(\alpha)} \otimes \boldsymbol{e}^{(\beta)} \end{aligned} \tag{35}$$

and the real vectors $\boldsymbol{e}^{(1)}(p)$, $\boldsymbol{e}^{(2)}(p)$ are determined by relations (28), (31). The relation

$$\Pi^*(-p) = \Pi(p) \tag{36}$$

is in accordance to the condition (30). Generators (35) realize Lee algebra of the group $U(2)$:

$$[\Pi_a, \Pi_b]_- = i\varepsilon_{abc}\Pi_c, \quad [\Pi_0, \Pi_a]_- = 0 \tag{37}$$

in accordance with the condition $p^2 = 0$ (on the "mass shell"). The determination of generators of polarization symmetry of free Maxwell's equations in the form (35) means that the vectors $\boldsymbol{e}^{(1)}$, $\boldsymbol{e}^{(2)}$ are compared to the vectors of polarization of electromagnetic field.

Based on the general approach to the description of the polarization symmetry of relativistic wave equations it is possible to obtain the expressions (35) in other way. In generalized Coulomb gauge Maxwell's equations may be written in form:

$$\Box A_\mu = 0,\ \partial_\mu A_\mu = 0,\ k_\mu A_\mu = 0,\ \mu = 1 \div 4, \tag{38}$$

where k_μ is a some constant unit time-liked vector $k^2 = k_\mu k_\mu = \boldsymbol{k}^2 - k_0^2 = 1$. In a momentum representation we have

$$p^2 A_\mu(p) = 0,\ pA(p) = 0,\ kA(p) = 0. \tag{39}$$

The first equation in (39) went to the limit on the value of momentum: $p^2 = 0$. The others equations introduce two (linear) limits on components of $A_\mu(p)$. That's why for all permitting values of p equations (39) have only two independent solutions. In other words the space of solutions of equations (39) $\Omega(p)$ for given p is a two-dimensional complex linear space. Let's determine the scalar product in $\Omega(p)$

$$(a,\, b) = a_\mu^* b_\mu \tag{40}$$

(and $a_4 = ia_0$, $a_4^* = ia_0^*$). Let's introduce an orthonormal basis in $\Omega(p)$ using a solutions of equations (39). Its elements are noted as $e^{(\alpha)} = (e_\mu^{(\alpha)})$, $\alpha = 1, 2$, such as $pe^{(\alpha)} = ke^{(\alpha)} = 0$, $e^{(\alpha)}e^{(\beta)} = \delta^{(\alpha)(\beta)}$. The transformations of polarization symmetry are determined as simultaneous and equals motions of the all spaces $\Omega(p)$ which are Hilbertian two-dimensional spaces.

Generators of the group of polarization symmetry are determined by expressions:

$$I_0 = e^{(\alpha)}(p)e^{*(\alpha)}(p),\ I_j = \frac{1}{2}(\sigma_j)_{\alpha\beta}e^{(\alpha)}(p)e^{(\beta)}, \tag{41}$$

where σ_j are Pauli matrices. Believing the condition of reality (30) is correct for a basis solutions $e^{(\alpha)}$ we have the following determination of generators:

$$\Pi_0 = \frac{p_0}{|\,\boldsymbol{p}\,|}I_0,\ \Pi_1 = \frac{p_1}{|\,\boldsymbol{p}\,|}I_1,\ \Pi_2 = \frac{p_2}{|\,\boldsymbol{p}\,|}I_2,\ \Pi_3 = \frac{p_3}{|\,\boldsymbol{p}\,|}I_3. \tag{42}$$

It is obviously that $\Pi_0^*(p) = -\Pi_0(p)$, $\Pi_a^*(p) = -\Pi_a(p)$, $a = 1 \div 3$.

It is possible to use a real 4-vectors $e^{(\alpha)}(p)$ with the property $e^{(\alpha)}e^{(\beta)} = \delta^{(\alpha)(\beta)}$, $ke^{(\alpha)} = pe^{(\alpha)} = 0$, $e^{(1)}(-p) = e^{(1)}(p)$, $e^{(2)}(-p) = -e^{(2)}(p)$ to the describtion of the basis states. The following expressions can be considered as a generators of polarization symmetry

$$\Pi_0(p) = \frac{p_0}{\mid \boldsymbol{p} \mid} e^{(\alpha)} \otimes e^{(\alpha)}, \quad \Pi_1(p) = \frac{1}{2}(\sigma_1)_{\alpha\beta} e^{(\alpha)} \otimes e^{(\beta)},$$
$$\Pi_2(p) = \frac{p_0}{2 \mid \boldsymbol{p} \mid}(\sigma_2)_{\alpha\beta} e^{(\alpha)} \otimes e^{(\beta)}, \quad \Pi_3(p) = \frac{p_0}{2 \mid \boldsymbol{p} \mid}(\sigma_3)_{\alpha\beta} e^{(\alpha)} \otimes e^{(\beta)}. \tag{43}$$

By choosing an ordinary Coulomb gauge ($k = (0, 0, 0, i)$) we get $A_4 = 0$ according to the condition $kA = 0$. The basis vectors together with $\boldsymbol{n} = \frac{\boldsymbol{p}}{|\boldsymbol{p}|}$ realize a right-hand thriple of vectors according to (29). The generators of polarization symmetry (43) are then in the form (35).

2.3. *Coordinate representation*

The concrete determination of the dependence of polarization vectors $\boldsymbol{e}^\alpha$ on the 4-momentum is necessary to define the generators Π nref1.25. The choice of the vectors $\boldsymbol{e}^{(1)}$, $\boldsymbol{e}^{(2)}$ is determined by the precision to the ($\boldsymbol{p}$-localized) group of rotation around the $\boldsymbol{p}$. The group of integro-differential transformations is isomorphic of the group $U(2)$ in a coordinate representation. A specific choice of $\boldsymbol{e}^{(1)}$ and $\boldsymbol{e}^{(2)}$ is determined by the reason of comfortable to the solving of the task.

For example the 4-vectors $e_\mu^{(1)}$, $e_\mu^{(2)}$ may be defined in the following form

$$e_\mu^{(1)} = \frac{\varepsilon_{\mu\nu\alpha\beta} V_\nu p_\alpha k_\beta}{\mid \varepsilon_{\mu\nu\alpha\beta} V_\nu p_\alpha k_\beta \mid}, \quad e_\mu^{(2)} = \frac{\varepsilon_{\mu\nu\alpha\beta} e_\nu^{(1)} p_\alpha k_\beta}{\mid \varepsilon_{\mu\nu\alpha\beta} e_\nu^1 p_\alpha k_\beta \mid}, \tag{44}$$

where V_ν is a time-like 4-vector. (It may be a vector of 4-momentum of the scattering particle in the studying of polarization of scattered vawe.) Taking $k = (0, 0, 0, i)$ we get

$$\boldsymbol{e}^{(1)} = \frac{[\boldsymbol{V}\boldsymbol{p}]}{\mid [\boldsymbol{V}\boldsymbol{p}] \mid}, \quad \boldsymbol{e}^{(1)} = \frac{[\boldsymbol{p}\boldsymbol{e}^{(1)}]}{\mid [\boldsymbol{p}\boldsymbol{e}^{(1)}] \mid} \tag{45}$$

in accordance to (44). In the system of reference where $\boldsymbol{V} = (0, 0, 1)$ we obtain by using (45)

$$\boldsymbol{e}^{(1)} = (p_1^2 + p_2^2)^{-\frac{1}{2}}(p_2, -p_1, 0),$$
$$\boldsymbol{e}^{(2)} = \mid \boldsymbol{p} \mid \boldsymbol{p}^{-2}(p_1^2 + p_2^2)^{-\frac{1}{2}}(-p_1p_3, -p_2p_3, p_1^2 + p_2^2). \tag{46}$$

It follows that $\boldsymbol{p}\delta\boldsymbol{A} = 0$ from the form of generators Π (35) and relations (28). The transformations of polarization symmetry affect on the transversal components of fields. That's why $\boldsymbol{A}$ (and $\boldsymbol{E}$, $\boldsymbol{H}$) means only its radiation part $\boldsymbol{A}^T$ ($\boldsymbol{E}^T$, $\boldsymbol{H}^T$ in correspondence).

The form of generators Π_0, Π_2 may be simplified by using the relations that follows from the conditions (28) (see[13]):

$$\begin{aligned} \boldsymbol{e}^{(1)} \otimes \boldsymbol{e}^{(1)} + \boldsymbol{e}^{(2)} \otimes \boldsymbol{e}^{(2)} + \boldsymbol{n}\boldsymbol{n} = 1, \\ \boldsymbol{e}^{(1)} \otimes \boldsymbol{e}^{(2)} - \boldsymbol{e}^{(2)} \otimes \boldsymbol{e}^{(1)} = [\boldsymbol{e}^{(1)}\boldsymbol{e}^{(2)}]^x = \frac{p^x}{|\boldsymbol{p}|}, \end{aligned} \tag{47}$$

where $p^x_{ab} = \varepsilon_{abc}p_c$. Taking in account that the components of $\boldsymbol{A}$ are transversal we can neglect by the expressions of type $\boldsymbol{n}\otimes\boldsymbol{n}$ in the definition of the generator Π_0. Then generators Π_0 and Π_2 will taken the form:

$$\Pi_0 = \frac{p_0}{2\mid \boldsymbol{p} \mid}, \quad \Pi_2 = -\frac{ip_0p^x}{2\boldsymbol{p}^2}. \tag{48}$$

The definition of the generator Π_2 in a coordinate representation is obvious:

$$\Pi_2(x) = -\frac{1}{2}\nabla^{-2}\partial_t\partial^x, \tag{49}$$

where "inverse Laplacian" ∇^{-2} is an integral operator with the nucleus $-(4\pi \mid \boldsymbol{x}-\boldsymbol{y} \mid^{-1})$. The Fourier mode of operator ∇^{-2} is the expression $\mid \boldsymbol{p} \mid^{-2}$. The operator with the Fourier mode $\mid \boldsymbol{p} \mid^{-1}$ is more less well-known. The action of this operator $F(x)$ is determined by relations (see [22,23]):

$$F\varphi(x) = (2\pi^2)^{-1}\int d^3y \mid \boldsymbol{x}-\boldsymbol{y} \mid^{-2} \varphi(y) \tag{50}$$

and

$$\mid \boldsymbol{x}-\boldsymbol{y} \mid^{-2} = \sqrt{\frac{\pi}{2}}(2\pi)^{-\frac{3}{2}}\int d^3p\boldsymbol{p}^{-1}exp(ip(x-y)). \tag{51}$$

It follows that

$$\Pi_0(x) = -i\partial_t F(x). \tag{52}$$

The generators Π_1 and Π_3 in the momentum representation will acquired the form:

$$\Pi_1(p) = [\mid \boldsymbol{p} \mid (p_1^2+p_2^2)]^{-1}D^{(1)}(p), \quad \Pi_3(p) = [\mid \boldsymbol{p} \mid (p_1^2+p_2^2)]^{-1}p_0D^{(3)}(p) \tag{53}$$

by using the definition (46). Where $D^{(1)}(p)$ and $D^{(3)}(p)$ are

$$D^{(1)}(p) = \frac{1}{2}\left[(p_1^2+p_2^2)\begin{pmatrix} 0 & 0 & p_2 \\ 0 & 0 & -p_1 \\ p_2 & -p_1 & 0 \end{pmatrix} + p_3\begin{pmatrix} -2p_1p_2 & (p_1^2-p_2^2) & 0 \\ (p_1^2-p_2^2) & 2p_1p_2 & 0 \\ 0 & 0 & 0 \end{pmatrix}\right],$$

$$D^{(3)}(p) = \begin{pmatrix} (p_1^2-p_2^2) & 2p_1p_2 & 0 \\ 2p_1p_2 & (p_2^2-p_1^2) & 0 \\ 0 & 0 & (p_2^2+p_1^2) \end{pmatrix}. \tag{54}$$

In the coordinate representation we have

$$\Pi_1(x) = D^{(1)}(x)F\nabla^{-2}_{(2)}, \quad \Pi_3(x) = D^{(3)}(x)F\nabla^{-2}_{(2)}\partial_t. \tag{55}$$

The expressions for $D^{(1)}(x)$ and $D^{(3)}(x)$ follow from (54) by replacing $\boldsymbol{p} \to -i\frac{\partial}{\partial \boldsymbol{x}}$ and $\nabla^{-2}_{(2)}$ is the inverse Laplacian in the plane (x_1, x_2).

Operators $\Pi(x)$ may be determined as follows

$$\begin{aligned} \Pi_m(x) = i\partial_t\phi^m D^m(x), \ m = (0,\, 2,\, 3), \\ \Pi_1 = \phi^1(x)D^1(x), \end{aligned} \tag{56}$$

where ϕ^1, ϕ^m are any integral operators (product of integral operators) and $D^0 = 1$, $D^2 = \partial^x$ are differential matrix operators. It follows from (49), (52), (55) that

$$\begin{aligned} \tilde{D}^{(1)}(x) = D^{(1)}(x), \ \tilde{D}^{(2)}(x) = -D^{(2)}(x), \\ \tilde{D}^{(3)}(x) = D^{(3)}(x), \ \tilde{D}^{(0)}(x) = D^{(0)}(x) \end{aligned} \tag{57}$$

and the notion $\tilde{}$ means an operation of transposition of the matrix. In the case of $\boldsymbol{e}^1$ is even function and $\boldsymbol{e}^2$ is odd function of momentum $\boldsymbol{p}$ we arrive by using (49), (52), (55) that

$$\begin{aligned} D^{(0)}(-x) = D^{(0)}(x), \ D^{(1)}(-x) = -D^{(1)}(x), \\ D^{(2)}(-x) = -D^{(2)}(x), \ D^{(3)}(-x) = -D^{(3)}(x). \end{aligned} \tag{58}$$

The properties (57), (58) of the operators D may be described in common

$$D^{\mu}_{jk}(-x) = D^{\mu}_{kj}(x). \tag{59}$$

There is a possibility to simplify the expressions for Π_3, Π_1. The generator Π_3 is represented by using (47) in form

$$\Pi_3 = \frac{p_0}{2\,|\,\boldsymbol{p}\,|}(2\boldsymbol{e}^{(1)} \otimes \boldsymbol{e}^{(1)} - 1 + \boldsymbol{n} \otimes \boldsymbol{n}). \tag{60}$$

Because fields are transversal the component $\boldsymbol{n} \otimes \boldsymbol{n}$ in the formula (60) vanishes and we obtain

$$\Pi_{(3)}(p) = \frac{p_0}{|\,\boldsymbol{p}\,|}\left[-\frac{1}{2} + \frac{1}{p_1^2 + p_2^2}\begin{pmatrix} 0 & 0 & 0 \\ 0 & 0 & 0 \\ 0 & 0 & \boldsymbol{p}^2 \end{pmatrix}\right]. \tag{61}$$

In the coordinate representation (61) has the form

$$\Pi_{(3)}(x) = -F^{-1}\partial_t\left[\frac{1}{2} + \nabla_{(2)}^{-2}\begin{pmatrix} 0 & 0 & 0 \\ 0 & 0 & 0 \\ 0 & 0 & \nabla^2 \end{pmatrix}\right]. \tag{62}$$

In the system of reference where $\boldsymbol{v} = (0,\,0,\,1)$ the expression for the generator Π_1 is simplified too. It is possible to replace $\boldsymbol{e}^{(2)}$ to $\breve{\boldsymbol{e}}^{(2)}$ according to the relation

$$\breve{\boldsymbol{e}^{(2)}}(p) = \frac{|\,\boldsymbol{p}\,|}{|\,\boldsymbol{p}^x\boldsymbol{v}\,|}\boldsymbol{v} = \frac{|\,\boldsymbol{p}\,|}{\sqrt{p_1^2 + p_2^2}}(0,\,0,\,1). \tag{63}$$

Then we arrive for transversal fields

$$\Pi_1(p) = \frac{1}{2}(\boldsymbol{e}^{(1)} \otimes \breve{\boldsymbol{e}}^{(2)} + \breve{\boldsymbol{e}}^{(2)} \otimes \boldsymbol{e}^{(1)}) = \frac{|\,\boldsymbol{p}^2\,|}{2\,|\,\boldsymbol{p}\,|\,(p_1^2 + p_2^2)}\begin{pmatrix} 0 & 0 & p_2 \\ 0 & 0 & -p_1 \\ p_2 & -p_1 & 0 \end{pmatrix}. \tag{64}$$

In the coordinate representation $\Pi_1(x)$ has the form

$$\Pi_1(p) = -\frac{i}{2}F^{-1}\nabla^2\begin{pmatrix} 0 & 0 & \partial_2 \\ 0 & 0 & -\partial_1 \\ \partial_2 & -\partial_1 & 0 \end{pmatrix}. \tag{65}$$

2.4. *Finite transformations*

The equations of electromagnetic field

$$p_0\boldsymbol{E}(p) = [\boldsymbol{p}\,\boldsymbol{H}(p)], \quad p_0\boldsymbol{H}(p) = -[\boldsymbol{p}\,\boldsymbol{E}(p)], \tag{66}$$

where $\boldsymbol{E}(p)$ is a complex intencity are invariance on the finite transformations of polarization symmetry. They are

$$\begin{aligned} \boldsymbol{E}(p) &\to \boldsymbol{E}'(p) = exp(i\Pi_\mu\theta_\mu)\boldsymbol{E}(p), \\ \boldsymbol{H}(p) &\to \boldsymbol{H}'(p) = exp(i\Pi_\mu^*\theta_\mu)\boldsymbol{H}(p), \end{aligned} \tag{67}$$

where θ_μ are the real parameters. It's taken into account that the trarsfomatbns of $\boldsymbol{H}(p)$ are generated by operators Π_μ^* in accordance with the form of matrices (16).

It may be shown by represanting of $\boldsymbol{E}(p)$ and $\boldsymbol{H}(p)$ in the form

$$\boldsymbol{E}(p) = E_\alpha e^{(\alpha)}(p), \quad \boldsymbol{H}(p) = H_\alpha e^{(\alpha)}(p) \tag{68}$$

and using the relations (35) and Maxwell's equations (66) that

$$\Pi_0^2 = 1, \ (2\Pi_1)^2 = 1, \ (2\Pi_2)^2 = 1, \ (2\Pi_3)^2 = 1. \tag{69}$$

Than

$$\boldsymbol{E}(p) \to (cos\,\frac{\theta_k}{2} + i\Pi_k sin\,\frac{\theta_k}{2})\boldsymbol{E}(p), \tag{70}$$

$$\boldsymbol{E}(p) \to (cos\,\theta_0 + i\Pi_0 sin\,\theta_0)\boldsymbol{E}(p). \tag{71}$$

In a coordinate representation the transformations of vectors $\boldsymbol{E}(p)$, $\boldsymbol{H}(p)$ are arriving from (70), (71) by replacing

$$\boldsymbol{E}(p) \to \boldsymbol{E}(x) \text{ and } \boldsymbol{H}(p) \to \boldsymbol{H}(x).$$

The operators $\Pi_0(p)$ and $Pi_\alpha(p)$ are replaced by their coordinate alalogs recived in previous section.

We can write using the definition of generators Π_2 (49) and free Maxwell's equations that

$$i\Pi_2\boldsymbol{E}(x) = \frac{1}{2}\nabla^{-2}[\partial\,\boldsymbol{E}] = \frac{1}{2}\nabla^{-2}\nabla^{-2}\boldsymbol{H} = -\frac{1}{2}\boldsymbol{H}, \ i\Pi_2\boldsymbol{H} = \frac{1}{2}\boldsymbol{E}. \tag{72}$$

In this case finite transformations of fields $\boldsymbol{E}$ and $\boldsymbol{H}$ have the well-known (local) form

$$\boldsymbol{E} \to \boldsymbol{E} cos\,\frac{\theta_2}{2} - \boldsymbol{H} sin\,\frac{\theta_2}{2}, \ \ \boldsymbol{H} \to \boldsymbol{H} cos\,\frac{\theta_2}{2} + \boldsymbol{E} sin\,\frac{\theta_2}{2}. \tag{73}$$

For the potential $\boldsymbol{A}$ we can write

$$i\Pi_2\boldsymbol{A}(x) = \frac{1}{2}\nabla^{-2} rot\,A = -\frac{1}{4\pi}\int \frac{d^3y}{\mid \boldsymbol{x}-\boldsymbol{y}\mid} rot\boldsymbol{E}(y), \tag{74}$$

and a finite transformations of dual symmetry of (transversal) potential of electromagnetic field have not a local character.

$$\boldsymbol{A} \to \boldsymbol{A}(x) cos\,\frac{\theta_2}{2} - \frac{1}{4\pi}\int \frac{d^3y}{\mid \boldsymbol{x}-\boldsymbol{y}\mid} rot\boldsymbol{E}(y) sin\,\frac{\theta_2}{2}, \tag{75}$$

But they can be written as local by introduction of the potential $\boldsymbol{B}$

$$\boldsymbol{B} = -\frac{1}{4\pi}\int \frac{d^3y}{\mid \boldsymbol{x}-\boldsymbol{y}\mid} rot\boldsymbol{E}(y), \quad -rot\boldsymbol{B} = \boldsymbol{E}. \tag{76}$$

Then (75) takes the form

$$\boldsymbol{A} \to \boldsymbol{A}(x) cos\,\frac{\theta_2}{2} - \boldsymbol{B}(x) sin\,\frac{\theta_2}{2}. \tag{77}$$

Introduction of the (second) potential $\boldsymbol{B}$ is not connected with the increasing of the number of degrees of freedom of electromagnetic field. The evident form of the finite transformations which realized by Π_1, Π_3 follows directly from (70), (71) by the placing (53) into (61), (60).

Using free-source Maxwell's equations and the relation (52) the finite transformations of the subgroup $U(1)$ defined by generator Π_0 can be written as

$$\begin{aligned} \boldsymbol{E} &\to \boldsymbol{E} cos\theta_2 + \frac{1}{2\pi^2}\int \frac{d^3y}{\mid \boldsymbol{x}-\boldsymbol{y}\mid^2} rot\boldsymbol{H}(y) sin\theta_0, \\ \boldsymbol{H} &\to \boldsymbol{H} cos\theta_2 - \frac{1}{2\pi^2}\int \frac{d^3y}{\mid \boldsymbol{x}-\boldsymbol{y}\mid^2} rot\boldsymbol{E}(y) sin\theta_0. \end{aligned} \tag{78}$$

The last transformations corresponds to the displacement of the Fourier components phase of the potential on the same value. Let's write the expression of electromagnetic potential in a real form:

$$\boldsymbol{A}(x) = \frac{1}{(2\pi)^{\frac{3}{2}}}\int \frac{d^3p}{\sqrt{2\mid \boldsymbol{p}\mid}} \boldsymbol{e}^{(\alpha)}(a_\alpha e^{ipx} + a^*_\alpha e^{-ipx}), \tag{79}$$

where $p = (\boldsymbol{p}, \mid \boldsymbol{p} \mid)$ end $\boldsymbol{e}^{(\alpha)}$ are the complex orths. The generator Π_0 induct the following transformations of $\boldsymbol{A}(x)$

$$\delta\boldsymbol{A}(x) = i\varepsilon_0\Pi(\partial)\boldsymbol{A}(x)\frac{1}{(2\pi)^{\frac{3}{2}}}\int \frac{d^3p}{\sqrt{2\mid \boldsymbol{p}\mid}} \boldsymbol{e}^{(\alpha)} i\varepsilon_0(a_\alpha e^{ipx} + a^*_\alpha e^{-ipx}), \tag{80}$$

and the finite transformation of the potential a_α has the form:

$$a_\alpha(p) \to a_\alpha(p)e^{i\lambda 0}$$

That means the well-known result that the directly measured quantity of electromagnetic field is only a difference of a phases. It corresponds to the invariance of Maxwell's equations on non-local transformations of fields (78).

Bibliography

1. V. I. Fushchich and A. G. Nicitin, *Symmetry of Maxwell's equatioms* (Navucova Dumka, Kiev, 1983).
2. G. I. Rainich, *Trans. Amer. Math. Soc.* **27**, 106 (1925).
3. C. Misner and J. A. Wheeler, it Ann. of Phys. **2**, 523 (1957).
4. V. I. Strazhew and L. M. Tomilchik, *Electrodinamic with magnetic charge* (Nauka i tecknika, Minsk, 1975).
5. V. I. Fushchich, *Lett. Nuovo Cim.* **11**, 508 (1974).
6. V. I. Fushchich and A. G. Nicitin, *Lett. Nuovo Cim.* **24**, 220 (1979).
7. V. A. Pletjuhov and V. I. Strazhew, *Izv. vuzov. Fizika* № 12, 39 (1981).

8. V. I. Strazhew, F. I. Fedorov and P. L. Shcol'nicov *Docl. AN BSSR* **26**, № 8 (1982).
9. V. I. Strazhew and P. L. Shcol'nicov, In *Problems of hight-energy physic and quantum theory of field* Protvino, IFVE V. 2, 200 (1983).
10. K. J. Molchanov and G. V. Scrotsky, *Quantum electrodynamic* № 3, 4 (1971).
11. M. Born and E. Volf, *Basics of optic* (Nauka, Moskva, 1973).
12. F. I. Fedorov, *Spectroscophy journal* **2**, № 6, 523 (1965).
13. F. I. Fedorov, *The theory of gyrotropy* (Nauka i tecknika, Minsk, 1976).
14. V. B. Berestetsky, E. M. Lifshits and L. P. Pitaevsky, *Relativistic quantum theory* (Nauka, Moskva, part 1, 1968).
15. F. I. Fedorov, *Docl. AN SSSR* **82**, 37 (1952).
16. L. G. Moroz and F. I. Fedorov, *Preprint Instituta fiziki i matematiki AN BSSR* № 3, 154 (1959).
17. A. A. Bogush and F. I. Fedorov, *Docl. AN BSSR* **12**, 21 (1968).
18. A. A. Bogush and L. G. Moroz, *Intrduction to the classical field's theory* (Nauka i tecknika, Minsk, 1968).
19. S. I. Kruglov, V. A. Pletjuhov and V. I. Strazhew, *About the polarization symmetry of Maxwell's equations* Preprint № 226 Inst. fiziki AN BSSR, Minsk, 1980.
20. F. I. Fedorov, *Lorentz group* (Nauka, Moskva, 1979).
21. V. I. Strazhew and P. L. Shcol'nicov, *Izv. vuzov. Fizika.* № 11, 74 (1982).
22. Ju. A. Brychkov and A. P. Prudnikov, *Integrals transformations of generalized functions* (Nauka, Moskva, 1977).
23. V. Kech and P. Todoresku, *Introduction to theory of generalized functions* (Mir, Moskva, 1978).

MULTIMONOPOLES AND CLOSED VORTICES IN $SU(2)$ YANG-MILLS-HIGGS THEORY

Yasha Shnir

Institute of Physics, University of Oldenburg, D-26111, Oldenburg, Germany
E-mail: shnir@marvin.physik.uni-oldenburg.de

We review classical monopole solutions of $SU(2)$ Yang-Mills-Higgs theory. New classes of static axially symmetric solutions which generalise 't Hooft - Polyakov monopole are described. The configurations of the first type are the chains of monopoles and antimonopoles in static equilibrium. The solutions of another type are closed vortices which are centred around the symmetry axis. Configurations of the third type are monopoles bounded with vortices. We suggest the classification of these solutions which is related to $2d$ Poincare index.

Introduction

There are two completely distinct periods in the long history of the monopole problem. Soon after the pioneering work by Dirac[1], the interest in this field of research faded away although some enthusiasts still worked on it[a]. Actually, for more then 40 years the monopole problem was considered to be a rather esoteric, fairly beyond mainstream of theoretical physics of that time. Such an attitude was caused in part by negative results of all the experiments searching for monopoles, and it was reinforced by rather cumbersome character of the theoretical constructions which were connected with the corresponding generalisation of quantum electrodynamics. As a matter of fact this problem still remains unsolved and, within an Abelian theory, a monopole looks like a stranger. Indeed, a magnetic monopole could be introduced into the Abelian electrodynamics

[a]For a complete discussion of 'stone age history' of the monopole problem, the properties of the Dirac monopole and extensive bibliography before 1974 see[2,3,4]. Note that in[4] the genesis of the monopole problem was traced up to the notes by Petrus Pelegrinius, written at the Crusades in 1269! We will not go into this fascinating story.

on the classical level if a vector potential is not defined globally or if there are singular objects in the theory (for related discussion see e.g.[2]). However, even in that case the quantum theory of the monopole is full of contradictions one can hardly avoid in the framework of an Abelian model.

The situation changes drastically if we take into account that the Abelian electrodynamics is not a lonely theory but a part of an unified model, i.e., the generator of the electromagnetic $U(1)$ subgroup has to be embedded into a non-Abelian gauge group of higher rank.

The modern era of the monopole theory started in 1974 when 't Hooft and Polyakov independently discovered monopole solutions of the $SO(3)$ Georgi–Glashow model[5,6]. Preskill pointed out that the essence of this breakthrough is that while a Dirac monopole *could be* incorporated in an Abelian theory, some non-Abelian models, like that of Georgi and Glashow, *inevitably contain* monopole-like solutions[11].

1. 't Hooft - Polyakov monopole

1.1. *SU(2) Yang-Mills-Higgs model*

We consider non-Abelian classical Lagrangian which describes coupled gauge and Higgs fields:

$$L = -\frac{1}{2}\mathrm{Tr}F_{\mu\nu}F^{\mu\nu} + \mathrm{Tr}D_\mu\phi D^\mu\phi - V(\phi) \tag{1}$$

Here, $F_{\mu\nu} = F^a_{\mu\nu}T^a$, $\phi = \phi^a T^a$ and we use standard normalisation of the generators of the gauge group: $\mathrm{Tr}(T^aT^b) = \frac{1}{2}\delta_{ab}$, $a, b = 1, 2, 3$, which satisfy the standard Lie algebra $[T^a, T^b] = i\varepsilon_{abc}T^c$.

In what follows we will choose the simplest non-trivial case of $SU(2)$ group[b], thus the generators could be taken in the fundamental representation, $T^a = \frac{\sigma^a}{2}$ or in the adjoint representation, $(T^a)_{bc} = -i\varepsilon_{abc}$ respectively. The covariant derivative of the Higgs field is defined as

$$D_\mu\phi = \partial_\mu\phi + ie[A_\mu, \phi], \quad \text{or} \quad D_\mu\phi^a = \partial_\mu\phi^a - e\varepsilon_{abc}A^b_\mu\phi^c\ , \tag{2}$$

and the potential of the scalar fields is taken to be $V(\phi) = \frac{\lambda}{4}(\phi^a\phi^a - a^2)^2$, where e and λ are gauge and scalar coupling constants respectively. The

[b]Note that the simply connected covering $SU(2)$ group is locally isomorphic to the $SO(3)$. In the monopole theory the global difference between these two groups is very important because the topological properties of the corresponding group spaces are different. For the sake of brevity we do not discuss this point and refer the reader to[11].

field strength tensor is

$$F_{\mu\nu} = \partial_\mu A_\nu - \partial_\nu A_\mu + ie[A_\mu, A_\nu] \equiv \frac{1}{ie}[D_\mu, D_\nu] \, . \tag{3}$$

The field equations corresponding to the Lagrangian (1) are

$$D_\nu F^{a\mu\nu} = -e\varepsilon_{abc}\phi^b D^\mu \phi^c \, ; \qquad D_\mu D^\mu \phi^a = -\lambda\phi^a(\phi^b\phi^b - a^2) \, . \tag{4}$$

The symmetric stress-energy tensor $T_{\mu\nu}$ which follows from the Lagrangian (1) is conserved by virtue of the field equations: $\partial_\mu T^{\mu\nu} = 0$. We can easily obtain the static Hamiltonian

$$\begin{aligned} E = \int d^3x\, T_{00} = \int d^3x \left[\frac{1}{4}F^a_{\mu\nu}F^{\mu\nu a} + \frac{1}{2}(D_\mu\phi^a)(D^\mu\phi^a) + \frac{\lambda}{4}(\phi^a\phi^a - a^2)^2\right] \\ = \int d^3x \frac{1}{2}\left[E^a_n E^a_n + B^a_n B^a_n + (D_n\phi^a)(D_n\phi^a)\right] + V(\phi) \, , \end{aligned} \tag{5}$$

where

$$E^a_n \equiv F^a_{0n}, \qquad \text{and} \qquad B^a_n \equiv \frac{1}{2}\varepsilon_{nmk}F^a_{mk} \tag{6}$$

are 'color' electric and magnetic fields. We see that the energy is minimal if the following conditions are satisfied

$$\phi^a\phi^a = a^2; \qquad F^a_{mn} = 0; \qquad D_n\phi^a = 0 \, . \tag{7}$$

These conditions define the vacuum.

The perturbative spectrum of the model can be found from analysing small fluctuations around the vacuum state. Let us suppose that the system under consideration is static, $E^a_n = 0$. Then the energy of the vacuum is equal to zero. We consider now a fluctuation χ of the scalar field ϕ around the trivial vacuum $|\phi| = a$ where only the third isotopic component of the Higgs field is non-vanishing:

$$\phi = (0, 0, a + \chi) \, . \tag{8}$$

Substitution of the expansion (8) into the Lagrangian (1) yields, up to terms of the second order

$$(D_n\phi^a)(D_n\phi^a) \approx (\partial_n\chi^a)(\partial_n\chi^a) + e^2a^2\left[\left(A^1_n\right)^2 + \left(A^2_n\right)^2\right] \, , \tag{9}$$

and $V(\phi) \approx \frac{\lambda}{2}a^2\chi^2$. Thus, the vacuum average of the scalar field is non-vanishing and the model describes spontaneous symmetry breaking. Further analysis shows that the perturbative spectrum consists of a massless photon A^3_μ corresponding to the unbroken $U(1)$ electromagnetic subgroup, massive vector fields $A^{\pm}_{\mu} = (1/\sqrt{2})(A^1_{\mu} \pm A^2_{\mu})$ with mass $m_v = ea$, and neutral scalars having a mass $m_s = a\sqrt{2\lambda}$.

Note that the electric charge of the massive vector bosons $A^{\pm}$ is given by the unbroken $U(1)$ subgroup. In general, this is a subgroup H of the gauge group G, the action of which leaves the Higgs vacuum invariant. Obviously, that is a little group of the rotation in isospace about the direction given by the vector ϕ^a. The generator of it, $(\phi^a T^a)/a$, must be identified with the operator of electric charge Q. Thus, the expression for the covariant derivative (2) can be written in the form

$$D_\mu = \partial_\mu + ieA^a_\mu T^a = \partial_\mu + iQA^{\text{em}}_\mu \tag{10}$$

that allows to define the 'electromagnetic projection' of the gauge potential

$$A^{\text{em}}_\mu = \frac{1}{a}A^a_\mu \phi^a, \qquad Q = e\frac{1}{a}\phi^a T^a. \tag{11}$$

Taking into account the definition of the generators T^a of the gauge group, we can easily see that the minimal allowed eigenvalues of the electric charge operator now are $q = \pm e/2$.

1.2. *Topological classification of the solutions*

The spectrum of possible solutions of the Yang-Mills-Higgs model is much richer than one would naively expect. There are stable soliton-like static solutions of the complicated system of field equations (4) having a finite energy density on the spatial asymptotic. An adequate description of these objects needs the extensive use of topological methods.

Indeed, the very definition (7) forces the classical vacuum of the Yang-Mills-Higgs theory to be degenerated. The condition $V(\phi) = 0$ means that $|\phi| = a$, i.e., the set of vacuum values of the Higgs field forms a sphere S^2_{vac} of radius a in $d = 3$ isotopic space. All the points on this sphere are equivalent because there is a well defined $SU(2)$ gauge transformation which connects them.

The solutions of the classical field equations map the vacuum manifold $\mathcal{M} = S^2_{\text{vac}}$ onto the boundary of 3-dimensional space, which is also a sphere S^2. These maps are charactered by a *winding number* $n = 0, \pm 1, \pm 2 \ldots$ which is the number of times S^2_{vac} is covered by a single turn around the spatial boundary S^2. The crucial point is that the solutions having a finite energy on the spatial asymptotic could be separated into different classes according to the behaviour of the field ϕ^a. The trivial case is that the isotopic orientation of the fields do not depend on the spatial coordinates and asymptotically the scalar field tends to the limit

$$\phi^a = (0, 0, a) \tag{12}$$

This situation corresponds to winding number $n = 0$.

To construct the solutions corresponding to the non-trivial minimum of the energy functional (5), we again consider the scalar field on the spatial asymptotic $r \to \infty$ taking values on the vacuum manifold $|\phi| = a$. However, we suppose that the isovector of the scalar field now is directed in the isotopic space along the direction of the radius vector on the spatial asymptotic[c]

$$\phi^a \stackrel{r\to\infty}{\longrightarrow} \frac{ar^a}{r} \tag{13}$$

This asymptotic behaviour obviously mixes the spatial and isotopic indices and defines a single mapping of the vacuum $\mathcal{M}$ onto the spatial asymptotic, a single turn around the boundary S^2 leads to a single closed path on the sphere S^2_{vac} and the winding number of such a mapping is $n = 1$.

1.3. *Definition of magnetic charge*

The condition of vanishing covariant derivative of the scalar field on the spatial asymptotic (7) together with the choice of the nontrivial hedgehog configuration implies that at $r \to \infty$

$$\partial_n \left(\frac{r^a}{r}\right) - e\varepsilon_{abc} A_n^b \frac{r^c}{r} = 0. \tag{14}$$

It provides an asymptotic form of the gauge potential

$$A_k^a(r) \stackrel{r\to\infty}{\longrightarrow} \frac{1}{e}\varepsilon_{ank}\frac{r_n}{r^2} \tag{15}$$

which corresponds to the non-Abelian magnetic field $B_n^a \stackrel{r\to\infty}{\longrightarrow} \frac{r_a r_n}{er^4}$. Therefore, the boundary conditions (13), (15) are compatible with the existence of a long-range gauge field associated with an Abelian subgroup which is unbroken in the vacuum. Since this field falls off like $1/r^2$, which characterizes the Coulomb-like field of a point charge, and since the electric components of the field strength tensor vanish, this configuration with a finite energy could be identified with a monopole.

---

[c]In the pioneering paper by Polyakov[6] this solution was coined a 'hedgehog'.

To prove it, we first have to define the electromagnetic field strength tensor. Recall that the unbroken electromagnetic subgroup $U(1)$ is associated with rotations about the direction of the isovector ϕ. Thus it would be rather natural to introduce the electromagnetic potential as a projection of the $SU(2)$ gauge potential A_μ^a onto that direction, see eq. (11). Furthermore, as was mentioned in the paper[7], a general solution of the equation $D_\mu\phi^a = 0$, for $\phi^a\phi^a = a^2$, can be written as

$$A_\mu^a = \frac{1}{a^2 e}\varepsilon_{abc}\phi^b\partial_\mu\phi^c + \frac{1}{a}\phi^a\Lambda_\mu \tag{16}$$

where Λ_μ is an arbitrary 4-vector. It can be identified with the electromagnetic potential because eq. (16) yields for $\phi^a\phi^a = a^2$:

$$\frac{\phi^a}{a}A_\mu^a = \Lambda_\mu \equiv A_\mu^{\rm em}.$$

Inserting Eq. (16) into the definition of the field strength tensor yields

$$F_{\mu\nu}^a = F_{\mu\nu}\frac{\phi^a}{a}, \quad \text{where} \quad F_{\mu\nu} = \partial_\mu A_\nu - \partial_\nu A_\mu + \frac{1}{a^3 e}\varepsilon_{abc}\phi^a\partial_\mu\phi^b\partial_\nu\phi^c \tag{17}$$

This gauge-invariant definition of the electromagnetic field strength tensor $F_{\mu\nu}$ suggested by 't Hooft[5], has a very deep meaning. It is rather obvious that in the topologically trivial sector (12) the last term in eq. (17) vanishes and then we have

$$F_{\mu\nu} = \partial_\mu A_\nu - \partial_\nu A_\mu.$$

This is precisely the case of standard Maxwell electrodynamics. Of course, in this sector there is no place for a monopole because the Bianchi identities are satisfied: $\partial^\mu \widetilde{F}_{\mu\nu} \equiv 0$.

However, for the configuration with non-trivial boundary conditions (13), (15), also the Higgs field gives a non-vanishing contribution to the electromagnetic field strength tensor (17). Then, the second pair of Maxwell equations becomes modified:

$$\partial^\mu \widetilde{F}_{\mu\nu} = k_\nu \tag{18}$$

Note, that if the electromagnetic potential $A_\mu^{\rm em}$ is regular, the magnetic, or topological current k_μ is expressed via the scalar field alone

$$k_\mu = \frac{1}{2}\varepsilon_{\mu\nu\rho\sigma}\partial^\nu F^{\rho\sigma} = \frac{1}{2a^3 e}\varepsilon_{\mu\nu\rho\sigma}\varepsilon_{abc}\partial^\nu\phi^a\partial^\rho\phi^b\partial^\sigma\phi^c. \tag{19}$$

Now we can justify the definition of magnetic charge[8]. According to Eq. (19)

$$g = \int d^3x k_0 = -\frac{1}{2ea^3}\int d^2S_n\, \varepsilon_{abc}\varepsilon_{mnk}\,\phi^a\partial_n\phi^b\partial_k\phi^c\ , \tag{20}$$

where the last integral is taken over the surface of the sphere S^2 on the spatial asymptotic. After some simple algebra we arrive at

$$g = \frac{1}{2ea^3} \int d^2\xi \, \varepsilon_{\alpha\beta}\varepsilon_{abc} \, \phi^a \partial_\alpha \phi^b \partial_\beta \phi^c = \frac{1}{e} \int d^2\xi \sqrt{\mathrm{g}} \, , \tag{21}$$

where $\mathrm{g} = \det(\partial_\alpha \hat{\phi}^a \partial_\beta \hat{\phi}^a)$ is the determinant of the metric tensor on the S^2_{vac} sphere in isospace which is parameterised by local coordinates ξ_α

Thus, the magnetic charge is proportional to an integer n which mathematicians refer to as the Brouwer degree. The geometrical interpretation of this integer is clear: it is the number of times the isovector ϕ^a covers the sphere S^2_{vac} while r^a covers the sphere S^2 on the spatial asymptotic once. Thus[8]:

$$g = \frac{4\pi n}{e}, \qquad n \in \mathbb{Z} \, , \tag{22}$$

where the factor 4π is due to integration over the unit sphere.

Another remark about the definition of the magnetic charge is that the Brouwer degree and homotopic classification are equivalent to the 3d Poincaré–Hopf index[8]. The latter is defined as a mapping of a sphere S^2 surrounding an isolated point r_0 where the scalar field vanishes, i.e., $\phi(r_0) = 0$, onto a sphere of unit radius S^2_ϕ (see e.g.[41]). In other words, the magnetic charge of an arbitrary field configuration can be defined as a sum of the Poincaré–Hopf indices i of non-degenerated zeros $r_0^{(k)}$ of the Higgs field:

$$g = \frac{4\pi}{e} \sum_k i(r_0^{(k)}) \, . \tag{23}$$

Indeed, if we consider a scalar field which is constant everywhere in space and satisfies the boundary condition (12), it has no zero at all. Thus, the 3d Poincaré–Hopf index, alias magnetic charge, is equal to zero. However, in the case of the hedgehog configuration which satisfies the boundary condition (13)

$$\phi^a = r^a h(r) \, , \tag{24}$$

where $h(r)$ is a smooth function having no zeros, there is a single zero at the origin. Thus, $i(0) = 1$ and that is a configuration of unit magnetic charge.

This approach allows one to identify monopoles according to the positions of zeros of the Higgs field. Such an identification is very useful from the point of view of constructing multi-monopole solutions which we will consider in the following section. But first we have to find a solution of the field equations (4), which would satisfy the boundary conditions (13) and (15).

1.4. *'t Hooft–Polyakov ansatz*

We showed that asymptotically, the monopole field configuration must satisfy the conditions (13) and (15). Now, we try to define the structure functions which form the radial shape of the monopole. As usual, this problem can be simplified if we take into account the constraints following from the symmetries of the configuration. Note, that we consider static fields. That condition leaves only rotational $SO(3)$ symmetry from the original Poincaré invariance of the Lagrangian (1). Therefore, the full invariance group of the system is $SO(3) \times SU(2)$, the product of spatial and group rotations. Moreover, the non-trivial asymptotic of the Higgs field (13) corresponds to the symmetry with respect to the transformation from the diagonal $SO(3)$ subgroup which mix spatial and group rotations. Thus, one can make the ansatz[5,6]:

$$\phi^a = \frac{r^a}{er^2}H(aer); \quad A_n^a = \varepsilon_{amn}\frac{r^m}{er^2}[1 - K(aer)]; \quad A_0^a = 0 \ , \tag{25}$$

where $H(aer)$ and $K(aer)$ are functions of the dimensionless variable $\xi = aer$.

Substituting the ansatz (25) back into Eq. (5) and varying the energy functional with respect to the functions H and K we obtain two equations

$$\xi^2\frac{d^2K}{d\xi^2} = KH^2 + K(K^2 - 1); \qquad \xi^2\frac{d^2H}{d\xi^2} = 2K^2H + \frac{\lambda}{e^2}H(H^2 - \xi^2) \ . \tag{26}$$

The functions K and H must satisfy the following boundary conditions:

$$\begin{aligned} &K(\xi) \to 1, \qquad H(\xi) \to 0 \quad \text{for} \quad \xi \to 0 \ ; \\ &K(\xi) \to 0, \qquad H(\xi) \to \xi \quad \text{for} \quad \xi \to \infty \ , \end{aligned} \tag{27}$$

which correspond to the asymptotic (13) and (15). Indeed, the substitution of the ansatz (25) into the expressions for the covariant derivative of the scalar field and the non-Abelian magnetic field yields

$$\begin{aligned} D_n\phi^a &= \frac{\delta_{an}}{er^2}KH + \frac{r^a r^n}{er^4}\left(\xi\frac{dH}{d\xi} - H - KH\right) \stackrel{r\to\infty}{\longrightarrow} 0; \\ B_n^a &= \frac{r_n r^a}{er^4}\left(1 - K^2 + \xi\frac{dK}{d\xi}\right) - \frac{\delta_{an}}{er^2}\xi\frac{dK}{d\xi} \stackrel{r\to\infty}{\longrightarrow} \frac{r_n r^a}{er^4} \ . \end{aligned} \tag{28}$$

Clearly, the magnetic charge could be calculated as an integral over the surface of the sphere S^2 on spatial infinity (compare with eq. (20)):

$$g = \frac{1}{a}\int d^2 S_n B_n = \frac{1}{a}\int d^2 S_n B_n^a \phi^a = \frac{1}{a}\int d^3 x B_n^a D_n \phi^a \ , \tag{29}$$

where we made use of the Bianchi identity for the tensor of non-Abelian magnetic field $D_n B_n^a = 0$. Substituting the ansatz (25) we obtain

$$g = \frac{4\pi}{e}\int\limits_0^\infty \frac{d\xi}{\xi^2}\left\{(K^2-1)(H-\xi H') - 2\xi K'KH\right\} = \frac{4\pi}{e}\int\limits_0^\infty d\xi \frac{d}{d\xi}\left\{\frac{1-K^2}{\xi}\right\} = \frac{4\pi}{e} \ . \tag{30}$$

Again we see that the boundary conditions (27) correspond to a monopole of unit magnetic charge.

Numerical solutions of the system (26) were discussed in the papers[9,10]. It turns out that the shape functions $H(\xi)$ and $K(\xi)$ approach rather fast to the asymptotic values. Thus, there is almost the Higgs vacuum outside of some region of the order of the characteristic scale R_c, which is called the *core* of the monopole. Inside the core, at distances shorter than the wavelength of the vector boson $m_v^{-1} \sim (ae)^{-1}$, the original $SU(2)$ symmetry is restored. However, outside the core this symmetry is spontaneously broken down to the Abelian electromagnetic subgroup. In this sense there is no difference between the 't Hooft–Polyakov and the Dirac monopole outside the core[12].

Unfortunately, the system of non-linear coupled differential equations (26) in general has no analytical solution. The only known exception is the very special case $\lambda = 0$[13–15]. This is the so-called *Bogomol'nyi–Prasad–Sommerfield (BPS) limit* which deserves a special consideration.

1.5. *BPS monopole*

In the BPS limit of vanishing Higgs potential the scalar field also becomes massless and the energy of the static field configuration is taking the form

$$E = \int \left\{\frac{1}{4}\mathrm{Tr}\left((\varepsilon_{ijk}F_{ij} \pm D_k\Phi)^2\right) \mp \frac{1}{2}\varepsilon_{ijk}\mathrm{Tr}\left(F_{ij}D_k\Phi\right)\right\} d^3 r \ . \tag{31}$$

Thus, the absolute minimum of the energy corresponds to the static configurations which satisfy the first order Bogomol'nyi equations[13–15]:

$$\varepsilon_{ijk}F_{ij} = \pm D_k\Phi \tag{32}$$

Substitution of the 't Hooft–Polyakov ansatz (25) yields the system of coupled differential equations of first order

$$\xi\frac{dK}{d\xi} = -KH; \qquad \xi\frac{dH}{d\xi} = H + (1 - K^2) , \tag{33}$$

which have an analytical solution in terms of elementary functions:

$$K = \frac{\xi}{\sinh\xi}; \qquad H = \xi\coth\xi - 1 . \tag{34}$$

Definitely, the solution to the first order BPS equation (32) automatically satisfies the system of field equations of the second order, (4).

In comparison with the 't Hooft–Polyakov solution, the behaviour of the Higgs field of the monopole in the BPS limit has changed drastically. As we can see from (34), alongside with the exponentially decaying component it also obtains a long-distance Coulomb tail

$$\phi^a \to a\hat{r}^a - \frac{r^a}{er^2} \quad \text{for} \quad r \to \infty \tag{35}$$

The reason for this is that in the limit $V(\phi) = 0$ the scalar field becomes massless. Because an interaction, which is mediated by a massless scalar field, always leads to attraction, the picture of the interaction between the monopoles is very different in the BPS limit, as compared with the naive picture based on pure electromagnetic interaction. Manton showed that monopole–monopole interaction is composed of two parts originated from the long-range scalar force and the standard electromagnetic interaction, which could be either attractive or repulsive[19]. Mutual compensation of both contributions leaves the pair of BPS monopoles static but the monopole and anti-monopole interact with double strength.

Many of the remarkable properties of the BPS equation (32) are connected with its property of integrability. As was pointed out by Manton[22], integrability of the BPS system is connected with a one-to-one correspondence between the system of BPS equations and the reduced equations of self-duality of the pure Euclidean Yang–Mills theory. Indeed, the Julia–Zee correspondence means that

$$\begin{aligned} D_n\phi^a &\rightleftharpoons D_n A_0^a \equiv F_{0n}^a, \\ B_n^a = D_n\phi^a &\rightleftharpoons \widetilde{F}_{0n} = F_{0n}^a \, . \end{aligned} \tag{36}$$

Therefore, if we suppose that all the fields are static, the Euclidean equations of self-duality $F_{\mu\nu}^a = \widetilde{F}_{\mu\nu}^a$ reduces to the equations (32) and the monopole solutions in the Bogomol'nyi limit could be considered as a special class of self-dual fields.

Of course, it would not be quite correct to make a direct identification between these fields and instantons, because the instanton configuration could be independent from Euclidean time only in the limit of infinite action. Nevertheless, this analogy opens a way to apply in the $d = 3+1$ monopole theory the same very powerful methods of algebraic and differential geometry which were used to construct multi-instanton solutions of the self-duality equations in $d = 4$[23]. In particular, in the case of the BPS monopole, the solution of the self-duality equations could be constructed on the ansatz of Corrigan and Fairlie[24].

The analogy between the Euclidean Yang–Mills theory and the BPS equations could be traced up to the solutions. It was shown[25,26], that the solutions of these equations exactly equal to an infinite chain of instantons directed along the Euclidean time axis t in $d = 4$.

2. Multimonopoles

So far, we have considered a single static monopole which has topological charge $n = 1$. An obvious generalisation would be a solution of the field equation with an arbitrary integer topological charge n. At this stage we have to consider two possibilities: a single 'fat' (possibly unstable!) monopole, having a charge $n > 1$, or a system of several monopoles with a total charge equal to n. Obviously, the second situation could be much more interesting, because in this case one would encounter the effects of interaction between the monopoles, their scattering and decay.

Over the last 20 years, the investigation of the exact BPS multimonopole configurations was definitely at the crossing of the most fascinating directions of modern field theory and differential geometry. This kind of research may have caused more enthusiasm from the side of the mathematicians, rather than the physicists. The point is that the Bogomol'nyi equation is 3-dimensional reduction of the integrable self-duality equations. Thus, its solution is simpler than the investigation of the multi-instanton configurations which arise in Yang–Mills theory in $d = 4$. On the other hand, there is a significant interest to the BPS configurations in supersymmetric theories and relation of these objects to electro-magnetic duality. The reader wishing to know more about the BPS multimonopoles should consult the classic book by Atiyah and Hitchin[23] and the original papers. Modern developments are discussed in a very good review by Sutcliffe[17]. Here, we briefly outline another development, related with recently discovered *essentially non-BPS* multimonopoles[29,40,18,37,38].

2.1. *Rebbi-Rossi multimonopoles, monopole-antimonopole chains and closed vortices*

We have seen that the configuration space of YMH theory consists of sectors characterized by the topological charge of the Higgs field. While the unit charge ‘t Hooft - Polyakov hedgehog solution (25) corresponds to a single covering of the vacuum manifols S^2_{vac} is by a single turn around the spatial boundary S^2, multimonopole configurations have to be characterised by n-fold covering of the vacuum manifold. This configuration is spherically symmetric. It was shown that $SU(2)$ monopoles with higher topological charge cannot be spherically symmetric[30] and posess at most axial symmetry[16,31–33] or no rotational symmetry at all[17,35].

In the BPS limit of vanishing Higgs potential spherically symmetric monopole and axially symmetric multimonopole solutions, which satisfy the first order Bogomol'nyi equations (32) are known analytically[16,31,32]. For these solutions all zeros of the Higgs field are superimposed at a single point. Multimonopole solutions of the Bogomol'nyi equations which do not possess any rotational symmetry[34], have recently been constructed numerically by making use of the rational map ansatz[17]. In these solutions the zeros of the Higgs field are no longer all superimposed at a single point but are located at several isolated points[d].

The energy of the BPS solutions satisfies exactly the lower energy bound given by the topological charge. As shown by Taubes using infinite dimensional Morse theory, in each topological sector there exist in addition smooth, finite energy solutions of the second order field equations, which do not satisfy the Bogomol'nyi equations but only the second order field equations (4)[20]. Consequently, the energy of these solutions exceeds the Bogomol'nyi bound.

The simplest such solution resides in the sector with zero topological charge and corresponds to a saddlepoint of the energy functional[20]. It possesses axial symmetry, and the two zeros of its Higgs field are located symmetrically on the positive and negative z-axis. This solution corresponds to a monopole and antimonopole in static equilibrium[29,36]. Such a configuration is a deformations of the topologically trivial sector.

Spherically symmetric non-Bogomol'nyi BPS $SU(N)$ monopoles were considered by Ioannidou and Sutcliffe[18]. Very recently we discovered new classical solutions, which are associated with monopole-antimonopole

[d]Static spherically symmetric $SU(N)$ non-Bogomol'nyi BPS monopoles were recently constructed for $N > 2$[18].

systems in $SU(2)$ YMH theory with the scalar field in the adjoint representation[37–39]. In these solutions, which generalizes both Rebbi-Rossi multimonopoles[33] and monopole-antimonopole pair solution[29,36], the Higgs field vanishes either at some set of discrete isolated points or at rings. The latter configurations corresponds to the closed vortices while the former are (multi)-monopole-antimonopole bound systems. There are also a third class of solutions which corresponds to a single (multi) monopole bounded with a system of vortex rings centered around the symmetry axis. We review these configurations below.

2.1.1. *Static axially symmetric ansatz*

Since the Higgs field is taking values in $su(2)$ Lie algebra, we may consider a triplet of unit vectors

$$\begin{aligned} \hat{e}_r^{(n,m)} &= [\sin(m\theta)\cos(n\varphi), \sin(m\theta)\sin(n\varphi), \cos(m\theta)]; \\ \hat{e}_\theta^{(n,m)} &= [\cos(m\theta)\cos(n\varphi), \cos(m\theta)\sin(n\varphi), -\sin(m\theta)]; \\ \hat{e}_\varphi^{(n)} &= [-\sin(n\varphi), \cos(n\varphi), 0] \end{aligned} \tag{37}$$

which describe both rotations in azimuthal angle and in polar angle as well.

Now we define the $SU(2)$ matrices $\tau_r^{(n,m)}$, $\tau_\theta^{(n,m)}$, and $\tau_\varphi^{(n)}$ as a product of these vectors with the usual Pauli matrices $\tau^a = (\tau_x, \tau_y, \tau_z)$ and parametrize the gauge potential and the Higgs field by the static, purely magnetic Ansatz

$$A_\mu dx^\mu = \left(\frac{K_1}{r}dr + K_2 d\theta\right)\tau_\varphi^{(n)} - n\sin\theta\left(K_3\tau_r^{(n,m)} + K_4\tau_\theta^{(n,m)}\right)d\varphi \tag{38}$$

$$\Phi = \Phi_1\tau_r^{(n,m)} + \Phi_2\tau_\theta^{(n,m)} \,. \tag{39}$$

which generalizes the spherically symmetric 't Hooft-Polyakov ansatz (25). The latter can be recovered if we impose the constrains $K_1 = K_3 = \Phi_2 = 0, K_2 = K_4 = 1 - K(\xi), \Phi_1 = H(\xi)$.

We refer to the integers m and n in (37), (38), (39) as θ winding number and φ winding number, respectively. Indeed, as the unit vector (37) parametrized by the polar angle θ and asimuthal angle φ covers the sphere S_2 once, the fields defined by the Ansatz (38), (39) wind n and m times around z-axis and ρ-axis respectively.

Profile functions $K_1 - K_4$ and Φ_1, Φ_2 depend on the coordinates r and θ only. Thus, this ansatz is axially symmetric in a sence that a spacial rotation around the z-axis can be compencated by an Abelian gauge transformation $U = \exp\{i\omega(r,\theta)\tau_\varphi^{(n)}/2\}$ which leaves the ansatz form-invariant.

Variation of the Lagrangian (1) with respect to the profile functions yelds a system of six second order non-linear partial differential equations in the coordinates r and θ which are solved numerically subject of the boundary conditions[37-39].

2.1.2. *Properties of the solutions*

We give here a description of the general properties of the solutions. Note that asymptotic behavior of the profile functions allows to check the equivalence between the topological charge Q and magnetic charge g. Indeed,

$$\begin{aligned}
Q &= \frac{1}{8\pi}\int\limits_{S_2} d^2\xi\ \varepsilon_{\alpha\beta}\varepsilon_{abc}\phi^a\partial_\alpha\phi^b\partial_\beta\phi^c = \\
&= \frac{2nm}{8\pi}\int d\theta d\varphi\ [\Phi_1\sin(m\theta) + \Phi_2\cos(m\theta)] = \frac{n}{2}\,[1-(-1)^m]\,; \\
g &= \frac{1}{4\pi}\int \mathrm{Tr}\,\varepsilon_{ijk}\left(F_{ij}D_k\hat{\Phi}\right)d^3r = \frac{1}{2\pi}\int\limits_{S_2} d\theta d\varphi\ \mathrm{Tr}(F_{\theta\varphi}\Phi) \\
&= \frac{n}{2}\int d\theta\,(\sin(m\theta) - \partial_\theta(\sin\theta K_3)) = \frac{n}{2}\,[1-(-1)^m]
\end{aligned} \tag{40}$$

where we used the definitions (20), (30) and substitute the asymptotic behavior of the provile functions K_3, K_4 and Φ_1, Φ_2. Thus, the configurations given by the axially symmetric Ansatz (38),(39) are either deformations of the topologically trivial sector (e.g. monopole-antimonopole pair, $n = 1$, $m = 2$) or deformations of core of charge n Rebbi-Rossi multimonopoles (e.g. separation of two zeros of the Higgs field of the charge 2 monopole).

2.1.3. *2d Poincare index of the vector field*

The axial symmetry of the ansatz (38), (39) means that we can choose any value of the azimuthal angle, for example $\varphi = 0$, that is consider the behavour of the fields on the xz plane. Then classification of the solutions can be constructed by making use of the *2d Poincare index* of the Higgs field on that plane which supplements 3d topological characteristic (23)[41].

By analogy with (23), 2d index is defined by consideration of a smooth two-dimensional vector field $\vec{v}(x) = (v_1, v_2)$ on a compact space X with an isolated zero at x_0. The function $f(x) = \frac{\vec{v}(x)}{|v(x)|}$ maps a small circle S^1 centered at x_0 into the unit sphere S^1_ϕ. The degree of this mapping is

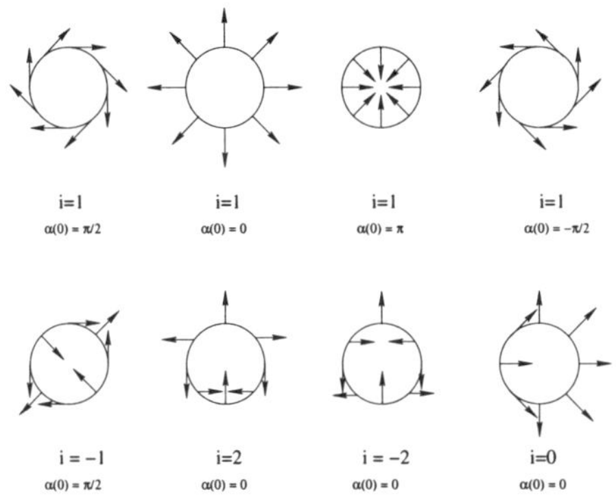

Figure 1. Indices of two dimensional vector field.

called the Poincare index i_{x_0} of vector field at the zero x_0. If the circle is parametrized by an angle $\alpha \in [0 : 2\pi]$, the index is

$$i_{x_0} = \frac{1}{2\pi} \int\limits_{S_1} \frac{\vec{v} \wedge d\vec{v}}{|v|^2} = \frac{1}{2\pi} \int d\alpha \, \frac{v_1 d_\alpha v_2 - v_2 d_\alpha v_1}{v_1^2 + v_2^2} \tag{41}$$

Note that different orientation of the vector field at the point $\alpha = 0$ on the small circle S_1 yields configurations which are labeled by the same index (see Fig. 1). Any perturbation of the contour without crossing a node of the 2-dimensional vector field does not change the index.

Let us consider a closed contour near spacial boundary of a plane which encircles some set of isolated zeros of a two-dimensional vector field

Mapping of a vector field $\vec{v}(x)$ at the spacial infinity into a unit circle yields the index i_∞. This index will not change upon a perturbation of the field, as long as no node crosses the contour during perturbation. The Poincare-Hopf theorem then can be formulated as a statement that the index of that field with respect to a contour near infinity is equal to the sum of the indices computed locally around each node of the field: $i_\infty = \sum\limits_k i(x_0^{(k)})$.

In the case under consideration we consider nodes of the Higgs field on the xz plane. The boundary condition yields $\Phi(r,\theta) \to \tau_r^{(n,m)}$ as $r \to \infty$. Thus, $\Phi_y = 0$ and coresponding components of two-dimensional vector field are $v_1 \equiv \Phi_z = \cos(m\theta), v_2 \equiv \Phi_x = \sin(m\theta)$ as $x > 0$ and $v_1 \equiv \Phi_z = \cos(m\theta), v_2 \equiv \Phi_x = (-1)^n \sin(m\theta)$ as $x < 0$ respectively. Then calculation of the index i_∞ according to eq. (41) yields

$$i_\infty = \frac{m}{2}\left[1 - (-1)^n\right] \tag{42}$$

This is a 2-dimensional topological characteristics of the axially symmetric solutions which is complementary to the 3-dimensional invariant (23). Note that it is "dual" to the topological charge (40) in a sence that permutation $n \rightleftharpoons m$ yields $i_\infty \rightleftharpoons Q$. However, the index i_∞ is not an invariant of the transformations between homotopically related configurations within the same topological sector. Corresponding deformations of 3-dimensional vector field may be related with vanishing of the projection of the field on xz plane on the circle at the spacial infinity. The crossing of this node yields jump of the 2-dimensional index i_∞ by some integer value.

2.1.4. *m-chains*

The solution of the first type are m-chains constructed in[37,38]. They are characterized by θ winding number $m > 1$ and φ winding number $n = 1, 2$.

Let us consider $n = 1$ configurations first. These m-chains possess m nodes of the Higgs field on the z-axis. Due to reflection symmetry, each node on the negative z-axis corresponds to a node on the positive z-axis. The nodes of the Higgs field $x_0(k)$ are associated with the location of the monopoles and antimonopoles and all of them are characterized by the index $i_(x_0^{(k)}) = 1$ (see Fig. 3). The index i_∞ of these m-chains is obviously equal to the winding number m whereas the topological charge is either unity (for odd m) or zero (for even m).

Indeed, for odd m ($m = 2k + 1$) the Higgs field possesses k nodes on the positive z-axis and one node at the origin. The node at the origin corresponds to a monopole if k is even and to an antimonopole if k is odd. For even m ($m = 2k$) there is no node of the Higgs field at the origin.

The $m = 1$ solution is the spherically symmetric 't Hooft-Polyakov monopole which we discussed above. The $m = 3$ (M-A-M) and $m = 5$ (M-A-M-A-M) chains represent saddlepoints with unit topological charge. The $m = 2$ chain is identical to the monopole-antimonopole (M-A) pair discussed in[29,36]. The M-A pair as well as the $m = 4$ (M-A-M-A) and $m = 6$ (M-A-M-A-M-A) chains form saddlepoints in the vacuum sector.

In Fig. 2 we present the dimensionless energy density and nodes of the Higgs field for the $n = 1$ solutions with θ winding number $m = 1, \ldots, 6$. The energy density of the m-chain possesses m maxima on the z-axis, and decreases with increasing ρ. The locations of the maxima are close to the nodes of the Higgs field. For a given m the maxima are of similar magnitude, but their height decreases with increasing m. Increasing of λ makes these maxima sharper and decreases the distance between the

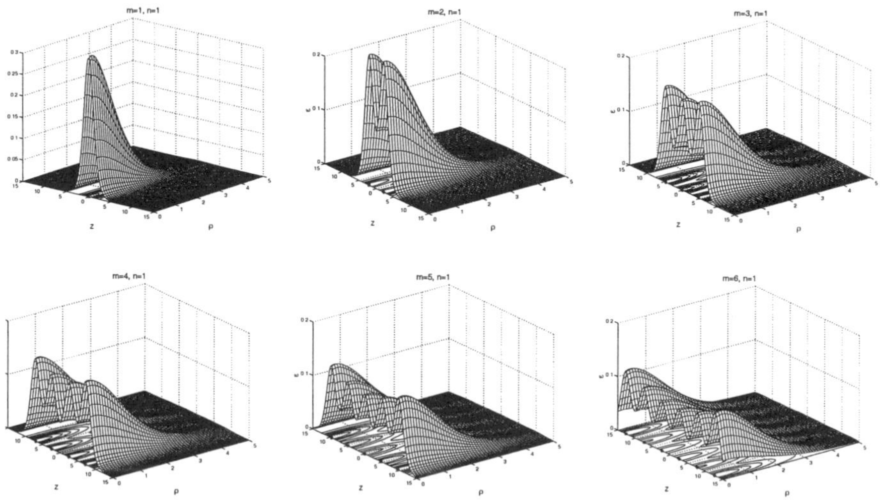

Figure 2. The rescaled energy density $E(\rho, z)$ is shown for the monopole-antimonopole chains with winding numbers $n = 1, m = 1, \ldots 6$ at $\lambda = 0$.

locations of the monopoles. We observe that for a given λ the distances between the corresponding nodes increase with increasing m.

We observe that the energy $E^{(m)}$ of an m-chain is always smaller than the energy of m single monopoles or antimonopoles (with infinite separation between them), i. e. $E^{(m)} < E_\infty = 4\pi\eta m$. On the other hand $E^{(m)}$ exceeds the minimal energy bound given by the Bogolmol'ny limit $E_{\rm min}/4\pi\eta = 0$ for even m, and $E_{\rm min}/4\pi\eta = 1$ for odd m. We observe an (almost) linear dependence of the energy $E^{(m)}$ on m. This can be modelled by taking into account only the energy of m single (infinitely separated) monopoles and the next-neighbour interaction between monopoles and antimonopoles on the chain. Defining the interaction energy as the binding energy of the monopole-antimonopole pair, $\Delta E = 2\ (4\pi\eta) - E^{(2)}$, we obtain as energy estimate for the m-chain $E^{(m)}_{\rm est}/4\pi\eta = m + (m-1)\Delta E$.

We interpret the m-chains as equilibrium states of m non-BPS monopoles and antimonopoles. As shown long ago[19], the force between BPS monopoles is given by twice the Coloumb force when the charges are unequal, and vanishes when the charges are equal, provided the monopoles well separated. Thus, monopoles and antimonopoles can only be in static

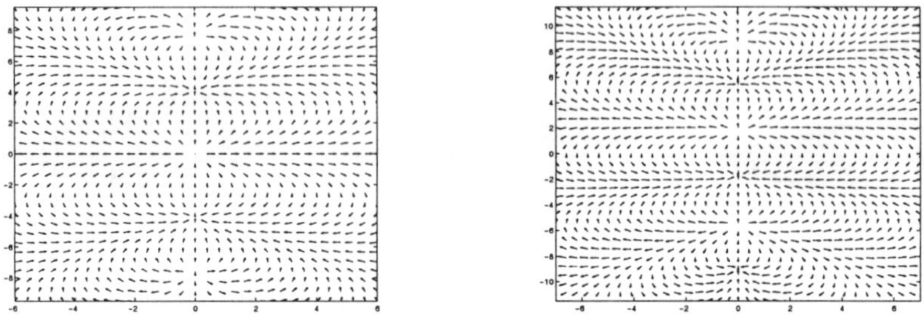

Figure 3. 2d Higgs field of monopole-antimonopole chains ($m = 5, n = 1$ and $m = 6, n = 1$) All the nodes correspond to the unit index.

equilibrium, if they are close enough to experience a repulsive force that counteracts the attraction. Indeed, is the separation between monopoles is not too large, both photon and scalar particles remain massive, the vector bosons $A_\mu^\pm$ are "defrosted" and all the particles from the spectrum contribute to the short-range Yukawa-type interactions.

In other words, m-chains are essentially non-BPS solutions. To see this in another way let us consider the limit $\lambda = 0$. Then the energy of these 'non-Bogomol'nyi BPS monopoles' can be written in the form (31) as a sum of the intergal of the square of the Bogomol'nyi equations and the term which is proportional to the topological charge and vanishes when m is even. Thus, in the latter case the energy is a measure for the deviation of the solution from selfduality.

Let us now consider chains consisting of multimonopoles with $n = 2$. Identifying the locations of the Higgs zeros on the symmetry axis with the locations of the monopoles and antimonopoles, we observe that when each pole carries charge $n = 2$, the zeros form pairs, where the distance between the monopole and the antimonopole of a pair is less than the distance to the neighboring monopole or antimonopole, belonging to the next pair.

The difference from the chains composed of pairs of monopole-antimonopole with winding number $n = 1$ we considered above, is that the index i_∞ of the m-chains composed of $n = 2$ multimonopoles is zero. Moreover, the index of each double node of the Higgs field, which is compatible with the symmetry conditions, is also zero. Another difference is that for $n = 2$ chains maxima of the energy density are no longer coincide with (double) zeros of the Higgs field. The latter still are placed on the z axis (cf. Fig. 4).

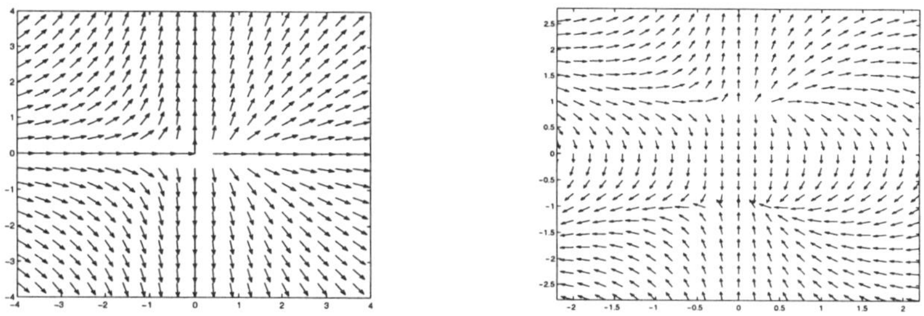

Figure 4. 2d Higgs field of a single 2-monopole and 2-monopole—2-antimonopole pair ($m = 1, n = 2$ and $m = 2, n = 2$) The double nodes correspond to zero indices.

2.1.5. *Closed vortices*

The boundary conditions imposed at the origin on the Higgs field (39) means that $\Phi_\rho(0,\theta) = 0$ and $\Phi(0,\theta) = \Phi_z(0,\theta)\tau_z^{(n,m)}$. Thus, the scalar field can either vanish there (for an odd θ winding number m), or to be directed along z axis (for an even m). In the former case there is a single n-monopole placed at the origin whereas the latter configuration in the limit of very large scalar coupling approaches the vacuum expectation value not only on the spacial boundary but also in the vicinity of the origin. Then the solutions with different winding number n link the trivial configuration $\Phi(0,\theta) = (0,0,\Phi_z(0,\theta))$ and its gauge rotated on the spacial infinity. For $n = 1,2$ these solutions are monopole-antimonopole chains we discussed above.

The situation changes dramatically if φ winding number $n > 2$. The solution of the second type, which appears in that case are not multimonopole chains but sustems of vortex rings, in the BPS limit either bounded with a single n-monopole placed at the origin, or without it[38]. One could expect that when charge of poles is increasing further beyond $n = 2$ the chain solutions consisting of multimonopoles with winding number n should exist, the monopoles and antimonopoles of the pairs should approach each other further, settling at a still smaller equilibrium distance.

Constructing solutions with charge $n = 3$ in the BPS limit ($\lambda = 0$) however, we do not find chains at all. Now there is no longer sufficient repulsion to balance the strong attraction between the 3-monopoles and 3-antimonopoles. Instead of chains, we now observe solutions with vortex rings, where the Higgs field vanishes on closed rings around the symmetry axis.

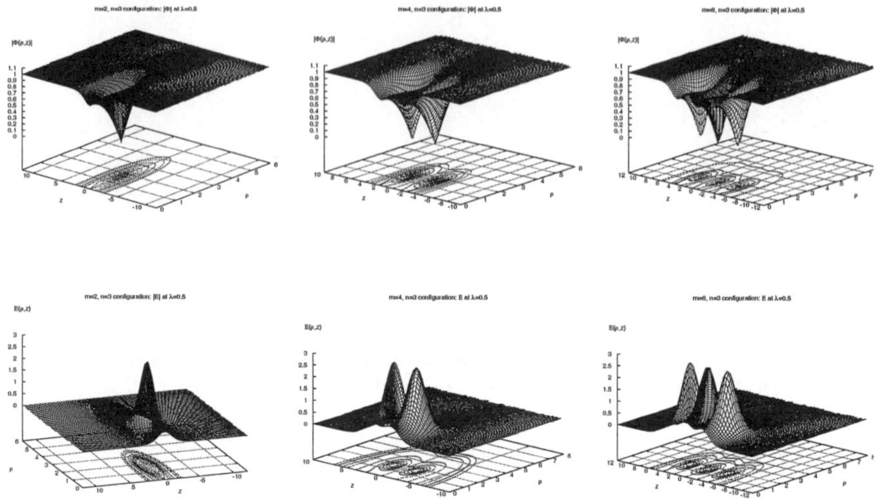

Figure 5. The rescaled energy density $E(\rho, z)$ and the modulus of the Higgs field $|\Phi(\rho, z)|$ are shown for the circular vortex solutions $m = 2, n = 3$ (single vortex), $m = 4, n = 3$ (double vortex) and $m = 6, n = 3$ (triple vortex) at $\lambda = 0.5$.

Such a situation is not a novelty, for examply the rings of zeros of a Higgs field are observed in Alice electrodynamics[43]. It is also known that closed knotted vortices arise in theories with a non-trivial Hopf number[44]. However the closed vortex solutions of Yang-Mills-Higgs theory were previously unknown.

While the dipole moment of monopole-antimonopole chains with an equal number of monopoles and antimonopoles has its origin in the magnetic charges of the configuration[36,37], the dipole moment of the closed vortices can be associated with loops of electric currents in analogy with the case of the sphaleron dipole moment[42]. Indeed, the electromagnetic current can be defined as

$$j_i^{em} = \partial_k F_{ik} \ ,$$

where F_{ik} is the 't Hooft field strength thensor (17) and there are closed rings of currents inside of the core of the configuration[39].

Other solutions with even θ winding number reside in the vacuum sector as well. For $m = 2k > 2$ solutions with zero scalar coupling it is now clear how they evolve, when the φ winding number is increased beyond $n = 2$. Starting from k pairs of physical dipoles, the pairs merge and form k vortex

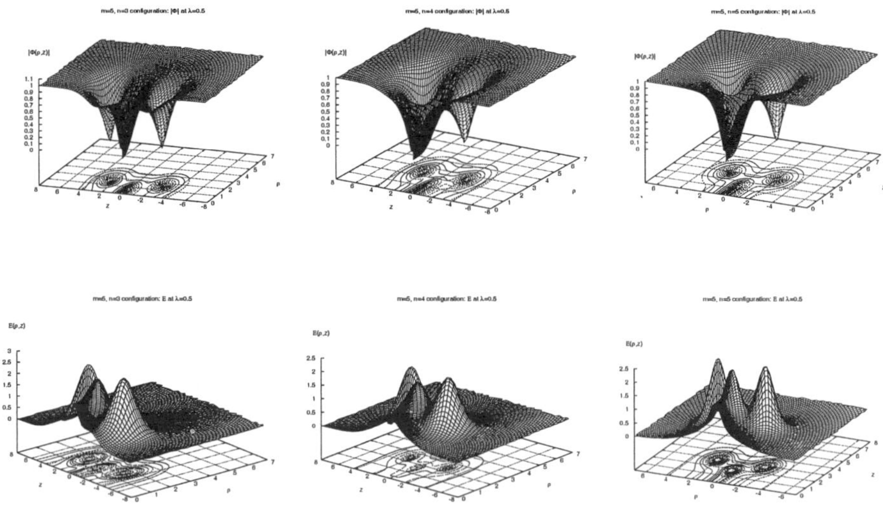

Figure 6. The rescaled energy density $E(\rho, z)$ and the modulus of the Higgs field $|\Phi(\rho, z)|$ are shown for the monopole-double vortex solution with winding numbers $m = 5, n = 3, 4, 5$ at $\lambda = 0.5$

rings, which carry the dipole strength of the solutions.

2.1.6. $n > 2$ *and odd* m*: Vortices bounded with monopole*

The solutions with odd θ winding number have an isolated zero of the Higgs field at the origin, thus their reside in the topological sector with charge n. For $m = 4k + 1$ the situation is somewhat similar to the above. Here a single n-monopole remains at the origin, whereas all other zeros form pairs, which for $n > 2$ approach each other, merge and form $2k$ rings carrying dipole strength. Since, however, a dipole on the positive axis and its respective counterpart on the negative axis have opposite orientation, their contributions cancel in the total magnetic moment. Thus the magnetic moment remains zero, as it must, because of the symmetry of the ansatz[37,39]. Non-zero scalar coupling does not change the situation but the picks of the energy density are getting sharper.

For $m = 4k - 1$, on the other hand, the situation is more complicated. Let us consider the simplest case $m = 3$ in the limit $\lambda = 0$. Again, we consider unphysical configurations with winding number n continuously vary between $n = 2$ chain solution and $n = 3$ configuration. The difference from the case of an even m is that in the initial state there are 3 poles on the z-axis, which cannot form pairs, such that all zeros belong to a pair, symmetrically located around the origin, thus the dipole moment of that configuration is zero. For $m = 3$, we observe in the BPS limit, that two vortices appear in the charge $n = 3$ solution, emerging from the upper and lower unpaired zero, respectively, carrying opposite dipole strength. Thus, the local $2d$ index at the origin is -1 whereas other four indices are 1, thus $i_\infty = 3$. Increasing of scalar coupling decreases the radius of the vortices as well the distance between then. However both rings remain individual.

This qualitative sketch of the properties of the new non-BPS axially symmetric multimonopoles and closed vortices is, of course, very superficial. We refer the reader to the papers[36–39]. Nevertheless, there are many questions which remains to be studied. This work is currently underway.

Acknowledgments

I have always considered it a privilege to have had the opportunity of my study at the Department of Theoretical Physics, University of Belarus. The time I have spent there has been very fruitful. I am very thankful to L.M.Tomilchik and E.A.Tolkachev, who were my teachers and advisors, for their valuable support, encouragement and guidance. They awaked my interest to the monopole problem. I would like to thank I.D. Feranchuk for asking me to write this brief notes for this collection dedicated to the anniversary of my Alma Mater.

This review is based on a work with Burkhard Kleihaus and Jutta Kunz[37,38,39]. I am ibdebted to Stephane Nonnemacher for helpful discussions, and in particular for clarifying the issue of the 2d index of vector fields. I am grateful to Ana Achucarro, Pierre van Baal, Falk Bruckmann, Adriano Di Giacomo, Dieter Maison, and Valentine Zakharov for very useful discussions and remarks. I would like to acknowledge the hospitality at the Service de Physique Théorique, CEA-Saclay and the Abdus Salam International Center for Theoretical Physics, where some parts of this work were carried out.

Bibliography

1. P.A.M. Dirac, *Proc. Roy. Soc.* **A133**, 60 (1931).
2. V.I. Strazhev and L.M. Tomilchik, *Electrodynamics with a magnetic charge*, Nauka i Tekhnika, Minsk, 1975 (in Russian).
3. G. Giacomelli et. al., *Magnetic monopole bibliography*, Preprint DFUB-2000-09, Bologna, 2000; hep-ex/0005041.

4. A.S. Goldhaber and W.P. Trower, *Magnetic Monopoles*, American Association of Physics Teachers, College Park, MD, 1990.
5. G. 't Hooft, *Nucl. Phys.* **B79**, 276 (1974).
6. A.M.Polyakov, *Pis'ma JETP* **20**, 430 (1974).
7. E. Corrigan, D. Olive, D. Fairlie and J. Nugts, *Nucl. Phys.* **B106**, 475 (1976).
8. J. Arafune, P.G.O. Freund and C.J.Goebel, *Journal of Math. Phys.* **16**, 433 (1975).
9. F.A. Bais and J.R. Primack, *Phys. Rev.* **D13**, 819 (1976).
10. T.W. Kirkman and C.K. Zachos, *Phys. Rev.* **D24**, 999 (1981).
11. J. Preskill, *Ann. Rev. of Nucl. and Part. Science* **34**, 461 (1984).
12. P. Goddard and D. Olive *Rep. Prog. Phys.* **41**, 1357 (1978).
13. E.B. Bogomol'nyi, *Sov. J. Nucl. Phys.* **24**, 449 (1976).
14. M.K. Prasad and C.M. Sommerfield, *Phys. Rev. Lett.* **35**, 760 (1975).
15. S. Coleman, S. Parke, A. Neveu and C. Sommerfield, *Phys. Rev.* **D15**, 544 (1977).
16. R.S. Ward, *Comm. Math. Phys.* **79**, 317 (1981).
17. P.M. Sutcliffe, *Int. J. of Mod. Phys.* **A12**, 4663 (1997).
18. Th. Ioannidou and P. Sutcliffe, *Phys. Rev.* **D60**, 105009 (1999).
19. N.S. Manton, *Nucl. Phys.* **B126**, 525 (1977).
20. C. H. Taubes, *Commun. Math. Phys.* **86**, 257 (1982); ibid **86**, 299 (1982).
21. C. H. Taubes, *Commun. Math. Phys.* **97**, 473 (1985).
22. N.S. Manton, *Nucl. Phys.* **B135**, 319 (1978).
23. M.F. Atiyah and N.J. Hitchin, *The geometry and dynamics of magnetic monopoles*, Princeton University Press, Princeton, 1988.
24. E. Corrigan and D. Fairlie, *Phys. Lett.* **67**, 69 (1977).
25. P. Rossi, *Nucl. Phys.* **B149**, 170 (1979).
26. P. Rossi, *Phys. Rep.* **86**, 317 (1982).
27. M.F. Atiyah and N.J. Hitchin, *Phys. Lett.* **A107**, 21 (1985).
28. S. Coleman *The magnetic monopole fifty years later* in Proceedings of 19th International School of Subnuclear Physis, 1983. Plenum Press. New York-London.
29. B. Rüber, *Eine axialsymmetrische magnetische Dipollösung der Yang-Mills-Higgs-Gleichungen*, MS Thesis, University of Bonn (1985).
30. E.J. Weinberg and A.H. Guth, *Phys. Rev.* **D14**, 1660 (1976).
31. P. Forgacs, Z. Horvath and L. Palla, *Phys. Lett.* **99B**, 232 (1981).
32. M.K. Prasad, *Comm. Math. Phys.* **80**, 137 (1981);
    M.K. Prasad and P. Rossi, *Phys. Rev.* **D24**, 2182 (1981).
33. C. Rebbi and P. Rossi, *Phys. Rev.* **D22**, 2010 (1980).
34. E. Corrigan and P. Goddard, *Comm. Math. Phys.* **80**, 575 (1981).
35. Th.A. Ioannidou and P. Sutcliffe, *Phys. Lett.* **B457**, 133 (1999).
36. B. Kleihaus and J. Kunz, *Phys. Rev.* **D61**, 025003 (2000).
37. B. Kleihaus, J. Kunz and Ya. Shnir, *Phys. Lett.* **B570**, 237 (2003).
38. B. Kleihaus, J. Kunz and Ya. Shnir, *Phys. Rev.* **D68**, 101701 (2003).
39. B. Kleihaus, J. Kunz and Ya. Shnir, *accepted by Phys. Rev. D, (2004)*.
40. B. Kleihaus, J. Kunz and D. H. Tchrakian, *Mod. Phys. Lett.* **A13**, 2523 (1998).

41. J.W. Milnor, *Topology from differentiable viewpoint*, The University Press of Virginia, 1965.
42. M. Hindmarsh and M. James, *Phys. Rev.* **D49**, 6109 (1994).
43. F.A. Bais and P. John, *Int. J. Mod. Phys.* **A10**, 3241 (1995); F.A. Bais and J. Striet, *Phys. Lett.* **B540**, 319 (2002).
44. L. D. Faddeev, and A. J. Niemi, *Nature* **387**, 56 (1997); A. J. Niemi, K. Palo and S. Virtanen, *Phys. Rev.* **D61**, 085020 (2000).

Part II

MATHEMATICAL METHODS IN THEORETICAL PHYSICS

FEDOROV'S COVARIANT APPROACH TO PARTICLE PHYSICS AND VECTOR PARAMETRIZATION OF QUANTUM GROUPS

Andrey A. Bogush
B.I.Stepanov Institute of Physics of National Academy of Sciences of Belarus
E-mail: bogush@dragon.bas-net.by

A brief analyzes of Fedorov's covariant approach to solving problems in the current field theory of elementary particles and their interactions is presented. The main attention is given to the applications of vector parametrization to the theory of physical symmetry groups elaborated by F.I. Fedorov. It is pointed out that this parametrization forms a foundation for elaborating some general finite transformation technique not only of geometrical, but also of dynamical (gauge) and, finally, of new quantum symmetries. The efficiency of this approach is demonstrated by listing the main applications of local dynamical transformations in gauge field theories. The vector-parametrical description of the simplest quantum groups (q-deformed $SL_q(2.C)$ and h-deformed $SL_h(2.C)$) is also given.

The work is dedicated to the memory of Academician Fedor Ivanovich Fedorov (19.06.1911–13.10.1994).

1. Introduction

The first research works, with which the development of the elementary particle physics in Belarus began, were carried out by Fedor Ivanovich Fedorov (19.06.1911–13.10.1994). Being a post-graduate student of Leningrad State University (1933-1936) and working under direct supervision of V.A.Fock, he applied a new method, the method of Fock's functionals to solving actual problems in quantum electrodynamics. In F.I. Fedorov thesis "On the application of the method of functionals to some problems of radiation theory" (1936, see[1]) the natural width of spectral lines and the cross sections of Compton scattering of gamma-quanta by electrons were calculated.

Later, at the Department of Theoretical Physics in Belarus State

University, the basic investigations in the field of the general theory of rdativistic wave equations for elementary particles with arbitrary spin and mass have been elaborated by F.I.Fedorov[2].

Starting from 1955, a wide systematic research in elementary particle physics and in the theory of fundamental interactions (including gravity) has been intensively developed in the Laboratory of Theoretical Physics at the Institute of Physics of Belarus Academy of Sciences (LTP IP BAS) up to now. This laboratory was also founded by Academician F.I.Fedorov and headed by him for 32 years (up to 1987).

It is quite natural, that a very close and effective collaboration between the theoreticians of the University and the Academy has been established.

2. Fedorov's covariant approach to the elementary particle theory and the vector parametrization of the physical symmetry groups

The main concepts and ideas of the general covariant approach to solving actual problems in the field theory of elementary particles and their interactions (1958–1968), are revealed in the following most important Fedorov's investigations (see monograph[3]).

a) The general method of projective operators in the theory of Dirac-like relativistic wave equations[4]

$$(\alpha_\mu \partial_\mu + \alpha_0)\psi(x) = 0 \qquad (1)$$

for free elementary particles with arbitrary spin and mass. Solutions of these equations have been obtained in covariant dyadic-matrix form by using of invariance properties of the matrices α_μ and α_0.

On this basis the covariant theory of the elementary particles with spin 0, 1/2, 1, 3/2 and 2 has been elaborated.

(1958–1968, A.I. Bedritski, L.G. Moroz, A.A. Bogush, A.I. Bolsun, V.K. Gronski, S.I. Lobko, et al.).

b) The universal non-linear matrix equations[5]

$$(\alpha_\mu \partial_\mu + \alpha_0)\psi(x) + \Lambda\psi(x)\psi(x) = 0 \qquad (2)$$

for interacting (self-interacting) fields. By now the well known schemes for the description of gravitational, electromagnetic, weak and strong interactions have been reformulated and analyzed on such a unique mathematical foundation.

(1968-1988, F.I. Fedorov, A.A. Bogush, L.F. Zhirkov, V.I. Kuvshinov, L.F. Babichev, A.A. Grigoriev, A.A. Babich).

c) The three-dimensional complex vector parametrization(VP) of the Lorentz group $SO(3.1)$[3,6,7] with the simplest composition law (group operation)

$$\mathcal{L}\mathcal{L}' = \mathcal{L}'' \leftrightarrow < \boldsymbol{q}, \boldsymbol{q}' >= \frac{\boldsymbol{q} + \boldsymbol{q}' + [\boldsymbol{q}\boldsymbol{q}']}{1 - \boldsymbol{q}\boldsymbol{q}'} = \boldsymbol{q}'' \tag{3}$$

for the complex vector parameters $\boldsymbol{q} = \boldsymbol{a} + i\boldsymbol{b}$ ($\boldsymbol{q}^* = \boldsymbol{a} - i\boldsymbol{b}$). This parametrization forms now a foundation of the general approach to the finite geometrical, dynamical and quantum transformations.

(1958–2003, F.I. Fedorov, A.A. Bogush, L.G. Moroz, E.E. Tcharev, V.S. Otchik, A.M. Fedorovych, A.V. Berezin, L.I. Komarov, Yu.A. Kurochkin, L.F. Zhirkov, V.I. Kuvshinov, N. Tho, E.A. Tolkachev, I.Ya. Pris', et al.).

Thus, F.I. Fedorov has worked out, in the framework of the covariant approach, a general mathematical foundation for the unified description of all the elementary particles, all the four fundamental interactions, as well as of all the kinds of the physical symmetries related to these particles and interactions generated by them.

On the basis and in the framework of the Fedorov's covariant approach, alongside with general theoretical investigations, the effective covariant methods of relativistic kinematics for description, calculation and analysis of the particular interaction processes of elementary particles have been worked out. They allow one to obtain some new physical results which would be impossible or difficult to get in other way. (See, for example, the review works: "Covariant description of spin properties of particles and its applications"[8], "Finite Lorentz transformations in quantum field theory"[9], "Covariant approach to calculation of the interaction processes of polarized particles"[10], "Diagonal spin basis and calculation of processes with polarized particles"[11], "On the principles and physical program of $\gamma\gamma$-colliders"[12] and monographs[3,13]).

As one of the consequences, the Belarusian physicists (IP NASB) have passed from the purely theoretical investigations to the direct participation in the realization of the experimental programs in high energy physics (see, for example, review work "On the contribution of Belarusian physicists in the development of experimental research in particle physics"[14]). That turned out to be possible thanks to the successful collaboration with scientists of the Joint Institute for Nuclear Research (Dubna, Russia).

3. The local vector parameters of the dynamical symmetry groups and non-linear field models

In the frames of the VP of $SU(2)$ $(SO(3.R))^3$ the transition from the global finite transformation $U = U(\boldsymbol{n}) \in SU(2)$, $(O = O(\boldsymbol{n}) \in SO(3.R))$ to the corresponding local (gauge) finite transformation $U(x)$ $(O(x))$ reduces solely to the obvious redefinition of the three-dimensional real vector parameter $\boldsymbol{n} \to \boldsymbol{n}(x)$[15-17].

As a result, when treating the group $SU(2)$ $(SO(3.R))$ as dynamical symmetry group in Yang-Mills gauge field theory starting from the known formula

$$\begin{aligned} \boldsymbol{b}'_\mu = \boldsymbol{b}'^a_\mu(x)t^a = U[\boldsymbol{n}(x)]\boldsymbol{b}^a_\mu(x)t^a U^{-1}[\boldsymbol{n}(x)] + \\ + (1/ig)\{\partial_\mu U[\boldsymbol{n}(x)]U^{-1}[\boldsymbol{n}(x)]\}, \end{aligned} \tag{4}$$

the following basic relations have been obtained[16,17]:

$$\boldsymbol{b}'_\mu(x) = O(\boldsymbol{n}(x))\boldsymbol{b}_\mu(x) + \boldsymbol{a}_\mu(x), \tag{5}$$

$$\boldsymbol{a}_\mu(x) = \boldsymbol{a}_\mu(\boldsymbol{n}(x)) = f(\boldsymbol{n}(x))\partial_\mu \boldsymbol{n}(x), \tag{6}$$

$$f(\boldsymbol{n}(x)) = 2(1 - \boldsymbol{n}^\times(x))/(1 + \boldsymbol{n}^2(x)). \tag{7}$$

These formulas define explicitly the transformation properties of the Yang-Mills vector potentials $\boldsymbol{b}_\mu(x) = (b^a_\mu(x))$ under the finite (rather than infinitesimal ones, as it has been done previously) gauge transformations $O(x) = O(\boldsymbol{n}(x))$. As a consequence, some important problems in the gauge field theory are considered on this basis.

First of all, in such a way an explicit expression (6) for the vacuum connection $\boldsymbol{a}_\mu(x)$ written in terms of local vector parameters $\boldsymbol{n}(x)$ is obtained[18].

Then, by taking into account that the multitude N of the local vector parameters $\boldsymbol{n}(x)$ forms a noncommutative group $N(x)$ with respect to the group operation (3), the nonlinear equation[18,19]

$$\partial_\mu \boldsymbol{a}_\mu(\pm\boldsymbol{n}(x)) = 0 \tag{8}$$

has been derived for the principal chiral fields $+\boldsymbol{n}^0(x)$ and $-\boldsymbol{n}^0(x)$ of the $SO(3.R)$ group. [This equation becomes solvable in the two-dimensional space-time and allows the existence of soliton-like solutions[20]].

As an analogy of the relations (4)–(7), i.e. starting from the evident formula

$$-i\{\partial_\mu U(\boldsymbol{n}(x))\}U^{-1}(\boldsymbol{n}(x)) = \boldsymbol{a}_\mu(x)\boldsymbol{t}, \quad (\boldsymbol{t} = \boldsymbol{\sigma}/2), \tag{9}$$

the differential Cartan form $dU(x)U^{-1}(x)$ has been also expressed in terms of local vector parameters $\boldsymbol{n}(x)$ of the gauge group $SU(2)$ (see (4))

$$dU(x)U^{-1}(x) = id\boldsymbol{n}'(x)\boldsymbol{t}, \quad d\boldsymbol{n}'(x) = f(\boldsymbol{n}(x))d\boldsymbol{n}(x). \tag{10}$$

The method of construction of the Cartan differential forms, worked out on this basis, which in following turned out to be general, is essentially based on the direct use of the specific properties of the VP[3]: the simple composition law (see (3)), as well as the naturalness conditions and linearity relations (see below (15) and (17)). These results were later extended to the wide set of gauge (supergauge) groups and effectively used in study of actual nonlinear field models (see, for example, the review paper "Local vector parameters of the groups, the Cartan forms and its applications to the gauge and chiral field theories" [21]).

4. On the quantum vector parameters and generators of the q-deformed group $SL_q(2.C)$

First of all, let us demonstrate explicitly that by extension of the Fedorov's vector parametrization (VP) (see[3]) on the q-deformed group $SL_q(2.C)$ (see, for example,[22]) not only the basic peculiarities and advantages of the VP are conserved, but also the new possibilities for simple and natural solution of specific problems in theory of quantum groups, including realization of the transition $SL(2.C) \to SL_q(2.C)$(see[23]), are opened.

1. Let us remind that the description of the classical group $SL(2.C)$ in the framework of VP (see[3]) is based on the following main relations:

a) the explicit very compact vector-parametrical expressions for the finite transformation 2×2-matrices

$$\begin{gathered} L = L(\boldsymbol{q}) = q_0(1 - i\boldsymbol{q}\boldsymbol{\sigma}), L^{-1} = L(-\boldsymbol{q}) = q_0(1 + i\boldsymbol{q}\boldsymbol{\sigma}); \\ q_0 = (1 + \boldsymbol{q}^2)^{-1/2}, \end{gathered} \tag{11}$$

where $\boldsymbol{\sigma} = (\sigma_1, \sigma_2, \sigma_3)$ are the known Pauli matrices with usual properties:

$$\sigma_a\sigma_b = \delta_{ab} + i\epsilon_{abc}\sigma_c, \quad ([\sigma_a, \sigma_b]_- = i\epsilon_{abc}\sigma_c); \tag{12}$$

b) the composition rules for the group vector parameters $\boldsymbol{q}$

$$LL' = L'' \to < \boldsymbol{q}, \boldsymbol{q}' > = (\boldsymbol{q} + \boldsymbol{q}' + [\boldsymbol{q}\boldsymbol{q}'])(1 - \boldsymbol{q}\boldsymbol{q}')^{-1} = \boldsymbol{q}'', \tag{13}$$

$$q_0q_0'(1 - \boldsymbol{q}\boldsymbol{q}') = q_0''; \tag{14}$$

c) the linearity relations:

$$LL^{(0)}L^{-1} = L'^{(0)} \to < \boldsymbol{q}, \boldsymbol{q}^{(0)}, -\boldsymbol{q} > = O(\boldsymbol{q})\boldsymbol{q}^{(0)} = \boldsymbol{q}'^{(0)}, \tag{15}$$

where (see[3])

$$O=O(\boldsymbol{q})=1+2\frac{\boldsymbol{q}^{\times}+\boldsymbol{q}^{\times}\boldsymbol{q}^{\times}}{1+\boldsymbol{q}^2}\in SO(3.C),\quad \boldsymbol{q}^{\times}=\begin{pmatrix}0&-q_3&q_2\\ q_3&0&-q_1\\ -q_2&q_1&0\end{pmatrix};\qquad (16)$$

$$O^{-1}=O^{-1}(\boldsymbol{q})=O(-\boldsymbol{q});$$

d) the naturalness relations:

$$L\to\boldsymbol{q},\quad L^{-1}\to-\boldsymbol{q},\quad L_0=I_2\to\boldsymbol{q}_0=0.\qquad (17)$$

In following an additional formula will be also used:

$$L\boldsymbol{\sigma}L^{-1}=L(\boldsymbol{q})\boldsymbol{\sigma}L(-\boldsymbol{q})=O(-\boldsymbol{q})\boldsymbol{\sigma}=\boldsymbol{\sigma}',\qquad (18)$$

which follows directly from the linearity relations (15).

Here and in the following the vector $\boldsymbol{q}=(q_1,q_2,q_3)$ defines the three-dimensional complex vector parameter of the group $SL(2.C)$.

It is worthy to note that the transition from the linear group $SL(2.C)$ to the related unitary group $SU(2)$ reduces here to the change of the complex quantity $\boldsymbol{q}$ by the three-dimensional real vector parameter $\boldsymbol{n}$.

2. It is known that the usual quantum deformation procedure is based, as a rule, on the independent introducing of the inverse quantum matrix L_q^{-1} with respect to the initial quantum matrix

$$L_q=\begin{pmatrix}\hat{a}&\hat{b}\\ \hat{c}&\hat{d}\end{pmatrix},\qquad (19)$$

in the following explicit form (see[22]):

$$L_q^{-1}=\begin{pmatrix}\hat{d}&-q^{-1}\hat{b}\\ -q\hat{c}&\hat{a}\end{pmatrix}.\qquad (20)$$

As distinct from this approach, let us realize the transitions from the classical to the q-deformed matrices, simultaneously for the initial and inverse ones, in the following two stages[23]:

$$L^{(0)}\to L_q^{(0)}\to L_q=\eta_q L_q^{(0)}\eta_q^{-1},$$

$$L^{(0)}=\begin{pmatrix}a&b\\ c&d\end{pmatrix},L_q^{(0)}=\begin{pmatrix}\hat{a}&\hat{b}\\ \hat{c}&\hat{d}\end{pmatrix},L_q(q^{1/2})=\begin{pmatrix}\hat{a}&q^{1/2}\hat{b}\\ q^{-1/2}\hat{c}&\hat{d}\end{pmatrix};\qquad (21)$$

$$L^{(0)-1}\to L'^{(0)-1}_q\to\eta_q^{-1}L'^{(0)-1}_q\eta_q,\ L^{(0)-1}=\begin{pmatrix}d&-b\\ -c&a\end{pmatrix},$$

$$L'^{(0)-1}_q=\begin{pmatrix}\hat{d}&-\hat{b}\\ -\hat{c}&\hat{a}\end{pmatrix},\ L_q^{-1}=L_q(-q^{-1/2})=\begin{pmatrix}\hat{d}&-q^{-1/2}\hat{b}\\ -q^{1/2}\hat{c}&\hat{a}\end{pmatrix},\qquad (22)$$

where, from the beginning, two different quantum deformation parameters, q and q^{-1}, are used.

Here on the first stage the both initial classical matrices $L^{(0)}$, $L^{(0)-1}$ are transformed into the matrices $L_q^{(0)}$ and $L_q^{(0)-1}$ of the same form, but with matrix elements $\hat{a}, \hat{b}, \hat{c}, \hat{d}$, which are now considered as quantum quantities (operators).

On the second stage the following matrices are introduced:

$$\eta_q = \eta(q^{1/4}) = \begin{pmatrix} q^{1/4} & 0 \\ 0 & q^{-1/4} \end{pmatrix}, \quad \eta_q^{-1} = \eta(q^{-1/4}) = \begin{pmatrix} q^{-1/4} & 0 \\ 0 & q^{1/4} \end{pmatrix}, \tag{23}$$

by means of them the matrices $L_q^{(0)}$ and $L_q^{(0)-1}$ are transformed finally to the quantum matrices $L_q(q^{1/2})$, $L_q(-q^{-1/2})$. The obtained in a such way resulting matrices, as distinct from the usual quantum matrices, (19), (20), both have the explicit quantum structure of the same type, but depend on the two different quantum deformation parameters q and q^{-1}, more precisely, on $q^{1/2}$ and $q^{-1/2}$ correspondingly (see[23]).

In the framework of the proposed approach the evident defining conditions:

$$L_q L_q^{-1} = L_q^{-1} L_q = I_2 \tag{24}$$

play the most crucial role and importance.

Namely, by substituting in these relations the explicit expressions for quantum matrices $L_q = L_q(q^{1/2})$ (see (21)) and $L_q^{-1} = L_q(-q^{-1/2})$ (see (22)) we can obtain in direct way the known basic relations[22] which characterize the quantum properties of the $q-$deformed group $SL_q(2.C)$ under consideration:

$$\hat{a}\hat{b}=q\hat{b}\hat{a}, \ \hat{b}\hat{d}=q\hat{d}\hat{b}, \ \hat{b}\hat{c}=\hat{c}\hat{b}, \ \hat{a}\hat{c}=q\hat{c}\hat{a}, \ \hat{c}\hat{d}=q\hat{d}\hat{c}, \ \hat{a}\hat{d}-\hat{d}\hat{a}=(q-q^{-1})\hat{b}\hat{c}; \tag{25}$$

$$D_q = \det{}_q L_q = \hat{a}\hat{d} - q\hat{b}\hat{c} = \hat{d}\hat{a} - q^{-1}\hat{b}\hat{c} = 1, \tag{26}$$

where $D_q = 1$ defines the quantum determinant of the q-deformed matrix $L_q \in SL_q(2.C)$.

3. Let us now rewrite the relations (21) and (22) in explicit vector-parametrical form (see[3,23] and (11)):

$$\begin{aligned} L = L(\boldsymbol{q}) = q_0(1 - i\boldsymbol{q\sigma}) \rightarrow L_q^{(0)} = L_q(\hat{\boldsymbol{q}}) = \hat{q}_0(1 - i\hat{\boldsymbol{q}}\boldsymbol{\sigma}) = \\ = (1 - i\hat{\bar{\boldsymbol{q}}}\boldsymbol{\sigma})\hat{\boldsymbol{q}}_0 = L_q(\hat{\bar{\boldsymbol{q}}}) \rightarrow L_q = L_q[\boldsymbol{q}(q^{1/2})] = L_q[\bar{\boldsymbol{q}}(q^{1/2})], \end{aligned} \tag{27}$$

$$\begin{aligned} L^{-1} = L(-\boldsymbol{q}) = q_0(1 + i\boldsymbol{q\sigma}) \rightarrow L_q'^{(0)-1} = L_q(-\hat{\boldsymbol{q}}) = \hat{q}_0(1 + i\hat{\boldsymbol{q}}\boldsymbol{\sigma}) = \\ = (1 + i\hat{\bar{\boldsymbol{q}}}\boldsymbol{\sigma})\hat{q}_0 = L_q(-\hat{\bar{\boldsymbol{q}}}) \rightarrow L_q^{-1} = L_q[-\boldsymbol{q}(q^{-1/2})] = L_q[-\bar{\boldsymbol{q}}(q^{-1/2})]. \end{aligned} \tag{28}$$

Hereafter, it follows that in the framework of the VP the transition from the classical group $SL(2.C)$ (see (11)) to the quantum group $SL_q(2.C)$ reduces to the simple quantum deformation of the initial classical vector parameters, as follows:

$$\boldsymbol{q} = (q_1, q_2, q_3) \to \begin{cases} \hat{\boldsymbol{q}} = (\hat{q}_1, \hat{q}_2, \hat{q}_3) \to \boldsymbol{q}(q^{1/2}), \\ \hat{\hat{\boldsymbol{q}}} = (\hat{\hat{q}}_1, \hat{\hat{q}}_2, \hat{\hat{q}}_3) \to \bar{\boldsymbol{q}}(q^{1/2}); \end{cases} \tag{29}$$

$$-\boldsymbol{q} \to \begin{cases} -\hat{\boldsymbol{q}} \to -\boldsymbol{q}(q^{-1/2}), \\ -\hat{\hat{\boldsymbol{q}}} \to -\bar{\boldsymbol{q}}(q^{-1/2}). \end{cases} \tag{30}$$

As a result, they are transformed into the three-dimensional quantum vector parameters of two kinds:"left" $\hat{\boldsymbol{q}}$ $(-\hat{\boldsymbol{q}}, \boldsymbol{q}(q^{1/2}), -\boldsymbol{q}(q^{-1/2}))$ and "right" $\hat{\hat{\boldsymbol{q}}}$ $(-\hat{\hat{\boldsymbol{q}}}, \bar{\boldsymbol{q}}(q^{1/2}), -\bar{\boldsymbol{q}}(q^{-1/2}))$, which are defined by the formulas

$$\hat{q}_0 \hat{\boldsymbol{q}} = \hat{\hat{\boldsymbol{q}}} \hat{q}_0, \quad \hat{q}_0 \boldsymbol{q}(q^{\pm 1/2}) = \bar{\boldsymbol{q}}(q^{\pm 1/2}) \hat{q}_0. \tag{31}$$

It follows from the last relations that there is necessary to introduce two kinds of three-dimensional quantum vector parameters since the quantum quantities $\hat{q}_0$ and $\hat{\boldsymbol{q}}$ not commute one with another.

In the beginning, taking into account that the Pauli matrices $\boldsymbol{\sigma} = (\sigma_a)$ conserve their form and properties (see (12)) after deformation, it is easy to establish that for all the introduced three-dimensional quantum vector parameters, similar to the classical case (13) and (14), the simplest composition rules are valid:

$$\begin{gathered} L_q^{(0)} L'^{(0)}_q = L''^{(0)}_q \leftrightarrow < \hat{\boldsymbol{q}}, \hat{\boldsymbol{q}}' > = \boldsymbol{q}'', \\ < \hat{\boldsymbol{q}}, \hat{\boldsymbol{q}}' > = (1 - \hat{\boldsymbol{q}} \hat{\boldsymbol{q}}')^{-1} (\hat{\boldsymbol{q}} + \hat{\boldsymbol{q}}' + [\boldsymbol{q}\boldsymbol{q}']); \end{gathered} \tag{32}$$

$$\hat{q}_0 (1 - \hat{\boldsymbol{q}} \hat{\boldsymbol{q}}') \hat{q}'_0 = \hat{q}''_0. \tag{33}$$

4. Now, let us write the similarity transformations $\eta_q L_q^{(0)} \eta_q^{-1}$ and $\eta_q^{-1} L'^{(0)-1}_q \eta_q$ in relations (21) and (22) also in the vector-parametrical form by using the classical matrices (see (23))

$$\begin{gathered} \eta_a = L(\boldsymbol{a}) = a_0 (1 - i \boldsymbol{a} \boldsymbol{\sigma}), \quad \eta_a^{-1} = L(-\boldsymbol{a}) = a_0 (1 + i \boldsymbol{a} \boldsymbol{\sigma}); \\ a_0 = \omega'_{(+)}, \ \boldsymbol{a} = (0, 0, a_3), \ a_3 = \omega'_{(-)} / \omega'_{(+)}, \ \omega'_{(\pm)} = (1/2)(q^{1/4} \pm q^{-1/4}). \end{gathered} \tag{34}$$

As a result, in accordance with general composition rules (32) and (33) and linearity relations (see (15)), we have:

$$\begin{gathered} L(\boldsymbol{a}) L_q(\hat{\boldsymbol{q}}) L(-\boldsymbol{a}) = L_q(< \boldsymbol{a}, \hat{\boldsymbol{q}}, -\boldsymbol{a} >) = L_q[\boldsymbol{q}(q^{1/2})], \\ < \boldsymbol{a}, \hat{\boldsymbol{q}}, -\boldsymbol{a} > = O(\boldsymbol{a}) \hat{\boldsymbol{q}} = \boldsymbol{q}(q^{1/2}); \end{gathered} \tag{35}$$

$$\begin{aligned} L(-\boldsymbol{a})L_q(-\hat{\boldsymbol{q}})L(\boldsymbol{a}) = L_q(< -\boldsymbol{a}, -\hat{\boldsymbol{q}}, \boldsymbol{a} >) = L_q[-\boldsymbol{q}(q^{-1/2})], \\ < -\boldsymbol{a}, -\hat{\boldsymbol{q}}, \boldsymbol{a} >= -O(-\boldsymbol{a})\hat{\boldsymbol{q}} = -\boldsymbol{q}(q^{-1/2}), \end{aligned} \tag{36}$$

where (see (16))

$$\begin{aligned} O=O(\boldsymbol{a})=\begin{pmatrix} \omega_{(+)} & i\omega_{(-)} & 0 \\ -i\omega_{(-)} & \omega_{(+)} & 0 \\ 0 & 0 & 1 \end{pmatrix}, \quad O^{-1}=\tilde{O}(\boldsymbol{a})=O(-\boldsymbol{a}); \\ \omega_{(\pm)} = (1/2)(q^{1/2} \pm q^{-1/2}). \end{aligned} \tag{37}$$

Hence, the explicit expressions for the three-dimensional vector parameters $\boldsymbol{q}(q^{1/2})$ (29) and $\boldsymbol{q}(q^{-1/2})$ (30) follow immediately:

$$O(\boldsymbol{a})\hat{\boldsymbol{q}} = \boldsymbol{q}(q^{1/2}) = \{\omega_{(+)}(\hat{q}_1 - i\alpha\hat{q}_2), \omega_{(-)}(\hat{q}_2 + i\alpha\hat{q}_1), \hat{q}_3\}, \tag{38}$$

$$O(-\boldsymbol{a})\hat{\boldsymbol{q}} = \boldsymbol{q}(q^{-1/2}) = \{\omega_{(+)}(\hat{q}_1 + i\alpha\hat{q}_2), \omega_{(-)}(\hat{q}_2 - i\alpha\hat{q}_1), \hat{q}_3\}, \tag{39}$$

where (see (34))

$$\alpha = -\omega_{(-)}/\omega_{(+)} = (1-q)/(1+q) = \alpha(q) = -\alpha(q^{-1}). \tag{40}$$

Respectively, substituting the expressions for the quantum matrices $L_q = L_q[\boldsymbol{q}(q^{1/2})] = L_q[\bar{\boldsymbol{q}}(q^{1/2})]$ (27) and $L_q^{-1} = L_q[-\boldsymbol{q}(q^{-1/2})] = L_q[-\bar{\boldsymbol{q}}(q^{-1/2})]$ (28) in defining conditions (24) and using only the composition rules (32) and (33), we find the basic relations (25) and (26) written in the following compact vector-parametrical form:

$$\begin{aligned} [\boldsymbol{q}(q^{1/2})\bar{\boldsymbol{q}}(q^{-1/2})] = \boldsymbol{q}(q^{1/2}) - \bar{\boldsymbol{q}}(q^{-1/2}), \\ [\boldsymbol{q}(q^{-1/2})\bar{\boldsymbol{q}}(q^{1/2})] = -\boldsymbol{q}(q^{-1/2}) + \bar{\boldsymbol{q}}(q^{1/2}); \end{aligned} \tag{41}$$

$$D_q = \det{}_q L_q = \hat{q}_0[1 + \boldsymbol{q}(q^{1/2})\bar{\boldsymbol{q}}(q^{-1/2})]\hat{q}_0 = 1. \tag{42}$$

It follows, in particular, from the formula (42), that quantum quantity $\hat{q}_0$ may be presented as

$$\hat{q}_0 = \pm[1 + \boldsymbol{q}(q^{1/2})\bar{\boldsymbol{q}}(q^{-1/2})]^{-1/2}. \tag{43}$$

Let us point out that the naturalness conditions (17) taken here the form

$$L_q \to \boldsymbol{q}(q^{1/2}), \quad L_q^{-1} \to -\boldsymbol{q}(q^{-1/2}), \quad L_{q0} = I_2 \to \hat{\boldsymbol{q}} = 0, \tag{44}$$

i.e., in the transition $L_q \to L_q^{-1}$ the change of sign of the vector parameter is accompanied by the change of the quantum deformation parameter: $q^{1/2} \to q^{-1/2}$.

5. The presented construction may be simply extended also on the cases, when in the theory of the quantum group $SL_q(2.C)$ (or $SU_q(2)$) the q-deformed Lee algebra is introduced and used (see, for example,[24]).

In our case it is sufficient to consider the above similarity transformations (35) and (36) on the initial quantum matrices $L_q^{(0)}$ (see (27)) and $L'^{(0)-1}_q$ (see (28)) with help of matrices $\eta_q = L(\boldsymbol{a})$ and $\eta_q^{-1} = L(-\boldsymbol{a})$ (34) [22]. Assuming that after these transformations the quantum vector parameters $\hat{\boldsymbol{q}}$ and $-\hat{\boldsymbol{q}}$ remain to be unchanged we conclude that the Pauli matrices are transformed according the formula (18) (see (34) and (37)):

$$\eta_q \boldsymbol{\sigma} \eta_q^{-1} = L(\boldsymbol{a})\boldsymbol{\sigma} L(-\boldsymbol{a}) = O(-\boldsymbol{a})\boldsymbol{\sigma} = \boldsymbol{\sigma}_q, \tag{45}$$

$$\eta_q^{-1} \boldsymbol{\sigma} \eta_q = L(-\boldsymbol{a})\boldsymbol{\sigma} L(\boldsymbol{a}) = O(\boldsymbol{a})\boldsymbol{\sigma} = \bar{\boldsymbol{\sigma}}_q. \tag{46}$$

Thus, we obtain two sets of the q-deformed Pauli matrices $\boldsymbol{\sigma_q} = (\sigma_{1q}, \sigma_{2q}, \sigma_{3q})$ and $\bar{\boldsymbol{\sigma}}_{\boldsymbol{q}} = (\bar{\sigma}_{1q}, \bar{\sigma}_{2q}, \bar{\sigma}_{3q})$:

$$\sigma_{1q} = \begin{pmatrix} 0 & q^{-1/2} \\ q^{1/2} & 0 \end{pmatrix}, \quad \sigma_{2q} = \begin{pmatrix} 0 & -iq^{-1/2} \\ iq^{1/2} & 0 \end{pmatrix}, \quad \sigma_{3q} = \sigma_3; \tag{47}$$

$$\bar{\sigma}_{1q} = \begin{pmatrix} 0 & q^{1/2} \\ q^{-1/2} & 0 \end{pmatrix}, \quad \bar{\sigma}_{2q} = \begin{pmatrix} 0 & -iq^{1/2} \\ iq^{-1/2} & 0 \end{pmatrix}, \quad \bar{\sigma}_{3q} = \sigma_3, \tag{48}$$

both of them satisfying the same relations (12), as in the case of usual Pauli matrices.

In order to reflect the quantum specific of the q-deformed Lie algebra for the group $SL_q(2.C)$, let us introduce the additional commutation relations which interconnect the two sets of the Pauli matrices $\boldsymbol{\sigma}_q$ (47) and $\bar{\boldsymbol{\sigma}}_q$ (48) (see (12)):

$$\begin{aligned} &[\sigma_{2q}, \bar{\sigma}_{3q}] = 2i\sigma_{1q}, \quad [\sigma_{3q}, \bar{\sigma}_{1q}] = 2i\bar{\sigma}_{2q}, \quad [\sigma_{1q}, \bar{\sigma}_{2q}] = 2i\omega_{(+)}\sigma_{3q}; \\ &[\bar{\sigma}_{2q}, \sigma_{3q}] = 2i\bar{\sigma}_{1q}, \quad [\bar{\sigma}_{3q}, \sigma_{1q}] = 2i\sigma_{2q}, \quad [\bar{\sigma}_{1q}, \bar{\sigma}_{2q}] = 2i\omega_{(+)}\sigma_{3q}. \end{aligned} \tag{49}$$

As a result, the quantum matrices L_q (27) and L_q^{-1}(28), written in the two different q-deformed bases, $\boldsymbol{\sigma}_q$ (47) and $\bar{\boldsymbol{\sigma}}_q$ (48), take now the following form:

$$L_q = \hat{q}_0(1 - i\hat{\boldsymbol{q}}\boldsymbol{\sigma}_q) = (1 - i\hat{\hat{\boldsymbol{q}}}\boldsymbol{\sigma}_q)\hat{q}_0, \tag{50}$$

$$L_q^{-1} = \hat{q}_0(1 + i\hat{\boldsymbol{q}}\bar{\boldsymbol{\sigma}}_q) = (1 + i\hat{\hat{\boldsymbol{q}}}\bar{\boldsymbol{\sigma}}_q)\hat{q}_0, \tag{51}$$

It is easy to verify, that after substituting these quantum matrices in the defining condition (24) and by using the relations (49), we go to the basic commutation relations (41) and (42) (see (25) and (26)).

Thus, we can conclude that the proposed approach to the description of the transition $SL(2.C) \to SL_q(2.C)$ ensures the adequate reflection at the quantum properties of the q-deformed group $SL_q(2.C)$. Simultaneously the new, previously not known possibilities of the VP method are unveiled.

5. On the vector parametrization of the h-deformed group $SL_h(2.C)$

The extensions of the Fedorov's vector parametrization (VP) on the dynamical (gauge) symmetry groups[15], and then, as was shown above, on the simplest quantum (q-deformed) groups[23] has opened, as has been pointed out already above, the new possibilities for the solutions of the problems in gauge field theory and, respectively, in the theory of the simplest quantum groups (see review work[25]).

It is known (see, for example,[26]), that in the last decade a series of various applications of quantum groups has appeared: representation theory, q-harmonic analysis and q-special functions, conformal field theories, the vertex and spin models, quantum optics. Of special interest are the attempts to construct the Yang-Mills theory, based on the q-deformed group $SU_q(2)$.

Moreover, in a recent work[26], an another quantum group, h-deformed (Jordanian) group $SU_h(2)$[27], is used as gauge symmetry group in the Yang-Mills theory, considered in the space with noncommutative geometry.

In this connection, we will also introduce the vector-parametrical description of the h-deformed group $SL_h(2.C)$[28], by analogy with the case of the q-deformed group $SL_q(2.C)$ given above (see sect. 1.4).

By the definition (see, for example,[26]), the quantum 2×2-matrix $L_h \in SL_h(2.C)$ and its inverse matrix are determined as follows:

$$L_h = \begin{pmatrix} \hat{a} & \hat{b} \\ \hat{c} & \hat{d} \end{pmatrix}, \quad L_h^{-1} = \begin{pmatrix} \hat{d} + h\hat{c} & -\hat{b} - h\hat{a} + h\hat{d} + h^2\hat{c} \\ -\hat{c} & \hat{a} - h\hat{c} \end{pmatrix}.$$

Here, as distinct from the inverse quantum matrix $L^{-1}(q)$ (19), the elements of the matrix L_h^{-1} have not multiplicative, but polynomial explicit structure.

Therefore, it is impossible to use here the same simple quantum deformation schema which has been realized above by considering the transition $SL(2.C) \rightarrow SL_q(2.C)$, and, as a result, to obtain the both quantum matrices, L_h and L_h^{-1}, simultaneously in the explicit quantum form.

So, let us consider the following two possibilities[28]:

$$L = \begin{pmatrix} a & b \\ c & d \end{pmatrix} \rightarrow L_h = \begin{pmatrix} \hat{a} & \hat{b} \\ \hat{c} & \hat{d} \end{pmatrix},$$
$$L^{-1} = \begin{pmatrix} d & -b \\ -c & a \end{pmatrix} \rightarrow L_h^{-1} = \begin{pmatrix} \hat{d} & -\hat{b} \\ -\hat{c} & \hat{a} \end{pmatrix} \rightarrow L_h^{-1}(-h) = \begin{pmatrix} \hat{d}(-h) & -\hat{b}(-h) \\ -\hat{c}(-h) & \hat{a}(-h) \end{pmatrix} \tag{52}$$

and respectively

$$L \rightarrow L_h \rightarrow L_h(h) = \begin{pmatrix} \hat{a}(h) & \hat{b}(h) \\ \hat{c}(h) & \hat{d}(h) \end{pmatrix}, \quad L^{-1} \rightarrow L_h^{-1} = \begin{pmatrix} \hat{d} & -\hat{b} \\ -\hat{c} & \hat{a} \end{pmatrix}, \tag{53}$$

where, from the beginning, two different quantum deformation parameters, $+h$ and $-h$, are used.

Now, by introducing the similarity transformations (see (21), (22) and (23))

$$L(h) L_h L(-h) = L_h(h), \quad L(-h) L_h^{-1} L(h) = L_h^{-1}(-h), \tag{54}$$

$$L = L(h) = \begin{pmatrix} 1 & h \\ 0 & 1 \end{pmatrix}, \quad L^{-1} = L(-h) = L^{-1}(h) = \begin{pmatrix} 1 & -h \\ 0 & 1 \end{pmatrix}, \tag{55}$$

we can obtain the following explicit expressions for the h-deformed elements of the quantum matrices L_h and $L_h^{-1}(-h)$:

$$\begin{gathered} \hat{a}(\pm h) = \hat{a} + (\pm h)\hat{c}, \quad \hat{b}(\pm h) = \hat{b} - (\pm h)(\hat{a} - \hat{d}) - h^2 \hat{c}, \\ \hat{c}(\pm h) = \hat{c}, \quad \hat{d}(\pm h) = \hat{d} - (\pm h)\hat{c}, \end{gathered} \tag{56}$$

which coincide with the known elements of the inverse quantum matrix L_h^{-1}.

The known basic relations, defining the quantum properties of the h-deformed group $SL_h(2.C)$[26],

$$\begin{gathered} [\hat{a}, \hat{c}]_- = h\hat{c}^2, \quad [\hat{b}, \hat{a}]_- = h(\hat{a}^2 - D_h), \quad [\hat{a}, \hat{d}]_- = -h(\hat{a} - \hat{d})\hat{c}, \\ [\hat{d}, \hat{c}]_- = h\hat{c}^2, \ [\hat{b}, \hat{d}]_- = h(\hat{d}^2 - D_h), \\ [\hat{b}, \hat{c}]_- = h(\hat{a}\hat{c} + \hat{c}\hat{d}), \ ([a, b]_\pm = ab \pm bc); \end{gathered} \tag{57}$$

$$D_h = \hat{a}\hat{d} - \hat{c}\hat{b} - h\hat{c}\hat{d} = 1, \quad D_h = \hat{a}\hat{d} - \hat{b}\hat{c} + h\hat{a}\hat{c} = 1 \tag{58}$$

where $D_h = 1$ is quantum determinant of the matrix L_h, we find directly, starting from the evident conditions (see (24)),

$$L_h L_h^{-1}(-h) = L_h^{-1}(-h) L_h, \quad L_h(h) L_h^{-1} = L_h^{-1} L_h(h) = 1. \tag{59}$$

2. Let us present the relations (52) and (53) in explicit three-dimensional vector-parametrical form (see (27), (28))[28]:

$$L = L(\boldsymbol{q}) = q_0(1 - i\boldsymbol{q}\boldsymbol{\sigma}) \to L_h = L_h(\hat{\boldsymbol{q}}) = \hat{q}_0(1 - i\hat{\boldsymbol{q}}\boldsymbol{\sigma}) = (1 - i\hat{\hat{\boldsymbol{q}}}\boldsymbol{\sigma})\hat{q}_0 = L_h(\hat{\hat{\boldsymbol{q}}}),$$

$$L^{-1} = L(-\boldsymbol{q}) = q_0(1 + i\boldsymbol{q}\boldsymbol{\sigma}) \to L_h^{-1} = L_h(-\hat{\boldsymbol{q}}) \to \tag{60}$$

$$\to L_h^{-1}(-h) = L_h[-\hat{\boldsymbol{q}}(-h)] = L_h[-\hat{\hat{\boldsymbol{q}}}(-h)];$$

$$\begin{gathered} L(\boldsymbol{q}) \to L_h(\hat{\boldsymbol{q}}) = L_h(\hat{\hat{\boldsymbol{q}}}) \to L_h(h) = L_h[\hat{\boldsymbol{q}}(h)] = L_h[\hat{\hat{\boldsymbol{q}}}(h)], \\ L^{-1} = L(-\boldsymbol{q}) \to L_h^{-1} = L_h(-\hat{\boldsymbol{q}}) = L_h(-\hat{\hat{\boldsymbol{q}}}). \end{gathered} \tag{61}$$

Here the usual Pauli matrices $\boldsymbol{\sigma} = (\sigma_1, \sigma_2, \sigma_3)$ with known properties (12) are used and, in analogy with the VP of group $SL_q(2.C)$ [23], the three-dimensional "left" $\hat{q}, \hat{q}(h)(-\hat{q}, -\hat{q}(-h))$ and "right" $\hat{\hat{q}}, \hat{\hat{q}}(h)(-\hat{\hat{\boldsymbol{q}}}, -\hat{\hat{q}}(-h)$ quantum vector parameters are introduced, which are interconnected due to the relations (see (31))

$$\hat{q}_0 \hat{\boldsymbol{q}} = \hat{\hat{\boldsymbol{q}}} \hat{q}_0, \quad \hat{q}_0 \hat{\boldsymbol{q}}(\pm h) = \hat{\hat{\boldsymbol{q}}}(\mp h) \hat{q}_0. \tag{62}$$

It follows from (60) and (61) that the transition $SL(2.C) \to SL_h(2.C)$ reduces in the framework of VP to the following simple changes of the vector-parameters related to the classical and quantum transformation matrices:

$$\boldsymbol{q} \to \hat{\boldsymbol{q}}(\hat{\hat{\boldsymbol{q}}}) \to \hat{\boldsymbol{q}}(h)[\hat{\hat{\boldsymbol{q}}}(h)], \quad -\boldsymbol{q} \to -\hat{\boldsymbol{q}}(-\hat{\hat{\boldsymbol{q}}}) \to -\hat{\boldsymbol{q}}(-h)[-\hat{\hat{\boldsymbol{q}}}(-h)]. \tag{63}$$

The explicit expressions for the quantum vector parameters $\hat{q}(\pm h)$ will be established later.

It is easy to see that the almost simple composition law of vector-parameters (see (32), (33)) and, as a consequence, the linearity relations (see (15)) remain to be valid for the introduced VP of the quantum group $SL_h(2.C)$:

$$L_h L'_h = L''_h \to \langle \hat{\boldsymbol{q}}, \hat{\boldsymbol{q}}' \rangle = \hat{\boldsymbol{q}}'',$$

$$\langle \hat{\boldsymbol{q}}, \hat{\boldsymbol{q}}' \rangle = (1 - \hat{\boldsymbol{q}}\hat{\boldsymbol{q}}')^{-1}(\hat{\boldsymbol{q}} + \hat{\boldsymbol{q}}' + [\hat{\boldsymbol{q}}\hat{\boldsymbol{q}}']), \tag{64}$$

$$\hat{q}_0 \hat{q}'_0 (1 - \hat{\boldsymbol{q}}\hat{\boldsymbol{q}}') = \hat{q}''_0; \tag{65}$$

$$LL_h^{(0)}L^{-1} = L_h^{(0)\prime} \to \langle \boldsymbol{q}, \hat{\boldsymbol{q}}^{(0)}, -\boldsymbol{q}\rangle = O(\boldsymbol{q})\hat{\boldsymbol{q}}^{(0)} = \hat{\boldsymbol{q}}^{(0)\prime}. \tag{66}$$

3. It is easy to show that they are the linearity relations (66) that allow to determine the explicit quantum structure of the introduced above h-deformed quantum vector parameters $\hat{\boldsymbol{q}}(\pm h)$ of the group $SL_h(2.C)$.

Indeed, by writing the similarity transformations (54) in the vector-parametrical form and by using the relations (66), we immediately find the evident relations

$$\begin{aligned} &L(\pm\boldsymbol{q})L_h(\pm\hat{\boldsymbol{q}}^{(0)})L(\mp\boldsymbol{q}) = L_h\left(\langle \pm\boldsymbol{q}, \pm\hat{\boldsymbol{q}}^{(0)}, \mp\boldsymbol{q}\rangle\right) = L_h[\pm\hat{\boldsymbol{q}}^{(0)}(\pm h)], \\ &\langle \pm\boldsymbol{q}, \pm\hat{\boldsymbol{q}}^{(0)}, \mp\boldsymbol{q}\rangle = \pm\langle \pm\boldsymbol{q}, \hat{\boldsymbol{q}}^{(0)}, \mp\boldsymbol{q}\rangle = \pm O(\pm\boldsymbol{q})\hat{\boldsymbol{q}}^{(0)} = \pm\hat{\boldsymbol{q}}^{(0)}(\pm h), \end{aligned} \tag{67}$$

from which it follows the basic formula

$$O(\pm\boldsymbol{q})\hat{\boldsymbol{q}} = \hat{\boldsymbol{q}}(\pm h). \tag{68}$$

Finally, by taking into account the explicit structure of the classical matrices $L(\pm h)$ (55) and O (O^{-1}) (16)

$$L(\pm h) = L(\pm\boldsymbol{q}) = (1 \pm i\boldsymbol{q}\boldsymbol{\sigma}) = 1 + (1/2)(\pm h)(\sigma_1 + i\sigma_2), \tag{69}$$

$$\boldsymbol{q} = (q_1, q_2, q_3) = (i\frac{h^2}{2}, -\frac{h}{2}, 0), \quad q_0 = 1, \quad \boldsymbol{q}^2 = 0, \tag{70}$$

$$O(\pm\boldsymbol{q}) = 1 + 2(\pm\boldsymbol{q}^{\times} + \boldsymbol{q}\cdot\boldsymbol{q}) = \begin{pmatrix} 1-\frac{h^2}{2} & -i\frac{h^2}{2} & \mp h \\ -i\frac{h^2}{2} & 1+\frac{h^2}{2} & \mp ih \\ \pm h & \pm ih & 1 \end{pmatrix}, \tag{71}$$

we obtain the following expressions for the three-dimensional quantum vector parameters $\hat{\boldsymbol{q}}(\pm h)$ of the group $SL_h(2.C)$:

$$\begin{aligned} \hat{q}_1(\pm h) &= (1-\frac{h^2}{2})\hat{q}_1 - i\frac{h^2}{2}\hat{q}_2 - (\pm h)\hat{q}_3, \\ \hat{q}_2(\pm h) &= -i\frac{h^2}{2}\hat{q}_1 + (1+\frac{h^2}{2})\hat{q}_2 - i(\pm h)\hat{q}_3, \\ \hat{q}_3(\pm h) &= (\pm h)(\hat{q}_1 + i\hat{q}_2) + \hat{q}_3. \end{aligned} \tag{72}$$

In the same way, starting from the evident conditions (57) and using only the composition rules (64) and (65), we obtain directly the basic commutation relations for the quantum vector parameters $\hat{\boldsymbol{q}}(\pm h)$ written in the following compact form (see (41):

$$\begin{gathered}\hat{\boldsymbol{q}} - \hat{\hat{\boldsymbol{q}}}(-h) = [\hat{\boldsymbol{q}}\hat{\hat{\boldsymbol{q}}}(-h)], \quad -\hat{\boldsymbol{q}}(-h) + \hat{\hat{\boldsymbol{q}}} = [\hat{\boldsymbol{q}}(-h)\hat{\hat{\boldsymbol{q}}}]; \\ \hat{\boldsymbol{q}}(h) - \hat{\hat{\boldsymbol{q}}} = [\hat{\boldsymbol{q}}(h)\hat{\hat{\boldsymbol{q}}}], \quad -\hat{\boldsymbol{q}} + \hat{\hat{\boldsymbol{q}}}(h) = [\hat{\boldsymbol{q}}\hat{\hat{\boldsymbol{q}}}(h)]\end{gathered} \tag{73}$$

as well as the explicit expressions for the quantum determinant (see (42), (43))

$$\begin{gathered}D_h = \hat{q}_0 \left\{1 + \tfrac{1}{2}[\hat{\boldsymbol{q}}\hat{\hat{\boldsymbol{q}}}(\pm h) + \hat{\boldsymbol{q}}(\pm h)\hat{\hat{\boldsymbol{q}}}]\right\} \hat{q}_0 = 1, \\ q_0 = \pm \left\{1 + \tfrac{1}{2}[\hat{\boldsymbol{q}}\hat{\hat{\boldsymbol{q}}}(\pm h) + \hat{\boldsymbol{q}}(\pm h)\hat{\hat{\boldsymbol{q}}}]\right\}^{1/2}.\end{gathered} \tag{74}$$

4. Now by considering the relations

$$L(\pm\boldsymbol{q})\hat{\boldsymbol{q}}\boldsymbol{\sigma}L(\mp\boldsymbol{q}) = \hat{\boldsymbol{q}}\left[\tilde{O}(\pm\boldsymbol{q})\boldsymbol{\sigma}\right] = \hat{\boldsymbol{q}}\left[O(\mp\boldsymbol{q})\boldsymbol{\sigma}\right] = \hat{\boldsymbol{q}}\hat{\boldsymbol{\sigma}}(\mp h), \tag{75}$$

which follow from formulas (67) (see (18)) we find two sets of h-deformed Pauli matrices

$$\hat{\boldsymbol{\sigma}}(\pm h) = O(\pm\boldsymbol{q})\boldsymbol{\sigma}, \tag{76}$$

which, in accordance with (71), have the following explicit form[28]:

$$\begin{gathered}\sigma_1(\pm h) = (1 - \tfrac{h^2}{2})\sigma_1 + i\tfrac{h^2}{2}\sigma_2 - (\pm h)\sigma_3, \\ \sigma_2(\pm h) = -i\tfrac{h^2}{2}\sigma_1 + (1 + \tfrac{h^2}{2})\sigma_2 - i(\mp h)\sigma_3, \\ \sigma_3(\pm h) = (\pm h)(\sigma_1 + i\sigma_2) + \sigma_3.\end{gathered} \tag{77}$$

Since the both sets of the obtained h-deformed Pauli matrices $\hat{\boldsymbol{\sigma}}(+h)$ and $\hat{\boldsymbol{\sigma}}(-h)$ satisfy the usual commutation relations (12), the following additional commutation relations are to be introduced:

$$\begin{gathered}[\hat{\sigma}_2(\pm h), \hat{\sigma}_3(\mp h)]_- = 2i\left[\hat{\sigma}_1(\pm h) + 2h\hat{\sigma}_3(\pm h)\right], \\ [\hat{\sigma}_3(\pm h), \hat{\sigma}_1(\mp h)]_- = 2i\left\{\hat{\sigma}_2(\pm h) + 2ih^2\left[\hat{\sigma}_1(\pm h) + i\hat{\sigma}_2(\pm h)\right]\right\}, \\ [\hat{\sigma}_1(\pm h), \hat{\sigma}_2(\mp h)]_- = 2i\left[(1 + 2h^2)\hat{\sigma}_3(\pm h) - 2ih\hat{\sigma}_2(h)\right],\end{gathered} \tag{78}$$

which define the specific quantum properties of the h-deformed Lie algebra $sl_h(2.C)$.

6. Conclusion

Finally, we can conclude that the Fedorov's covariant approach and, in particular, the elaborated by him vector parametrization has allowed to solve in simple and natural way a wide series of actual problems in the modern particle physics.

This review shows that at present the activity of the Fedorov's scientific school in Belarus is fruitfully continued.

Bibliography

1. F. I. Fedorov, *Sci. Scripta of Leningrad University. Phys. Sci. Series*, No 146, Iss. 8, 33 (1952).
2. F. I. Fedorov, *Doklady Akademii Nauk SSSR*, **65**, No 6, 813 (1949); **79**, No 5, 787 (1951); **82**, No 1, Pp.37-40. (1952).
3. F. I. Fedorov, *Lorentz Group* (Nauka, Moscow, 1979); (Editorial URSS, Moscow, 2003).
4. F. I. Fedorov, *JETP*, **35**, Iss. 2, 493 (1958).
5. F. I. Fedorov, *Doklady Akademii Nauk SSSR*, **179**, No 4, 802 (1968).
6. F. I. Fedorov, *Doklady Akademii Nauk BSSR*, **5**, No 3, 101 (1961); **5**, No 5, 194 (1961).
7. F. I. Fedorov, *Doklady Akademii Nauk SSSR*, **143**, No 1, 56 (1962).
8. A. A. Bogush, F. I. Fedorov, *Vesti AN BSSR. Ser. phys.-tekhn. nav*, No 2, 26 (1962).
9. A. A. Bogush, F. I. Fedorov, *Rep. Math. Phys.* **11**, No 1, 37 (1977)
10. A. A. Bogush, L. G. Moroz, S. M. Sikach, F. I. Fedorov. In book:*Problems on High Energy Physics and Field Theory.* (Nauka, M., 1989), p.308.
11. M. V. Galynski, S. M. Sikach, *Physics of Elem. Part. and At. Nucl.*, Dubna. **29**, iss.3, 1113 (1998).
12. E. A. Kuraev, M. V. Galynski, M. I. Levchuk, *Physics of Elem. Part. and At. Nucl.*, Dubna, **31**, iss.1, 1955 (2000).
13. A. A. Bogush, (Preprint No 727. Institute of Physics of NASB. Minsk, 2000).
14. A. V. Berezin, Yu. A. Kurochkin, E. A. Tolkachev, *Quaternions in Relativistic Physics* (Nauka i tekhnika, Minsk, 1979); (Editorial URSS, Moscow, 2003).
15. A. A. Bogush, *Doklady Akademii Nauk BSSR*, **24**, No 12, 1073 (1980).
16. A. A. Bogush, *Introduction to the Field Theory of Elementary Partiles* (Nauka i tekhnika, Minsk 1981).
17. A. A. Bogush, *Introduction to the Gauge Field Theory of Electroweak Interactions* (Nauka i tekhnika, Minsk 1987); (Editorial URSS, Moscow, 2003).
18. A. A. Bogush, L. F. Zhirkov. in *Cov. Meth.in Theor. Phys. Elem. Particle Phys. and Relat. Theory*, (Institute of Physics AN BSSR, Minsk, 1981), p. 5.
19. A. A. Bogush, L. F. Zhirkov, A. M. Fedorovych, in *Group Theor. Methods in Physics. Vol. 1*, (Harv.A. Publ. London. N.-Y., 1987), p.271.
20. E. V. Doktorov, in *Cov. Meth.in Theor. Phys. Elem. Particle Phys. and Relat.*

Theory, (Institute of Physics AN BSSR, Minsk, 1981), p. 64.
21. V. I. Kuvshinov, N. V. Tho, *Physics of Elem. Part. and At. Nucl.*, Dubna. **25**, 603 (1994).
22. S. P. Vokos, B. Zumino and J. Wess, *Z . Phys.* **48A**, N 1, 65 (1990).
23. A. A. Bogush, *Dokl. NAN Belarusi*, **44**, N 1, 46 (2000).
24. R. J. Finkelstein, *J. Math. Phys*, **37**, N 2, 59 (1996).
25. A. A. Bogush, in *Quantum Systems. New Trends and Methods, III Intern. Workshop*. (Met. Minsk, 2001), p. 79.
26. L. Mesref, *Int. Journ. Mod. Phys. A.* **17**, No 31, 4777 (2002).
27. C. H. Ohn, *Lett. Math. Phys.* **25**, 85 (1992).
28. A. A. Bogush, *Dokl. NAN Belarusi*, **V.47**, No 6, 32 (2003).

CHARACTERIZING AND DESIGNING LOCALIZED ELECTROMAGNETIC FIELDS

Georgy N. Borzdov

Department of Theoretical Physics, Belarusian State University,
4, Fr. Skorina Ave., 220050, Minsk, Belarus
E-mail: BorzdovG@bsu.by

An approach to characterizing and designing localized electromagnetic fields in complex media and free space, based on the use of differentiable manifolds, differentiable mappings, and the rotation group, is discussed. Families of exact time–harmonic solutions to Maxwell's equations — standing waves defined by spherical harmonics, and localized fields defined by the rotation group — are presented.

1. Introduction

During the last two decades, a considerable amount of attention has been devoted to new solutions of the wave equation, that describe scalar and vector fields with a high degree of two–dimensional and three–dimensional spatial localization (2–D and 3–D localized fields). In particular, 2–D localized fields include Bessel[1–4], Bessel–Gauss[5] and Weber[6] beams which possess high power confinement near the axis and maintain it over a long propagation range. Bessel beams are a typical example of so–called nondiffracting (or diffraction–free) beams conserving their shape across any plane orthogonal to the direction of propagation, i.e., having a transverse intensity distribution independent of the propagation distance. Although, the total energy flux of a Bessel beam is infinite, a truncation of this exact solution can give an approximate solution with finite energy flux and good propagation properties[6].

In 1983, Brittingham[7] proposed the problem of searching for 3–D localized electromagnetic waves being nondispersive for all time and moving at light velocity in straight lines. This work lent impetus to a search localized wave solutions both in electrodynamics and acoustics. The obtained solutions include, in particular, focus wave modes[7,8], modulated, moving Gaussian pulses[9], electromagnetic missiles[10], acoustic[11,12] and electromagnetic[13,14] directed energy pulse trains (ADEPT's and EDEPT's), moving modified Bessel–Gauss pulses[15], moving pulses with Gaussian localization in both longitudinal and transverse directions[16], etc. Such moving localized fields can be realized physically, they appear to be launchable from finite–aperture antennas[11–14,17,18]. An optical implementation of localized wave fields also has attracted considerable interest (see, for example,[19,20] and references therein). In particular, it was shown[20] that wavelength dispersion of the cone angle of an axicon and circular diffraction gratings can be used to generate a good approximation to focus wave modes in the optical region. Bessel beams can be obtained from a Gaussian beam in the conditions of internal conic refraction[21].

Plane wave expansions provide a very fruitful technique for designing and investigating various diffraction–free beams and 3–D localized fields. In recent works[22,23], we have introduced a specific type of electromagnetic field in free space and complex media — plane wave superpositions defined by a set of orthonormal scalar functions on a 2–D or 3–D manifold (beam manifold $\mathcal{B}$). The proposed mathematical technique makes it possible to obtain a set of orthonormal beams, normalized to either the energy flux through a given plane or the total energy transmitted through this plane, and various families of localized fields. As an illustration, the fields defined by the spherical harmonics are being treated.

The families of orthonormal electromagnetic beams form convenient functional bases which can be very useful in the characterization of both complex mediums and fields[24]. In particular, this provides a means to generalize the free–space techniques[25–28] for characterizing complex mediums to the case of incident beams.

The families of localized electromagnetic fields defined by the spherical harmonics and called, for the sake of brevity, storms, whirls and tornadoes, also possess very interesting properties [22,23]. They have a small (about several wavelengths) clearly defined core region with maximum intensity of field oscillations and unique spatial distributions of energy densities, energy fluxes and polarization states. For an electromagnetic storm, the time–average energy flux vector $\mathbf{S}$ is identically zero at all points. Whirls and tornadoes have circular and spiral energy flux lines, respectively. The solutions, which describe electromagnetic whirls moving without dispersion with speed $0 < V < c$ (c is the speed of light in vacuum), finite–energy evolving electromagnetic whirls, and weak gravitational orthonormal beams and localized fields [22,23], have been found as well.

The mathematical formalism proposed in[22] can be applied to any scalar, vector, or tensor field that is linear. By now this formalism has been illustrated for the cases of electromagnetic waves in free space and complex media[22–24,23], weak gravitational fields[23], elastic waves in isotropic and anisotropic solids[29–31], and sound waves in an ideal liquid[32]. In particular, it has been shown [33] that the presented localized fields can be combined as constructive elements to obtain a complex field structure with desirable properties, such as one–, two–, and three–dimensional field gratings. They also can be used to induce chiral inhomogeneities in isotropic mediums[34]. All this makes them promising tools to control laser and particle beams.

Recently, we proposed[33] another technique for obtaining localized exact solutions of linear field equations, that can be considered as a generalization of the earlier presented approach[22]. It has extended potential for characterizing and designing various localized fields. In this chapter, we discuss its main notions and tools on an example of localized electromagnetic fields in complex media and free space.

2. Geometric Parametrization of Plane–wave Superpositions

Electromagnetic homogeneous plane waves (eigenwaves), propagating in a complex medium (anisotropic, chiral, or bianisotropic) or the source-free space, can be written as

$$F(\mathbf{x}) = F(\mathbf{r}, t) = F(0)e^{i\Phi}, \tag{1}$$

where

$$\Phi = \mathbf{x} \cdot \mathbf{K} = \mathbf{k} \cdot \mathbf{r} - \omega t, \tag{2}$$

F is the four–dimensional electromagnetic field tensor, $\mathbf{x} = \mathbf{r} + ct\mathbf{e}_4$ is the four-dimensional position vector, and $\mathbf{K} = \mathbf{k} + \mathbf{e}_4\omega/c = k(\hat{\mathbf{k}} + \mathbf{e}_4)$ is the four-dimensional wave vector. Here, t and $\mathbf{r}$ are the time and the three-dimensional radius vector in a Lorentz frame L with the basis $(\mathbf{e}_i)$, ω and $\mathbf{k}$ are the angular frequency and the three-dimensional wave vector in this frame, c is the velocity of light in vacuum, $\hat{\mathbf{k}} = \mathbf{k}/k$ is the unit wave normal, $k = \omega/c = 2\pi/\lambda$ is the wavenumber, and λ is the wave length.

To compose an electromagnetic field of eigenwaves, one must specify their propagation directions, frequencies or wave numbers, polarizations, intensities, and initial phases. To this end, one can use the four-dimensional wave vector $\mathbf{K}$, the field tensor F, and the so-called beam manifold (BM) $\mathcal{B}$ [22,23]. This approach provides Lorentz covariant parametrization for an electromagnetic plane-wave superposition as

$$F(\mathbf{x}) = \int_{\mathcal{B}} e^{i\Phi(\mathbf{x},b)} F(b)u(b)d\mathcal{B}, \tag{3}$$

where

$$\Phi(\mathbf{x}, b) = \mathbf{x} \cdot \mathbf{K}(b) + \Phi_0(b) \tag{4}$$

is the phase of a partial eigenwave, $\Phi_0(b)$ is the initial phase, $F(b)$ is the normalized real or complex amplitude specifying the eigenwave polarization, $u = u(b)$ is a scalar real function on $\mathcal{B}$, specifying the eigenwave magnitude, and $d\mathcal{B}$ is the infinitesimal element of $\mathcal{B}$.

The image $\mathcal{K}$ of $\mathcal{B}$ under the mapping $b \mapsto \mathbf{K}(b)$ is the subset $\{\mathbf{K}(b);\ b \in \mathcal{B}\}$ of the four-dimensional wave vector space. It is called below the wave vector manifold (WVM). Any point $\mathbf{K}(b) = \mathbf{k}(b) + \mathbf{e}_4\omega(b)/c$ of the WVM satisfies the dispersion equation for the corresponding complex medium[22,35]. In vacuum, this equation reduces to $[\mathbf{K}(b)]^2 = 0$, i.e., $[\mathbf{k}(b)]^2 = [\omega(b)/c]^2$.

By changing initial phases of infinitesimal eigenwaves, one can obtain an infinite family of different, but closely related fields that can be treated as different phase states of the same plane-wave superposition. All members of this family are described by the same functions $\mathbf{K} = \mathbf{K}(b)$, $F = F(b)$, $u = u(b)$ on $\mathcal{B}$, but they are distinguished by different initial phase functions (IPFs) $\Phi_0 = \Phi_0(b)$ on the same BM $\mathcal{B}$. The functions $\mathbf{K} = \mathbf{K}(b)$ and $\Phi_0 = \Phi_0(b)$ in Eq. (4) can be set independently from each other. Alternatively, one can use the parametrization

$$\Phi(\mathbf{x}, b) = \mathbf{K}(b) \cdot [\mathbf{x} - \mathbf{x}_p(b)]\,, \tag{5}$$

where $\mathbf{x}_p = \mathbf{x}_p(b)$ is a vector real functions on $\mathcal{B}$, specifying the IPF as $\Phi_0 = -\mathbf{K}(b) \cdot \mathbf{x}_p(b)$. The image $\mathcal{Z}$ of $\mathcal{B}$ under the mapping $b \mapsto \mathbf{x}_p(b)$ is the subset $\{\mathbf{x}_p(b);\ b \in \mathcal{B}\}$ of the Minkowski space. Since $\Phi(\mathbf{x}_p(b), b) = 0$ for any $b \in \mathcal{B}$, $\mathcal{Z}$ is called below the zero phase manifold (ZPM). Both manifolds $\mathcal{B}$ and $\mathcal{Z}$ are the Lorentz invariant geometrical characteristics of each field under consideration. However, in contrast to $\mathcal{Z}$, $\mathcal{B}$ is not related directly with the space-time, it is merely a parametrization manifold. The four-dimensional formalism is especially useful in analysis of moving localized electromagnetic fields[22,23].

Using the relations $\mathbf{x} = \mathbf{r} + ct\mathbf{e}_4$, $\mathbf{x}_p = \mathbf{r}_p + ct_p\mathbf{e}_4$, and $\mathbf{K} = \mathbf{k} + \mathbf{e}_4\omega/c$, one can present the eigenwave phase as

$$\Phi(\mathbf{x}, b) = \Phi(\mathbf{r}, t, b) = \mathbf{k}(b) \cdot [\mathbf{r} - \mathbf{r}_p(b)] - \omega(b)\,[t - t_p(b)]\,. \tag{6}$$

In the frame L, the ZPM is described by a vector function $\mathbf{r}_p = \mathbf{r}_p(b)$ and a scalar function $t_p = t_p(b)$, specifying the spatial and the temporal eigenwave shifts, respectively. If the field is time-harmonic (THF) in some Lorentz frame L, it is convenient to set the ZPM so that to obtain $t_p(b) \equiv 0$.

Let $\mathbf{v}_a \in V$ be an auxiliary vector related with the contravariant $\mathbf{K}$ and the covariant K wave vectors by the normalization condition $\mathbf{v}_a \cdot \mathbf{K} = \mathbf{v}_a \rfloor K = 1$. The eigenwave 2-form F and 2-vector $\mathbf{F}$ can be written as[25,26]

$$F = K \wedge f, \quad \mathbf{F} = \mathbf{K} \wedge \mathbf{f}. \tag{7}$$

Here, $\wedge$ and $\rfloor$ are the exterior and interior products, $f = \mathbf{v}_a \rfloor F$ and $\mathbf{f}$ are the covariant and contravariant polarization vectors satisfying the condition $\mathbf{K} \rfloor f \equiv \mathbf{f} \rfloor K = 0$. To set the normalized amplitude function $F = F(b)$ in Eq. (3), it is sufficient to set either the covariant $f = f(b)$ or contravariant $\mathbf{f} = \mathbf{f}(b)$ vector function on the BM. Like $\mathcal{Z}$, the image $\mathcal{P}$ of $\mathcal{B}$ under the mapping $b \mapsto \mathbf{f}(b)$ is the subset $\{\mathbf{f}(b);\ b \in \mathcal{B}\}$ of the Minkowski space. Since $\mathcal{P}$ provides the Lorentz invariant geometrical characteristic of partial eigenwaves polarization states, it is called below the polarization manifold (PM). The auxiliary vector $\mathbf{v}_a$ can be chosen so that to obtain $\mathbf{f}$ with a specific physical meaning in the frame L. In particular, to describe the eigenwave polarization in terms of the electric field in the frame L, we chose $\mathbf{v}_a = -\mathbf{e}_4/k$. Then, $\mathbf{f} = \mathbf{E}/k$, and we obtain

$$\mathbf{F} = (\hat{\mathbf{k}} + \mathbf{e}_4) \wedge \mathbf{E}. \tag{8}$$

The infinitesimal eigenwave with the wave vector $\mathbf{K}(b)$ has the amplitude $F(b)u(b)d\mathcal{B}$, i.e., the polarization and the magnitude of this wave are set by two independent factors — the 2-form $F(b)$ and the scalar $u(b)$, respectively. The magnitude function (MF) $u = u(b)$ plays the major role in setting the contribution of each each eigenwaves to the total field, even though $d\mathcal{B}$ also may depend on b.

In the frame of the described covariant approach, a superposition of homogeneous eigenwaves is defined by the functions $\mathbf{K} = \mathbf{K}(b)$, $\mathbf{x}_p = \mathbf{x}_p(b)$ [or $\Phi_0 = \Phi_0(b)$], $\mathbf{f} = \mathbf{f}(b)$, and $u = u(b)$ on the beam manifold $\mathcal{B}$. All these functions as well as the related geometrical objects — the beam manifold $\mathcal{B}$ and its images $\mathcal{K}$, $\mathcal{P}$, and $\mathcal{Z}$ — are Lorentz invariant characteristics of the field. Except for the only condition $\mathbf{K}(b) \cdot \mathbf{f}(b) = 0$ imposed on the points of

$\mathcal{K}$ and $\mathcal{P}$, the described design characteristics can be set independently from each other in any convenient reference frame. This approach can be applied to localized fields in linear homogeneous complex media and free space. For the sake brevity, we illustrate it below on two examples of time-harmonic electromagnetic fields in free space.

3. Standing Waves Defined by the Spherical Harmonics

If the field under consideration is time-harmonic in some reference frame L, instead of the four-dimensional vector functions $\mathbf{K} = \mathbf{K}(b)$, $\mathbf{f} = \mathbf{f}(b)$, and $\mathbf{x}_p = \mathbf{x}_p(b)$, one can describe this field by the three-dimensional vector functions $\mathbf{k} = \mathbf{k}(b)$, $\mathbf{E} = \mathbf{E}(b)$, and $\mathbf{r}_p = \mathbf{r}_p(b)$. Correspondingly, instead of the manifolds $\mathcal{K}$, $\mathcal{P}$, $\mathcal{Z}$ in the Minkowski space, one can characterize the THF by their sections $\mathcal{K}_3$, $\mathcal{P}_3$, and $\mathcal{Z}_3$, lying in the Euclidean subspace in the frame L.

By way of illustration, let us consider the three-dimensional standing electromagnetic waves defined by the spherical harmonics (Y_j^s) as[22,23]

$$\mathbf{E}_j^s(\mathbf{r},t) = u_0 e^{-i\omega t} \int_0^{2\pi} d\varphi \int_0^{\pi} e^{i\mathbf{k}(\theta,\varphi)\cdot\mathbf{r}}\, Y_j^s(\theta,\varphi) \times \mathbf{E}(\theta,\varphi) \sin\theta d\theta, \tag{9}$$

where

$$Y_j^s(\theta,\varphi) = N_{js} P_j^{|s|}(\cos\theta) e^{is\varphi}, \tag{10}$$

$$N_{js} = \sqrt{\frac{(2j+1)(j-|s|)!}{4\pi(j+|s|)!}}, \tag{11}$$

and $P_j^s(\cos\theta)$ is the spherical Legendre function, u_0 is some normalizing constant factor. For these waves, the beam manifold $\mathcal{B}$ is the unit sphere S^2, and $\mathrm{d}\mathcal{B} = \sin\theta \mathrm{d}\theta \mathrm{d}\varphi$.

To set the functions $\mathbf{k} = \mathbf{k}(\theta,\varphi)$ and $\mathbf{E} = \mathbf{E}(\theta,\varphi)$, one can use the spherical basis vectors

$$\mathbf{e}_r(\theta,\varphi) = \mathbf{e}_R(\varphi)\sin\theta + \mathbf{e}_3\cos\theta, \tag{12}$$

$$\mathbf{e}_M(\theta,\varphi) = \mathbf{e}_R(\varphi)\cos\theta - \mathbf{e}_3\sin\theta, \tag{13}$$

$$\mathbf{e}_A(\varphi) = -\mathbf{e}_1\sin\varphi + \mathbf{e}_2\cos\varphi, \tag{14}$$

where

$$\mathbf{e}_R(\varphi) = \mathbf{e}_1\cos\varphi + \mathbf{e}_2\sin\varphi. \tag{15}$$

In Refs.[22,23], we have treated the standing waves with $\mathbf{k}(\theta, \varphi) = k\mathbf{e}_r$ and two different polarization states, namely, E_M fields with the meridional orientation $[\mathbf{E}(\theta, \varphi) = \mathbf{e}_M]$ and E_A fields with the azimuthal orientation $[\mathbf{E}(\theta, \varphi) = \mathbf{e}_A]$ of the eigenwave electric field. They are formed from plane waves of all possible propagation directions.

For these fields, the WVM $\mathcal{K}_3$ is the sphere S_k^2 with a radius $k = \omega/c$ in the three-dimensional wave vector space, and $\hat{\mathbf{k}}(\theta, \varphi) \equiv \mathbf{e}_r(\theta, \varphi)$. For various types of time-harmonic orthonormal beams also treated in Refs.[22,23], the WVM is some domain of S_k^2. The PM $\mathcal{P}_3$ for E_M and E_A fields can be conveniently illustrated as the meridional and azimuthal unit vector fields on the surface of this sphere, respectively.

The ZPM $\mathcal{Z}_3$ is described by the relations

$$t_p \equiv 0, \qquad \mathbf{r}_p(\theta, \varphi) = -\frac{s\varphi}{k}\mathbf{e}_r(\theta, \varphi), \tag{16}$$

where $\theta \in [0, \pi]$ and $\varphi \in [0, 2\pi]$. For the electromagnetic storms $(s = 0)$, the ZPM shrinks to the point $\mathbf{r}'_p = 0$.

Let us now consider magnitude functions of the standing waves under the consideration. It follows from Eqs. (9) and (10) that the MF of the wave $\mathbf{E}_j^s$ has the form

$$u = u_0 N_{js} P_j^{|s|}(\cos\theta) \tag{17}$$

and can be graphically illustrated by a parameterized surface as

$$\mathbf{r}'_u = (u/u_0)\mathbf{e}_r(\theta, \varphi) = N_{js} P_j^{|s|}(\cos\theta)\mathbf{e}_r(\theta, \varphi), \tag{18}$$

where $\theta \in [0, \pi]$ and $\varphi \in [0, 2\pi]$. This surface is axially symmetric. It should be mentioned that the MF u [Eq. (17)] may take negative values at some intervals of θ (see, for example, Ref.[33]). Alternatively, one can replace u by its absolute value $|u|$ and introduce an additional phase shift π whenever u becomes negative.

4. Fields Defined by the Rotation Group

Let us confine our consideration to the time-harmonic superpositions of linearly polarized homogeneous plane waves in free space, that can be parameterized by points $b \in \mathcal{B}$ of a two-dimensional BM as follows:

$$\mathbf{W}(\mathbf{r}, t) = e^{-i\omega t} \int_{\mathcal{B}} e^{i\mathbf{k}(b)\cdot[\mathbf{r} - \mathbf{r}_p(b)]} \mathbf{W}(b) u(b) d\mathcal{B}. \tag{19}$$

The unit real vectors $\mathbf{W}(b) = \mathbf{E}(b)$ and $\mathbf{W}(b) = \mathbf{B}(b)$ specify the polarization in terms of the electric and the magnetic fields, respectively, whereas the eigenwave magnitude is defined by the MF $u = u(b)$. The normalized amplitudes $\mathbf{E}(b)$ and $\mathbf{B}(b)$ of each eigenwave are related as

$$\mathbf{B}(b) = \hat{\mathbf{k}}(b) \times \mathbf{E}(b). \tag{20}$$

For the plane-wave superposition under consideration, the orthonormal triads $[\hat{\mathbf{k}}(b), \mathbf{E}(b), \mathbf{B}(b)]$ of any pair of partial eigenwaves are related by a rotation operator. Let us denote $\hat{\mathbf{k}}_0 = \hat{\mathbf{k}}(b_0)$, $\mathbf{E}_0 = \mathbf{E}(b_0)$, and $\mathbf{B}_0 = \mathbf{B}(b_0)$, where $b_0 \in \mathcal{B}$ is an arbitrarily given point of the BM. Then, we can set the triads of all eigenwaves by a rotation operator function $G = G(b)$ as

$$\hat{\mathbf{k}}(b) = G(b)\hat{\mathbf{k}}_0, \quad \mathbf{E}(b) = G(b)\mathbf{E}_0, \quad \mathbf{B}(b) = G(b)\mathbf{B}_0, \tag{21}$$

where $G(b_0) = \mathbf{1}$ is the unit dyadic. In its turn, the rotation operator G can be parameterized by the Euler angles or the Fedorov complex vector parameter[36]. This approach makes it possible to describe the beam by the evolution operator

$$\mathcal{F}(\mathbf{r}, t) = e^{-i\omega t} \int_{\mathcal{B}} e^{i\mathbf{k}(b)\cdot[\mathbf{r}-\mathbf{r}_p(b)]} u(b) G(b) d\mathcal{B}, \tag{22}$$

that defines both the electric and magnetic fields as

$$\mathbf{E}(\mathbf{r}, t) = \mathcal{F}(\mathbf{r}, t)\mathbf{E}_0, \quad \mathbf{B}(\mathbf{r}, t) = \mathcal{F}(\mathbf{r}, t)\mathbf{B}_0. \tag{23}$$

It should be noted that Eqs. (22) and (23) can be readily applied to superpositions of eigenwaves with elliptic or circular polarization. To this end, it is sufficient to set the normalized amplitudes $\mathbf{E}_0$ and $\mathbf{B}_0$ as

$$\mathbf{E}_0 = \mathbf{a} + i\mathbf{b}, \quad \mathbf{B}_0 = \hat{\mathbf{k}}_0 \times \mathbf{E}_0, \tag{24}$$

where $\mathbf{a}$ and $\mathbf{b}$ are the major and the minor semiaxes, $\mathbf{a}\cdot\hat{\mathbf{k}}_0 = \mathbf{b}\cdot\hat{\mathbf{k}}_0 = \mathbf{a}\cdot\mathbf{b} = 0$, $\mathbf{a}^2 + \mathbf{b}^2 = 1$, and $\mathbf{a}^2 - \mathbf{b}^2 \geq 0$. Beams in isotropic mediums, including chiral ones, can be treated similarly, except that the eigenwaves in an isotropic chiral medium are circularly polarized ($\mathbf{a}^2 = \mathbf{b}^2$). Superpositions of eigenwaves with different polarizations can be treated by making use of the exponential evolution operators[37,38].

The operator G of rotation through the angle φ around the unit vector $\mathbf{n}$ can be written as

$$G \equiv \exp(\varphi\mathbf{n}^\times) = \mathbf{n}\otimes\mathbf{n} + (\mathbf{1} - \mathbf{n}\otimes\mathbf{n})\cos\varphi + \mathbf{n}^\times \sin\varphi, \tag{25}$$

where $\mathbf{n}^\times$ is the antisymmetric tensor dual to $\mathbf{n}$ ($\mathbf{n}^\times\mathbf{E} = \mathbf{n} \times \mathbf{E}$). Let us parameterize the operator G by three rotation angles ($\psi_j, j = 1, 2, 3$) as

$$G = \exp(\psi_3\mathbf{d}^\times)\exp(\psi_2\mathbf{e}_A^\times)\exp(\psi_1\mathbf{e}_3^\times), \tag{26}$$

where

$$\mathbf{d} = \exp(\psi_2 \mathbf{e}_A^{\times}) \exp(\psi_1 \mathbf{e}_3^{\times}) \mathbf{d}_0, \tag{27}$$

$\mathbf{e}_A = \mathbf{e}_A(\psi_1)$, and $\mathbf{d}_0$ is a unit vector.

In particular, if $\mathbf{d}_0 = \mathbf{e}_3$, the operator G can be written as

$$G = \mathbf{g}_1 \otimes \mathbf{e}_1 + \mathbf{g}_2 \otimes \mathbf{e}_2 + \mathbf{d} \otimes \mathbf{e}_3, \tag{28}$$

where

$$\mathbf{g}_1 = \mathbf{e}_M(\psi_2, \psi_1) \cos\psi_3 + \mathbf{e}_A(\psi_1) \sin\psi_3, \tag{29}$$

$$\mathbf{g}_2 = -\mathbf{e}_M(\psi_2, \psi_1) \sin\psi_3 + \mathbf{e}_A(\psi_1) \cos\psi_3, \tag{30}$$

$$\mathbf{d} = \mathbf{e}_r(\psi_2, \psi_1) = \mathbf{e}_R(\psi_1) \sin\psi_2 + \mathbf{e}_3 \cos\psi_2. \tag{31}$$

The operator function $G = G(b)$ can be uniquely defined by three real scalar functions $\psi_= \psi_j(b)$, $j = 1, 2, 3$. In particular, for the E_M and E_A fields defined by the spherical harmonics, $\mathbf{E}_0 = \mathbf{e}_1$ and $\mathbf{E}_0 = \mathbf{e}_2$, respectively. In both cases, $\hat{\mathbf{k}}_0 = \mathbf{e}_3$, $\psi_1 = \varphi$, $\psi_2 \doteq \theta$, and $\psi_3 \equiv 0$.

All three-dimensional standing waves defined by the spherical harmonics have the same BM (the unit sphere $\mathcal{B} = S^2$) as well as the same WVM (the sphere $\mathcal{K}_3 = S_k^2$). E_M and E_A waves have different PMs ($\mathcal{P}_3$ is set by the meridional and the azimuthal vectors, respectively) and their electric and magnetic fields are related by the duality transformation. The THF defined by the rotation group possess a much larger diversification potential. In particular, they include a multitude of localized fields with topologically different beam manifolds, for example, spherical and toroidal BMs[33].

5. Complex Field Structures

The presented electromagnetic fields have a very small (about several wavelengths) core region with maximum intensity of field oscillations and unique space distributions of polarization states, energy densities, and energy fluxes. Outside the core, the intensity of oscillations rapidly decreases in all directions. The three-dimensional localization makes it possible to use these fields as structural elements to form various complex electromagnetic fields. By way of illustration, let us consider a field defined as

$$\mathbf{W}'(\mathbf{r}, t) = \sum_{n=M_1}^{N_1} \sum_{m=M_2}^{N_2} \sum_{l=M_3}^{N_3} \mathbf{W}(\mathbf{r} - \mathbf{a}_{nml}, t - \tau_{nml}), \tag{32}$$

$$\mathbf{a}_{nml} = n\mathbf{a}_1 + m\mathbf{a}_2 + l\mathbf{a}_3, \tag{33}$$

$$\tau_{nml} = n\tau_1 + m\tau_2 + l\tau_3, \tag{34}$$

where $\mathbf{a}_j$ and τ_j are some given spatial and temporal shifts. This field can be treated as a different state of $\mathbf{W}(\mathbf{r}, t)$, obtained by the magnitude function transformation transformation $u(b) \mapsto u(b)|h(b)|$ and the phase transformation $\Phi(\mathbf{r}, t, b) \mapsto \Phi(\mathbf{r}, t, b) + \Delta\Phi(b)$. Here, $|h(b)|$ and $\Delta\Phi(b)$ are defined as

$$h(b) \equiv |h(b)| e^{\Delta\Phi(b)} = h_1(b) h_2(b) h_3(b), \tag{35}$$

where

$$\begin{aligned} h_j(b) &= \sum_{m=M_j}^{N_j} e^{im\Phi_j(b)} \\ &= \frac{\exp\left[iM_j\Phi_j(b)\right] - \exp\left[i\left(N_j + 1\right)\Phi_j(b)\right]}{1 - \exp\left[i\Phi_j(b)\right]}, \end{aligned} \tag{36}$$

$$\Phi_j(b) = \omega(b)\tau_j - \mathbf{k}(b) \cdot \mathbf{a}_j. \tag{37}$$

These fields have the same WVM $\mathcal{K}_3$ and the same PM $\mathcal{P}_3$.

The great diversity of the presented localized fields with different geometries of core regions provides a great scope for combining them, as constructive elements, into various complex geometrical structures, such as one-, two-, or three-dimensional gratings[33]. Designing localized electromagnetic fields to control motion and state of charged and neutral particles (atoms and molecules) is one of possible applications of the presented techniques. In Ref.[33], this is illustrated on an example of relativistic electrons.

6. Conclusions

An approach to characterizing and designing localized electromagnetic fields, based on the use of differentiable manifolds, differentiable mappings, and the group of rotation, is discussed. The beam manifold $\mathcal{B}$, the wave vector manifold $\mathcal{K}$, the polarization manifold $\mathcal{P}$, the zero phase manifold $\mathcal{Z}$, and the magnitude function u are the key elements of the Lorentz covariant description. Every localized field under consideration is characterized by the differentiable mappings $\mathcal{B} \to \mathcal{K}$, $\mathcal{B} \to \mathcal{P}$, $\mathcal{B} \to \mathcal{Z}$, and $\mathcal{B} \to R$, given as $\mathbf{K} = \mathbf{K}(b)$, $\mathbf{f} = \mathbf{f}(b)$, $\mathbf{x}_p = \mathbf{x}_p(b)$, and $u = u(b)$. They specify wave vectors, polarizations, initial phases, and magnitudes of eigenwaves constituting the field. If the field is time-harmonic in a frame L, one can characterize it by the corresponding sections $\mathcal{K}_3$, $\mathcal{P}_3$, and $\mathcal{Z}_3$ of $\mathcal{K}$, $\mathcal{P}$, and $\mathcal{Z}$. By way of illustration, families of exact time-harmonic solutions to Maxwell's equations in the source-free space — fields defined by the rotation group — are treated. The proposed approach provides a broad spectrum of tools to design localized fields, i.e., to build-in symmetry properties of oscillating electric and magnetic fields, to govern the distributions of their energy densities (both size and form of localization domains), and to set the structure of time-average energy fluxes[33].

Localized fields of one or more types can be combined as constructive elements to obtain a complex field structure with desirable properties, such as one-, two-, or three-dimensional field gratings. The proposed approach can be used in designing localized electromagnetic fields to govern motion and state of charged and neutral particles.

Bibliography

1. J. Durnin, *J. Opt. Soc. Am.* **A4**, 651 (1987).
2. J. Durnin, J. J. Miceli, Jr. and J. H. Eberly, *Opt. Lett.* **13**, 79 (1988).
3. A. M. Belsky, *Vestnik Belarusian State University*, Ser. 1, No. 2, 8 (1995).
4. A. M. Belsky, *Optics of coherent light beams* (Belarusian State University, Minsk, 2000).
5. F. Gori, G. Guattari and C. Padovani, *Opt. Commun.* **64**, 491 (1987).
6. F. Bloisi and L. Vicari, *Opt. Commun.* **75**, 353 (1990).
7. J. N. Brittingham, *J. Appl. Phys.* **54**, 1179 (1983).
8. A. Sezginer, *J. Appl. Phys.* **57**, 678 (1985).
9. R. W. Ziolkowski, *J. Math. Phys.* **26**, 861 (1985).
10. T. T. Wu, *J. Appl. Phys.* **57**, 2370 (1985).
11. R. W. Ziolkowski, D. K. Lewis and B. D. Cook, *Phys. Rev. Lett.* **62**, 147 (1989).
12. R. W. Ziolkowski and D. K. Lewis, *J. Appl. Phys.* **68**, 6083 (1990).
13. R. W. Ziolkowski, *Phys. Rev.* **A39**, 2005 (1989).
14. A. M. Shaarawi, I. M. Besieris and R. W. Ziolkowski, *J. Appl. Phys.* **65**, 805 (1989).
15. P. L. Overfelt, *Phys. Rev.* **A44**, 3941 (1991).
16. A. P. Kiselev and M. V. Perel, *J. Math. Phys.* **41**, 1934 (2000).
17. R. W. Ziolkowski, I. M. Besieris and A. M. Shaarawi, *J. Opt. Soc. Am.* **A10**, 75 (1993).
18. R. W. Ziolkowski, *Phys. Rev.* **A44**, 3960–3984 (1991).
19. P. Saari and K. Reivelt, *Phys. Rev. Lett.* **79**, 4135 (1997).
20. K. Reivelt and P. Saari, *J. Opt. Soc. Am.* **A17**, 1785 (2000).
21. N. S. Kazak, N. A. Khilo, A. A. Ryzhevich and E. G. Kantranji, *Optical Journal* **67**, 54 (2000).
22. G. N. Borzdov, *Phys. Rev.* **E61**, 4462 (2000).

23. G. N. Borzdov, *Phys. Rev.* **E63**, 036606 (2001).
24. G. N. Borzdov, *Arch. Elektron. Übertrag.* **55**, 224 (2001).
25. G. N. Borzdov, *Opt. Commun.* **94**, 159 (1992).
26. G. N. Borzdov, *J. Math. Phys.* **34**, 3162 (1993).
27. G. N. Borzdov, in *Advances in complex electromagnetic materials*, Ed. A. Priou, A. Sihvola, S. Tretyakov and A. Vinogradov,(Kluwer, Dordrecht, 1997), p. 71.
28. G. N. Borzdov, in *Electromagnetic fields in unconventional materials and structures*, Ed. O. N. Singh and A. Lakhtakia, (Wiley, New York, 2000), p. 83.
29. G. N. Borzdov, *J. Phys.* **A34**, 6249 (2001).
30. G. N. Borzdov, *J. Phys.* **A34**, 6259 (2001).
31. G. N. Borzdov, *J. Phys.* **A34**, 6269 (2001).
32. G. N. Borzdov, *J. Phys.* **A34**, 6281 (2001).
33. G. N. Borzdov, *Phys. Rev.* **E65**, 066612 (2002).
34. G. N. Borzdov, in *Advances in electromagnetics of complex media and metamaterials*, Ed. S. Zouhdi, A. Sihvola and M. Arsalane, (Kluwer, Dordrecht, 2002), p. 363.
35. F. I. Fedorov, *Theory of gyrotropy* (Nauka i Tekhnika, Minsk, 1976).
36. F. I. Fedorov, *Lorentz group* (Nauka, Moscow, 1979).
37. L. M. Barkovsky, G. N. Borzdov and F. I. Fedorov, *Zh. Prikl. Spektrosk.* **39**, 996 (1983).
38. L. M. Barkovsky and A. N. Furs, *Operator methods of description of optical fields in complex media* (Belarusian Science, Minsk, 2003).

OPERATOR METHOD FOR NONPERTURBATIVE DESCRIPTION OF QUANTUM SYSTEMS

Ilya Feranchuk[1] and Alexey Ivanov[2]

[1]*Belarusian State University, 4, Fr. Skorina Ave., 220050, Minsk, Belarus*
E-mail: fer@open.by

[2]*Belarusian National Technical University, 65 Fr. Skorina Ave., 220013, Minsk, Belarus*
E-mail: alivanov@tut.by

A nonperturbative method for describing quantum systems — the operator method (OM) and the conception of a uniformly suitable estimation (USE) are considered for a series of real physical systems. It is shown that the OM zeroth-order approximation permits one to find the analytical approximation for eigenfunctions and eigenvalues with high accuracy within the entire range of the Hamiltonian parameters and any quantum numbers. The OM subsequent approximations converge rapidly to the exact solutions of the Schrödinger equation. The generalization of OM for quantum statistics is also developed.

1. Introduction

There is no doubt at present that the whole variety of atomic and molecular physical phenomena are determined by an electromagnetic interaction and quantum mechanics laws. Therefore any states of atomic and molecular systems can be described by the corresponding solution of the Schrödinger equation with definite Hamiltonian. Besides, quantum description is universal and simple enough, because the wave function of any complex system is the solution of linear equation with a mathematical structure which is often similar for systems with different physical properties. That's why one can say, as Wigner once said[1], there is a special role for mathematics in quantum mechanics. In particular, the development of a new effective method of approximate solution of the Schrödinger equation provides essential progress in describing a number of real systems.

It is necessary to point out, that a lot of such methods have arisen from generalization of some unconventional approaches proposed in describing real physical systems (see, e.g. papers[2–8]).

New methods of solution of the Schrödinger equation are of great interest not only for analytical description of quantum system but for the construction of an effective algorithm for numerical computation of their characteristics as well. The thing is that despite considerable increase in modern computer power it is still insufficient for solving the Schrödinger equation for most systems. The above-mentioned difficulty is of principal meaning when considering the systems with infinite number of degrees of freedom. At present the only method of such a numerical investigation is based on Feynman formulation of quantum mechanics[9]. Unfortunately, the volume of calculations in terms of this method remains so great, that sufficient accuracy can be obtained only for simple models of the quantum field theory in so-called numerical experiments[10].

What requirements should the mathematical method satisfy to be widely adopted? In our opinion, the first one is universality, that is, the possibility of representation of the main computational procedure in such a form, that is not bound up with special properties of a concrete system. We mean, of course, that any universal method can not provide a ready-made formula for solution of the Schödinger equation with an arbitrary Hamiltonian, but gives a prescription for the calculation of successive approximation to it. First, it requires that the zeroth-order approximation of the method should describe qualitatively the most essential properties of the considered physical system, and, second, the consequent approximations should converge uniformly to the exact solution. Finally, the important condition is that the calculation algorithm for the successive approximations should be rather simple to be carried out on a computer.

There are lots of approximate schemes which satisfy these requirements based on the perturbation theory and its modifications[11–13]. However, the nontrivial problems with some small parameter are almost exhausted at present. At the same time the practical value of problems with arbitrary Hamiltonian parameters is increasing. Therefore the necessity arises to develop the mathematical methods for physical theories without small parameters. At present there are many nonperturbative methods for describing of quantum systems (see, e.g.[14–17]).

Sufficient advantage of a nonperturbative approximate method is the possibility to obtain such an approximation that is suitable for a wide range of Hamiltonian parameters (or even for a whole range of them). We call such an approximation the uniformly suitable estimation (USE). Let us clarify, what we mean by the USE.

Let us consider some physical variable with eigenvalues $F_n(\lambda)$ depending on the quantum number n and the physical parameter λ (the totality of the parameters and quantum numbers is implied for the systems with many degrees of freedom). Let us introduce

Definition 1: the function $F_n^{(0)}(\lambda)$ is the USE for $F_n(\lambda)$ if the following inequality holds in the entire range of variation of the values n and λ

$$\left| \frac{F_n^{(0)}(\lambda) - F_n(\lambda)}{F_n(\lambda)} \right| \leq \xi^{(0)}. \tag{1}$$

Here the parameter $\xi^{(0)} < 1$ is assumed to be independent of n and λ, and defines the accuracy of the USE. We also consider

Definition 2: there are the sequence of functions $F_n^{(s)}(\lambda)$ and the method for their calculation corresponding to the decreasing sequence of the parameters $\xi^{(s)}$; $s = 0, 1, 2, \ldots$, so that

$$\lim_{s \to \infty} F_n^{(s)}(\lambda) = F_n(\lambda). \tag{2}$$

It seems that Definition 1 is not constructive because the exact values $F_n(\lambda)$ are unknown in the general case. However, there are several possibilities to estimate the value $\xi^{(0)}$. In particular, one can compare the asymptotic series for $F_n(\lambda)$ in various limit cases either of the parameter λ or of the quantum number n with the corresponding expansions of the function $F_n^{(0)}(\lambda)$ taking into account that inequality (1) should hold for all cases simultaneously. Besides, an estimation for $\xi^{(0)}$ can be found as the difference between $F_n^{(1)}(\lambda)$ and $F_n^{(0)}(\lambda)$.

Considerable advantage of the USE for various applications in comparison to the asymptotic expansions is the possibility to investigate the qualitative peculiarities of the quantum system in the intermediate range of the parameters connected with Hamiltonian and the external conditions. At the same time, rather high precision of the USE in the zeroth order is of great use for practical calculations and it defines the rate of convergence of the consecutive approximations to the exact solution.

To obtain the USE we use the operator method (OM) of approximate solution of the Schrödinger equation. The OM was introduced in[18,19], then some of its applications for specific systems were considered in[20–35]. This method was developed by Fernandez, Meson and Castro[36–41], Gerry and Silverman[42], Witschel[43] and others[44,45]. Several ideas of the OM were introduced earlier in[46–52].

2. General scheme of the operator method

We consider here only the basic expressions defining the consecutive approximations of the OM for the eigenvalues E_n and eigenvectors $|\psi_n\rangle$ of a Hamiltonian $\hat{H}$ of some quantum system (the index n may involve the whole set of quantum state numbers)

$$\hat{H}\,|\psi_n\rangle = E_n\,|\psi_n\rangle\,. \tag{3}$$

Let us introduce the complete set of state vectors $|n,\omega\rangle$, depending on the same quantum numbers and arbitrary parameters ω. In the canonical perturbation theory (CPT) the eigenfunctions of some part $\hat{H}_0$ of the full Hamiltonian (zeroth-order approximation Hamiltonian) are used as such a complete set. The Schrödinger equation for that part is assumed to be exactly solvable. In terms of the OM, the vectors $|n,\omega\rangle$ are considered as a set of variational wave functions. Thus, the choice of these vectors is rather arbitrary and is defined by qualitative features of the system under consideration and by the possibility of rather simple calculation of the matrix elements of the total Hamiltonian $\hat{H}$. Then we can represent the solution of equation (3) in the form of the expansion

$$|\psi_n\rangle = |n,\omega\rangle + \sum_{k\neq n} C_{nk}\,|k,\omega\rangle\,, \tag{4}$$

and choose the normalization of the exact solution as the following

$$\langle\psi_n|\,n,\omega\rangle = 1. \tag{5}$$

As a result, the initial eigenvalues problem (3) exactly reduces to the system of infinite number of nonlinear algebraic equations for E_n and coefficients C_{nk}[30]

$$E_n = H_{nn} + \sum_{k\neq n} C_{nk}H_{nk}; \quad H_{nk}(\omega) = \langle n,\omega|\,\hat{H}\,|k,\omega\rangle\,;$$
$$C_{nm} = -[H_{mm} - E_n]^{-1}\big[H_{mn} + \sum_{k\neq m,n} C_{nk}H_{mk}\big]; \quad m\neq n. \tag{6}$$

As shown in[30], the most effective method for solving this system of equations is an iterative scheme, in which the exact solutions are calculated as the limit of the sequences (and not by summation of series, as in the CPT)

$$E_n = \lim_{s\to\infty} E_n^{(s)}; \quad C_{nk} = \lim_{s\to\infty} C_{nk}^{(s)}, \tag{7}$$

and the consecutive approximations in (2.1) are calculated by means of the recurrence relations

$$E_n^{(s)}(\omega) = H_{nn} + \sum_{k\neq n} C_{nk}^{(s-1)} H_{nk};$$
$$C_{nm}^{(s)} = -[H_{mm} - E_n^{(s-1)}]^{-1}[H_{mn} + \sum_{k\neq m,n} C_{nk}^{(s-1)} H_{mk}];$$
$$E_n^{(0)} = E_n^{(1)} = H_{nn}; \quad C_{nk}^{(0)} = 0. \tag{8}$$

As shown for various physical systems[20−35], the recurrence relations (8) converge rapidly to the exact solutions, and the choice of the parameter ω influences only the rate of this convergence. For example, the second iteration leads to the following approximation

$$E_n^{(2)}(\omega) \simeq E_n^{(0)}(\omega) + \Delta E_n(\omega) = H_{nn} - \sum_{m\neq n} [H_{mm} - H_{nn}]^{-1} H_{nm} H_{mn}. \tag{9}$$

Strictly speaking, this expression is suitable if the diagonal matrix elements $H_{kk}(\omega) \neq H_{nn}(\omega)$, for the considered value of the parameter ω and for all states from the set $|k,\omega\rangle$ in expansion (4).

If for some concrete values m and ω_m $H_{mm}(\omega_m) = H_{nn}(\omega_m)$, one should extract the state $|m,\omega_m\rangle$ from expansion (4) and solve the secular equation in the OM zeroth order approximation analogously to the canonical perturbation theory in case of degenerate states[54]. Some applications of the OM for physical systems with such features were considered in papers[31,32].

For systems with several degrees of freedom the state vector $|\psi_n, j\rangle$ in equation (3) may also depend on some additional index j. The energy level E_n is supposed to be independent of this index. Actually it means that the wave function is also the eigenvector of some operator

$$\hat{J}\,|\psi_n, j\rangle = j\,|\psi_n, j\rangle\,, \quad [\hat{J}, \hat{H}] = 0, \tag{10}$$

which commutes with the Hamiltonian. As shown in[30], in such a case the projection operator $\hat{T}_j$ for the state with fixed quantum number j should be used in series (4) in order to solve both equations (3) and (10) simultaneously. It leads to unessential modification of recurrence equations (8) but does not change the properties of the OM consecutive approximations.

In spite of the formal similarity of formula (9) and the expression obtained in the second order of the CPT, the former one maintains some essential differences. First, the matrix elements of the operator $\hat{H}$ are calculated with the set of state vectors $|n,\omega\rangle$ for the whole Hamiltonian (and not for its unperturbed part as in the CPT). According to[30], this circumstance is essential for the convergence of the consequent approximations of the OM. In particular, this convergence depends on the numerical dimensionless parameter, defined by the ratio of the diagonal and nondiagonal matrix elements of the total Hamiltonian

$$\xi(\omega) \simeq \left|\frac{H_{nk}}{H_{nn}}\right|_{max} ; \quad k \neq n. \tag{11}$$

This value is independent of the physical parameter of the Hamiltonian and defines both the convergence of the consequent approximations of the OM, and the accuracy of the USE for the zeroth order approximation of the OM[30] in accordance with the definition of the USE.

Another specific feature of the approximation (9) is that it depends on the undefined parameter ω, which should be chosen so that to provide the best accuracy of the USE in the zeroth order approximation of the OM. It was shown in[30], that the best zeroth order approximation of the OM for the energy level with the quantum number n

$$E_n \simeq E_n^{(0)} = E_n^{(0)}(\omega_n), \tag{12}$$

is achieved when $\omega = \omega_n$ is chosen as the solution of the equation

$$\frac{\partial E_n^{(0)}(\omega)}{\partial \omega} = 0. \tag{13}$$

It should be underlined, that equation (13) is not the variational principle for the excited states. The point is that the artificial parameter ω defines actually the representation for the wave functions, and hence, the exact eigenvalues of the Hermitian Hamiltonian should not depend on its choice. The exact condition is

$$\frac{\partial E_n(\omega)}{\partial \omega} = 0. \tag{14}$$

Therefore, equation (13) can be considered as the OM zeroth order approximation for the exact condition (14).

It is essential that the optimal value of the parameter ω_n depends on the quantum number of the considered state. This means that the orthonormal set of basic vectors used in expansion (4) should be chosen for every state in different ways. As a result, the consecutive approximations of the OM (8) have "local" character in the space of quantum numbers, i.e., they are calculated independently with different sets of state vectors for every state of the system. On the contrary, in terms of the CPT the corrections for all quantum numbers are defined by the same spectrum and basic state vectors of the unperturbed Hamiltonian.

3. How and why does the general scheme work?

Let us consider this question for well-known model system, which is widely used for approbation of various approximate theories – the quantum anharmonic oscillator. This system is described by the following Hamiltonian

$$\hat{H} = \frac{1}{2}(\hat{p}^2 + \hat{x}^2) + \mu\hat{x}^2 + \lambda\hat{x}^4. \tag{15}$$

It is well known, that the CPT expansion for this system has zero radius of convergence with respect to the parameter λ[55]. Therefore, the QAO is widely used for approbation of various nonperturbative methods, and all approximations can be compared with the detailed numerical analysis of this problem in[55]. In the problem of the QAO the most adequate choice of the full set of state vectors for the OM algorithm is the set of eigenfunctions of the harmonic oscillator with an arbitrary frequency, playing a role of the parameter ω. At the same time, it is convenient to perform all the necessary calculations of the matrix elements in an algebraic form, using the representation of the secondary quantization[18]. In such a case, the variational parameter ω can be introduced directly into the Hamiltonian by means of creation $\hat{a}^+$ and annihilation $\hat{a}$ operators

$$\begin{gathered} \hat{x} = \frac{1}{\sqrt{2\omega}}(\hat{a} + \hat{a}^+), \qquad \hat{p} = -i\sqrt{\frac{\omega}{2}}(\hat{a} - \hat{a}^+), \\ \hat{n} = a^+a; \quad \left[a, a^+\right] = 1; \quad \hat{n}\left|n\right\rangle = n\left|n\right\rangle. \end{gathered} \tag{16}$$

Some fairly simple algebraic operations with expressions (6)–(13) lead to the following results

$$\begin{aligned} E_n^{(0)} &= \frac{1}{4\omega_n}(\omega_n^2 + 1 + 2\mu)(2n+1) + \frac{3\lambda}{4\omega_n^2}(1 + 2n + 2n^2); \\ \omega_n^3 &- \omega_n(1+2\mu) - \frac{6\lambda(2n^2+2n+1)}{2n+1} = 0; \end{aligned} \tag{17}$$

$$\begin{aligned} \Delta E_n^{(2)} = &- \frac{(n+1)(n+2)[\omega_n(1 + 2\mu - \omega_n^2) + 2\lambda(2n+3)]^2}{16\omega_n^2[\omega_n(\omega_n^2 + 1 + 2\mu) + 3\lambda(3+2n)]} \\ &- \frac{\lambda^2(n+1)(n+2)(n+3)(n+4)}{32\omega_n^2[\omega_n(\omega_n^2 + 1 + 2\mu) + 3\lambda(5+2n)]}. \end{aligned} \tag{18}$$

It can be shown by strict comparison with the numerical results[55] that simple analytical expressions (17) define the USE for QAO with $\xi^{(0)} \simeq 0.03$. Figure 1 and Fig. 2 illustrate this statement. They show the results of calculation of the QAO energy levels in dependence on the parameter λ and quantum number n, obtained by means of various methods: numerical calculations (scattered)[55]; the CPT; expansion in the strong coupling limit; the algebraic formulae in the OM zeroth and second orders. The absolute values and relative errors of various calculation methods in dependence on the parameter λ are compared in Fig. 1, and Fig. 2 shows the relative error in dependence on the quantum number n. One can see that only approximation for the functions $E_n(\lambda)$ obtained in terms of the OM does satisfy the definition for the USE approximated energy levels of the QAO.

4. Generalization for quantum statistics

Let us consider the partition function of some quantum system

$$Z(\beta) = \sum_{n=0}^{\infty} g_n \exp[-\beta E_n], \tag{19}$$

where E_n is the energy level corresponding to a quantum number n and degeneracy g_n.

To obtain the USE for this function one has to solve two problems: 1) to construct the USE for the energy spectrum of the considered system that is valid for any quantum numbers, and 2) to obtain the approximate estimation for the summation on quantum numbers suitable in the whole range of the temperature. It should be pointed out, that for single-dimensional systems the latter problem is less important. It is sufficient to use expression for the energy levels, obtained in terms of the OM, to derive the USE for thermodynamic values and to carry out direct numerical summation in (19). However, for systems with many degrees of freedom such summation is rather complicated numerical problem, therefore we shall discuss the possibility of constructing the USE for summation on quantum states.

First, let us show that the partition function can be identically represented in the operator form as an average over some state vector. For that purpose one can use the basic set of the state vectors considered formally as the eigenfunctions of some excitation number operator

$$\hat{n}\,|n\rangle = n\,|n\rangle\,. \tag{20}$$

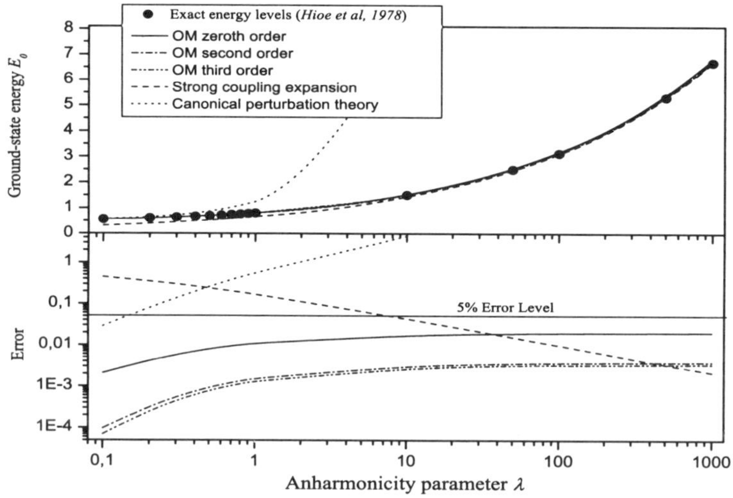

Figure 1. Various estimations for the ground-state energy of the quantum anharmonic oscillator in dependence on the anharmonicity parameter λ.

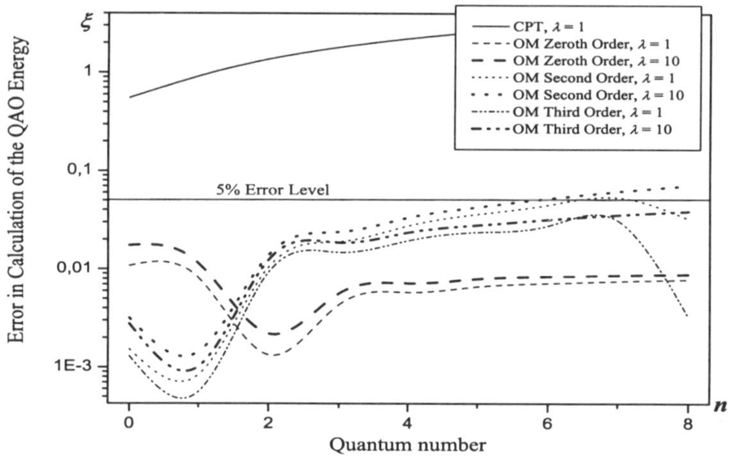

Figure 2. Errors in calculation of the energy levels of the QAO in terms of various approximations considered as the functions of quantum numbers.

Then one can rewrite equation (19) as

$$Z(\beta) = \langle \beta^* | \exp[-\beta E(\hat{n}) + 2\beta^* \hat{n} - \ln N(\beta^*)] \, |\beta^*\rangle \,. \tag{21}$$

Here $|\beta^*\rangle$ is the normalized "trial" state vector that depends on an arbitrary parameter β^* having the physical meaning of the effective inverse temperature for the equilibrium system of the excitations considered in (19). The actual value of β^* should be defined from the condition of the best approximation for the partition function of the real system. By definition, vector $|\beta^*\rangle$ is of the following form

$$|\beta^*\rangle = \sqrt{N} \sum_{n=0}^{\infty} \sqrt{g_n} \exp[-\beta^* n] \, |n\rangle \,; \quad N = \left[\sum_{n=0}^{\infty} g_n \exp(-2\beta^* n) \right]^{-1} . \tag{22}$$

It is necessary to stress that the operational representation for the partition function, analogous to (21), can be obtained by choosing a more complicated trial vector instead of (22). Such a vector should take into account specific features of the problem under consideration. In every particular case, such a choice may lead to a more accurate zeroth order approximation, but a simple vector (22) used in this paper allows us to construct the universal scheme for calculating the USE that is valid for an arbitrary quantum system.

Now, we can use the cumulant expansion (CE) for calculating the average value (21). Recall that the CE is valid for an arbitrary exponential operator when averaging over the normalized state vector[53]

$$\left\langle \exp \hat{A} \right\rangle = \exp \left[\sum_{n=1}^{\infty} \frac{K_n}{n!} \right], \tag{23}$$

where the cumulants K_n are expressed in terms of the moments of the operator $\hat{A}$. This expansion as a whole is an accurate one but at the same time every cumulant corresponds to the partial summation of a usual power series. As follows from[53], the first few terms in equation (23) are given by

$$K_1 = \left\langle \hat{A} \right\rangle; \; K_2 = \left\langle \hat{A}^2 \right\rangle - \left\langle \hat{A} \right\rangle^2; \; K_3 = \left\langle \hat{A}^3 \right\rangle - 3 \left\langle \hat{A} \right\rangle \left\langle \hat{A}^2 \right\rangle + 2 \left\langle \hat{A} \right\rangle^3 . \tag{24}$$

The general procedure of the CE can be applied to the partition function in the form (21). Then successive approximations for the partition function will be calculated taking into account the corresponding cumulants. In particular, usage of the first and second cumulants leads to the following approximation

$$\begin{aligned} Z \simeq Z(\beta, \beta^*) =& Z_0(\beta, \beta^*) Z_1(\beta, \beta^*) = \\ &= \exp\left[\langle A(\hat{n})\rangle - \ln N + \frac{1}{2}\left(\langle A^2(\hat{n})\rangle - \langle A(\hat{n})\rangle^2\right)\right], \\ A(\hat{n}) =& -\beta E(\hat{n}) + 2\beta^* \hat{n}. \end{aligned} \tag{25}$$

It is obvious that with any fixed number of cumulants in formula (25) the partition function $Z(\beta, \beta^*)$ depends on the artificial value β^*, which can be considered as a variational parameter, leading to the best approximation of the CE in some order. Particularly, the zeroth order approximation $Z_0(\beta, \beta^*)$ and the first correction $Z_1(\beta, \beta^*)$ are defined by the formulae

$$\begin{aligned} Z_0 &= \exp\left[-\beta \bar{E}(\beta^*) + 2\beta^* \bar{n} - \ln N(\beta^*)\right]; \\ Z_1 &= \exp\{\frac{1}{2}\left[\beta^2(\overline{E^2} - \bar{E}^2) - 4\beta\beta^*(\overline{En} - \bar{E}\bar{n}) + 4\beta^{*2}(\overline{n^2} - \bar{n}^2)\right]\}. \end{aligned} \tag{26}$$

Here, all values are averaged over the "trial" distribution function corresponding to the ensemble of the excitations (19) with the degeneracy of states g_n and the effective inverse temperature β^*, as for example

$$\bar{E}(\beta^*) = N\sum_{n=0}^{\infty} g_n E_n e^{-2\beta^* n}; \; \bar{n}(\beta^*) = N\sum_{n=0}^{\infty} g_n n e^{-2\beta^* n} = \frac{N\partial N}{2\partial\beta^*}. \tag{27}$$

Then the free energy calculated with the considered accuracy takes the following form

$$F(\beta) = -\frac{1}{\beta}\ln Z(\beta) \simeq F^{(0)} + F^{(1)} + \cdots = -\frac{1}{\beta}(\ln Z_0(\beta^*, \beta) + \ln Z_1(\beta^*, \beta) + \cdots). \tag{28}$$

To obtain the estimation for the partition function of the given energy spectrum, the variational parameter β^* considered as the function of the real temperature β can be defined from the condition of the best approximation in the zeroth order of the CE. For this purpose one has to solve the equation

$$\frac{\partial Z_0}{\partial \beta^*} = 0; \Rightarrow \beta\frac{\partial \bar{E}}{\partial \beta^*} - 2\bar{n} - 2\beta^*\frac{\partial \bar{n}}{\partial \beta^*} + \frac{\partial N}{N\partial \beta^*} = 0. \tag{29}$$

One can use the energy spectrum, obtained in terms of the OM to calculate the moments $\bar{E}, \overline{E^2}$ with the distribution function (27). Then the totality of the expressions (8), (13), (26) and (29) lead to the USE for the thermodynamic characteristics of a system with the accuracy to the terms of second order in approximation of the OM and the CE. In some cases, the additional approximation in these equations can also be applied. This approximation leads to the worst accuracy, but permits one to simplify significantly the evaluations. It is based on the following.

Recall that condition (13) corresponds to the optimal choice of the parameter ω_n for the best approximation for the energy level with quantum number n in the zeroth order of the OM. But the convergence of the successive approximations of the OM exists for all values of ω[30]. At the same time, the partition function and the free energy are the integral characteristics in the space of quantum numbers of a system. Thus, it is possible to expect that in order to construct the USE for these values it is sufficient to choose the parameter ω as the same value for all quantum numbers, and to use it for defining the best approximation condition for the free energy in the zeroth order of the OM and the CE. This means that for the calculation of the moments of the energy spectrum one has to take into account only the explicit dependence of the approximate energy spectrum on the quantum number n and disregard a more complicated dependence related to the choice of ω_n from the condition (13). Hence, the values

$$\bar{E}(\beta^*,\omega) \simeq N\sum_{n=0}^{\infty} g_n[E_n^{(0)}(\omega)+\Delta E_n(\omega)]\exp[-2\beta^* n];$$
$$\overline{E^2}(\beta^*,\omega) \simeq N\sum_{n=0}^{\infty} g_n[E_n^{(0)}(\omega)]^2\exp[-2\beta^* n] \qquad (30)$$

depend on both the parameter β^* and the variational parameter ω, which should be taken from the minimum condition for the free energy. As a result, instead of a system of algebraic equations (26) for ω_n it is required to solve only one equation for the parameter ω

$$\frac{\partial}{\partial\omega}\sum_{n=0}^{\infty} g_n[E_n^{(0)}(\omega)]\exp[-2\beta^* n] = 0. \qquad (31)$$

The physical meaning of this equation is that the variational parameter ω should be chosen from the condition of the best approximation of the energy level with maximum occupancy corresponding to a given temperature.

5. Statistics of quantum anharmonic oscillator

The problem considered in this section illustrates the possibilities of the discussed method in the case when the approximate calculations are used for obtaining the USE for both the eigenvalues and summation over the states (see section 4 for general procedure). For that purpose, we use the equilibrium statistical system of the quantum anharmonic oscillators (QAO).

Let us consider the QAO with the Hamiltonian (15) and parameter $\mu = 0$. This can be done without loss of generality, since the parameter μ can always be excluded from the Hamiltonian by the scaling transformation of the coordinate. We thus consider

$$\hat{H} = \frac{1}{2}(\hat{p}^2 + \hat{x}^2) + \lambda\hat{x}^4. \tag{32}$$

The USE for this system is defined by formulae (20). When obtaining the USE for the partition function and the free energy it is interesting to compare the results, which can be obtained by means of approximate calculations in two different ways: 1) direct numerical summation over the QAO states taking into account the OM energy levels and 2) evaluation of the USE in analytical form by means of the approximate summation over the states based on the cumulant expansion. To estimate the accuracy of our results we shall use direct numerical calculation of the partition function $((Z_A^{num})$ for the QAO ($g_n = 1$) taking into account 8 energy levels of the QAO, obtained numerically in paper[55]. It is essential to stress, that in real calculations with any fixed number of the QAO levels in the summation over the states, the numerical calculations lose their accuracy in the limit case of the high temperature ($\beta \to 0$), when the whole energy spectrum becomes essential. In these cases one can use for comparison the asymptotic expressions for the partition function which can be obtained analytically:

$$\begin{aligned} Z_A(\beta,\lambda) &= \sum_{n=0}^{\infty} \exp[-\beta E_n(\lambda)] \simeq 1 + e^{-\beta E_0(\lambda)}, \quad \beta \gg 1; \\ Z_A(\beta,\lambda) &\simeq \frac{e^{-\beta/2}}{1-e^{-\beta}}\left[1 + \frac{3\lambda e^{-\beta}}{(1-e^{-\beta})^2}\right], \quad \frac{\lambda}{(1-e^{-\beta})^2} \ll 1; \\ Z_A(\beta,\lambda) &\simeq \sum_{n=0}^{\infty} \exp[-\beta\lambda^{1/3} b_n], \quad \lambda \gg 1. \end{aligned} \tag{33}$$

Here b_n are known numerical coefficients of the asymptotic expansion for the QAO energy levels in the strong coupling limit[55]. The analytical approximate expression for them in the OM zeroth order was obtained in paper[30]

$$b_n \simeq (\frac{3}{4})^{4/3}[\frac{1 + 2n + 2n^2}{(1+2n)^2}]^{1/3}. \tag{34}$$

Figure 3 represents the results of comparison of the free energy $F_A^{num} = -1/\beta \ln(Z_A^{num})$ with various approximate expressions for it, that are defined as the following

$$F_A^{(0)}(\beta,\lambda) = -\frac{1}{\beta}\ln\left[\sum_{n=0}^{\infty}\exp[-\beta E_n^{(0)}(\lambda)]\right];$$
$$F_A^{(1)}(\beta,\lambda) = -\frac{1}{\beta}\ln\left[\sum_{n=0}^{\infty}\exp[-\beta(E_n^{(0)}+\Delta E_n)]\right];$$
$$F_A^{(01)}(\beta,\lambda) = -\frac{1}{\beta}\ln\left[\sum_{n=0}^{\infty}\exp[-\beta E_n^{(0)}](1-\beta\Delta E_n)\right], \quad (35)$$

where the energy is calculated in different orders of the OM. We also compare our approximations with the results of the canonical thermodynamic perturbation theory and the direct summation over the exact eigenvalues for the first 8 levels of QAO calculated numerically in[55].

The presented results show that the zeroth order approximation $F_A^{(0)}$ satisfies definitions 1 and 2 and defines the USE for the pointed values with a relative error $\xi^{(0)} \simeq 0.1$. If one takes into account the subsequent iteration of the OM, the accuracy of the estimation improves. At the same time, it is necessary to stress that the expansion of the exponent in a rather small value ΔE_n used in expression $Z_A^{(01)}(\beta,\lambda)$ breaks the USE conditions in the low temperature limit ($\beta \to \infty$). This means that the improvement of the accuracy of the zeroth order approximation for the free energy in the subsequent iterations of the considered method has in general nonadditive character, unlike the thermodynamical CPT.

Let us consider now the approximate expression for the partition function obtained in terms of the zeroth order approximation of both the OM and the CE

$$Z_A^{(0C)}(\beta,\lambda) = \exp[\varphi(\beta,\beta^*,\lambda,\omega)];$$
$$\varphi(\beta,\beta^*,\lambda,\omega) = -\beta\overline{E_n^{(0)}(\beta^*,\omega,\lambda)} + 2\beta^*\bar{n}(\beta^*) - \ln N;$$
$$N = (1-q); \bar{n} = \frac{q}{(1-q)}; \overline{n^2} = \frac{q+q^2}{(1-q)^2}; q = e^{-2\beta^*};$$
$$\overline{E_n^{(0)}} = \frac{1}{4\omega}(\omega^2+1)(2\bar{n}+1) + \frac{3\lambda}{4\omega^2}(1+2\bar{n}+2\overline{n^2}). \quad (36)$$

Variational parameters q and ω are defined from the minimum conditions for function φ, which lead to the following equations

$$\frac{\partial\varphi}{\partial q} = 0, \Rightarrow -\beta\frac{\omega^2+1}{2\omega(1-q)} + \frac{3\lambda(1+q)}{\omega^2(1-q)^2} + \frac{2\ln q}{1-q} + 1 = 0;$$
$$\frac{\partial\varphi}{\partial\omega} = 0, \Rightarrow \omega^3 - \omega - 6\lambda\frac{1+q}{1-q} = 0. \quad (37)$$

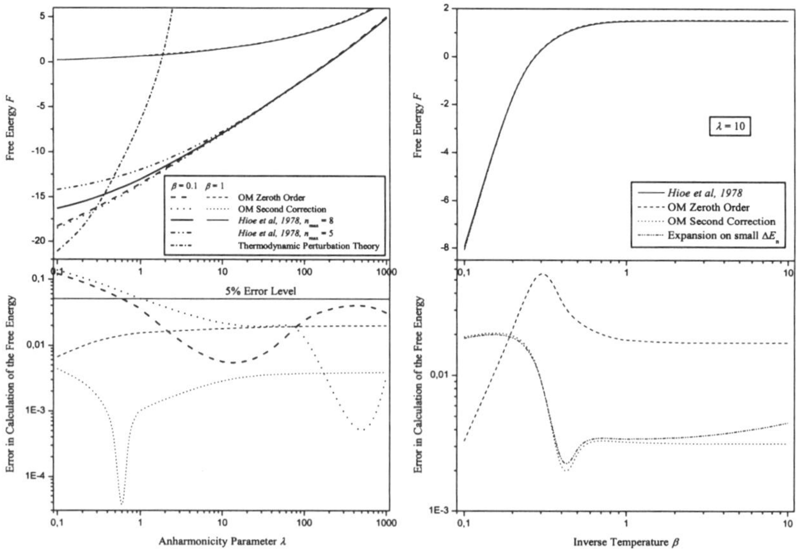

Figure 3. The free energy of the QAO using the OM zeroth and second orders and errors of these approximations in dependence on the anharmonicity parameter λ (left panel) and on the inverse temperature β (right panel) in comparison with the results of the numerical calculations. Lines, corresponding to $\beta = 1$ on the upper part of the left panel, and all lines on the upper part of the right panel almost coincide. Results obtained in terms of the thermodynamic perturbation theory are also represented on the upper part of the left panel (they are not uniformly suitable and can not be shown on the right panel).

It is convenient not to solve numerically equations (37), but to consider $\lambda(q,\omega)$ and $\beta(q,\omega)$ as functions of variables q and ω. Together with definition (36) these functions assign in parametric form the functions $Z_A^{(0C)}(\beta,\lambda)$ and $F_A^{(0C)}(\beta,\lambda) = -1/\beta \ln Z_A^{(0C)}(\beta,\lambda)$. The results of such a calculation and its comparison with the numerically calculated free energy are shown in Fig.4. As can be seen, the application of the CE to summation over the states keeps the zeroth order approximation of the method under consideration satisfying the criteria of the USE with a relative error $\xi^{(0)} \simeq 0.1$.

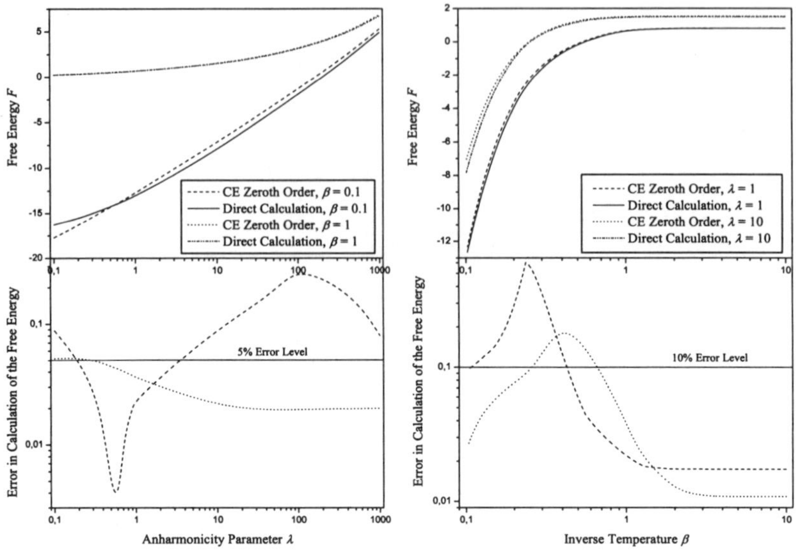

Figure 4. Approximation for the free energy of the QAO using the CE zeroth order and error of such an approximation in dependence on the anharmonicity parameter λ for various values of inverse temperature β (left panel) and on the inverse temperature β for various values of the anharmonicity parameter λ (right panel). Lines 3 and 4 on the upper part of the left panel almost coincide.

6. Conclusions

The review on the OM is given. The general procedure of the OM and its applications to a series of model and physical systems is presented. Besides, there are lots of problems, that were considered in terms of the OM (e.g. system of coupled anharmonic oscillators[32], the USE for the density matrix, solving differential equations etc.) which are not considered in this paper because of the lack of space. The generalization of the OM to the case of quantum statistics based on the cumulant expansion is also given. It is shown, that the obtained approximations describe correctly the essential peculiarities of the considered systems in the zeroth orders, and the following approximations converge to the exact solutions, i.d. the presented approximate scheme satisfies the criteria of the USE.

Bibliography

1. E. P. Wigner, *Symmetries and Reflections* (Indiana University Press, Bloomington/London, 1970).
2. P. Antosik, J. Mikusinski and R. Sikorski, *Theory of Distributions* (Elsevier Sci., Amsterdam, 1973).
3. K. Huang, C. N. Yang and J. M. Luttinger, *Phys. Rev* **105**, 776 (1957).
4. N. N. Bogoliubov, B. I. Sadovnikov and A. S. Shumovsky, *Mathematical Methods for Statistical Mechanics of Model Systems* (Nauka, Moskow, 1989).
5. N. N. Bogoliubov, *Physica S* **26**, 1 (1960).
6. M. L. Goldberger and K. M. Watson, *Collision Theory* (Wiley, New York/London, 1964).
7. V. G. Baryshevsky, *Nuclear Optics of Polarized Media* (Belarusian University Press, Minsk, 1976).
8. N. H. March, W. H. Young and S. Sampanthar, *The Many-Body Problem in Quantum Mechanics* (Cambridge University Press, Cambridge, 1967).
9. R. P. Feynman and A. R. Hibbs, *Quantum Mechanics and Path Integrals* (McGraw-Hill. New York, 1965).
10. M. Creutz, *Quarks, Gluons and Lattice* (Cambridge University Press, Cambridge, 1983).
11. P. M. Morse and G. Feshbach, *Methods of Theoretical Physics* (McGraw-Hill, New York, 1953).
12. L. I. Komarov, S. T. Zavtrak and I. D. Feranchuk, *Theoret. and Math. Phys.* **47**, 55 (1981).
13. B.-G. Englert, *Semiclassical Theory of Atoms* (Springer-Verlag, Berlin, 1988).
14. *Statistical Mechanics and Field Theory: Mathematical Aspect; Proc. Internat. Conf.* (Springer-Verlag, Berlin, 1985).
15. G. N. Hardy, *Divergent Series* (Oxford University Press, Oxford, 1949).
16. B. Simon, *Ann. of Phys. (N.Y.)* **58**, 79 (1970).
17. V. I. Yukalov and E. P. Yukalova, *Chaos, Solitons and Fractals* **14**, 839 (2002).
18. I. D. Feranchuk and L. I. Komarov, *Phys. Lett. A* **88**, 211 (1982).
19. I. D. Feranchuk and L. I. Komarov, *J. Phys. C:* **15**, 1965 (1982).
20. I. D. Feranchuk, S. I. Fisher and L. I. Komarov, *J. Phys. C: Solid State Phys.* **17**, 4309 (1984).
21. I. D. Feranchuk and L. I. Komarov, *Methods of Atomic Calculations* (Institute of Spectroscopy, Moskow, 1983).
22. I. D. Feranchuk and L. I. Komarov, *J. Phys. A: Math. Gen.* **17**, 3111 (1984).
23. C. Z. An and I. D. Feranchuk, *Mol. Phys.* **64**, 589 (1988).
24. I. D. Feranchuk and L. X. Hai, *Phys. Lett. A* **137**, 385 (1989).
25. L. I. Komarov and T. S. Romanova, *J. Phys. B: At. Mol. Phys.* **18**, 859 (1985).
26. I. D. Feranchuk, S. I. Fisher and L. I. Komarov, *J. Phys. C: Solid State Phys.* **18**, 5083 (2985).
27. C. Z. An, I. D. Feranchuk, L. I. Komarov and L. S. Nakhamchik, *J. Phys.*

A: Math. Gen. **19**, 1583 (1986).
28. C. Z. An, I. D. Feranchuk and L. I. Komarov, *Phys. Lett. A* **125** 123 (1987).
29. I. D. Feranchuk, L. I. Komarov and I. V. Nichipor, *J. Phys. A: Math. Gen.* **20**, 3849 (1987).
30. I. D. Feranchuk, L. I. Komarov, I. V. Nichipor and A. P. Ulyanenkov, *Ann. of Phys. (N.Y.)* **238**, 370 (1995).
31. I. D. Feranchuk, L. I. Komarov and A. P. Ulyanenkov, *J. of Phys. A: Math. and Gen.* **29**, 4035 (1996).
32. I. D. Feranchuk and A. L. Tolstik, *J. of Phys. A: Math. and Gen.* **32**, 2115 (1999).
33. I. D. Feranchuk, L. I. Gurskii, L. I. Komarov, O. M. Lugovskaya, F. Burgäzy and A. P. Ulyanenkov, *Acta Cryst.* **A58**, 370 (2002).
34. I. D. Feranchuk, A. P. Ulyanenkov and V. S. Kuzmin, *Chem. Phys.* **157**, 61 (1991).
35. I. D. Feranchuk and A. A. Ivanov, in *Fluctuating Paths and Fields*, Eds. W. Janke, A. Pelster, H.-J. Schmidt and M. Bachmann (World Scientific, Singapore, 2001), p. 327.
36. F. M. Fernandez and E. A. Castro, *Phys. Lett. A* **88**, 4 (1982).
37. F. M. Fernandez and E. A. Castro, *Phys. Lett. A* **91**, 339 (1982).
38. F. M. Fernandez, A. M. Meson and E. A. Castro, *Phys. Lett. A* **104**, 401 (1984).
39. F. M. Fernandez, A. M. Meson and E. A. Castro, *Phys. Lett. A* **111**, 104 (1985).
40. F. M. Fernandez, A. M. Meson and E. A. Castro, *Phys. Lett. A* **112**, 107 (1985).
41. F. M. Fernandez, E. A. Castro and A. M. Meson, *Mol. Phys.* **58**, 365 (1986).
42. C. C. Gerry and S. Silverman, *Phys. Lett. A* **95**, 481 (1983).
43. W. Witschel, *Phys. Lett. A* **97**, 315 (1983).
44. H. Mitter and K. Yamazaki, J. Phys. A: Math. Gen. **17**, 1215 (1984).
45. B. Bidyut, B. Sikha and R. Raykumar, *Fizika* **16**, 161 (1984).
46. I. G. Holliday and P. Suranyi, *Phys. Lett. B* **85**, 421 (1979).
47. W. E. Caswell, *Ann. Phys. (N.Y.)* **123**, 153 (1979).
48. I. G. Holliday and P. Suranyi, *Phys. Rev. D* **21**, 1529 (1980).
49. I. K. Dmitrieva and G. I. Plindov, Phys. Lett. A **79**, 47 (1980).
50. K. Bhattacharyya, *J. Phys. B: At. Mol. Phys.* **14**, 783 (1981).
51. J. Austin and J. Killingbeck, *J. Phys. A: Math. Gen.* **15**, L433 (1982).
52. C. K. Au, G. W. Roders and J. Aharonov, *Phys. Lett. A* **95**, 287 (1983).
53. H. Cramer, *Mathematical Methods of Statistics* (Princeton University Press, Princeton, 1951).
54. L. D. Landau and E. M. Lifsitz, *Quantum Mechanics. Nonrelativistic Theory* (Nauka, Moskow, 1985).
55. F. T. Hioe, D. MacMillen and F. T. Montroll, *Phys. Rep.* **43**, 307 (1978).
56. *Handbook of Mathematical Functions*, Eds. M. Abramowitz and A. Stegun. (National Bureau of Standards, USA, 1964).
57. I. D. Feranchuk, L. I. Komarov and A. P. Ulyanenkov, *J. Phys. A: Math. Gen* **29**, 4035 (1996).

AROUND OBSERVABLES AND RADIATIVE CORRECTIONS

I. Akushevich[a]
North Carolina Central University, Durham, NC 27707, USA
and
Jefferson Lab, Newport News, VA 23606, USA

A. Ilyichev, N. Shumeiko
National Center of Particle and High Energy Physics,
153 M. Bogdanovich Str., 220040 Minsk, Belarus
E-mail: ily@hep.by

Recent developments in the field of radiative effects in polarized lepton-nuclear scattering are reviewed. The processes of inclusive, semi-inclusive, diffractive and elastic scattering are considered. The explicit formulae obtained within the covariant approach are discussed. FORTRAN codes POLRAD, RADGEN, HAPRAD, DIFFRAD, MASCARAD, ELRADGEN and EXCLURAD created on the basis of the formulae are briefly described. Data analysis applications of the current experiments on lepton-nuclear scattering at CERN, DESY, SLAC and TJNAF are illustrated by numerical results.

1. Introduction

Starting from the late 1960s the main source of information on the internal nature of nucleon is lepton nuclei scattering. Nowadays an experimental design of modern detectors opens a possibility for a large set of experimental data within a short period of time. As a result the calculation of background radiative effects or the procedure of radiative corrections (RC) becomes quite important thing.

While considering lepton nucleon scattering it should be noted that

[a]on leave of absence from the National Center of Particle and High Energy Physics, 153, M. Bogdanovich Str., 220040 Minsk, Belarus

the main contribution to the observables gives subprocess with one photon exchange graphs that is called Born process. However all sets of experimental data consist of not only such contribution, but of other processes appearing during the measurement which are called background ones because of their small contributions to the observables. Therefore the main task of data processing in experiments on lepton nucleon scattering is the extraction of the one-photon exchange contribution. It leads to the necessity to cancel the background events from all data sets. When it cannot be made by experimental methods the correspond contributions can be calculated in the frame of some theory. Usually a such kind of contribution depends on the quantities (the structure function, asymmetries), that are observables in the given experiment. As a result we come to the necessity of the iteration procedure application for the extraction of the observables on Born level from the full set of experimental data. This procedure is usually called as the procedure of RC of experimental data.

There are two basic methods of calculation of model independent QED radiative correction. First one is connected with introducing of an artificial parameter separating the momentum phase space into soft and hard parts. One can find a classical review introducing this formalism in ref.[1,2]. However the presence of the artificial parameter is a disadvantage of this method. For the soft photon part the calculation is performed in such approximation when the photon energy is considered to be small with respect to all momenta and masses in the problem. So this parameter should be chosen as small as possible to reduce the region evaluated approximately. From the other side it cannot be chosen too small because of possible numerical instabilities in calculating of hard-photon emission. Calculation within the approach of Mo and Tsai was performed only for the case of unpolarized deep inelastic scattering. In the end of seventies Bardin and Shumeiko developed a covariant approach [3] of extraction and cancellation of infrared divergence without introducing this artificial parameter. Then this approach was applied to the calculation of electromagnetic correction to the lepton current for deep inelastic scattering (DIS)[4–12] as well as for some approximation and taking into account the contribution of net soft photon emission. The exact expressions for the lowest order QED RC to polarized DIS of polarized leptons by polarized light nuclei including targets of spin 1 such as deuterium are presented in[13]. Basing on these formulae FORTRAN code POLRAD[14] are constructed. Besides, some others tasks such as Monte Carlo generator RADGEN[15] was made, RC both to diffractive vector meson electroproduction[16,17] and to spin-density matrix elements in the exclusive vector meson production[18] by this method were solved. Recently Bardin-Shumeiko approach has being applied to experiment on polarized elastic electron-proton scattering[19–21] and exclusive pion electroproduction[22] at TJNAF. The comprehensive analysis of some results obtained by[1,2] and[3] was made in[23,15].

The formulae of[3] as well as[24] allow along with using the parton model to estimate the value of electromagnetic correction to the hadronic current in unpolarized[25] and polarized [26,27] lepton nucleon DIS.

The calculation of RC within the structure function method was suggest by Kuraev and Fadin[28] and applied for DIS[29,30]. This method was used for the calculation of Compton tensor with heavy photon in the case of unpolarized[31] and longitudinal-polarized[32] fermions. The calculation of LO and NLO corrections to radiative tail from elastic peak in DIS can be found in[33].

In the article which is basing on Ref.[7,11,15–22]some radiative effects in polarized lepton by polarized light nuclei scattering calculated within Bardin Shumeiko covariant approach is reviewed.

2. Corrections to the cross sections and asymmetries in DIS

2.1. *Semi-inclusive physics*

We calculate the radiative corrections to data of semi-inclusive polarized experiments

$$e(k_1) + N(p) \longrightarrow e'(k_2) + h(p_h) + X(p_x) \tag{1}$$

when a hadron h is detected in coincidence with the outgoing lepton. In this case the cross section depends on five kinematic variables which can be chosen as

$$x,\ y; \phi_h, t = (q - p_h)^2, \quad z = p_h p/pq, \quad q = k_1 - k_2, \tag{2}$$

where x and y are usual scaling variables, ϕ_h is an angle between planes $(\vec{k}_1, \vec{k}_2)$ and $(\vec{q}, \vec{p}_h)$ in the rest frame $(p = (M, \vec{0}))$. Instead of $t-$dependence we will also consider the cross section as a function of transversal momentum p_t defined in (A.8) of[11].

Notice that there are two tasks for this process. The first task corresponds the description of semi-inclusive DIS in the frame of quark-parton model. In this way the cross section depends on three variables x, y and z. However for the real situation the integration region is limited by the experimental cuts. To deal with them on the angles of registered hadron a special procedure was developed and cross section describing such situation is defined by five variables: three standard x, y, z and two additional p_t, ϕ_h. The relation between these two cross sections can be established as follows:

$$\frac{d\sigma_R}{dxdydz} = \int \frac{1}{2\pi}\frac{d^3k}{2k_0}\frac{d\tilde{z}}{dz}dp_t d\phi_h \frac{d\sigma}{dxdydzdp_t d\phi_H}, \\ \times\theta(\sin^2\vartheta_{max} - \sin^2\vartheta)\theta(\sin^2\vartheta - \sin^2\vartheta_{min}) \qquad (3)$$

if θ–function are removed. Here ϑ is the angle between the beam direction ($\vec{k}_1$) and hadron momentum in the lab frame.

Within quark-parton model for the Born cross section of longitudinal polarized semi-inclusive DIS we have

$$\sigma_0^s \equiv \frac{d^3\sigma_0}{dxdydz} = \frac{2\pi\alpha^2}{Sxy}[F_0^u\Sigma^+(x,z) + P_L P_N F_0^p \Sigma^-(x,z)] \qquad (4)$$

where $F_0^u = 2(1/y - 1 - \mu_N x) + y$, $F_0^p = y - 2$, $\mu_N = M^2/S$ and the quantities

$$\Sigma^{+(-)}(x,z) = \sum_q e_q^2[f_q^+(x) \pm f_q^-(x)]D_q^H(z) \qquad (5)$$

is the ordinary for QPM combination of the distribution functions $f_q^{+(-)}(x)$ for the quark of flower q polarized (anti)parallel to the nucleon polarization, and of the fragmentation functions $D_q^h(z)$ of the quark q into the hadron h, e_q being the quark charge in units of elementary charge.

As it was mentioned above there are three radiative tails give contribution to DIS: inelastic, elastic and quasielastic. However, in semi-inclusive process the transfer energy is above pion threshold, so in RC calculation one have to take into account only σ^{in} and σ^v. The lowest order QED correction was calculated in Ref.[7] and presented in[14] as the sum of factorizing and non-factorizing infrared free parts

$$\sigma_{EM} \equiv \frac{d^3\sigma_{EM}}{dxdydz} = \frac{\alpha}{\pi}\delta_{VR}\sigma_0^s + \sigma_R^F, \qquad (6)$$

where the first one is the sum of infrared divergence separated from radiative tail and the additional virtual photon contribution: $\delta_{VR} = \delta_V + \delta_R^{IR}$.

The finite part of (4) has the form

$$\sigma_R^F = \frac{2\alpha^3}{S}\int\limits_{t_{1d}}^{t_{1u}} dt_1 \int\limits_{t_{2d}}^{t_{2u}} dt_2 \{\frac{y^2}{x_t^2 y_t}[F_R^u\Sigma^+(\tilde{x},\tilde{z}) + P_L P_N F_R^p \Sigma^-(\tilde{x},\tilde{z})] - \\ - \frac{\theta(t_{1i} - t_1)}{xy}[F_{IR}^u\Sigma^+(x,z) + P_L P_N F_{IR}^p\Sigma^-(x,z)]\}, \qquad (7)$$

The definition of the rest quantities can be found in[14].

In order to write five-dimensional cross section it is useful to take the some scalar products with p_h in the form

$$\frac{1}{2}V_{1,2} = k_{1,2}p_h = a^{1,2} + b\cos\phi_h, \tag{8}$$
$$\frac{1}{2}\mu R = kp_h = R(a^k + b^k(\cos\phi_h\cos\phi_k + \sin\phi_h\sin\phi_k)).$$

where $\mu = kp_h/kp$ and ϕ_k is the rest frame angle between planes $(\vec{k}_1, \vec{k}_2)$ and $(\vec{q}, \vec{k})$. We will use $a^{\pm} = a^2 \pm a^1$. The explicit expressions for coefficients are given in[11].

As a result the five-dimensional cross section of unpolarized DIS on the Born level has the following dependence on ϕ_h:

$$\sigma_0 = \frac{d\sigma_0}{dxdydzdp_t^2d\phi_h} = \frac{N}{Q^4}(A + \cos\phi_h A^c + \cos 2\phi_h A^{cc}), \tag{9}$$

where $N = \alpha^2 y S_x/\sqrt{\lambda_Q}$. The coefficients A, A^c and A^{cc} do not depend on ϕ_h more and they have the form

$$\begin{aligned}
A \ &= 2Q^2\mathcal{H}_1 + (SX - M^2Q^2)\mathcal{H}_2 + (4a^1a^2 + 2b^2 - M_h^2Q^2)\mathcal{H}_3 \\
&\quad + (2Xa^1 + 2Sa^2 - zS_xQ^2)\mathcal{H}_4, \\
A^c \ &= 2b(2a^+\mathcal{H}_3 + S_p\mathcal{H}_4), \\
A^{cc} &= 2b^2\mathcal{H}_3.
\end{aligned} \tag{10}$$

The model for structure functions $\mathcal{H}_i$ can be constructed on basis of results of the paper Mulders and Tangerman[35].

When integrated over the kinematic variables ϕ_h and p_t the cross section (9) coincides with the unpolarized part of semi-inclusive cross section calculated within quark parton model (4).

The cross section that takes into account radiative effects can be written as

$$\sigma_{obs} = \sigma_0 e^{\delta_{inf}}(1 + \delta_{VR} + \delta_{vac}) + \sigma_F. \tag{11}$$

Here the corrections δ_{inf} and δ_{vac} come from radiation of soft photons[12] and effects of vacuum polarization[b]. The correction δ_{VR} is infrared free sum of factorized parts of real and virtual photon radiation. These quantities

[b]There are explicit formulae for leptonic contribution to vacuum polarization effect (see [13] for example) and parameterization of hadronic one [34].

are given by the following expressions

$$\delta_{VR} = \frac{\alpha}{\pi}\left(\frac{3}{2}l_m - 2 - \frac{1}{2}\ln^2\frac{X'}{S'} + \mathrm{Li}_2\frac{S'X' - Q^2p_x^2}{S'X'} - \frac{\pi^2}{6}\right),$$

$$\delta_{inf} = \frac{\alpha}{\pi}(l_m - 1)\ln\frac{(p_x^2 - (M+m_\pi)^2)^2}{S'X'},$$

$$\delta_{vac} = \delta_{vac}^{lept} + \delta_{vac}^{hadr}, \tag{12}$$

where $S' = X + Q^2 - V_2$, $X' = S - Q^2 - V_1$, $l_m = \ln Q^2/m^2$ and Li_2 is Spence function or dilogarithm.

The contribution of radiative tail has the form

$$\sigma_F = -\frac{\alpha N}{2\pi}\int_0^{2\pi} d\phi_k \int_{\tau_{min}}^{\tau_{max}} d\tau \sum_{i=1}^{4}\sum_{j=1}^{3}\theta_{ij}(\tau,\phi_k) \times \int_0^{R_{max}} dR R^{j-2}\left[\frac{\mathcal{H}_i}{(Q^2+R\tau)^2} - \delta_j\frac{\mathcal{H}_i^0}{Q^4}\right]. \tag{13}$$

Here $2M^2\tau_{max,min} = S_x \pm \sqrt{\lambda_Q}$ and $R_{max} = (M_x^2 - (M+m_\pi)^2)/(1+\tau-\mu)$, δ_j=1 for j=1 and δ_j=0 otherwise. The explicit formulae for functions $\theta(\tau,\phi_k)$ can be found in[11]. The structure functions $\mathcal{H}_i^0$ have to be calculated for Born kinematics, but $\mathcal{H}_i$ is calculated in terms of tilde variables introducing by

$$\tilde{Q}^2 = (q-k)^2 = Q^2 + R\tau,\ \tilde{W}^2 = (p+q-k)^2 = W^2 - R(1+\tau),$$
$$\tilde{t} = (q-k-p_h)^2 = t + R(\tau-\mu),\ \tilde{M}_x^2 = \tilde{p}_x{}^2 = M_x^2 + R(1+\tau-\mu). \tag{14}$$

Table 1. The results for RC factors to three dimensional semi-inclusive cross section obtained by FORTRAN codes SIRAD and HAPRAD (see text for further explanations). Kinematic points are taken from the Table 1 in ref..

x	y	Q^2 GeV^2	z	SIRAD	HAPRAD					
					without cuts			with cuts		
					$\mathcal{G}_1$	$\mathcal{G}_1'$	$\mathcal{G}_2$	$\mathcal{G}_1$	$\mathcal{G}_1'$	$\mathcal{G}_2$
0.038	0.677	1.33	0.25	1.029	1.033	1.024	0.982	1.041	1.025	0.985
0.062	0.567	1.82	0.35	0.996	0.989	0.989	0.947	0.989	0.980	0.951
0.092	0.529	2.52	0.45	0.970	0.961	0.961	0.934	0.961	0.956	0.936
0.131	0.499	3.38	0.55	0.945	0.936	0.933	0.912	0.934	0.931	0.906
0.198	0.476	4.88	0.65	0.918	0.902	0.902	0.889	0.897	0.897	0.881

2.2. *FORTRAN codes POLRAD 2.0 and HAPRAD*

Using the results presented in the previous subsections two FORTRAN code POLRAD 2.0[14] and HAPRAD[11] have been created. Here some results obtained by these codes are presented.

The program POLRAD 2.0 consists of two patches. The first one which itself is called POLRAD describes RC to inclusive DIS.

The second patch of POLRAD 2.0 which is called as SIRAD allows us to calculate RC to three-dimension cross section $d\sigma/dxdydz$ of semi-inclusive DIS.

To estimate the RC to the cross section $d^5\sigma/dxdydzdp_t^2d\theta_h$ the special FORTRAN code HAPRAD was developed. Now we show that numerical results for RC to $d\sigma/dxdydz$ reproduced by these two codes which coincide with a good accuracy. It can be seen in the Table 1 where we represent RC to the cross section as it follows from the runs of the codes POLRAD 2.0 and HAPRAD. Since HAPRAD allows us to take into account the kinematic cuts and to use different models for p_t^2-slope, three fits for p_t^2-distribution and the cases with and without experimental cuts were considered (see[11] for detail). We can conclude from this analysis that neither important differences between SIRAD and HAPRAD results, if exponential model for p_t^2-distribution is used, nor dependence on slope parameter model and applying of geometrical cuts are found. However RC takes a negative shift in the model based on power functional. As it is shown in[11] RC depends on steepness of p_t^2 distribution. It is a reason why models like $\delta(p_t^2)$ (QPM, POLRAD 2.0) give larger RC. Within practical RC procedure in concrete measurement of $d\sigma/dxdydz$ the model can be fixed only if the information about p_t^2-distribution is additionally considered.

2.3. *Monte Carlo generator RADGEN 1.0*

In this subsection we present a Monte Carlo generator RADGEN 1.0[15] for the events with a possible radiation of a real photon in DIS on polarized and unpolarized targets. The events are generated in accordance with their contribution to the observed total cross section given by

$$\sigma_{obs} = \sigma_{non-rad}(\Delta) + \sigma_{in}(\Delta) + \sigma_q + \sigma_{el}. \tag{15}$$

The first term $\sigma_{non-rad}(\Delta)$ contains not only the contribution from the Born process but the contributions from loop corrections (σ_v) and from multiple soft photon production with a total energy not exceeding a cut-off parameter Δ. In handling the radiative corrections two approaches have been used, the one developed by Mo and Tsai[1,2] and the other given by Bardin and Shumeiko[3].

Kinematics of the Born (or one photon exchange) process is completely defined by the scattering angle θ and the energy E' of the scattered lepton, two variables which are usually measured. All other inclusive kinematic variables can be expressed in the terms of θ and E'. Having in mind the laboratory frame, i.e. the frame with the target nucleus at rest the relevant variables are

$$\begin{aligned} Q^2 = 4EE' \sin^2 \frac{\theta}{2}, \qquad \nu = E - E', \qquad x = \frac{Q^2}{2M\nu}, \\ y = \frac{\nu}{E}, \qquad W^2 = Q^2(1/x - 1) + M^2, \end{aligned} \tag{16}$$

where E is the beam energy.

The event registered in the detector with certain values of E' and θ for the scattered lepton can either be a non-radiative or a radiative event, i.e. an event containing a real hard radiated photon. For the radiative event there are three additional variables necessary to fix the kinematics of the radiated real photon apart from E' and θ. A possible choice is the photon energy E_γ and the two angles θ_γ and ϕ_γ, where θ_γ is the angle between the real and the virtual photon momenta $\vec{k}$ and $\vec{q} = \vec{k}_1 - \vec{k}_2$ and ϕ_γ is the angle between the planes defined by the momenta $(\vec{k}_1, \vec{k}_2)$ and $(\vec{k}, \vec{q})$. For the events with the radiation of a real photon the kinematic variables describing the virtual photon and which is used to generate the hadronic final state is derived from eq.(16) because the substitution $q \to q - k$ has to be made in their definition. The variables obtained after this substitution will be referred to as 'true' ones:

$$\begin{aligned} W^2_{true} &= W^2 - 2E_\gamma(\nu + M - \sqrt{\nu^2 + Q^2} \cos\theta_\gamma), \; \nu_{true} = \nu - E_\gamma, \\ Q^2_{true} &= Q^2 + 2E_\gamma(\nu - \sqrt{\nu^2 + Q^2} \cos\theta_\gamma), \qquad x_{true} = \frac{Q^2_{true}}{2M\nu_{true}}. \end{aligned} \tag{17}$$

For non-radiative events the true kinematics exactly coincide with (16).

In case of the elastic and quasielastic radiative processes the mass squared of the hadronic final state is fixed imposing additional constraints on the true kinematic variables. Indeed the following ranges are allowed:

$$\begin{cases} x \le x_{true} \le 1 & \text{for } \sigma_{in} \\ x_{true} = 1 & \text{for } \sigma_q \\ x_{true} = M_A/M & \text{for } \sigma_{el} \end{cases} \qquad Q^2_{min} \le Q^2_{true} \le Q^2_{max} \tag{18}$$

where

$$Q^2_{max,min} = Q^2 \frac{2(1-x_r)(1 \pm \sqrt{1+\gamma^2}) + \gamma^2}{\gamma^2 + 4x_r(1-x_r)}, \qquad (19)$$

and

$$\gamma^2 = 4M^2x^2/Q^2, \qquad x_r = x/x_{true}. \qquad (20)$$

The Monte-Carlo procedure for the generation of events with a possible photon radiation is the following:
The procedure starts off with the generation of the kinematics of the scattered lepton and the calculation of an event weight from these kinematics. Then the appropriate scattering channel (non-radiative; elastic, quasielastic or inelastic radiative tail) has to be chosen according to their contribution to the total observed cross section (see (15)). If a radiative channel is selected the radiated photon has to be generated and the values of the kinematic variables have to be re-computed to obtain the true values. For each event the weight has to be recalculated. The new weight is defined as the ratio of the radiatively corrected and the Born cross section. After this recalculation the weighted sum of all events (generated originally in accordance with the Born cross section) gives the observed cross section.

The recalculation of the weight requires the knowledge of the cross section integrated over the photon momentum. This is done differently in two theoretical approaches. In the Mo-Tsai approach an additional parameter Δ dividing the integration region into a soft and a hard photonic part is introduced. There is no such parameter in the Bardin-Shumeiko approach. However, in both approaches a minimal photon energy E^{γ}_{min} is adapted in generating a radiated photon. If the photon energy is above this value the kinematics of the photon is calculated and the event becomes a radiative one. The actual value of E^{γ}_{min} depends on the aim of the photon generator. In general photons should be detected in the calorimeter. So the energy threshold of the calorimeter sets the value for E^{γ}_{min}.

The unpolarized generator is based on the FORTRAN code FERRAD35[37]. The code calculates the radiative correction to deep inelastic scattering of unpolarized particles in accordance with the analytical formulae given by Mo and Tsai[1,2].

The polarized generator is constructed utilizing the FORTRAN code POLRAD 2.0[14,13].

To estimate the value of radiative effects numerically Monte-Carlo event samples of 100k events each have been generated for a ^{3}He target and kinematic cuts relevant for the HERMES experiment[38] have been

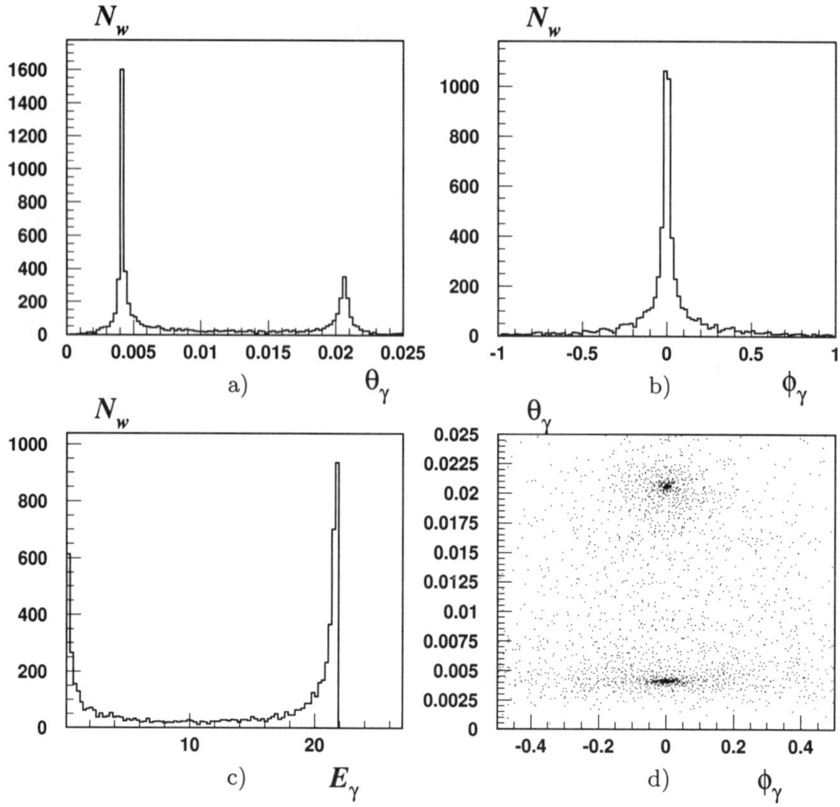

Figure 1. *The distribution of the radiation angles θ_γ a) and ϕ_γ b) and of the energy c) of the radiated photon for $x = 0.1$ and $y = 0.8$. The two-dimensional distribution d) shows θ_γ vs ϕ_γ.*

applied (Q^2 >1 GeV^2, W^2 >4 GeV^2, y <0.85, 0.037< θ <0.14rad). The distributions of the radiation angles θ_γ and ϕ_γ as defined in the so-called Tsai-system in which the z-axis is along the direction of the virtual photon and the y-axis is normal to the scattering plane (see Ref.[1,2]) and of the energy E_γ are displayed in Fig. 1. These distributions are generated for a certain point in the kinematic plane of the scattered lepton, i.e. for $x = 0.1$ and $y = 0.8$. The two peaks seen in the θ_γ distribution of Fig. 1 correspond to the s- and p-peak emerging from collinear radiation along the direction of the incoming and outgoing lepton, respectively.

3. Corrections to other processes

3.1. *Diffractive vector meson electroproduction and code DIFFRAD*

The measurement of the cross section of the exclusive vector meson electroproduction

$$e(k_1) + p(p) \longrightarrow e'(k_2) + \vec{v}(p_h) + p(p_2), \tag{21}$$

can provide information on the hadronic component of the photon and on nature of diffraction. During several years the diffractive production of the vector meson has been the subject of the muonproduction[39–41] and electroproduction[42–44] experiments. Data analysis of these experiments is considerably affected by the QED radiative effects. At practice RC to the processes of electroproduction are taken into account using codes originally developed for the inclusive case (see ref.[45], for example).

The purpose of this subsection is to present the electromagnetic correction to experimentally observed cross sections that was calculate by Bardin-Shumeiko approach in ref.[16].

We consider RC to three and four dimensional cross sections $\sigma = d\sigma/dxdydtd\phi_h$ and $\bar{\sigma} = d\sigma/dxdydt$. They are related as

$$\bar{\sigma} = \int\limits_0^{2\pi} d\phi_h \, \sigma \, . \tag{22}$$

The four differential Born cross section can be presented in the form

$$\sigma_0 = \frac{\alpha}{4\pi^2 xy}\Big(y^2\sigma_T + 2\big(1 - y - \frac{1}{4}y^2\gamma^2\big)(\sigma_L + \sigma_T)\Big), \tag{23}$$

where x and y are usual scaling variable, σ_T and σ_L are differential cross sections of the photoproduction, $\gamma^2 = Q^2/\nu^2$ and ν is the virtual photon energy.

For the observed cross section of the vector meson electroproduction we obtain

$$\sigma_{obs} = \sigma_0 e^{\delta_{inf}}(1 + \delta_{VR} + \delta_{vac}) + \sigma_F. \tag{24}$$

As earlier the correction δ_{vac} comes from the effects of vacuum polarization by leptons and hadrons.

The contribution of the infrared finite part can be written in terms of POLRAD 2.0 notation[13,14]:

$$\sigma_F = -\frac{\alpha^2 y}{16\pi^3} \int_0^{2\pi} d\phi_k \int_{\tau_{min}}^{\tau_{max}} d\tau \sum_{i=1}^{2} \sum_{j=1}^{3} \theta_{ij} \int_0^{v_m} \frac{dv}{f} R^{j-2} \left[\frac{\mathcal{F}_i}{\tilde{Q}^4} - \delta_j \frac{\mathcal{F}_i^0}{Q^4} \right], \quad (25)$$

where $R = v/f$, $f = 1 + \tau - kp_h/kp$ and $2M\tau_{max,min} = S_x \pm \sqrt{\lambda_q}$; as usual $\delta_j = 1$ for $j = 1$ and $\delta_j = 0$ otherwise. The quantities θ_{ij} depend only on the kinematic invariants and the integration variables τ and ϕ_k. The explicit expressions for them can be found in[16]. θ-functions are kinematical ones without any integrals for unpolarized case, and only one integration are left in the case of polarization part.

The dependence on the photoproduction cross sections is included in $\mathcal{F}_i$:

$$\begin{aligned} \mathcal{F}_1 &= (S_x - R)\sigma_T^R, \quad \mathcal{F}_2 = \frac{2\tilde{Q}^2}{S_x - R}(\sigma_T^R + \sigma_L^R), \\ \mathcal{F}_1^0 &= S_x \sigma_T, \qquad\qquad \mathcal{F}_2 = 2x(\sigma_T + \sigma_L). \end{aligned} \quad (26)$$

The quantities $\sigma_{T,L}$ have to be calculated for Born kinematics, but $\sigma_{T,L}^R$ is calculated in terms of so called true kinematics. It means that they have to be calculated for the tilde variables instead of standard Q^2, W^2 and t which have being already defined by Eq. (14).

The important point is the dependence of the results on the maximal inelasticity v_m. The inelasticity is calculated in terms of the measured momenta, so it is possible to make a cut on the maximal value of this quantity. If this cut is not applied the maximal inelasticity is defined by kinematics only. Below we give the formulae for v_m in terms of the kinematic invariants

$$4Q^2 v_m = \left(\sqrt{\lambda_q} - \sqrt{t_q^2 + 4m_v^2 Q^2} \right)^2 - (S_x - 2Q^2 + t_q)^2 - 4M^2 Q^2. \quad (27)$$

The FORTRAN code DIFFRAD created on the basis of the presented formulae calculates the lowest order RC to the diffractive vector meson electroproduction. The higher order effects are approximated by the procedure of exponentiation. The formulae for the cross section are given in a covariant form, so the code can be run both for the fixed target experiments and for the experiments at collider.

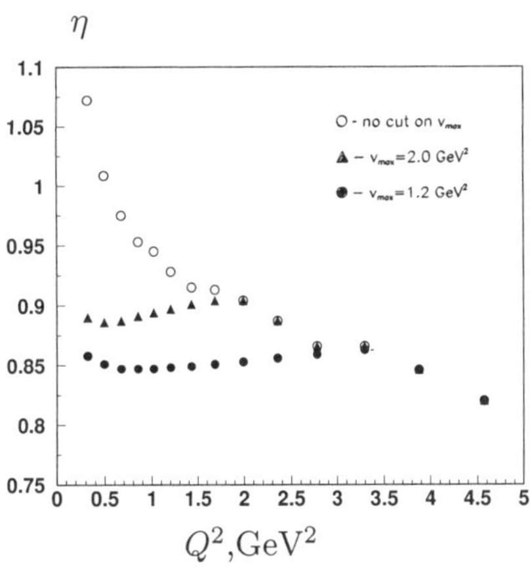

Figure 2. *RC factor under kinematic conditions of HERMES; $\sqrt{S} = 7.9$ GeV, $t = -0.11$ GeV^2.*

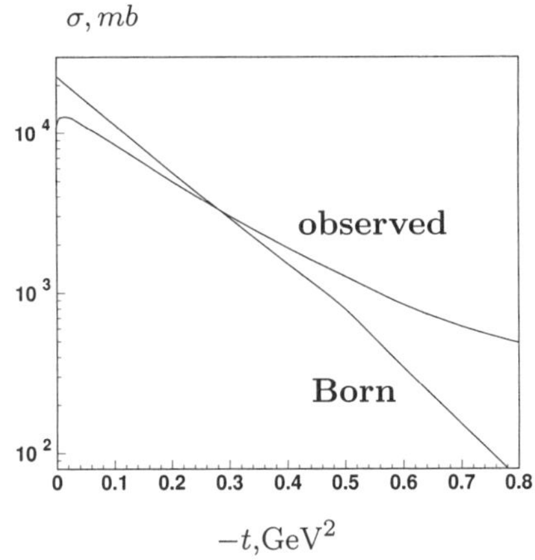

Figure 3. *Observable and Born cross sections as a function of t under kinematic conditions of HERA; $\sqrt{s}$=300 GeV; W=70 GeV; Q^2=4 GeV^2.*

Below we give numerical results for RC factor

$$\eta = \frac{\bar{\sigma}_{obs}}{\bar{\sigma}_0} = \int_0^{2\pi} d\phi_h \sigma_{obs} \left[\int_0^{2\pi} d\phi_h \sigma_0 \right]^{-1} \tag{28}$$

obtained within the kinematic regions of recent HERMES and HERA experiments for exclusive $\rho(770)$ electroproduction.

The dependence on the v_m cut in the region of HERMES kinematics is analyzed in Fig. 8. The usage of the cut changes RC factor for small Q^2 and does not influence on RC for larger values of Q^2. In the case of the cut usage we have to define v_m as a minimum of the value of the cut and v_m given by the kinematic restrictions (27). For fixed values of $-t$ the cut influence on v_m and RC till certain value of Q^2 only.

The Q^2-dependence shown in this figure is typical for the situation when the inelasticity cut is not applied. If this cut is used then the rise of η when Q^2 goes down would be suppressed. Notice that the t-dependence is rather important too. Fig. 3 illustrates this last property. η crosses unity for $-t \sim$ 0.25–0.3 GeV^2 and rises with increasing $|t|$. The large positive correction in this case is a result of the large phase space for photon radiation. The cut applied to the inelasticity (or $E - p_z$) can again reduce the value of the RC factor for large $|t|$. As a consequence of the t-dependence of η, the observed slope parameter also receives large RC.

Now as application of the obtained result let us consider the process

$$e(k_1) + p(p) \rightarrow e(k_2) + p(p') + \rho(p_V), \quad \rho \rightarrow \pi^+(p_+) + \pi^-(p_-), \qquad (29)$$

that can be viewed as an off-diagonal Compton scattering analytically continued in the virtuality of the photon γ^* to the vector meson mass $\gamma^* p \rightarrow Vp$ and gives access to the whole set of the corresponding helicity amplitudes.

The process (29) is analyzed experimentally by means of spin-density matrix elements. When measured, they give an indication of vector meson internal constituents motion and its spin-angular structure. The angular distribution of unpolarized vector meson decay is parameterized by fifteen matrix elements $r_{ij}^{\alpha}, r_{ij}^{\alpha\beta}$. For a long time it was believed, that their behavior complies with the s-channel helicity conservation (SCHC) hypothesis, which means that the helicity of the virtual photon is conserved in the s-channel process $\gamma^* p \rightarrow \rho p$. In this case ten matrix elements (which corresponds to the case when photon and vector meson have different helicities) are equal to zero. But in the recent measurements r_{00}^5 has been observed to be non zero[46–48], what has been considered as an indication to SCHC violation.

The procedure of the experimental data analysis is based on the correlation of the lepton scattering, vector meson production and decay planes, which are affected by RC. Hence it is topical to look at whether the measured r_{00}^5 can, at least partly, be the result that RC coming from non-observed QED effects and real photon emission was underestimated. In any case in order to make the data processing of the corresponding experiments[46–48] to be consistent, RC should be taken into account.

As it was shown in[18] the value of the QED corrections ($\Delta r = r_{obs} - r_{Born}$) to the matrix elements are defined by the following quantity:

$$I_1 = \int\limits_0^{2\pi} d\phi_h \, \cos\phi_h \, \delta(\phi_h) \left[\int\limits_0^{2\pi} d\phi_h \, (1 + \delta(\phi_h)) \right]^{-1}, \quad n = 1, ..., 4, \qquad (30)$$

where $\delta(\phi_h) = \sigma_{obs}/\sigma_0 - 1$.

For the majority of the matrix elements vanishing in the SCHC limit, radiative corrections turn out to be not greater then 1%. However there are two of them, Re r_{10}^{04} and r_{00}^5, which RC appears to be substantial. One can see, that corrections ΔRe r_{10}^{04} and $\Delta\mathrm{r}_{00}^5$ may reach $\sim 20\%$.

The last result is interesting from point of view of the found SCHC violation: the radiative correction procedure reduces the observed effect.

3.2. *Elastic ep scattering and code MASCARAD*

Precise polarization measurements of nucleon form factors in electron scattering is an essential component of new-generation electron accelerators such as CEBAF[49]. This unprecedented precision requires knowledge of higher-order electromagnetic effects at a per-cent level. The purpose of this subsections is to analyze radiative corrections in elastic electron proton scattering and present computational techniques that could be used in experiments at Jefferson Lab and other electron accelerator laboratories.

The observed cross section of the process

$$e(k_1) + N(p) \longrightarrow e'(k_2) + N(p_2), \tag{31}$$

is described by one non-trivial variable, which is usually chosen to be squared of momentum transferred. There are two ways to reconstruct the variables, when both lepton and nucleon final momenta are measured. In the first way it will be denoted as $Q_l^2 = -(k_1 - k_2)^2$, and for second one it is $Q_h^2 = -(p_2 - p)^2$. It is clear that there is no difference between these definition at the Born level. However when the radiation of the lepton is only considered at the level of RC the symmetry is breaking. Here we will deal with both situations.

In the first case the structure of bremsstrahlung cross section looks like

$$\frac{d\sigma}{dQ_l^2} \sim \alpha^3 \int \frac{d^3k}{k_0} \sum \mathcal{K}\mathcal{F}^2(Q_h^2)\mathcal{A} \tag{32}$$

where $\mathcal{K}$ is a kinematic coefficient calculatable exactly in the lowest order. It depends on photon variables. $\mathcal{F}^2$ is a bilinear combination of nucleon formfactors dependent on Q_h^2 only, which is function of photon momentum. Usually only the final momenta are measured in the part of the space. It is controlled by the function of acceptance $\mathcal{A}$, which is 1 or 0 in depending whether the final particles make it in detectors or not. The integral (32) should not be analytically calculated for two reasons. The first one is dependence of formfactors on Q_h^2. We avoid to use some specific model for them. The second one is acceptance usually very complicated function of kinematic variable dependent on photon momentum.

For the second version of reconstruction of transfer momentum squared the structure of the cross section is

$$\frac{d\sigma dQ_h^2}{\sim} \alpha^3 \sum \mathcal{F}^2(Q_h^2) \int \frac{d^3k}{k_0} \mathcal{K}\mathcal{A} \tag{33}$$

In this case the squared formfactor is not dependent on photon momentum and for 4π kinematics ($\mathcal{A} = 1$) this integral can be calculated analytically.

In the experimental conditions at JLab[49], both final electron and proton were detected in order to reduce background. However elastic scattering kinematics was restored by the final proton kinematics, while electron momentum was integrated over. Therefore, formalism related with Eq.(33) applies for this case.

Polarization effects are described by polarization four-vectors of the lepton (ξ) and proton (η) which can be expanded over the measured momenta k_1, p_1 and p_2. The polarization vector of the lepton as well as polarization vector of the proton can be expand over four momenta of scattering particles

$$\eta = 2a_\eta k_1 + b_\eta q + c_\eta(p_1 + p_2), \tag{34}$$

where the expansion coefficients defined by the different polarization states of the target.

As a result the Born cross section for the process (31) can be written in the form

$$\frac{d\sigma_0}{dQ^2} = \frac{2\pi\alpha^2}{S^2Q^4}\sum_{i=1}^{4}\theta_B^i\mathcal{F}_i, \tag{35}$$

where

$$\theta_1^0 = 2Q^2, \qquad \theta_3^0 = -\frac{2Q^2}{M}(Q^2a_\eta + (2S - Q^2)c_\eta),$$
$$\theta_2^0 = \frac{(S^2 - Q^2S - M^2Q^2)}{M^2}, \quad \theta_4^0 = \frac{Q^4}{M^3}(2S - Q^2)(a_\eta - 2b_\eta).$$

The structure functions $\mathcal{F}_i^2$ are bilinear combinations of the nucleon formfactors dependent on Q_h^2 only

$$\mathcal{F}_1 = 4\tau_p M^2G_M^2, \qquad \mathcal{F}_3 = -2M^2G_EG_M,$$
$$\mathcal{F}_2 = 4M^2\frac{G_E^2 + \tau_pG_M^2}{1+\tau_p}, \quad \mathcal{F}_4 = -M^2G_M\frac{G_E - G_M}{1+\tau_p}.$$

where $\tau_p = Q^2/4M^2$.

According to[19] the cross section that takes into account radiative effects within the leptonic variables can be written as

$$d\sigma_{obs} = d\sigma_0 e^{\delta_{inf}}(1 + \delta_{VR} + \delta_{vac}) + d\sigma_F. \tag{36}$$

Here the corrections δ_{inf} and δ_{vac} come from radiation of soft photons and effects of vacuum polarization. The correction δ_{VR} is an infrared-free sum of factorized parts of real and virtual photon radiation. The infrared-free contribution of bremsstrahlung process can be presented as integral over three variables: inelasticity $v = \Lambda^2 - M^2$ ($\Lambda = p + k_1 - k_2$), $\tau = kq/kp$ and angle ϕ_k between planes ($\mathbf{q}$, $\mathbf{k}$) and ($\mathbf{k_1}$, $\mathbf{k_2}$)

$$d\sigma_F = -\frac{\alpha^3}{2S^2} dQ_l^2 \int\limits_0^{v_m} dv \int\limits_{\tau_{min}}^{\tau_{max}} \frac{d\tau}{1+\tau} \int\limits_0^{2\pi} d\phi_k \times \sum_i \left[\sum_{j=1}^{3} \mathcal{A} R^{j-2} \theta_{ij} \frac{\mathcal{F}_i}{Q_h^4} - 4F_{IR}^0 \theta_i^B \frac{\mathcal{F}_i^0}{RQ_l^4} \right]. \quad (37)$$

Here $R = v/(1+\tau)$ and $\mathcal{A}$ is integrated over the ϕ acceptance function. The explicit expression for functions $\theta_{ij}(\tau)$ are presented in[14,13]. However there they are integrated over ϕ_k. We keep this integration because of possible dependence of acceptance function on this angle. The explicit expressions of the θ-functions in (37) are discussed in Appendix of[19].

Since within the hadronic variables the integration over the phase space of the real photon can be performed analytically, the observable cross section for this situation has more simple form

$$d\sigma_{obs} = \frac{\alpha^3}{4} \frac{dQ_h^2}{S^2 Q_h^4} \sum_{i=1}^{4} [\theta_i^F + 4(\delta^{el} + \delta_{vac}^l + \delta_{vac}^h)\theta_i^0] \mathcal{F}_i + d\sigma_0. \quad (38)$$

The explicit expressions for θ_i^F and δ^{el} can be found in[20].

On the basis of formalism presented in this subsections the FORTRAN code MASCARAD developed. This code uses Monte-Carlo methods to calculate RC to the observable quantities in polarized *ep* scattering measurements.

Here we consider RC to some observables within two polarization measurements:

(1) The initial proton is polarized and the final electron is measured to reconstruct Q^2. As a result there are four experimental situations for asymmetry definition: the target is polarized along (perpendicular) to beam or $\vec{q}$ ($q = p_2 - p$). Corresponding polarization 4-vectors are denoted as η_L (η_T) or η_L^q (η_T^q).
(2) Polarization and momentum of the final proton are measured. Two polarization states should be considered: the final proton is polarized along (η'_L) and perpendicular (η'_T) to $\vec{q}$.

It is natural that for the fist type of measurement we have to calculate RC within the leptonic and for the second type within hadronic variables.

Both the spin averaged and spin dependent parts of the cross section (σ^u and σ^p) can be presented as

$$\sigma_{obs}^{u,p} = (1+\delta)\sigma_0^{u,p} + \sigma_R^{u,p}. \tag{39}$$

Both the factorized correction δ and unfactorized cross section coming from bremsstrahlung process contribute to cross section.

Absolute and relative corrections to asymmetry can be defined as (see 39)

$$\Delta A_i = A_i - A_{i0} = \frac{(1+\delta)\sigma_0^p + \sigma_R^p}{(1+\delta)\sigma_0^u + \sigma_R^u} - \frac{\sigma_0^p}{\sigma_0^u},$$
$$\Delta_i = \frac{A_i - A_{i0}}{A_{i0}} = \frac{\delta_p - \delta_u}{1+\delta+\delta_u} \tag{40}$$

where index i runs over all considered cases: $i = L, T, qL, qT$; $\delta_{u,p} = \sigma_R^{u,p}/\sigma_0^{u,p}$. Here the correction δ is usually large because of contributions of leading logarithms. However it exactly canceled in numerator of expression for correction to asymmetry. That is a reason why the correction to cross section can be large while the correction to the asymmetry is relatively small.

Born and observed asymmetries are presented in Fig. 4. There were no cuts for the missing mass. As a result due to unfactorizing properties hard photon contributions gives different contributions to spin averaged and spin dependent part of cross sections. For the longitudinal asymmetries $\delta_p > \delta_u$ and there are positive contributions to RC. Opposite situation with the transverse asymmetries.

For the second type of measurement let us define the relative corrections to the observables in the current experiments:

$$\delta_u = \frac{\sigma_u^{obs}}{\sigma_u^b} - 1, \qquad \delta_{L,T} = \left[\frac{\sigma_p^{obs}}{\sigma_u^{obs}} - \frac{\sigma_p^b}{\sigma_u^b}\right]\left[\frac{\sigma_p^b}{\sigma_u^b}\right]^{-1} = \frac{\sigma_u^b}{\sigma_u^{obs}}\frac{\sigma_p^{obs}}{\sigma_p^b} - 1. \tag{41}$$

The first quantity δ_u is the relative correction to the unpolarized part of cross section. The $\delta_{L,T}$ give contribution to polarization asymmetries measured by rotating the polarization states of the initial protons. The correction to the unpolarized part of cross section is presented in Fig. 5a. Its behavior is quite typical. For the very hard inelasticity cut ($u_m \ll Q^2$) the positive contribution is suppressed due to real bremsstrahlung, so there is only negative loop correction contributing to cross section. Different ending values for the curves corresponds to different kinematically allowed regions.

Algorithm of radiative event simulation described in Section 2.3 can be applied to elastic scattering as well. The Monte Carlo generator based on

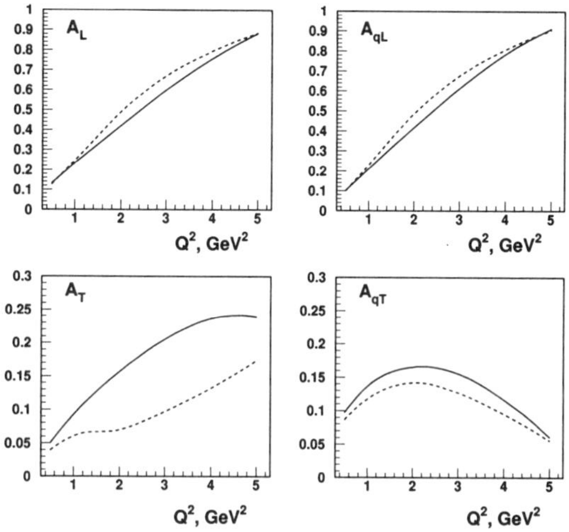

Figure 4. *Born (solid line) and observed (dashed line) asymmetries vs* Q^2. *No kinematic cuts on inelasticity were used.* $E = 4\ GeV$.

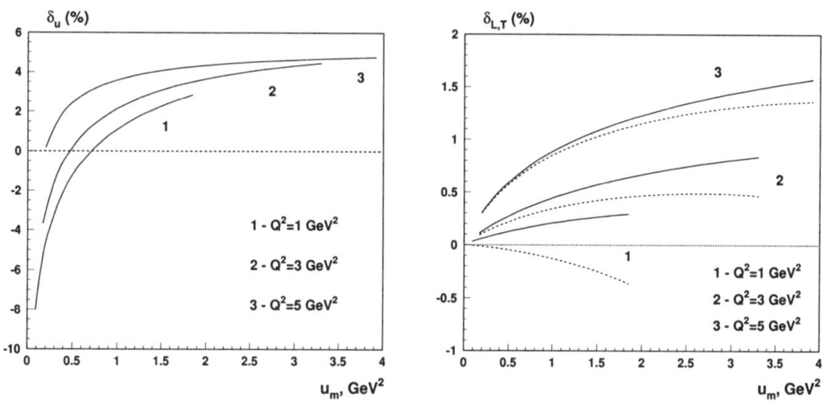

Figure 5. *Radiative corrections to the unpolarized cross section (left plot) and polarization asymmetries (right plot) defined in (12). Solid and dashed lines corresponds to longitudinal and transverse cases.* $S{=}8\ GeV^2$.

it and above formulae is named ELRADGEN. The strategy for simulation of one event is defined as follows.

- For the fixed initial energy and Q^2 the non–radiative and radiative parts of the total cross section are calculated. Note, that together with Born subprocess, the contributions of loop effects and soft photon radiation are also included into the non–radiative part.
- The channel of scattering is simulated for this event in accordance with partial contributions of these two positive parts into the total cross section.
- For any process ϕ is simulated uniformly from 0 to 2π.
- For the radiative event the kinematic variables t, v and ϕ_k are simulated in accordance with their calculated distribution.
- The four–momenta of all final particles in the required system form are calculated.
- If the initial Q^2 has not a fixed value but is instead simulated according to some probability distribution (for example, the Born cross section) then the cross sections have to be stored for reweighting. The Q^2 distribution is simulated over the Born cross section, and realistic observed Q^2 distribution is calculated as sum of weights, they are ratios of the total and Born cross sections.

An example of the simulated distributions at CEBAF energies can be found in[21].

4. Exclusive Pion Production

Recent measurements with the CLAS detector at Jefferson Lab/Hall B [50] covered a wide kinematic range in four momentum transfer Q^2 and invariant mass W, as well as to measure the full angular range of the resonance decay into the nucleon-pion final state:

$$
\begin{aligned}
e(k_1) + N(p) &\longrightarrow e'(k_2) + \pi^+(p_h) + n(p_u),\\
e(k_1) + N(p) &\longrightarrow e'(k_2) + p(p_h) + \pi^0(p_u).
\end{aligned}
$$

$(p^2 = M^2,\ p_h^2 = m_h^2,\ p_u^2 = m_u^2)$, where $p_h(p_u)$ denotes the momentum of the detected (undetected) hadron. Adequate calculation of RC becomes important in interpreting the measured observables such as unpolarized coincidence cross sections and polarization asymmetries.

The Born cross section for this process ($\sigma_0 = d\sigma/dW^2dQ^2d\Omega_h$) can be presented in terms of the photoabsorption cross sections (or the response functions), where each term corresponds to the certain polarization states of the virtual photon[51],

$$\sigma_0 = \alpha\frac{(W^2 - M^2)}{2S^2Q^2(1-\epsilon)}(\sigma_T + \epsilon\sigma_L + \epsilon\cos 2\phi_h\,\sigma_{TT} \\ +\sqrt{\epsilon(1+\epsilon)/2}\cos\phi_h\,\sigma_{LT} + h_e\sqrt{\epsilon(1-\epsilon)/2}\sin\phi_h\,\sigma'_{LT}), \quad (42)$$

where h_e is the longitudinal polarization degree of the incoming leptons, ϕ_h is detected proton angle in the center–of–mass of the final hadrons and ϵ (Ref. [51]) describes the virtual photon polarization.

Formalism RC calculation in exclusive pion electroproduction on the proton is presented in[22]. Exact formulae for observe differential cross section (over Q^2, W and two angles of a detected hadron) are obtained in the form traditional for the covariant formalism:

$$\sigma_{obs} = \sigma_0 e^{\delta_{inf}}(1 + \delta_{VR} + \delta_{vac}) + \sigma_F. \quad (43)$$

The finite part has the following structure

$$\sigma_F = -\frac{\alpha^3}{2^9\pi^4 S^2 W^2}\int d\Omega_k \frac{dv}{f}\sum_i\left[\frac{\sqrt{\lambda_W}}{Q_h^4}\theta_i\mathcal{H}_i - \frac{4F_{IR}\sqrt{\lambda_W^0}}{Q^4}\theta_i^0\mathcal{H}_i^0\right] \quad (44)$$

where

$$\begin{aligned}\lambda_W^0 &= (W^2 - m_h^2 - m_u^2)^2 - 4m_h^2W^2,\\ \lambda_W &= (W^2 - m_h^2 - m_u^2 - v)^2 - 4m_h^2W^2.\end{aligned} \quad (45)$$

The integration is carried out over invariant mass of unobserved states $v = (p+k_1-k_2-p_h)^2 - m_u^2$ and photonic solid angle Ω_k. Sum over i corresponds to contribution all five structure function $\mathcal{H}_i$ that can be presented by Born photoabsorption cross sections. Functions θ_i and F_{IR} have the same sense as their counterparts described above. Their explicit form, properties and definition of kinematical factors can be found in[22]. In addition, two types of leading log formulae are considered in this papers. Note, that is not so trivial because of complicated phase space. Distinguishing feature of the paper is explicit integration of the exclusive RC cross section over hadronic angles and, as a result, demonstration of transferring to inclusive RC.

The results were implemented in FORTRAN code EXCLURAD. Numerical analysis was performed for kinematics of current and future experiments at JLab. New feature of the analysis is an investigation of beam asymmetry (Fig. 6).

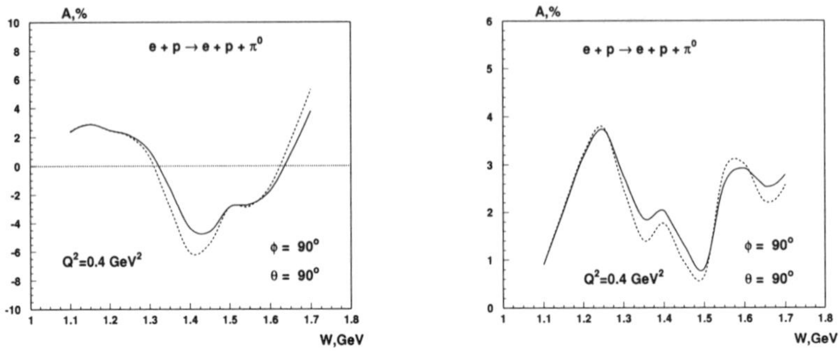

Figure 6. W-dependence of the beam polarization asymmetry in neutral (left) and charged (right) pion production. The solid (dashed) curve denote the asymmetry with (without) RC. MAID2000 was used to compute the structure functions.

Conclusion

Radiative effects in lepton scattering on polarized and unpolarized nucleons and light nuclei have been briefly reviewed. The processes of inclusive, semi-inclusive, diffractive and elastic scattering are considered. The explicit expressions for QED and electroweak lowest-order correction within Bardin-Shumeiko approach have been presented. Besides the next order correction in LO has been estimated.

Basing on the presented formulae the FORTRAN codes have been created which allow one to provide the procedure of RC of experimental data in the current and future experiments as well. More popular of them are POLRAD[14] and RADGEN[15]. The version 2.0 of FORTRAN code POLRAD provides complete RC procedure in inclusive and semi-inclusive experimental setup. Monte Carlo generator RADGEN creating from FORTRAN code POLRAD 2.0 provides event simulation with taking into account radiative effects in various experimental setup on inclusive DIS. Codes DIFFRAD, MASCARAD an EXCLURAD calculates RC in diffractive, elastic and exclusive processes.

The obtained results and FORTRAN codes constructed on their basis are widely used in experimental collaborations at CERN, DESY, SLAC and TJNAF. Many of them can be download from http://www.jlab.org/RC or obtained directly from the authors.

Acknowledgement

One of us (I. A.) thanks the US Department of Energy for support under contract DE-AC05-84ER40150.

Bibliography

1. L.W. Mo, Y.S. Tsai, *Rev. Mod. Phys.* **41**, 205 (1969).
2. Y.S. Tsai, *Preprint* (SLAC, PUB - 848, 1971).
3. D.Yu. Bardin, N.M.Shumeiko, *Nucl. Phys.* **B127**, 242 (1977).
4. A.A. Akhundov, D.Yu. Bardin, N.M. Shumeiko, *Yad. Fiz.* **26**, 1251 (1977).
5. A.A. Akhundov, D.Yu. Bardin, N.M. Shumeiko, *Yad. Fiz.* **44**, 1517 (1986).
6. A. Akhundov, D. Bardin, L. Kalinovskaya, T. Rieman, *Fortschr. Phys.* 1996. **44**, 373 (1996).
7. A.V. Soroko, N.M. Shumeiko, *Yad. Fiz.* **49**, 1348 (1989).
8. T.V. Kukhto, N.M. Shumeiko, *Yad. Fiz.* **36**, 707 (1982).
9. I. Akushevich, D. Ryckbosch, N. Shumeiko, A. Tolkachev, *J. Phys.* **G26**, 1389 (2000).
10. A.V. Soroko, N.M. Shumeiko, *Yad. Fiz.* **53**, 1015 (1991).
11. I. Akushevich, N. Shumeiko, A. Soroko, *Eur. Phys. J.* **C10**, 681 (1999).
12. N.M. Shumeiko, *Yad. Fiz.* **29**, 1571 (1979).
13. I. Akushevich, N. Shumeiko, *J. Phys.* **G20**, 513 (1994).
14. I. Akushevich, A. Ilyichev, N. Shumeiko, A. Soroko, A. Tolkachev, *Comput. Phys. Commun.* **104**, 201 (1997).
15. I.Akushevich, H.Boettcher, D.Ryckbosch, *Workshop "Monte Carlo Generators for HERA Physics"*, (Hamburg, 1998/99).
16. I. Akushevich, *Eur. Phys. J.* **C8**, 457 (1999).
17. I. Akushevich, *In Monte Carlo generators for HERA physics* (Hamburg 1998/1999), p. 547.
18. I.Akushevich, P.Kuzhir, *Phys. Lett.* **B474**, 411 (2000).
19. A.Afanasev, I.Akushevich, N.Merenkov, *Phys. Rev.* **D64**, 113009 (2001).
20. A.Afanasev, I.Akushevich, A. Ilyichev, N. Merenkov, *Phys. Lett.* **B514**, 269 (2001).
21. A.Afanasev, I.Akushevich, A. Ilyichev, B. Nicziporuk, *Czech. J. Phys.*, **53 Suppl. B**, B449 (2003).
22. A. Afanasev, I. Akushevich, V. Burkert and K. Joo, *Phys. Rev.* **D66**, 074004 (2002)
23. P.P.Kuzhir, N.M.Shumeiko, *Yad. Fiz.* **55**, 1958 (1992).
24. T.V.Kukhto, N.M.Shumeiko, S.I.Timoshin, *J. Phys.* **G13**, 725 (1987).
25. D.Yu. Bardin, N.M.Shumeiko, *Yad. Fiz.* **29**, 969 (1979).
26. N.M.Shumeiko, S.I.Timoshin, *J. Phys.* **G17**, 1145 (1991).
27. N.M.Shumeiko, S.I.Timoshin, V.A.Zykunov, *Yad. Fiz.* **58**, 2021 (1995).
28. E.A. Kuraev, V.S. Fadin, *Yad. Fiz.* **41**, 733 (1985).
29. E.A. Kuraev, N.P. Merenkov, V.S. Fadin *Yad. Fiz.* **47**, 1593 (1988).
30. A.Afanasev, I.Akushevich, N.Merenkov, *Zh.Eksp.Teor.Fiz* **98**, 462 (2004).
31. E.A. Kuraev, N.P. Merenkov, V.S. Fadin *Yad. Fiz.* **45**, 782 (1987).

32. I.V. Akushevich, A.V. Arbuzov, E.A. Kuraev *Phys. Lett.* **B432**, 222 (1998).
33. I. Akushevich, E. Kuraev, B. Shaikhatdenov *Phys.Rev.* **D62**, 53016 (2000).
34. H. Burkhardt, B. Pietrzyk, *Phys. Lett.* **B356**, 398 (1995).
35. P.J. Mulders, R.D. Tangerman, *Nucl. Phys.* **B461**, 197 (1996), Erratum – ibid. **B484**, 538 (1996).
36. K. Ackerstaff et al., HERMES Collab. *Phys. Rev. Lett.* **58** 5519 (1998).
37. FERRAD 3.5 . This code was first created by J. Drees for EMC and further developed by M. Dueren.
38. K. Ackerstaff et al. [HERMES Collaboration], *Phys. Lett.* **B404**, 383 (1997).
39. J.J. Aubert et al. EMC Collaboration, *Phys. Lett.* **B161**, 203 (1985).
40. M. Arneodo et al. NMC Collaboration, *Nucl. Phys.* **B429**, 503 (1994).
41. M.R. Adams et al. E665 Collaboration, *Z. Phys.* **C74**, 237 (1997).
42. S. Aid et al. H1 Collaboration, **Nucl. Phys. B468**, 3 (1996).
43. M.Derrick et al. ZEUS Collaboration, *Phys. Lett.* **B356**, 601 (1995).
44. M.Tytgat, HERMES Collaborationm, Proceedings of the DIS98 Workshop, Brussels, Belgium, April 1998, World Scientific.
45. K.Kurek, *DESY 96-209*, hep-ph/9606240 (1996).
46. J. Breitweg et al. ZEUS Collaboration, *Eur.Phys.J.* **C6**, 603 (1999).
47. C. Adloff et al. H1 Collaboration, *Eur.Phys.J.* **C13**, 371 2000.
48. J. Breitweg et al. ZEUS Collaboration, *Eur.Phys.J.* **C12**, 393 (2000).
49. M. K. Jones et al. Jefferson Lab Hall A Collaboration, *Phys. Rev. Lett.* **84**, 1398 (2000).
50. K. Joo et al. CLAS Collaboration, *Phys. Rev.* **C68**, 032201 (2003).
51. N. Dombey, *Rev. Mod. Phys.* **41**, 236 (1969).
52. D. Drechsel, O. Hanstein, S. S. Kamalov and L. Tiator, *Nucl. Phys.* **A645** 145 (1999)

PROCESSES OF LEPTON AND BOSONS BIRTH IN LINEAR PHOTON COLLIDER

Tatiana V. Shishkina
Department of Theoretical Physics, Belarusian State University, 4 Fr. Skorina Ave., 220050, Minsk, Belarus;
National Center of Particle and High Energy Physics, 153 Bogdanovich Str., 220040 Minsk, Belarus
E-mail: shishkina@bsu.by, shishkina@hep.by

The main possibilities of investigating lepton and boson production by the interaction of polarized photons are analyzed. The main contribution of high order electroweak effects is also considered.

1. Introduction

The Standard Model (SM) makes possible to describe all experimental data up to now with a typical precision about one per mille. Nevertheless, the Model is not the final theory valid up to very high scales and in the linear collider that can run at centre of mass energies about 1 TeV one can hope to see that finally deviations in precision measurements occur typically for two reasons.

If the new physics occurs in loop diagrams their effect is usually suppressed by a loop factor $\alpha/4\pi$ and very high precision is required to obscure it. If the new physics is already at the Born level but at very high masses the effects are suppressed by the propagator factor $\frac{s}{s-m_{N_P}^2-\imath m_{N_P}\Gamma}$, so it is important to work at the highest possible energies.

Linear lepton colliders will provide the opportunity to investigate photon collisions at energies and luminosities close to those of in e^+e^- collisions[1]. The possibility to transform the future linear e^+e^--colliders into the $\gamma\gamma$ and γe -colliders with approximately the same energies and luminosities is shown. The basic e^+e^--colliders can be transformed into the $e\gamma$- or $\gamma\gamma$-colliders. The intense γ -beams for photon colliders are suggested to be obtained by Compton scattering of laser light which is focused on electrons beams of basic e^+e^--accelerators.

The electron and photon linear colliders of next generation will attack unexplored higher energy region where new behaviour can turn up. In this area the photon colliders have a number of advantages.

– The first of the above advantages is connected with the better signal/background ratio at both e^+e^-- and $e\gamma/\gamma\gamma$-colliders in comparison with hadron ones.

– The production cross sections at photon colliders are usually larger than those at electron colliders.

– The photon colliders permit to investigate both of the problems of new physics and those ones of "classical" hadron physics and QCD. So it is exclusively important task to use possibilities of $\gamma\gamma$-colliders to realize the experiments of the next generation.

In the SM the couplings of the gauge bosons and fermions are constrained by the requirements of gauge symmetry. In the electroweak sector this leads to trilinear and quartic interactions between the gauge bosons with completely specified couplings. The trilinear and quartic gauge boson couplings probe different aspects of the weak interactions. The trilinear couplings directly test the non-Abelian gauge structure, and possible deviations from the SM forms have been extensively studied. In contrast, the quartic couplings can be regarded as a more direct way of consideration of electroweak symmetry breaking or, more generally, on new physics which couples to electroweak bosons. In this respect it is quite possible that the quartic couplings deviate from their SM values while the triple gauge vertices do not. For example, if the mechanism for electroweak symmetry breaking doesn't reveal itself through the discovery of new particles such as the Higgs boson, supersymmetric particles or technipions it is possible that anomalous quartic couplings could provide the first evidence of new physics in this sector of electroweak theory.

The production of several vector bosons is the best place to search directly for any anomalous behaviour of triple and quartic couplings. We wait for colliders with higher center of mass energy in order to produce a final state with three or more gauge bosons and to test the quartic gauge-bosons couplings. The future project TESLA[2] is the real candidate for investigation of process three boson production.

Previously the cross sections for triple gauge bosons production in framework of the SM presented for e^+e^--colliders and hadronic colliders. By using of transforming a linear e^+e^- collider in a $\gamma\gamma$ collider, one can obtain very energetic photons from an electron or positron beams. Such machines as TESLA which will reach a center of mass energy $\sim 1000GeV$ with high luminosity ($\sim 10^{33}cm^{-2}s^{-1}$) will be able to study multiple vector boson production with high statistics.

For obvious kinematic reasons, processes where at least one of the gauge bosons is a photon have the largest cross sections. So the photon linear colliders have the great physical potential[2,3] (Higgs and SUSY particles searching, study of anomalous gauge boson couplings and hadronic structure of photons etc.). Performing of this set of investigations requires a fine measurement of the luminosity of photon beams. For this purpose some of the well-known and precisely calculated reactions (see, for example, $\gamma\gamma \to 2f, 4f$[4-9]) are traditionally used.

It was shown that it is convenient to use the events of $\gamma\gamma \to l^+l^-$ process for measuring the luminosity of the $J = 2$-beams (J is the total angular momentum of initial photon couple). Here l is the unpolarized light lepton (e or μ). It is the dominating QED process on $J = 2$ beams and its events are easily detected. The difficulties appear in the calibration of photon beams of similar helicity (the total helicity of $\gamma\gamma$-system $J=0$) since the small magnitude of cross sections of the most QED processes. For example, the leading term of cross section of $\gamma\gamma \to l\bar{l}$ scattering on $J=0$-beams is of order α/π (≈ 0.002). The exclusive reaction $\gamma\gamma \to l^+l^-\gamma$ provides the unique opportunity to measure the luminosity of $J=0$ beams on a linear photon collider. One of the main purposes of the linear photon collider is the s-channel of the Higgs boson production at energies about $\sqrt{s} = 120GeV$ [2]. That is the reason of using this value of c.m.s. energy in our analysis.

2. Two lepton production with photon in $\gamma\gamma$-collisions

The two various helicity configuration of the $\gamma\gamma$-system leads to the different spectra of final particles and requires the two mechanisms of beam calibration. We have analyzed[4] the behaviour of the $\gamma\gamma \to l^+l^-\gamma$ reaction on beams with various helicities as a function of the parameters of detectors, and performed the detail comparison of cross section on $\gamma^+\gamma^+$- ($J = 0$) and $\gamma^+\gamma^-$-beams ($J = 2$). Since experimental beams are partially polarized the ratio of cross sections of $\gamma\gamma \to l^+l^-\gamma$ scattering on $J = 0$ to $J = 2$-beams should be high for the effective luminosity measurement. We have outlined the conditions that greatly restrict the observation of the process on "$J = 2$" beams, remaining the "$J = 0$" cross section almost unchanged.

Finally we estimate the precision of luminosity measurement.

Consider the process

$$\gamma(p_1, \lambda_1) + \gamma(p_2, \lambda_2) \to f(p_1{}', e_1{}') + \bar{f}(p_2{}', e_2{}') + \gamma(p_3, \lambda_3), \qquad (1)$$

where λ_i and $e_i{}'$ are the photon and the fermion helicities.

We denote the c.m.s. energy squared by $s = (p_1 + p_2)^2 = 2\,p_1 \cdot p_2$, the final-state photon energy by w. For the differential cross section the normalized final-state photon energy (c.m.s. is used) $x = w/\sqrt{s}$ is introduced. The differential cross section $d\sigma/dx$ appears to be the energy spectrum of final-state photons.

The matrix elements are obtained using two methods: the massless helicity amplitudes[10] for the fast estimations and the exact covariant analysis including finite fermion mass. Since final-state polarizations will not be measured we have summarized over all final particles helicities. The integration over the phase space of final particles is performed numerically using the Monte-Carlo method.

The calculations have been performed for various experimental restrictions on the parameters of final particles. Events are not detected if energies and angles are below the corresponding threshold values. The considering restrictions on the phase-space of final particles (the cuts) are denotes as follows:

- Minimum final-state photon energy: ω_{cut},
- Minimum fermions energy: $E_{f,cut}$,
- Minimum angle between any final and any initial particles (polar angle cut): Θ_{cut},
- Minimum angle between any pair of final particles: φ_{cut}.

Consider the energy spectrum of final photons. In Fig. 1 the spectra for $J=0$ and $(J=2)$ are presented (the $(J=2)$-cross section is scaled on factor 0.1 for the convenience). The differential cross section $d\sigma/dx$ on $J = 2$ beams decreases while one on $J = 0$ beams raises with increasing of the final-state photon energy. This leads to the conclusion that if one increases the threshold on w, the process on $J=2$ beams will be greatly restricted, but the rate of $J = 0$ events remains almost unchanged.

The ratio of events on $J = 0$ and $J = 2$ beams strongly depends on the experimental cuts. We obtained the region (the configuration of cuts) where the processes on the both $J = 0$ and $J = 2$ beams have the cross sections close by each other. That is the region of small polar angle cut, high collinear angle cut as well as high minimal energy of final-state photons. At these parameters the total cross sections of $\gamma\gamma \to f\bar{f}\gamma$ in experiments using $\gamma^+\gamma^+$- and $\gamma^+\gamma^-$- beams appear to be the same order of magnitude.

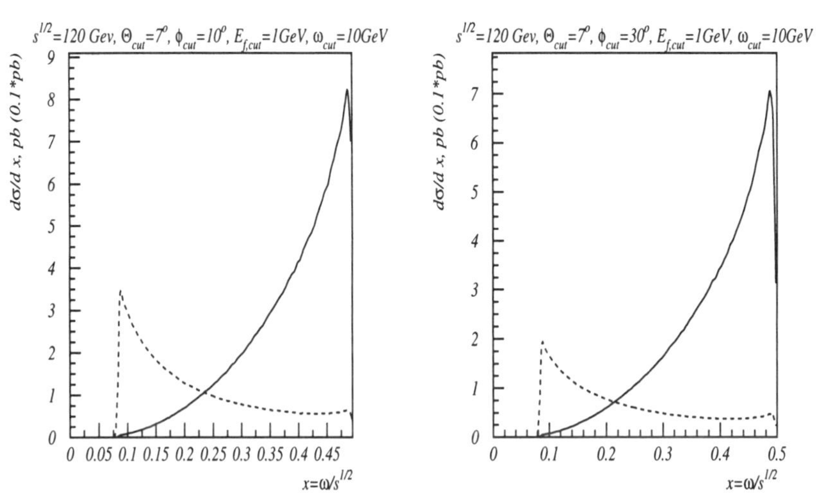

Figure 1. Final-state photon energy spectrum for $J=0$ (solid) and $(J=2)*0.1$ (dotted) at $\sqrt{s} = 120GeV$ and cuts: $\Theta_{min} = 7^o$, $\varphi_{min} = 10^o$ (left) and $\varphi_{min} = 30^o$ (right), $E_{f,min}=1GeV$, $\omega_{min}=10GeV$.

The mass contribution is small in the great part of phase space of final particles. The most significant contribution is for the $J = 0$ energy spectra. The high value of the contribution corresponds to regions where the differential cross section is minimal. The mass contribution to the total cross section is below the 1% level at any realistic set of cuts. It means that the helicity amplitudes is a good approach for study the $\gamma\gamma \to l^+l^-\gamma$ process.

2.1. *Luminosity measurement of $J=0$ beams.*

For analysis the precision of luminosity measurement[4] that can be achieved using the reaction $\gamma\gamma \to f\bar{f}\gamma$, the most interest are offered by the two kinds of measurement. The first one is the measuring of beams luminosity with the wide energy spectrum. The second one is the same measurement for the narrow band around the energy of Higgs boson production.

We use for consideration the following TESLA project parameters[2]:

1. luminosity

$$\mathcal{L}(\sqrt{s'} > 0.8\sqrt{s'_{\max}}) = 5.3 \cdot 10^{33} cm^{-2}s^{-1},$$
$$\mathcal{L}(m_H \pm 1GeV) = 3.8 \cdot 10^{32} cm^{-2}s^{-1};$$

2. polarization $\mathcal{P} \approx 90\%$.

Our calculations allow to choose the set of cuts with the high $J=0$ cross section and high ratio $\sigma_{J=0}/\sigma_{J=2}$: $\omega_{cut}=20GeV$, $E_{f,cut}=5GeV$, $\Theta_{cut}=6°$, $\varphi_{cut}=30°$. For these cuts the total cross sections have the following values:

$$\sigma(J=0) = 0.82pb,$$
$$\sigma(J=2) = 1.89pb.$$

So for the precision of luminosity measurement in a 2 years run $(2\cdot10^7\text{s})$ one can obtain:

$$\frac{\Delta\mathcal{L}}{\mathcal{L}}\left(\sqrt{s'} > 0.8\sqrt{s'_{\max}}\right) = 0.35\%,$$
$$\frac{\Delta\mathcal{L}}{\mathcal{L}}(m_H \pm 1GeV) = 1.3\%.$$

3. Lepton-antilepton production in high energy polarized photons interaction

The luminosity measurement at $J=2$ beams will be performed using the reaction $\gamma\gamma \to l^+l^-$. It has the great cross section that provides the number of events enough for the 0.1% precision of luminosity determination.

The main task is to calculate the cross section with maximal precision. For realization of this purpose we have calculated the complete one-loop QED radiative corrections to cross section of $\gamma\gamma \to l^+l^-$ process including the hard photon bremsstrahlung.

The major feature of $\gamma\gamma \to f\bar{f}$ process is the small value of cross section if the total angular momentum of $\gamma\gamma-$beams equals zero.

We analyze both the angular spectra and the invariant distributions of final particles. The angular spectrum of final leptons is calculated in form $d\sigma/d\cos\Theta(p_l, p_\gamma)$. It is more convenient to use Lorentz-invariant results for the experimental reasons. Therefore we analyze the process $\gamma\gamma \to f\bar{f}[+\gamma]$ including $O(\alpha)$-corrections using the method of covariant calculations. The invariant differential cross section is calculated in the form $d\sigma/d(p_l - p_\gamma)^2$ and can be used in the arbitrary experimental configuration.

The cross section of process (1) to be calculated is

$$d\sigma = \frac{1}{2s}\left|M_{fi}^{\lambda_1,\lambda_2,e_1{}',e_2{}',[\lambda_k]}\right|^2 \cdot d\phi,$$

where

$$\int A d\phi_{2[3]} = \frac{1}{(2\pi)^2}\cdot\frac{d^3p_1'}{2\mathcal{E}_1'}\cdot\frac{d^3p_2'}{2\mathcal{E}_2'}[\cdot\frac{d^3k}{(2\pi)^3\,2\omega}]\cdot\delta\left(p_1+p_2-p_1'-p_2'[-k]\right).$$

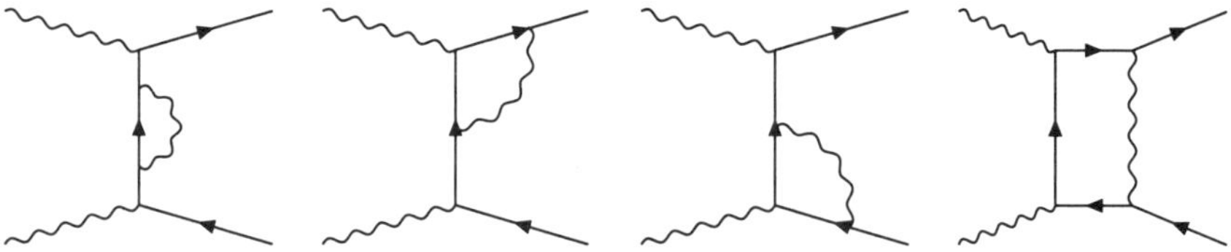

+ crossed graphs

Figure 2. QED loop corrections.

The matrix elements are obtained using the method of helicity amplitudes [10]:

$$\left|M_2^{+-+-}\right|^2 = 4e^4 \frac{u}{t} = 4e^4 \frac{1+\cos\Theta_{2,2'}}{1-\cos\Theta_{2,2'}} = 4e^4 \frac{s+t}{-t}, \tag{2}$$

$$\left|M_2^{+--+}\right|^2 = 4e^4 \frac{t}{u} = 4e^4 \frac{1-\cos\Theta_{2,2'}}{1+\cos\Theta_{2,2'}} = 4e^4 \frac{-t}{s+t}. \tag{3}$$

The set of invariants

$$\begin{array}{llll} s = (p_1+p_2)^2, & t = ({p_2}'-p_2)^2, & u = ({p_2}'-p_1)^2, & y = -t/s, \\ \upsilon = 2\,{p_1}'\!\cdot k\,, & \nu = 2\,{p_2}'\!\cdot k\,, & z = 2\,p_1\!\cdot k\,, & z' = 2\,p_2\!\cdot k \end{array}$$

are introduced.

It is essential feature of this process that M_2^{++xx} and M_2^{--xx} amplitudes at the Born approximation have the order $\left(m^2/s\right)^2$ and are negligible at high energies.

The integration over $d\phi$ for the $\gamma\gamma \to l^+l^-$ process is performed as follows:

$$\int A d\phi_2 = \frac{1}{8\pi s}\int A\, dt_{(2)} \simeq \frac{1}{16\pi}\int A\, d\cos\Theta_{2,2'}. \tag{4}$$

The QED loop corrections are represented by diagrams on Fig. 2. We can factorize them upon the Born cross section as follows:

$$d\sigma^{+-xx} = d\sigma_{(2)}^{+-xx} \cdot \frac{\alpha}{2\pi} \cdot \delta_V^{+-xx}, \tag{5}$$

$$\delta_V^{+-+-} = 2\ln\frac{s}{\lambda^2}\left(1-\ln\frac{s}{m^2}\right)+\ln^2\frac{s}{m^2}+\ln\frac{s}{m^2}+\ln^2\frac{-u}{s}+ \\ +\frac{s^2}{u^2}\ln^2\frac{-t}{s}+\left(1-\frac{2s}{u}\right)\ln\frac{-t}{s}+\frac{4\pi^2}{3}-3; \tag{6}$$

$$\delta_V^{+--+} = 2\ln\frac{s}{\lambda^2}\left(1-\ln\frac{s}{m^2}\right)+\ln^2\frac{s}{m^2}+\ln\frac{s}{m^2}+\ln^2\frac{-t}{s}+ \\ +\frac{s^2}{t^2}\ln^2\frac{-u}{s}+\left(1-\frac{2s}{t}\right)\ln\frac{-u}{s}+\frac{4\pi^2}{3}-3. \tag{7}$$

Here we have introduced the finite photon mass λ to remove the infrared divergence.

The real photon emission for this process is a pure QED reaction. It is indistinguishable from the $\gamma\gamma \to l^+l^-$ process in the infrared (IR) limit and it's singularities cancel ones caused by loop corrections.

Using the method of helicity amplitudes[10], one can calculate

$$\left|M^{+--++}\right|^2 = e^6\frac{4\,p_1'\cdot p_2'\,(p_2'\cdot p_2)^2}{p_1'\cdot k\; p_2'\cdot k\; p_1'\cdot p_1\; p_2'\cdot p_1}, \tag{8}$$

The other non-vanishing amplitudes are obtained from $|M^{+--++}|$ by using C, P, Bose and crossing (between final and initial particles) symmetries:

$$d\sigma^{-\lambda_1,-\lambda_2,-e_1',-e_2',-\lambda_3} = d\sigma^{\lambda_1,\lambda_2,e_1',e_2',\lambda_3},\ (P)$$

$$d\sigma^{+-+--} = d\sigma^{+--++}|_{1\leftrightarrow 2},\ (P+Bose)$$

$$d\sigma^{+-+-+} = d\sigma^{+--++}|_{1'\leftrightarrow 2'},\ (C)$$

$$d\sigma^{+--+-} = d\sigma^{+--++}|_{\substack{1\leftrightarrow 2\\ 1'\leftrightarrow 2'}},\ (CP+Bose)$$

$$d\sigma^{+++--} = d\sigma^{+--++}|_{\substack{3\leftrightarrow 2\\ 1'\leftrightarrow 2'}},\ (C+crossing)$$

$$d\sigma^{++-+-} = d\sigma^{+++--}|_{1'\leftrightarrow 2'}.\ (C)$$

The last couple of substitutions leads to the non-divergent leading term of $\gamma^+ - \gamma^+$ – scattering.

It is convenient to perform the integration over the phase-space of the final particles numerically. But the Monte-Carlo methods of numerical analysis require to eliminate all the divergences in the integration expressions.

The "forward-backward" divergences can be deleted by imposing cuts on the scattering angle (in calculation of $d\sigma/d\cos\Theta$) or on the $(p_i\cdot p_f)$-invariants (for $d\sigma/dy$). Another singularities should be extracted as a single expression $|M|^2_{sub}$. After this term has been subtracted the matrix element doesn't contain divergences and can be integrated numerically. The singular term $|M|^2_{sub}$ should be integrated analytically.

The infrared behaviour of helicity amplitudes can be found by covariant expanding (8) of matrix elements into a series around pole $\omega_{\gamma_{real}} \to 0$:

$$\begin{aligned} \left|M^{+-+-}\right|^2_{IR} &= 16e^6 \frac{s}{\upsilon\nu}\frac{u}{t}\left(1-\frac{\nu}{s}-\frac{\upsilon}{s}\right) + 8e^6 \frac{s}{\upsilon\nu}\frac{(\upsilon-z)u+(\nu-z)t}{t^2}, \\ \left|M^{+--+}\right|^2_{IR} &= 16e^6 \frac{s}{\upsilon\nu}\frac{t}{u}\left(1-\frac{\nu}{s}-\frac{\upsilon}{s}\right) - 8e^6 \frac{s}{\upsilon\nu}\frac{(\upsilon-z)u+(\nu-z)t}{u^2}. \end{aligned} \tag{9}$$

The first term of each expression has the usual IR-singularity and the rest one is divergent in the massless limit.

The divergences caused by $p_f{\cdot}k \to 0$ can be extracted using the method of peaking approximation:

$$\begin{aligned} \left|M^{+-+-}\right|^2_{peak} &= 8e^6 \frac{s}{\upsilon\nu}\frac{u}{t}\left(1-\frac{\nu}{s}+\frac{\nu^2}{s^2}-\frac{\upsilon}{s}+\frac{\upsilon^2}{s^2}\right), \\ \left|M^{+--+}\right|^2_{peak} &= 8e^6 \frac{s}{\upsilon\nu}\frac{t}{u}\left(1-\frac{\nu}{s}+\frac{\nu^2}{s^2}-\frac{\upsilon}{s}+\frac{\upsilon^2}{s^2}\right). \end{aligned} \tag{10}$$

Each formula of Eqs. (9) and (10) can be combined into the united expression:

$$\begin{aligned} \left|M^{+-+-}\right|^2_{sub} &= 8e^6 \frac{s}{\upsilon\nu}\left(\frac{u}{t}\left(1-\frac{\nu}{s}+\frac{\nu^2}{s^2}-\frac{\upsilon}{s}+\frac{\upsilon^2}{s^2}\right)+\frac{tu'-ut'}{t^2}\right), \\ \left|M^{+--+}\right|^2_{sub} &= 8e^6 \frac{s}{\upsilon\nu}\left(\frac{t}{u}\left(1-\frac{\nu}{s}+\frac{\nu^2}{s^2}-\frac{\upsilon}{s}+\frac{\upsilon^2}{s^2}\right)-\frac{tu'-ut'}{u^2}\right). \end{aligned} \tag{11}$$

This directly leads to eqs. (9) in the IR-limit. And it differs from (10) on the term that vanishes in the peaking limits due to

$$(\upsilon-z)u+(\nu-z)t = (tu'-ut') \overset{peak}{\longrightarrow} 0. \tag{12}$$

The second term in (11) is only a IR-divergent one. To simplify further calculations we introduce arbitrary value $\bar{\upsilon}$ as an upper bound for it's integration (and subtraction). Neither numerical no analytical part of the result does not depend on $\bar{\upsilon}$ if it is chosen in the region $m^2 \ll \bar{\upsilon} \ll s$ (or $m^2 \ll \bar{\upsilon} \ll (s+t)$ in case of y-dependent differential cross-section).

The IR-divergences can be factorized upon matrix element in a covariant path that after squaring gives

$$|M^{\lambda}|^2 = 4e^2|M^{Born}|^2 \left(\frac{s'}{v\nu} - \frac{m^2}{v^2} - \frac{m^2}{\nu^2}\right). \tag{13}$$

The m^2–dependent terms form (13) do not appear in helicity amplitude expressions since setting mass to zero but they should be included in calculations for proper cancelation of divergences.

The result of analitical integration over the phase space of final photon for the"soft" and "collinear" parts of bremsstrahlung is

$$\delta_R^{\Theta} = 2\ln\frac{s}{\lambda^2}\left(\ln\frac{s}{m^2} - 1\right) - \ln^2\frac{s}{m^2} - \ln\frac{s}{m^2} - \frac{4\pi^2}{3} + \frac{13}{2}. \tag{14}$$

Combining loop correction expressions (6), (7) and the bremsstrahlung contribution (14) one can obtain

$$\delta_R^{\Theta} + \delta_V^{+-+-} = \ln^2(1-y) + \frac{1}{(1-y)^2} ln^2 y + \left(1 + \frac{2}{1-y}\right)\ln y + \frac{7}{2}, \tag{15}$$

$$\delta_R^{\Theta} + \delta_V^{+--+} = \ln^2(y) + \frac{1}{y^2}\ln^2(1-y) + \left(1 + \frac{2}{y}\right)\ln(1-y) + \frac{7}{2}. \tag{16}$$

Here y is the function of angle between initial and final particles:

$$y = \frac{1}{2}\left(1 - \cos\Theta_{2,2'}\right).$$

The integration results for the invariant-dependent spectra are so complicated that can't be outlined here.

The final state polarization can scarcely be measured at experiment. That is the reason for summarizing over the helicities of all final particles.

The major feature of $\gamma\gamma \to l^+l^-$ process is the small value of cross section if the total angular momentum of $\gamma\gamma$-beams equals zero. This polarization selectivity can be useful at the experiment.

Since masses don't appear in eqs. (15), (16), the relative value of the c.m.s.-angle dependent QED-correction has the order about 1%. These corrections don't depend on the energy of scattering photons. The invariant expression $d\sigma/dy$ includes $\ln m$ terms that leads to the corrections of order $2 - 20\%$, raising with energy.

The achieved precision is sufficient for the huge variety of experiments at the photon collider.

4. Production of four leptons in $\gamma\gamma$ electroweak interaction

Since the high accuracy and relatively clean environment are provided by a linear collider, a precision calculation of background including $\gamma\gamma \to (...) \to 4l$ process is necessary.

Total cross sections of such interactions have been already calculated and analyzed[12] about 30 years ago and were found to be large enough. However these calculations have used the low energy approximation and obtained results are not applicable for modern high energy investigations.

There was applied the algorithm ALPHA[7] for automatic computations of scattering amplitude. However modern high energy experiments require the cross section calculation and analysis at fixed polarization states of initial and final particles that ALPHA method couldn't provide.

Our purpose was investigation of process $\gamma\gamma \to 4l$ using parameters of linear colliders of new generation such as TESLA[2].

There are six topographically different Feynman diagrams of electroweak interactions for describing of this process (see Fig. 3). Using C-, P- and crossing symmetries one can built 40 different diagrams.

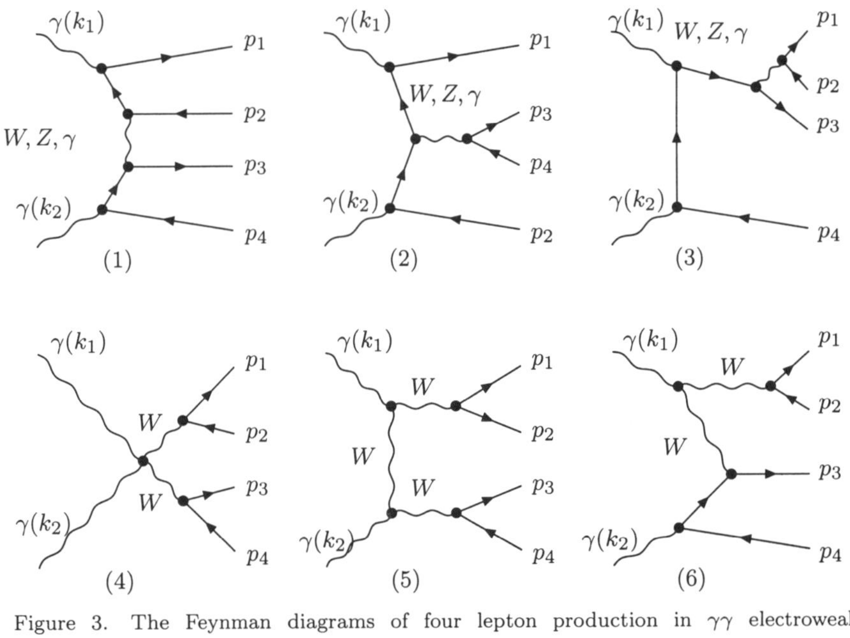

Figure 3. The Feynman diagrams of four lepton production in $\gamma\gamma$ electroweak interaction.

The diagrams containing charged current exchange are excluded because only processes with four charged leptons in the final state are considered.

The corresponding cross section has the form:

$$\sigma = \frac{1}{4(k_1 k_2)} \int |M|^2 d\Gamma. \tag{17}$$

Here $M = \sum_{i=1}^{3} M_i$, M_i are the matrix elements of remaining diagrams (1)–(3) in Fig. 3 and $d\Gamma$ is the phase space element.

Squared matrix element is constructed using the method of helicity amplitudes[10] as well as the method of precision covariant calculation. The first one allows to calculate matrix element directly for each definite polarization state of initial and final particles. Amplitude constructed by this method contains only invariants without any bispinor, so many difficulties at squaring and numerical integration are excluded.

The explicit form of all matrix elements constructed using helicity amplitude method are recently published[9].

The method of precision covariant calculations allows to obtain matrix element without any approximations and was used for verifying results in each step of our construction and calculation.

For the investigation of total and differential cross section the method of Monte-Carlo integration was applied. If two or more produced particles propagate into very close direction, the square of matrix element becomes very large (so-called collinear peaks problem). For the achievement of high precision the Monte-Carlo generator was adopted. Despite regular distributions of kinematic variables (such distributions usually applied into Monte-Carlo generators) we have used irregular one, which is very closed to matrix element behavior. This proximity can be achieved by choosing of several free parameters available in the distribution function. So the adopted method of Monte-Carlo integration leads to results with very small numerical error, nearly $(0.1\% - 0.2\%)$.

The magnitude of differential cross section increases significantly at polar angle tends to 0 or π. The cross section raises with decreasing of interaction particles energy. It has symmetric (asymmetric) form in case of similar (different) spin configuration of scattering photons. The cross section dependence on angle between initial photon momentum and a set of final particles directions has the same form due to electromagnetic interaction results are only presented.

5. The SM and anomalous amplitudes of $\gamma\gamma \to W^+W^-$

Future high-energy linear e^+e^--colliders in $\gamma\gamma$-mode could be a very useful instrument to explore the mechanism of symmetry breaking in electroweak interaction. $\gamma\gamma$-colliders give the opportunity for mesuarement of a light Higgs productions on resonance and $H\gamma\gamma$ coupling because of the processes $\gamma\gamma \to H \to ZZ, b\bar{b}, WW$. In the case of heavy mass Higgs bosons (over $1\,TeV$) a huge background from $\gamma\gamma \to W_T^+W_T^-$ complicates a search of the Higgs signal.

Another way to probe the symmetry breaking is examination of the self couplings of the W, Z bosons in non-minimal gauge models. Since the cross section of $\gamma\gamma \to W^+W^-$ is large the WW-production would be provided mainly by $\gamma\gamma$ scattering[13]. So the Born cross section $\sigma(\gamma\gamma \to WW)$ is about $110pb$ at $1\,TeV$ on unpolarized γ-beams at that all contributions of different polarization sets of W-bosons are considered. Corresponding cross section of WW-production in e^+e^- collisions is an order of magnitude smaller and amount to $10pb$ at the same circumstances. In addition to the $\gamma\gamma \to W^+W^-$ process one needs to consider a reaction of $\gamma\gamma \to W^+W^-Z$ because of its cross section becomes about 5% – 10% of the $\sigma(\gamma\gamma \to WW)$ at energies $\sqrt{s} \geq 500\,GeV$. The anomalous tri-linear[14] γWW, ZWW and quartic[15] $\gamma\gamma WW$, γZWW, $ZZWW$ couplings induce deviations of Born cross sections $\sigma(\gamma\gamma \to WW)$ and $\sigma(\gamma\gamma \to WWZ)$ from Standard Model values.

In order to evaluate contributions of anomalous couplings a cross section of $\gamma\gamma \to W^+W^-$ must be calculated with a high precision. Therefore one needs to calculate the main contribution of high order electroweak effects: one-loop correction, real photon and Z-boson emission.

The amplitude of $\gamma\gamma \to W^+W^-$ are defined as follows

$$M = G_v \epsilon_\mu(k_1)\epsilon_\nu(k_2)\epsilon_\alpha(p_+)\epsilon_\beta(p_-)M_T^{\mu\nu\alpha\beta}, \tag{18}$$

where k_1, k_2, p_+, p_- are 4-momenta of the γ, γ, W^+, W^- respectively, and $\epsilon_\mu(k_1)$, $\epsilon_\nu(k_2)$, $\epsilon_\alpha(p_+)$, $\epsilon_\beta(p_-)$ – 4-vectors of polarizations accordingly. The constant G_v is equal to $e^3 \cot\theta_W$. The total amplitude M_T in Born approximation is defined by three main amplitudes M_i $(i = 1, 3)$:

$$M_T^{\mu\nu\alpha\beta} = \sum_{i=1}^{3}, M_i^{\mu\nu\alpha\beta}. \tag{19}$$

These three amplitudes are constructed from a trilinear boson $\Gamma_3^{\mu\alpha\beta}$, a quartic boson $\Gamma_4^{\mu\nu\alpha\beta}$ vertices and a boson propagator $D_{\alpha\beta}(p)$, where p is

4-momentum of a virtual W-boson:

$$M_1^{\mu\nu\alpha\beta} = \Gamma_3^{\mu\alpha\xi}(-k_1, p_+) D_{\xi\lambda}(p_+ - k_1) \Gamma_3^{\nu\beta\lambda}(-k_2, p_-), \tag{20}$$

$$M_2^{\mu\nu\alpha\beta} = \Gamma_3^{\mu\beta\xi}(-k_1, p_-) D_{\xi\lambda}(p_- - k_1) \Gamma_3^{\nu\alpha\lambda}(-k_2, p_+), \tag{21}$$

$$M_3^{\mu\nu\alpha\beta} = \Gamma_4^{\mu\nu\alpha\beta}. \tag{22}$$

Consider the anomalous quartic boson vertices only. For this purpose the following 6-dimensional $SU(2)_c$ Lagrangians[16,17] have been chosen

$$\begin{aligned} \mathcal{L}_0 &= -\frac{e^2}{16\Lambda^2} a_0 F^{\mu\nu} F_{\mu\nu} \bar{W}^\alpha \bar{W}_\alpha, \\ \mathcal{L}_c &= -\frac{e^2}{16\Lambda^2} a_c F^{\mu\alpha} F_{\mu\beta} \bar{W}^\beta \bar{W}^\alpha, \\ \tilde{\mathcal{L}}_0 &= -\frac{e^2}{16\Lambda^2} \tilde{a}_0 F^{\mu\alpha} \tilde{F}_{\mu\beta} \bar{W}^\beta \bar{W}^\alpha, \end{aligned} \tag{23}$$

where the triplet of gauge bosons $\bar{W}_\mu$ and the field-strength tensors

$$F_{\mu\nu} = \partial_\mu A_\nu - \partial_\nu A_\mu, \quad W^i_{\mu\nu} = \partial_\mu W^i_\nu - \partial_\nu W^i_\mu, \quad \tilde{F}_{\mu\nu} = \frac{1}{2}\epsilon_{\mu\nu\rho\sigma} F^{\rho\sigma}$$

are introduced. The scale Λ keeps the coupling constant a_i dimensionless. In our calculations Λ are fixed by value of M_W ($\sim$ 80 GeV). As one can see the operators $\mathcal{L}_0$ and $\mathcal{L}_c$ are C-, P-, CP-invariant. $\tilde{\mathcal{L}}_0$ is the P- and CP-violating operator. The anomalous quartic boson vertices depend on the 4-momenta k_1, k_2 of the photons involved in the vertex.

Total cross section of $\gamma\gamma \to W^+W^-$ is calculated as

$$\sigma_{\lambda_1\lambda_2\lambda_3\lambda_4} = \frac{1}{2S} \int |M_{\lambda_1\lambda_2\lambda_3\lambda_4}|^2 d\Gamma^{(2)}, \tag{24}$$

where $M_{\lambda_1\lambda_2\lambda_3\lambda_4}$ have been defined in eq. (18), $d\Gamma$ is two-particle phase space of the bosons.

Through the method of Monte-Carlo integration one may obtain numerical values of the cross section[18]. For these purposes the events generator has been built.

Figure 4 shows the dependence of three total cross sections $\sigma(W^+W^-)$ on anomalous parameters. Graphs presented on the picture allow to value contributions of anomalous parts to the cross section. The fact that the minima of the curves are close to the SM point $a_i = 0$ means that the interference between the anomalous and the standard part of the matrix

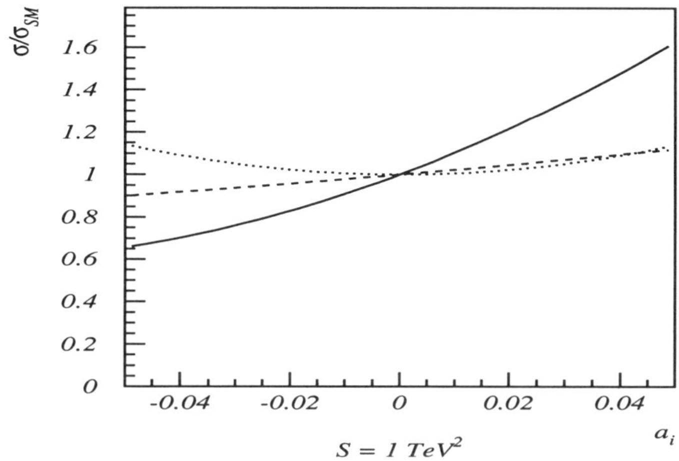

Figure 4. The comparison of the dependency of the cross sections $\sigma(W^+W^-)$. Solid line presents a_0-, dashed line – a_c-, dotted line – $\tilde{a}_0$-dependence

element is very small. Though the region of a_i is small, it's about 0.05, the cross section with anomalous constants may reach the values of 1.6σ. That stands for the evidence of great sensitivity of the cross section $\sigma(W^+W^-)$ to anomalous couplings. Taking into account a luminosity $\mathcal{L}$ of photons about $100 fb^{-1}/year$, energy $\sqrt{s} \sim 1\,TeV$ that corresponds to TESLA[2], statistical error will be equal to 0.05%. Therefore we can estimate bounds imposed on

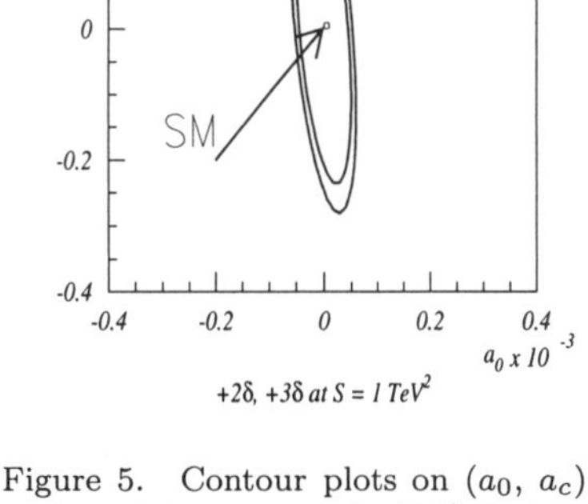

Figure 5. Contour plots on $(a_0,\, a_c)$ for $+2\delta$, $+3\delta$ deviations of $\sigma(W^+W^-)$

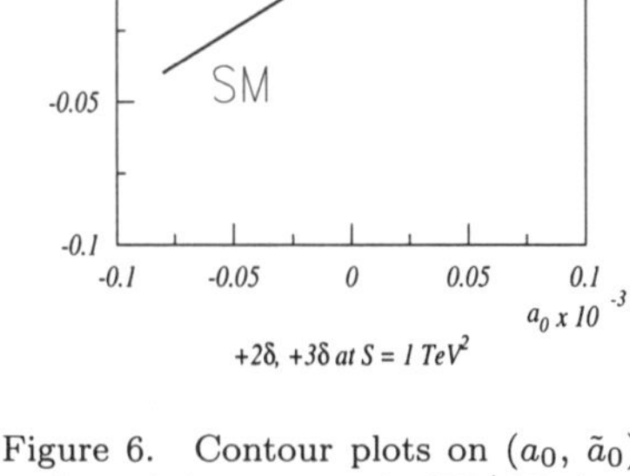

Figure 6. Contour plots on $(a_0,\, \tilde{a}_0)$ for $+2\delta$, $+3\delta$ deviations of $\sigma(W^+W^-)$

anomalous couplings to compare anomalous effect with statistical error. Contour plots on two non-zero anomalous constant, Figs. 5, 6, clearly demonstrate these bounds.

5.1. *The first-order radiative correction to anomalous constants*

Obviously a precision analysis of $\gamma\gamma \to WW$ at the future high energy $\gamma\gamma$-colliders is impossible without calculation of whole set the first-order $\mathcal{O}(\alpha)$ radiative corrections, including real photon emission as well as a set of one-loop diagrams of $\gamma\gamma \to W^+W^-$ that are presented in Fig. 1 and Fig. 2 of Ref.[19]. Inclusive cross section of the W pair production in the $\gamma\gamma$ collisions to the third order in α is given by the sum of Born cross section, interference term between the Born and one-loop amplitudes and cross section of the $WW\gamma$ production. In case of energy of $\gamma\gamma$-interaction exceeds threshold of three boson production these process ($\gamma\gamma \to W^+W^-Z$) must be considered as radiative correction:

$$
\begin{gathered}
d\sigma(\gamma\gamma \to W^+W^-) = d\sigma^{Born}(\gamma\gamma \to W^+W^-)+ \\
+\frac{1}{S}Re(M^{Born}M^{1-loop*})d\Gamma^{(2)} + d\sigma^{soft}(\gamma\gamma \to W^+ \\
+W^-\gamma) + d\sigma^{hard}(\gamma\gamma \to W^+W^-\gamma) + d\sigma^{Z}(\gamma\gamma \to W^+W^-Z).
\end{gathered} \tag{25}
$$

Since one-loop and soft photon emission amplitudes are IR-divergent and only their sum is IR-finite it is convenient to consider soft and hard photon emissions separately. $d\sigma^{soft}$ can be presented by factorizable expression

$$
d\sigma^{soft}(\gamma\gamma \to W^+W^-\gamma) = d\sigma^{Born}(\gamma\gamma \to W^+W^-)R^{soft}(\omega), \tag{26}
$$

where ω is soft photon energy cutoff, $\beta = \sqrt{1-4m_W^2/s}$. The differential cross section of hard photon emission is given by

$$
\begin{gathered}
d\sigma^{hard}(\gamma\gamma \to W^+W^-\gamma) = \\
= d\sigma(\gamma\gamma \to W^+W^-\gamma) - d\sigma^{soft}(\gamma\gamma \to W^+W^-\gamma)
\end{gathered} \tag{27}
$$

and can not be factorized. $d\sigma^{soft}$ and $d\sigma^{hard}$ present two contributions independent from infrared divergence as well as from soft photon cutoff.

One-loop amplitude M^{1-loop} were built due to usage of SCA (Algebra of Symbolic Calculations) programs (*MATHEMATICA*, *REDUCE*...) and transformations of scalar and tensor integrals to one-, two-, ..., six-point integrals. Cross section of WWZ production on $\gamma\gamma$ beams are obtained through application of the Monte-Carlo method of numerical integration and exact covariant expressions of $\gamma\gamma \to W^+W^-Z$ amplitudes. Cross section of $\gamma\gamma \to W^+W^-$ including the lowest-order QED radiative correction and three-boson production one can see in refs.[18–20].

Contour plot on a_0 and a_c of $\sigma(WW)$ with the lowest order QED correction are presented in Figs. 5, 6.

6. Conclusion

We have analyzed the possible usage of $\gamma\gamma \to f\bar{f}\gamma$ reaction for the luminosity measurement at $J = 0$ beams on linear photon collider. The achievable precision of the luminosity measuring is considered and calculated. The optimal conditions for that measurement are found (for the high magnitude of $J = 0$ cross section and small $J = 2$ background). The first-order QED correction to $\gamma\gamma \to l\bar{l}$ cross section is calculated and analyzed at $J = 2$-beams.

The considered process gives the excellent opportunity for luminosity measurements with substantial accuracy.

The cross section of process of four charged lepton production are obtained in frame of method of helicity amplitudes as well as method of covariant calculations. Comparative analysis are performed. Monte-Carlo generator was developed to calculate differential and total cross section of four lepton production.

The investigation of the sensitivity of process of $\gamma\gamma \to W^+W^-$ and $\gamma\gamma \to W^+W^-Z$ to genuine anomalous quartic couplings a_0, a_c and $\tilde{a}_0$ was performed at center-of-mass energy $\sqrt{s} = 1\,TeV$. It was discovered that two-boson production has great sensitivity to anomalous constants a_c and a_0 but process $\gamma\gamma \to W^+W^-Z$ is more suitable for study of $\tilde{a}_0$.

The fact that the minimum of the curves are close to the SM point $a_i = 0$ demonstrates the small value of the anomalous and the standard part interference. The first-order radiative correction to cross section $\sigma(\gamma\gamma \to W^+W^-)$ has significant magnitude and its calculation increases the precision of the a_0 and a_c measurement.

Bibliography

1. I. F. Ginzburg, G. L. Kotkin, V. G. Serbo and V. I. Telnov, *Nucl. Instr. Meth.* **205**, 47 (1983);
   I. F. Ginzburg, G. L. Kotkin, S. L. Panfil, V. G. Serbo and V. I. Telnov, *Nucl. Instr. Meth.* **219**, 5 (1984).

2. B. Badelek *et al.*, *TESLA Technical Design Report Part VI: The Photon Collider at TESLA*, *DESY-01-011E*, hep-ex/0108012.
3. E. A. Kuraev, M. V. Galynskii, M. I. Levchuk, *Phys. Part. Nucl.* **31**, 76 (2000), *Fiz. Elem. Chast. Atom. Yadra* **31**, 155 (2000);
V. Telnov, *Nucl. Instr. Meth.* **A494**, 35 (2002), hep-ex/0207093.
4. T. V. Shishkina, V. V. Makarenko, hep-ph/0212409;
V. Makarenko, K. Mönig, T. Shishkina, *Eur. Phys. J.* **C30**, d01, 011 (2003), *LC-PHSM-2003-016*, hep-ph/0306135;
V. V. Makarenko, T. V. Shishkina, hep-ph/0310104.
5. V. V. Makarenko, T. V. Shishkina, *Proceedings of 8th International School-Seminar On The Actual Problems Of Microwold Physics*, Gomel (2003).
6. A. Denner, S. Dittmaier, *Eur. Phys. J.* **C9**, 425 (1999), hep-ph/9812411.
7. M. Moretti, *Nucl. Phys.* **B484**, 3 (1997), hep-ph/9604303;
M. Moretti, hep-ph/9606225.
8. C. Carimalo, W. da Silva, F. Kapusta, *Nucl. Phys. Proc. Suppl.* **82**, 391 (2000), hep-ph/9909339.
9. T. Shishkina , I. Sotsky, hep-ph/0312208.
10. P. De Causmaecker *et al.*, *Nucl. Phys.* **B206**, 53 (1982);
F.A. Berends *et al.*, *Nucl. Phys.* **B206**, 61 (1982).
11. D.L. Borden, D.A. Bauer, D.O. Caldwell, *Phys. Rev.* **D48**, 4018 (1993);
J.F. Gunion, H.E. Haber, *Phys. Rev.* **D48**, 5109 (1993).
12. L.N.Lipatov and G.V. Frolov *ZHETV Pis. Red.* **10**, 399 (1969), *JETP Letters* **10**, 254 (1969);
V.G. Serbo, *ZHETF Pis. Red.* **12**, 33 (1970);
H. Cheng and T.T. Wu, *Phys. Rev.* **D1**, 3414 (1970).
13. K. Hagiwara *et al.*, *Nucl. Phys.* **B282**, 253 (1987).
14. G. Belanger *et al.*, *Eur. Phys. J.* **C13**, 283 (2000), hep-ph/9908254.
15. G. Gounaris et al., *"Physics at LEP2"* v.**1**, 525 (1992), hep-ph/9601233.
16. A. Denner, S. Dittmaier, M. Roth, D. Wackeroth, *Nucl. Phys.* **B560**, 33 (1999), hep-ph/9904472.
17. A. Denner, S. Dittmaier, M. Roth, D. Wackeroth, *Eur. Phys. J.* **C20**, 201 (2001), hep-ph/0104057.
18. I. Marfin, V. Mossolov, T. Shishkina, hep-ph/0305153.
19. I. Marfin, V. Mossolov, T. Shishkina, *LC-PHSM-2003-085*, hep-ph/0304250.
20. I. Marfin, V. Mossolov, T. Shishkina, *DESY-PROC-2003-2*, 384, *PIC-2003-FRAP11*, *eConf C030626*:FRAP11 (2003), hep-ph/0308317.

ALGEBRAIC METHOD WITH THE USE OF MANY-PARTICLE COULOMB GREEN FUNCTION FOR ATOMIC CALCULATIONS

Le Van Hoang

Department of Physics, HCMC University of Pedagogy
280 An Duong Vuong St., Ward 5, Ho Chi Minh City, Vietnam
E-mail: hoanglv@hcm.fpt.vn

On the basis of the relation between the Coulomb Green function and the Green function of a harmonic oscillator, the algebraic representation of the first one in the form of annihilation and creation operators is established. The many-particle Coulomb Green function is established in the algebraic representation in the same way. These results allow us to construct a purely algebraic method for atomic calculations and thus to reduce rather complicated calculations of matrix elements to algebraic procedures of transforming the products of annihilation and creation operators to a normal form. The effectiveness of the constructed method is demonstrated by two example problems: (i) non-relativistic radiation shifts of the levels of hydrogen-like atoms (ii) the ground state of hydrogenic molecule. The calculation algorithm of this algebraic approach is simple and suitable for symbolic calculation programs, such as Mathematica, that widely extend the application area of the Coulomb Green function.

1. Introduction

The Coulomb Green function (CGF) in the closed form was first constructed by Hostler in 1964[1] and has been used till now in many applications of atomic calculations (see[2,3]). The use of CGF widely enlarges the application area of analytical methods in perturbation theory for atomic calculations. However, researchers have to face up to an enormous amount of integrations of special functions. That is the reason why the other representations of CGF have been continuatively constructed[3]. In this context, the Sturmian expansion of CGF has been done and used for various applications[4–10]. From our point of view, the application problem of CGF is still open for further investigation (see also[3]).

Our developing trend in this area is related to the algebraic method established by Professor Komarov L. I.[11,12] for atomic calculations. In this algebraic approach, by using the connection between hydrogen-like atoms with harmonic oscillator, the hydrogen-like atom wave-functions are constructed via annihilation and creation operators by its action on the vacuum state. All physical operators are expressed via the annihilation and creation operators, and one can reduce rather complicated calculations of matrix elements into the pure algebraic transformations[11]. See also the work of Professor Kleinert H[13], in which the topic is well considered by the group theory. However, for concrete applications, especially for developing to the relativistic case, the calculation technique is given clearly in[11,12]. Visibly, it prompts us to construct a representation of CGF via the annihilation and creation operators. The operator of CGF should be established in a normal form that is suitable for applying the algebraic method.

This paper frames some of our results[14] related to the mentioned-above purpose. In Section 2, we establish the connection between the Coulomb Green function and the Green function for harmonic oscillator. The found connection is considered a basis for further constructing algebraic form of CGF in Section 4. The representation of CGF in the form of annihilation and creation operators, called the operator form of CGF, is applied for algebraic calculations of the non-relativistic radiation shift of the hydrogen-like atom energy[15]. Section 3 remains some points of the algebraic method of using the annihilation and creation operators for use in the other sections. In Section 5, the many-particle Coulomb Green function (MCGF) is given in the separated form of the CGF for each electron. The result allows us to construct operator form of MCGF suitable for algebraic calculations of many-electron atom problems. The application example is demonstrated via the problem of the hydrogenic molecule.

The concrete calculations given in this paper are for instances of the algebraic approach only. Applications for the most actual problems will be given in other works (see[16]). Moreover, this paper deals only with non-relativistic problems. For the results of development in relativistic case, see the works[17–19].

2. Connection between the Coulomb Green function and the Green function of harmonic oscillator

By path-integral technique, the above-mentioned connection has been established in[20] (see also lecture of Professor Kleinert H[21]) and generalized in[22,23]. The purpose of these cited works is to develop the Feynman calculation technique for new areas, such as the Coulomb problem, Mic-Kepler problem, and Aharonov-Bohm effect problem. Differently, in this paper we focus on the connection as itself and its further application in concrete calculations. Therefore, we will approve this connection by direct transformation of variables.

Let us consider an equation for the Green function of harmonic oscillator in space of two-dimensional complex coordinates ξ_s, ξ_s^*:

$$\left\{-\frac{1}{2}\frac{\partial^2}{\partial\xi_s\,\partial\xi_s^*}+\frac{1}{2}\omega^2\xi_s\,\xi_s^*-Ze^2\right\}U\left(\xi,\eta;Ze^2\right)=-i\delta^4(\xi-\eta), \tag{1}$$

where repeating indices denote summation over them ($s = 1,2$); ω is a positive number; $U\left(\xi,\eta;Ze^2\right)$ is the Green function of harmonic oscillator in ξ-space; the notation in the right side of (1) means:

$$\delta^4(\xi-\eta)=\delta(\xi_1^{'}-\eta_1^{'})\,\delta(\xi_1^{"}-\eta_1^{"})\,\delta(\xi_2^{'}-\eta_2^{'})\,\delta(\xi_2^{"}-\eta_2^{"})$$

with $\xi_s^{'}=\mathrm{Re}\,\xi_s$, $\xi_s^{"}=\mathrm{Im}\,\xi_s$. We change the variables in (1) using the KS transformation[24]:

$$\begin{cases} x_\lambda=\xi_s^*\,(\sigma_\lambda)_{st}\xi_t \\ \phi=\arg(\xi_1),\ 0\leqslant\phi\leqslant 2\pi \end{cases} \tag{2}$$

with σ_λ ($\lambda = 1,2,3$) are Pauli matrices.ϕ is assumed as "extra" variable in meaning that all wave-functions in ξ space satisfied condition:

$$\frac{\partial\psi(\xi)}{\partial\phi}=0, \tag{3}$$

after transformation (2), can be reduced to being dependent only on variables $\mathbf{r}$. They are called "physical" wave-functions, and in further consideration, we will deal only with them. From (2), we obtain

$$d\xi_1^{'}\,d\xi_1^{"}\,d\xi_2^{'}\,d\xi_2^{"}=\frac{1}{8r}dx_1\,dx_2\,dx_3\,d\phi \tag{4}$$

and, moreover:

$$\delta^4(\xi-\eta)=8r\,\delta(\mathbf{r}-\mathbf{r}')\,\delta(\phi-\phi') \tag{5}$$

with $r=\sqrt{(x_1)^2+(x_2)^2+(x_3)^2}$. Here $(\mathbf{r}',\phi')$ are variables obtained from variable η correspondently. Equation (1) after transformation (2) has the form:

$$\left\{-\frac{1}{2}r\left(\frac{\partial}{\partial x_\lambda}+A_\lambda(\mathbf{r})\frac{\partial}{\partial\phi}\right)\left(\frac{\partial}{\partial x_\lambda}+A_\lambda(\mathbf{r})\frac{\partial}{\partial\phi}\right)-\frac{1}{8r}\frac{\partial^2}{\partial\phi^2}\right.$$
$$\left.+\frac{1}{2}\omega^2r-Ze^2\right\}U\left(\mathbf{r},\phi,\mathbf{r}',\phi';Ze^2\right)=-8ir\delta(\mathbf{r}-\mathbf{r}')\,\delta(\phi-\phi'). \tag{6}$$

Here $A_\lambda(\mathbf{r})$ is related to the vector-potential of the Dirac monopole[22]. Operator in left side of (6) depends on ϕ via operator $\partial/\partial\phi$ only, thus follows the fact that the variable ϕ is contained in the Green function via the term $\phi - \phi'$ only. From this circumstance, we can assume that the function:

$$K(\mathbf{r}, \mathbf{r}'; E) = \frac{1}{8} \int_0^{2\pi} d\phi' \, U\left(\mathbf{r}, \mathbf{r}', \phi - \phi'; Ze^2\right) \tag{7}$$

is independent from variable ϕ. Integrating both sides of equation (6) over ϕ', we obtain:

$$\left\{ -\frac{1}{2} \frac{\partial^2}{\partial x_\lambda \partial x_\lambda} - \frac{Ze^2}{r} - E \right\} K\left(\mathbf{r}, \mathbf{r}'; E\right) = -i\delta(\mathbf{r} - \mathbf{r}'), \tag{8}$$

where $E = -\omega^2/2$. From (8) we can find that $K(\mathbf{r}, \mathbf{r}'; E)$ is the Coulomb Green function in the "energy" representation and, certainly, the found expression (7) describes the connection between the Green functions. For the "physical" wave-functions, we can ignore integration over ϕ', and (7) allows us to use the Green function of harmonic oscillator instead of CGF in concrete calculations.

3. Algebraic Method using annihilation and creation operators

In this section, we review some main points in the work of Professor Komarov about the algebraic method of solving Schrödinger equation based on the connection between the Coulomb problem and the problem of harmonic oscillator[11] (see also[13]), which will be used in later sections. Readers interested in relativistic case can see also[12,17].

By analogy given in Section 2, one can find the same connection between the Schrödinger equations. Thus the wave-functions of hydrogen-like atoms can be extracted from the wave-functions of harmonic oscillator in ξ space by demanding them to be satisfied the condition (3). From the other side, it is well known that the wave-functions of harmonic oscillator in ξ space can be represented in the form:

$$|n_1, n_2, n_3, n_4\rangle = \frac{1}{\sqrt{n_1! n_2! n_3! n_4!}} (\hat{a}_1^+)^{n_1} (\hat{a}_2^+)^{n_2} (\hat{b}_1^+)^{n_3} (\hat{b}_2^+)^{n_4} |0(\omega)\rangle \tag{9}$$

where the annihilation and creation operators are defined as follows:

$$\hat{a}_s(\omega) = \sqrt{\frac{\omega}{2}}\left(\xi_s + \frac{1}{\omega}\frac{\partial}{\partial \xi_s^*}\right), \quad \hat{a}_s^+(\omega) = \sqrt{\frac{\omega}{2}}\left(\xi_s^* - \frac{1}{\omega}\frac{\partial}{\partial \xi_s}\right)$$
$$\hat{b}_s(\omega) = \sqrt{\frac{\omega}{2}}\left(\xi_s^* + \frac{1}{\omega}\frac{\partial}{\partial \xi_s}\right), \quad \hat{b}_s^+(\omega) = \sqrt{\frac{\omega}{2}}\left(\xi_s - \frac{1}{\omega}\frac{\partial}{\partial \xi_s^*}\right) \qquad (10)$$

and the vacuum state is determined by the equations:

$$\hat{a}_s\,|0(\omega)\rangle = 0\,, \quad \hat{b}_s\,|0(\omega)\rangle = 0 \quad (s = 1,2), \quad \langle 0(\omega) \mid 0(\omega)\rangle = 1. \qquad (11)$$

Operators (10) satisfy the following standard commutative relations:

$$\left[\hat{a}_s(\omega), \hat{a}_t^+(\omega)\right] = \delta_{st}\,, \quad \left[\hat{b}_s(\omega), \hat{b}_t^+(\omega)\right] = \delta_{st} \qquad (12)$$

(the other commutators are equal to zero). From (9), we will now establish the state-vectors for hydrogen-like atom in discrete region of energy. For this purpose, the state-vector (9) is required to satisfy condition (3). In representation of operators (10), condition (3) has the equation form:

$$\left(\hat{a}_s^+\hat{a}_s - \hat{b}_s^+\hat{b}_s\right)\,|\psi\rangle = 0. \qquad (13)$$

Besides, it is well known that for hydrogen-like atoms, the orbital momentum and its projection in the z axis are integrals of motion. Thus we can build the wave-functions belonging to eigen-values of the operators $\hat{L}^2$, $\hat{l}_3$ i.e. satisfying the following equations:

$$\hat{L}^2\,|\psi\rangle = \left\{\frac{1}{4}N(N+2) - M^+M\right\}|\psi\rangle = l(l+1)|\psi\rangle, \qquad (14)$$

$$\hat{l}_3\,|\psi\rangle = \frac{1}{2}(\hat{a}_1^+\hat{a}_1 - \hat{a}_2^+\hat{a}_2 - \hat{b}_1^+\hat{b}_1 + \hat{b}_2^+\hat{b}_2)\,|\psi\rangle = m\,|\psi\rangle. \qquad (15)$$

For convenience, in (14) and henceforth, we use notations:

$$N + 2 = \hat{a}_s^+\hat{a}_s + \hat{b}_s^+\hat{b}_s + 2\,, \quad n_\lambda^a = (\sigma_\lambda)_{st}\hat{a}_s^+\hat{a}_t\,, \quad n_\lambda^b = (\sigma_\lambda)_{st}\hat{b}_t^+\hat{b}_s$$
$$M = \hat{a}_s\hat{b}_s\,, \quad M^+ = \hat{a}_s^+\hat{b}_s^+\,, \quad m_\lambda = (\sigma_\lambda)_{st}\hat{a}_t\hat{b}_s\,, \quad m_\lambda^+ = (\sigma_\lambda)_{st}\hat{a}_s^+\hat{b}_t^+. \qquad (16)$$

All operators corresponding to physical quantities after transforming into ξ space, can be expressed via combinations of annihilation and creation operators written in (16). We write, for example, some of them:

$$r = \xi_s^*\xi_s = \frac{1}{2\omega}(M + M^+ + N + 2),$$
$$x_\lambda = \xi_s^*(\sigma_\lambda)_{st}\xi_t = \frac{1}{2\omega}(m_\lambda + m_\lambda^+ + n_\lambda^a + n_\lambda^b), \qquad (17)$$
$$r\mathbf{p}^2 = -\frac{\partial^2}{\partial \xi_s \partial \xi_s^*} = \frac{\omega}{2}(N + 2 - M - M^+), \quad rp_\lambda \Rightarrow -\frac{i}{2}(m_\lambda - m_\lambda^+).$$

The Schrödinger equation for the hydrogen-like atom with nuclear charge Ze and energy $E = \omega^2/2$ by itself can be written in the form:

$$\left\{-\frac{1}{2}\omega\left(\hat{a}_s^+\hat{a}_s + \hat{b}_s^+\hat{b}_s + 2\right) - Ze^2\right\}|\psi\rangle = 0 \quad . \tag{18}$$

The way to construct solution of system of equations (13)–(15), (18) is straightforward. We write, for example, the hydrogen-like atom ns state-vectors:

$$|ns\rangle = \frac{1}{\sqrt{n!(n+1)!}}(\hat{a}_s^+\hat{b}_s^+)^n|0(\omega)\rangle = \frac{1}{\sqrt{n!(n+1)!}}(M^+)^n|0(\omega)\rangle. \tag{19}$$

For the other state wave-functions, one can find in the work[11]. The generalized form for wave functions in discrete energy region can be found in[12,13]. Presently, the problems of multi-photon ionization involve great interest for theoretical investigation[10]. For applying the algebraic approach to these problems, there is the need of constructing algebraic solution of equation (13)–(15), (18) for the continuous energy region. One can find these solutions in[16].

4. Coulomb Green function in the operator representation

When solving a Schrödinger equation such as:

$$(\hat{H} - E)\,\psi(\mathbf{r}) = -\hat{V}\psi(\mathbf{r}) \tag{20}$$

one can reduce to the integral equation:

$$\psi(\mathbf{r}) = -\int G(\mathbf{r},\mathbf{r}';E)\,\hat{V}\psi(\mathbf{r}')\,d\mathbf{r}' \tag{21}$$

with the use of the Green function, and then apply the method of perturbation theory onto this. However, in concrete calculations, it is more convenient to use the Green operator defined via the following formulation:

$$\hat{G}(\mathbf{r};E)\,\varphi(\mathbf{r}) = \int G(\mathbf{r},\mathbf{r}';E)\,\varphi(\mathbf{r}')\,d\mathbf{r}'. \tag{22}$$

Thus equation (21) can be rewritten in the form:

$$\psi(\mathbf{r}) = -(\hat{H}(\mathbf{r}) - E)^{-1}\hat{V}\psi(\mathbf{r}).$$

Here we use a formal notation $\hat{G}(\mathbf{r};E) = (\hat{H}(\mathbf{r}) - E)^{-1}$ with the meaning that the Green operator is inverse operator of the left part of the equation (20). Using an integral representation of an inverse operator:

$$\hat{A}^{-1} = -i\lim_{\varepsilon\to+0}\int_0^{+\infty} dt\, e^{-\varepsilon t + i\hat{A}\,t}, \tag{23}$$

we obtain:

$$\hat{G}(\mathbf{r};E) = -i \lim_{\varepsilon \to +0} \int_0^{+\infty} dt\, e^{-\varepsilon t + i(\hat{H}(\mathbf{r}) - E)t}. \tag{24}$$

In the case of the Coulomb Green operator, from (22) and from the correlation (7) between the Green functions, we can write:

$$\hat{K}(\mathbf{r};E)\,\varphi(\mathbf{r}) = \int K(\mathbf{r},\mathbf{r}';E)\,\varphi(\mathbf{r}')\,d\mathbf{r}' =$$
$$= \int \frac{1}{8}\int_0^{2\pi} d\phi'\, U\left(\mathbf{r},\mathbf{r}',\phi-\phi';Ze^2\right)\varphi(\mathbf{r}')\,d\mathbf{r}'.$$

Changing variables into the complex space ξ (η) and substituting (4), we have:

$$\hat{K}(\mathbf{r};E)\,\varphi(\mathbf{r}) = \int U\left(\xi,\eta;Ze^2\right)\,\varphi(\eta)\,\eta_s\eta_s^*\,d^4\eta = \hat{U}(\xi,Ze^2)\,\xi_s\xi_s^*\varphi(\xi).$$

Here we use notation $\hat{U}(\xi, Ze^2)$for the Green operator of harmonic oscillator, and the correlation between the Green operators, at last, is established as follows:

$$\hat{K}(\mathbf{r};E) = \hat{U}(\xi,Ze^2)\,\xi_s\xi_s^*. \tag{25}$$

This connection prompts us to rewrite the equation in the other way, such as:

$$\psi(\mathbf{r}) = -(r\hat{H}(\mathbf{r}) - Er)^{-1} r\hat{V}\psi(\mathbf{r}). \tag{26}$$

Here the Green operator will be used in concrete calculations in the form:

$$\hat{U} = -i \lim_{\varepsilon \to +0} \int_0^{+\infty} dt\, e^{-\varepsilon t + ir(\hat{H}(\mathbf{r}) - E)t}. \tag{27}$$

We note that all operators in (26)–(27) are written in r space just for convenience. This notation is formal only, which later on has to be conceived via the representation of operators (16)–(17) as:

$$r(\hat{H}(\mathbf{r}) - E) = \frac{1}{2}r\mathbf{p}^2 - Er - Ze^2 =$$
$$= \left(\frac{\omega}{4} - \frac{E}{2\omega}\right)(N+2) - \left(\frac{\omega}{4} + \frac{E}{2\omega}\right)(M + M^+) - Ze^2.$$

In case of $E = -\omega^2/2$, the Green operator (27) has a simple form:

$$\hat{U}(Z) = -i \lim_{\varepsilon \to +0} \int_0^{+\infty} dt\, e^{-\varepsilon t + iZe^2 t} \exp\left\{ \frac{i\omega t}{2}(N+2) \right\} \tag{28}$$

and is suitable for acting to the state-vectors such as (19).

In case of $E = \pm\nu^2/2$, $\nu \neq \omega$ $(\nu \geq 0)$, which is often met in atomic calculations, the exponential factor in (27) has to be introduced into a normal form i.e. in which annihilation operators are placed in right side, and creation operators in the left one. This normal form is need for algebraic calculations. For this purpose, we note that operators (16) generate a closed algebra according to the commutative correlations among them. Indeed, using (12), we can easily obtain all these commutative correlations. We write, for example, the following:

$$[M\,,\,M^+] = N+2\,, \quad [M\,,\,N+2] = 2M\,, \quad [N+2\,,\,M^+] = 2M^+ \tag{29}$$

only. For other correlations, one can see in[11,12]. With the use of (29), we find the mentioned-above exponential factor:

$$\exp\left\{-it\left(\frac{1}{2}r\mathbf{p}^2 \pm \frac{1}{2}\nu^2 r\right)\right\} = \\ = \exp\left\{-\frac{it}{4}\left(\left(-\omega \pm \frac{\nu^2}{\omega}\right)(M+M^+) + \left(\omega \pm \frac{\nu^2}{\omega}\right)(N+2)\right)\right\}$$

in a normal form as follows:

$$\begin{aligned}\exp\left\{-it\left(\frac{1}{2}r\mathbf{p}^2 + \frac{1}{2}\nu^2 r\right)\right\} &= \exp\left\{-\frac{(\nu^2-\omega^2)(1-\exp(-i\nu\, t))}{(\nu+\omega)^2 - (\nu-\omega)^2\exp(-i\nu\, t)}M^+\right\} \\ &\times \exp\left\{-\frac{i\nu\, t}{2}(N+2) + (N+2)\ln\frac{4\nu\omega}{(\nu+\omega)^2-(\nu-\omega)^2\exp(-i\nu\, t)}\right\} \\ &\times \exp\left\{-\frac{(\nu^2-\omega^2)(1-\exp(-i\nu\, t))}{(\nu+\omega)^2-(\nu-\omega)^2\exp(-i\nu\, t)}M\right\}\end{aligned} \tag{30}$$

$$\begin{aligned}\exp\left\{-it\left(\frac{1}{2}r\mathbf{p}^2 - \frac{1}{2}\nu^2 r\right)\right\} &= \exp\left\{\frac{(\nu^2+\omega^2)(1-\exp(-\nu\, t))}{(\omega-i\nu)^2-(\omega+i\nu)^2\exp(-\nu\, t)}M^+\right\} \\ &\times \exp\left\{-\frac{\nu\, t}{2}(N+2) + (N+2)\ln\frac{4i\nu\omega}{(\omega-i\nu)^2-(\omega+i\nu)^2\exp(-\nu\, t)}\right\} \\ &\times \exp\left\{\frac{(\nu^2+\omega^2)(1-\exp(-\nu\, t))}{(\omega-i\nu)^2-(\omega+i\nu)^2\exp(-\nu\, t)}M\right\}\end{aligned}$$

Operator (27) in the specific case (28) or in generalized casewith the normal form of exponential factor (30) is considered the operator representation of CGF.

4.1. *Application Example: Non-relativistic radiation shifts of hydrogen-like atom energy levels*

The Schrödinger equation for an electron in the Coulomb field of a nuclear with charge Ze, interacting with the quantum electromagnetic field can be written as follows:

$$\left\{ \frac{1}{2} r\mathbf{p}^2 + r\sum_{\mathbf{k}s} \omega_k C^+_{\mathbf{k}s} C_{\mathbf{k}s} - Er - \frac{1}{2}\delta m\, r\mathbf{p}^2 - Ze^2 \right.$$
$$\left. -e\sqrt{\frac{2\pi}{\Omega}} \sum_{\mathbf{k}s} \frac{e^{(s)}_\lambda}{\sqrt{\omega_k}} \left(C_{\mathbf{k}s} e^{i\mathbf{k}\mathbf{r}} + C^+_{\mathbf{k}s} e^{-i\mathbf{k}\mathbf{r}} \right) r p_\lambda \right\} |\psi\rangle = 0. \quad (31)$$

Here $C^+_{\mathbf{k}s}$, $C_{\mathbf{k}s}$ are the operators of creation and annihilation of a photon with momentum $\mathbf{k}$, energy $\omega_k = |\mathbf{k}|$ and given polarization state s; the polarization unit vectors satisfy the following correlations:

$$e^{(s)}_\lambda e^{(s')}_\lambda = \delta_{ss'}\,, \quad \sum_s e^{(s)}_\lambda e^{(s)}_\mu = \delta_{\lambda\mu} - \frac{k_\lambda k_\mu}{k^2},$$

and Ω is a normalized volume. To eliminate the principal divergent terms, we introduce the mass renormalization term into equation (31) (see[15]):

$$\delta m = \frac{8\pi e^2}{3\Omega} \sum_{\mathbf{k}} \frac{1}{\omega_k(\omega_k + k^2/2)}. \quad (32)$$

As mentioned above, all the operators related to electron should be understood in the meaning of transformation (17).

As far as we know, in most calculations, the interaction of an electron with the quantum electromagnetic field is described by the "dipole" approximation: $e^{i\mathbf{k}\mathbf{r}} \approx 1$. In this case, the mass renormalization term has another form:

$$\delta m = \frac{8\pi e^2}{3\Omega} \sum_{\mathbf{k}} \frac{1}{\omega_k^2}. \quad (33)$$

Firstly, we will solve equation (31) in "dipole" approximation by the method of perturbation theory, assuming that the interaction between electron and quantum field is small and the electron charge e can be considered a small parameter. Then, in zero order approximation, we have

$$\left|\psi^{(0)}\right\rangle = |0_f\rangle\, \varphi^{(0)}(\xi), \tag{34}$$

where $|0_f\rangle$ is the vacuum state of the quantum electromagnetic field; the electron wave-function $\varphi^{(0)}(\xi)$ is defined by equation:

$$\left\{\frac{1}{2} r\mathbf{p}^2 - Er - Ze^2\right\}\varphi^{(0)}(\xi) = 0 \quad . \tag{35}$$

(δm in (31) is assumed to be small with order of e^2). Putting $E = -\omega^2/2$ i.e. considering any discrete state of hydrogen-like atoms, we lead (35) into equation (18) and thus can define the value of ω and construct the state-vector in the form (19). For the first order correction to the wave-function we obtain:

$$\left|\psi^{(1)}\right\rangle = \sum_{\mathbf{k}s} C^{+}_{\mathbf{k}s} |0_f\rangle\, \varphi_{\mathbf{k}s}(\xi), \tag{36}$$

where the functions $\varphi_{\mathbf{k}s}(\xi)$ are defined by the equation:

$$\left\{\frac{1}{2} r\mathbf{p}^2 + \frac{1}{2}\nu_k^2\, r - Ze^2\right\}\varphi_{\mathbf{k}s}(\xi) = e\sqrt{\frac{2\pi}{\Omega\omega_k}}\, e_\lambda^{(s)} rp_\lambda \varphi^{(0)}(\xi) \tag{37}$$

with the notation $\nu_k^2 = \omega^2 + 2\omega_k$. Equation (37) has the same form as (20) and can be solved by the Green function method. Using the operator form of CGF, we obtain solutions of (37) as follows:

$$\varphi_{\mathbf{k}s}(\xi) = ie\sqrt{\frac{2\pi}{\Omega\omega_k}}\, e_\lambda^{(s)}$$

$$\times \lim_{\varepsilon\to 0}\int_0^{+\infty} dt\, \exp\left\{-\varepsilon t - it\left(\frac{1}{2} r\mathbf{p}^2 + \frac{1}{2}\nu_k^2\, r - Ze^2\right)\right\} rp_\lambda \varphi^{(0)}(\xi). \tag{38}$$

Here the Coulomb Green operator has the form of (27)–(30), operator rp_λ is understood corresponding to correlations (17). Finally, for the second-order correction to the energy, we find the following expression (the first order one is equal to zero):

$$E^{(2)} = \delta m \frac{\langle\varphi^{(0)}|r\mathbf{p}^2|\varphi^{(0)}\rangle}{2\,\langle\varphi^{(0)}|\,r\,|\varphi^{(0)}\rangle} - \frac{1}{\langle\varphi^{(0)}|\,r\,|\varphi^{(0)}\rangle}\frac{2\pi i e^2}{\Omega}\sum_{\mathbf{k}s}\frac{e_\lambda^{(s)} e_\mu^{(s)}}{\omega_k}\lim_{\varepsilon\to 0}$$

$$\times \int_0^{+\infty} dt \exp\left\{itZe^2 - \varepsilon t\right\}\left\langle\varphi^{(0)}\right| rp_\lambda \exp\left\{-it\left(\frac{r\mathbf{p}^2}{2} + \frac{\nu_k^2 r}{2}\right)\right\} rp_\mu \left|\varphi^{(0)}\right\rangle.$$

For concrete result, we will take into account the ns state i.e. consider the state-vector (19). Putting this state into equation (18), we have $\omega = Ze^2/(n+1)$ (n+1 is the principal quantum number). From the commutative relations (29), one can find for further use:

$$\begin{aligned} N(M^+)^n|0(\omega)\rangle &= 2n(M^+)^n|0(\omega)\rangle\,, \\ M(M^+)^n|0(\omega)\rangle &= n(n+1)(M^+)^{n-1}|0(\omega)\rangle. \end{aligned} \tag{39}$$

Using (39), we calculate, for example:

$$\left\langle \varphi^{(0)} \middle| r\mathbf{p}^2 \middle| \varphi^{(0)} \right\rangle = \omega(n+1) \quad , \quad \left\langle \varphi^{(0)} \middle| r \middle| \varphi^{(0)} \right\rangle = \frac{1}{\omega}(n+1).$$

The other matrix elements can be obtained by analogy using commutative relations (29) and the found formulae (39). Finally, we have:

$$\begin{aligned} E^{(2)} &= \frac{1}{2}\omega^2\delta m - \frac{\omega}{n+1}\frac{\pi i e^2}{3\Omega}\sum_{\mathbf{k}}\int_0^{+\infty} dt\, \exp\left\{itZe^2 - i\nu_k t\right\} \\ &\times \sum_{s=0}^{n} \frac{n!(n+1)!}{(s!)^2(n-s)!(n+3-s)!} \exp\left\{-i\nu_k t(n+1-s)\right\} \\ &\times \left(A_k(t)\right)^{2(n+2-s)} \left(B_k(t)\right)^{2s} \left((n+3)(n+2) - \frac{1}{B_k^2(t)} s(s+1)\right) \end{aligned} \tag{40}$$

with

$$\begin{aligned} A_k(t) &= \frac{4\omega\nu_k}{(\nu_k+\omega)^2 - (\nu_k-\omega)^2 \exp\left\{-i\nu_k t\right\}}, \\ B_k(t) &= \frac{\left(\nu_k^2-\omega^2\right)\left(1-\exp\left\{-i\nu_k t\right\}\right)}{(\nu_k+\omega)^2 - (\nu_k-\omega)^2 \exp\left\{-i\nu_k t\right\}}. \end{aligned}$$

As it is well known, the formula (40) involves a shift ($\operatorname{Re} E^{(2)}$) and a radiation width ($\Gamma = -2\operatorname{Im} E^{(2)}$) of energy level. From the calculation technique point of view, there is no principal problem to further analyze equation (40) to extract the principal divergency (in the high energy area of photons) and to obtain concrete numerical results. In other works we will give concrete result for non-relativistic contribution of the Lamb shift. Here we just want to show the way of using operator form of CGF in algebraic calculation, and it is pleasant to say that in our algebraic approach, the refusal of "dipole" approximation does not actually increase the amount of algebra and does not lead to any principal difficulty in calculation. Indeed, by analogy, we obtain for this case:

$$E^{(2)} = \delta m \frac{\langle \varphi^{(0)} | r\mathbf{p}^2 | \varphi^{(0)} \rangle}{2\langle \varphi^{(0)} | r | \varphi^{(0)} \rangle} - \frac{1}{\langle \varphi^{(0)} | r | \varphi^{(0)} \rangle} \frac{2\pi i e^2}{\Omega}$$
$$\times \sum_{\mathbf{k}s} \frac{e_\lambda^{(s)} e_\mu^{(s)}}{\omega_k} \lim_{\varepsilon \to 0} \int_0^{+\infty} dt \exp\{itZe^2 - \varepsilon t\}$$
$$\times \langle \varphi^{(0)} | r p_\lambda e^{i\mathbf{kr}} \exp\{-it(\tfrac{1}{2} r\mathbf{p}^2 + \tfrac{1}{2}\nu_k^2 r)\} e^{-i\mathbf{kr}} r p_\mu | \varphi^{(0)} \rangle. \quad (41)$$

Expression (41) of energy correction is different from the one in "dipole" approximation only in the mass renormalization term which is now considered by (32). Besides, the exponential factor in (41) now is as follows:

$$e^{i\mathbf{kr}} \exp\left\{-it\left(\frac{1}{2} r\mathbf{p}^2 + \frac{1}{2}\nu_k^2 r\right)\right\} e^{-i\mathbf{kr}}. \quad (42)$$

We can reduce (42) into the normal form suitable for algebraic calculation. Indeed, using formula:

$$e^A B e^{-A} = B + [A, B] + \frac{1}{2!}[A, [A, B]] + \frac{1}{3!}[A, [A, [A, B]]] + \ldots$$

(with A, B are two non-commutative operators) we have:

$$e^{i\mathbf{kr}} N e^{-i\mathbf{kr}} = N + \frac{i}{\omega} k_\lambda (m_\lambda - m_\lambda^+) + \frac{k^2}{2\omega^2}(M + M^+ + N + 2),$$
$$e^{i\mathbf{kr}}(M + M^+) e^{-i\mathbf{kr}} =$$
$$= \left(1 - \frac{k^2}{2\omega^2}\right)(M + M^+) - \frac{i}{\omega} k_\lambda (m_\lambda - m_\lambda^+) - \frac{k^2}{2\omega^2}(N + 2)$$

and lead operator (42) to the form:

$$e^{i\mathbf{kr}} \exp\{-it(\tfrac{1}{2} r\mathbf{p}^2 + \tfrac{1}{2}\nu_k^2 r)\} e^{-i\mathbf{kr}} =$$
$$\exp\{-\tfrac{it}{4\omega}(2i\omega\, k_\lambda (m_\lambda - m_\lambda^+) + (\nu_k^2 + k^2 + \omega^2)(N + 2)$$
$$(\omega^2 - \nu_k^2 + k^2)(M + M^+))\}.$$

Noticing that the operators:

$$N + 2, \quad C = (\nu_k^2 + k^2 - \omega^2) M + 2i\omega\, k_\lambda m_\lambda,$$
$$C^+ = (\nu_k^2 + k^2 - \omega^2) M^+ - 2i\omega\, k_\lambda m_\lambda^+$$

generate a closed algebra according to the commutative relations:

$$[C, N + 2] = 2C, \quad [N + 2, C^+] = 2C^+,$$
$$[C, C^+] = \left((\nu_k^2 + k^2 - \omega^2)^2 + 4\omega^2\right)(N + 2)$$

we find a normal form of the operator (42):

$$
\begin{aligned}
&e^{i\mathbf{kr}} \exp\left\{-it\left(\frac{1}{2}r\mathbf{p}^2 + \frac{1}{2}\nu_k^2 r\right)\right\} e^{-i\mathbf{kr}} \\
&= \frac{16\,\omega^2\nu_k^2 \exp\{-i\nu_k t\}}{\left((\nu_k+\omega)^2 + k^2 - ((\nu_k-\omega)^2+k^2)\exp\{-i\nu_k t\}\right)^2} \\
&\times \exp\left\{-\frac{(1-\exp\{-i\nu_k t\})\,\left((\nu_k^2+k^2-\omega^2)M^+ - 2i\omega\, k_\lambda m_\lambda^+\right)}{(\nu_k+\omega)^2+k^2-((\nu_k-\omega)^2+k^2)\,\exp\{-i\nu_k t\}}\right\} \\
&\times \exp\left\{-\frac{i\nu_k t}{2}N + N\ln\frac{4\omega\nu_k}{(\nu_k+\omega)^2+k^2-((\nu_k-\omega)^2+k^2)\,\exp\{-i\nu_k t\}}\right\} \\
&\times \exp\left\{-\frac{(1-\exp\{-i\nu_k t\})\,\left((\nu_k^2+k^2-\omega^2)M + 2i\omega\, k_\mu m_\mu\right)}{(\nu_k+\omega)^2+k^2-((\nu_k-\omega)^2+k^2)\,\exp\{-i\nu_k t\}}\right\}. \qquad (43)
\end{aligned}
$$

Using (43), we can calculate all matrix elements in (41) by pure algebraic transformation. Note that expressions (41) and (32) do not have any divergency, therefore, in summation, over$\mathbf{k}$ there is no need for the upper bound on the photon energy.

5. Many-particle Coulomb Green Function

When considering the many-electron atom problems as well as the problem of atom or molecule interaction, we can construct Hamiltonian in two parts. The first of them is the following:

$$\hat{H} = \sum_k \hat{H}_k\,, \qquad \hat{H}_k = -\frac{1}{2}\Delta_k - \frac{Z_k e^2}{r_k}\,, \quad (k = 1, 2, ..., N) \qquad (44)$$

where Δ_k is Laplace operator corresponding to $\mathbf{r}_k$- coordinate; Z_k $(k = 1, 2, ..., N)$ are different positive numbers connected with effective charges of electrons in the independent particle model with the hydrogenic wave-functions[25], but in the conventional description of atoms they are the same and equal to nuclear charge. The second remaining part of Hamiltonian is considered a perturbation term$\hat{V}$.

The Green function, according to Hamiltonian (44), is called many-particle Coulomb Green function (MCGF). The many-particle Coulomb Green operator can be written by analogy of the operator of CGF:

$$\hat{K}(\mathbf{r}_1, \mathbf{r}_2, ..., \mathbf{r}_N; E) = \frac{1}{i}\lim_{\varepsilon\to 0}\int_0^{+\infty} dt\, e^{-\varepsilon\, t + i(\hat{H}-E)\,t} \quad . \qquad (45)$$

We will show that (45) can be expressed via the operator of CGF for each$\mathbf{r}_k$, thus suitable for using in algebraic calculations. Indeed, from (45), it is easy to approve that:

$$\hat{K}(\mathbf{r}_1, \mathbf{r}_2, ..., \mathbf{r}_N; E) = \tfrac{1}{i} \lim_{\varepsilon \to 0} \int\limits_0^{+\infty} dt_1 \int\limits_0^{+\infty} dt_2 ... \int\limits_0^{+\infty} dt_N$$

$$\times \delta(t_2 - t_1)\, \delta(t_3 - t_1) ... \delta(t_N - t_1) \prod_{k=1}^{N} e^{-\varepsilon\, t_k + i(\hat{H}_k - E_k)\, t_k} \tag{46}$$

where $E = \sum\limits_{k=1}^{N} E_k$. By substituting into (46) the Furrier integral form of delta-function:

$$\delta(s - t) = \frac{1}{2\pi} \int\limits_{-\infty}^{+\infty} d\alpha\, e^{-i\alpha(s-t)}$$

we obtain:

$$\hat{K}(\mathbf{r}_1, \mathbf{r}_2, ..., \mathbf{r}_N; E) = \tfrac{1}{i(2\pi)^{N-1}} \int\limits_{-\infty}^{+\infty} d\alpha_1 \int\limits_{-\infty}^{+\infty} d\alpha_2 \cdots \int\limits_{-\infty}^{+\infty} d\alpha_N$$

$$\times \delta(\alpha_1 + \alpha_2 + \ldots + \alpha_N) \lim_{\varepsilon \to 0} \int\limits_0^{+\infty} dt_1 \int\limits_0^{+\infty} dt_2 \cdots \int\limits_0^{+\infty} dt_N$$

$$\times \prod_k \exp\left\{-\varepsilon\, t_k + i(\hat{H}_k + \alpha_k - E_k) t_k\right\}$$

$$= \tfrac{1}{i(2\pi)^{N-1}} \int\limits_{-\infty}^{+\infty} d\alpha_1 \int\limits_{-\infty}^{+\infty} d\alpha_2 \cdots \int\limits_{-\infty}^{+\infty} d\alpha_N$$

$$\times \delta(\alpha_1 + \alpha_2 + \ldots + \alpha_N) \prod_{k=1}^{N} \hat{K}\, (\mathbf{r}_k; E_k - \alpha_k). \tag{47}$$

In (47), $\hat{K}\, (\mathbf{r}_k; E_k - \alpha_k) = \lim\limits_{\varepsilon \to 0} \left(\hat{H}_k + \alpha_k - E_k + i\varepsilon\right)^{-1}$ is the Green operator of CGF corresponding to$\mathbf{r}_k$-coordinate and the energy$E_k - \alpha_k$. Thus we have reduced the operator of MCGF to the separated operators of CGF. From (47), it is easy to see that if each $\hat{K}\, (\mathbf{r}_k; E_k - \alpha_k)$ has a normal form, then $\hat{K}(\mathbf{r}_1, \mathbf{r}_2, ..., \mathbf{r}_N; E)$ has one also.

5.1. *Application example: Ground state of hydrogenic molecule*

The Schrödinger equation for a hydrogenic molecule can be written as follows:

$$\left\{\frac{1}{2}\mathbf{p}_1^2 - \frac{e^2}{r_1} + \frac{1}{2}\mathbf{p}_2^2 - \frac{e^2}{r_2} + V(R, \mathbf{r}_1, \mathbf{r}_2)\right\} \Psi\, (\mathbf{r}_1, \mathbf{r}_2) = E(R) \Psi\, (\mathbf{r}_1, \mathbf{r}_2) \tag{48}$$

($\hbar = m = c = 1$) with R is a distance between the molecule nucleus; $\mathbf{r}_1, \mathbf{r}_2$ are respectively coordinates of each electron. Assuming that $R >> r_1$, $R >> r_2$, the potential:

$$V(R, \mathbf{r}_1, \mathbf{r}_2) = \frac{e^2}{R} - \frac{e^2}{|\mathbf{R} - \mathbf{r}_1|} - \frac{e^2}{|\mathbf{R} + \mathbf{r}_2|} + \frac{e^2}{|\mathbf{R} + \mathbf{r}_2 - \mathbf{r}_1|}$$

can be expanded in the Taylor power series as follows:

$$\begin{aligned} V(\mathbf{R}, \mathbf{r}_1, \mathbf{r}_2) &= (2z_1 z_2 - x_1 x_2 - y_1 y_2) \frac{e^2}{R^3} \\ &+ \frac{3}{2} \left(z_1 \left(2z_1 z_2 + 2x_1 x_2 + 2y_1 y_2 - x_2^2 - y_2^2 \right) \right. \\ &\left. - z_2 \left(2z_1 z_2 + 2x_1 x_2 + 2y_1 y_2 - x_1^2 - y_1^2 \right) \right) \frac{e^2}{R^4} + \cdots \end{aligned} \tag{49}$$

For demonstration of algebraic calculation, we consider only the first term in expansion (49). This term is corresponding to the following energy correction:

$$\Delta E(R) = a\, e^2 / R^6.$$

The other terms in (49) can be considered absolutely by the same way. The equation (48) now becomes:

$$\begin{aligned} &\left\{ \left(\frac{1}{2} r_1 \mathbf{p}_1^2 - r_1 E_1(R) - e^2 \right) r_2 + \left(\frac{1}{2} r_2 \mathbf{p}_2^2 - r_2 E_2(R) - e^2 \right) r_1 + \right. \\ &\left. + r_1 r_2 (2z_1 z_2 - x_1 x_2 - y_1 y_2) \frac{e^2}{R^3} \right\} \Psi(\mathbf{r}_1, \mathbf{r}_2) = 0, \end{aligned} \tag{50}$$

with $E(R) = E_1(R) + E_2(R)$ (For the ground state $E_1(R) = E_2(R) =$ $= E(R)/2$. By doing the formal change (17) for each of the coordinates, we transform (50) into an equation describing the motion of two "particles" in the space of complex coordinates ξ. Henceforth, all operators in equation (50) should be understood via the formal changes (17). We solve now equation (50) by the perturbation method, assuming the parameter $\beta = e^2 / R^3$ to be small. Then, the wave-function in zero order approximation can be found in the form:

$$\left| \Psi^{(0)} (\mathbf{r}_1, \mathbf{r}_2) \right\rangle = |\Psi(\mathbf{r}_1)\rangle \, |\Psi(\mathbf{r}_2)\rangle,$$

where $|\Psi(\mathbf{r})\rangle$ is the ground state wave-vector defined by equation:

$$\left\{ \frac{1}{2} r \mathbf{p}^2 - \frac{1}{2} r E^{(0)}(R) - e^2 \right\} \Psi(\mathbf{r}) = 0. \tag{51}$$

Introducing parameter ω by the equation$E^{(0)} = -\omega^2$, then putting it into (51), this equation becomes equation (18) in the specific case $Z = 1$. We obtain for the ground state:

$$|\Psi\rangle = |0(\omega)\rangle\,, \quad \omega = e^2. \tag{52}$$

Using the operator of two-particle CGF of the form (47), from (50) we find the first order correction to the wave-function in ground state as follows:

$$\left|\Psi^{(1)}\right\rangle = \frac{i\beta}{2\pi}\int_{-\infty}^{+\infty} d\alpha \lim_{\varepsilon\to 0}\int_0^{+\infty} dt \exp\left\{-it\left(\frac{1}{2}r_1\mathbf{p}_1^2 + \frac{1}{2}\omega^2 r_1 - \alpha r_1 - e^2\right) + \varepsilon t\right\}$$
$$\times \int_0^{+\infty} ds \,\exp\left\{-i\,s\left(\frac{1}{2}\,r_2\,\mathbf{p}_2^2 + \frac{1}{2}\,\omega^2\,r_2 - \alpha\;r_2 - e^2\right) + \varepsilon\,s\right\}$$
$$\times\, r_1 r_2(2z_1z_2 - x_1x_2 - y_1y_2)\left|\Psi^{(0)}\right\rangle \tag{53}$$

Taking into account that the wave-function in ground state depends only on the distances r_1, r_2(not on the vectors $\mathbf{r}_1$, $\mathbf{r}_2$), we find the first order energy correction equal to zero. The second-order correction to energy in the following expression:

$$E^{(2)} = \beta\left\langle\Psi^{(0)}\middle|r_1r_2\middle|\Psi^{(0)}\right\rangle^{-1}\left\langle\Psi^{(0)}\right|\, r_1r_2(x_1x_2 + y_1y_2 - 2z_1z_2)\left|\Psi^{(1)}\right\rangle$$

after substitution of (53) leads to:

$$E^{(2)} = \tfrac{3i}{\pi}\beta^2\left\langle\Psi(\mathbf{r})|r|\Psi(\mathbf{r})\right\rangle^{-2}\int_{-\infty}^{+\infty} d\alpha\left\{\lim_{\varepsilon\to 0}\int_0^{+\infty} d\,t\;\exp\left\{ie^2t - \varepsilon t\right\}\right.$$
$$\left.\times\left\langle\Psi(\mathbf{r})\right|\,rx\;\exp\left\{-it\,\left(\tfrac{1}{2}\,r\,\mathbf{p}^2 + \tfrac{1}{2}\,\omega^2\,r - \alpha\;r\right)\right\}rx|\Psi(\mathbf{r})\rangle\right\}^2. \tag{54}$$

Here in (54), we eliminate the electron indices because they have no effect on calculation of matrix elements. Putting:

$$\nu^2 = \frac{1}{2}\omega^2 - \alpha \quad \text{if} \quad \omega^2 \geq 2\alpha\,, \qquad \nu^2 = \alpha - \frac{1}{2}\omega^2 \quad \text{if} \quad \omega^2 < 2\alpha,$$

we lead the exponential factor in (54) into the form (30) that permits of using the algebraic calculation. For the ground state we have, for example:

$$\langle 0(\omega)|\, r\, |0(\omega)\rangle = \frac{1}{\omega},$$

$$\langle 0(\omega)|\; rx\, \exp\left\{-it\left(\frac{1}{2}\,r\mathbf{p}^2 + \frac{1}{2}\,\nu^2\, r\right)\right\} rx|0(\omega)\rangle =$$

$$2^{11}\nu^4\omega^2\frac{(\omega+\nu)^2e^{-2i\nu\, t} + (3\nu^2 - 2\omega^2)e^{-3i\nu\, t} + (\omega-\nu)^2e^{-4i\nu\, t}}{\left((\omega+\nu)^2 - (\omega-\nu)^2e^{-i\nu\, t}\right)^6}. \quad (55)$$

Substituting (55) into (54) and taking integration over α, we obtain, for instance, the numerical result:

$$E^{(2)} = -6.499026\; e^2/R^6$$

which coincides with the well known data[26].

6. Conclusions and Outlook

Within the concern of this paper, we have demonstrated the way of using operator representation of the Coulomb Green function as well as of the many-particle Coulomb Green function for algebraic atomic calculations. The calculation method is simple, straightforward and thus reduces rather complicated integrations of special functions to simple, pure algebraic procedures of transforming the products of the annihilation and creation operators to a normal form. This method can be recommended for most problems in atomic physics. For example, the following modern physics problems have stimulated us to develop new trend of applying CGF and MCGF: (i) atomic characteristics of atoms and molecules in electromagnetic field (dynamical polarizability, magnetizability[4–6], hyper-polarizability[9], and so on; (ii) the multi-photon process (multi-photon ionization)[7,8], which recently involves great interest related to the new success in laser technology, which can enable experiments of multi ionization of heavy elements to be done in the near future[10]; and (iii) the QED effects in atomic energy levels, e.g. the Lamb shift[27,28]. The experiment precision for the $1s - 2s$ transition in hydrogen has already reached very high accuracy and is likely to be improved in the near future. However, the theoretical description of the system till now is not on adequate level[29]. Therefore, calculations of the two-loop-self-energy-correction is now one of the most interested problems[30].

In this context, we believe that the algebraic method described in this paper will be found adequate applications. This relates to an aspect of our investigation in Physics Department of HCMC University of Pedagogy. In the end, I would like to say that the main ideas described in this paper appear through the long-term cooperation with Professor Komarov L. I.

Acknowledgments

I would like to thank Prof. Barkovsky L. M. and Prof. Feranchuk I. D. for inviting me to write this paper, which is related to my research when working in the Department of Theoretical Physics – Belarus University. I would like to thank the involved people who provided me with scientific activities in Belarus. My special thanks to Prof. Komarov L. I. and Prof. Feranchuk I. D. for their cooperation, the ideas and many fruitful discussions, Prof. Yunhevich A. V. (Lab. of Physico-Chemistry of Surface) for stimulation of developing the method for the problems of NanoPhysics during my working in his laboratory (1991-1995), Prof. Khrustaliev O. A. (Moscow University), Acad. Prof. Tolmilchik L. M., Prof. Tolkachev E. A. (Belarus Academy of Science), Prof. Zavtrak S. T., Prof. Tikhomirov V. V. (Institute of Nuclear Physics) for useful comments on my works when being the opponents of my Ph.D and Dr.Sci. dissertations. And finally, thanks to my colleagues: Dr. Romanova T. S., Dr. Le Anh Thu, Dr. Viloria Tony, Dr. Nguyen Thu Giang, Dr. Ly Xuan Hai for cooperation.

Bibliography

1. L. Hostler, *J. Math. Phys.* **5**, 591 (1964).
2. A. G. Makhanhek and V. S. Korolkov, *Analytical Methods in Perturbation Theory of Quantum Mechanics* (Nauka I Tekhnika, Minsk, 1982).
3. A. Maquet, V. Veniard and T. A. Marian, *J.Phys.B: At.Mol.Opt. Phys.* **31**, 3743 (1998).
4. R. Szmutkowski, *J.Phys.B: At.Mol.Opt. Phys.* **30**, 825 (1997).
5. R. Szymanowski, V. Veniard, R. Taieb and A. Maquet, *EuroPhys. Lett.* **37**, 391 (1997).
6. R. Szmutkowski, *J.Phys.B: At.Mol.Opt. Phys.* **35**, 1379 (2002) .
7. E. Karule and B. Moine, *J.Phys.B: At.Mol.Opt. Phys.* **36**, 1963 (2003).
8. P. Koval, S. Fritzche and A. Surzhykov, *J.Phys.B: At.Mol.Opt. Phys.* **36**, 873 (2003).
9. N. L. Manakov, S. Marmo and E. Pronin, *JETP* **98**, 254 (2004).
10. N. L. Manakov, M. V. Frodov, B. Borca and Anthony F. Starace, *J.Phys.B: At.Mol.Opt. Phys.* **36**, R49 (2003).
11. L. I. Komarov and T. S. Romanova, *Izv.AkadN BSSR (ser.Fis.-Mat.Nauk)* **2**, 98 (1982).
12. L. I. Komarov and T. S. Romanova, *J. Phys. B: At. Mol. Opt. Phys.* **18**, 859 (1985).

13. H. Kleinert *In Lectures in Theor.Phys.*, Ed. W. Brittin and A. Barut (Gordon and Brach, NY, 1968) p. 427.
14. Le V. Hoang, L. I. Komarov and T. S. Romanova, *J. Phys.A: Math.Gen.* **22** 1543 (1989).
15. J. D. Bjorken and S. Drell, *Relativistic Quantum Mechanics* (McGraw-Hill, NY, 1964).
16. Le V. Hoang, *Prod.Conf.Theo.Phys.* **28**, (Hanoi, 2003), p. 108.
17. Le V. Hoang, Ly Xuan Hai, L. I. Komarov and T. S. Romanova, *J.Phys.B: At.Mol.Opt. Phys.* **25**, 6461 (1992).
18. Le Anh Thu, Le V. Hoang, L. I. Komarov and T. S. Romanova, *J.Phys.B: At.Mol.Opt. Phys.* **27**, 4083 (1994).
19. Le Anh Thu, Le V. Hoang, L. I. Komarov and T. S. Romanova, *J.Phys.B: At.Mol.Opt. Phys.* **29**, 2897 (1996).
20. R. Ho and A. Inomata, *Phys. Rev. Lett.* **48**, 231 (1982).
21. H. Duru and H. Kleinrt, *Phys. Lett. B* 84, **185** (1979).
22. Le V. Hoang, *Izv.AkadN BSSR (ser.Fis.-Mat.Nauk)* **2**, 76 (1992).
23. Le V. Hoang and Nguyen Thu Giang, *J. Phys. A: Math. Gen.* **26**, 3333 (1993).
24. P. Kustaanheimo and E. Stiefel, *J. für. Math.* **218**, 204 (1964)
25. M. Kregar, *Phys. Scripta.* **31**, 246 (1985).
26. J. Baras, *Ban der Walls force* (Nauka, Moskva, 1988).
27. T. Appelquist and S. T. Brodsky, *Phys. Rev. A* **2**, 2293 (1970)
28. K. Pachuki and D. Jentschura, *Phys. Rev. Lett.* **91**, 113005 (2003)
29. V. A. Yerokhin, P. Indelicato and V. M. Shabaev, *Eur.Phys.J.D.* **25**, 203 (2003)
30. V. A. Yerokhin, P. Indelicato and V. M. Shabaev, *Phys. Rev. Lett.* **91**, 73001 (2003)

THREE-DIMENSIONAL SOLITARY WAVES AND VORTICES IN A DISCRETE NONLINEAR SCHRÖDINGER LATTICE

P.G. Kevrekidis[1], B.A. Malomed[2], D.J. Frantzeskakis[3] and R.C̃arretero-González[4]

[1] *Department of Mathematics and Statistics, University of Massachusetts, Amherst MA 01003-4515, USA*

[2] *Department of Interdisciplinary Studies, Faculty of Engineering, Tel Aviv University, Tel Aviv 69978, Israel*

[3] *Department of Physics, University of Athens, Panepistimiopolis, Zografos, Athens 15784, Greece*

[4] *Nonlinear Dynamical Systems Group, Department of Mathematics, San Diego State University, San Diego CA, 92182*

In a benchmark dynamical-lattice model in three dimensions (3D), the discrete nonlinear Schrödinger (DNLS) equation, we find solutions in the form of discrete-vortex solitons with various values of the topological charge S. Stability regions for the vortices with $S = 0, 1, 3$ are discussed. The $S = 2$ vortex is unstable, spontaneously rearranging into a stable one with $S = 3$. We also examine a two-component extension of the model, and find a novel class of stable structures, consisting of vortices in the different components, oriented perpendicular to each other. Self-localized states of the proposed types can be observed experimentally in Bose-Einstein condensates loaded into optical lattices.

1. Introduction

Intrinsic localized modes (ILMs) in nonlinear dynamical lattices are the theme of a large volume of theoretical and experimental literature. These *discrete breathers* (as they are also known) have attracted attention due to their inherent ability to concentrate and, potentially, transport the vibrational energy in a coherent fashion (see, e.g.,[1]). Settings in which these objects, strongly localized in space and periodic in time, appear range from arrays of nonlinear-optical waveguides[2] to Bose-Einstein condensates (BECs) in optical-lattice (OL) potentials[3], and from various models based on nonlinear springs[4] to Josephson-junction ladders[5] and dynamical models of the DNA double strand[6].

A benchmark model, which generically emerges in the description of envelope waves in nonlinear lattices, is the discrete nonlinear Schrödinger (DNLS) equation[7]. It finds its most straightforward physical realizations in two of the above-mentioned settings, viz., arrays of optical waveguides[8,9], and networks of BEC drops trapped in OLs (with $\sim 10^3$ atoms in each droplet) in the tight-binding approximation[3]. While the former system may be, effectively, 1- and 2-, but not 3-dimensional, the latter one can be experimentally created in three dimensions (3D) as well[10], which suggests a direct (and truly unique) physical implementation of the 3D DNLS model.

In spite of the physical relevance of the NLS equation in the 3D case, very few works attempted to establish (computationally) localized solutions in this system, and those were actually done in the absence of OLs (see, e.g., Ref.[11]). Only very recently, a possibility of the existence of stable 3D solitons in continuum NLS equations including OL potentials has been demonstrated[12]. A problem of fundamental interest, both in its own right and in view of the possibility of the experimental realization in the BEC-OL setting, is the search for 3D solitons in the DNLS model proper. Especially attractive is the possibility of supporting ILMs with *intrinsic vorticity* (topological charge), which is an entirely new class of discrete breathers in 3D that has never been considered before.

In this work, we demonstrate that the discreteness indeed makes it possible to stabilize, in the DNLS model *with attraction*, not only ordinary 3D ILMs, but also vortex solitons that are strongly unstable in the continuum limit. These include not only fundamental discrete vortices, with the topological charge $S = 1$, whose persistence and stability in the 3D case may already be surprising, but also higher-order vortices, such as ones with $S = 3$. We also extend the considerations to multi-component DNLS models, that allow for the existence and stability of still more exotic configurations. In particular, we introduce a novel type of a compound vortex, consisting of two vortices with the same value of S in the two components, whose orientations are *perpendicular* to each other. Such solutions are stable too, in certain parametric intervals.

2. The Model

The DNLS equation on the cubic lattice with a coupling constant C is[7]

$$i\frac{d}{dt}\phi_{l,m,n} + C\Delta_2\phi_{l,m,n} + |\phi_{l,m,n}|^2\,\phi_{l,m,n} = 0, \tag{1}$$

where the discrete Laplacian is $\Delta_2\phi_{l,m,n} \equiv \phi_{l+1,m,n} + \phi_{l,m+1,n} + \phi_{l,m,n+1} + \phi_{l-1,m,n} + \phi_{l,m-1,n-1} + \phi_{l,m,n-1} - 6\phi_{l,m,n}$. We seek for ILM solutions of the form $\phi_{l,m,n} = \exp(i\Lambda t)u_{l,m,n}$, where $-\Lambda$ is the frequency (chemical potential, in the BEC context), and nonlinear eigenfunctions $u_{l,m,n}$ obey the stationary equation,

$$\Lambda u_{l,m,n} = C\Delta_2 u_{l,m,n} + |u_{l,m,n}|^2 u_{l,m,n}. \tag{2}$$

Solutions to Eq. (2) (generally, complex ones) are obtained by means of Newton iterations, and their stability is examined in the framework of linearized equations. To this aim, perturbed solutions are substituted into Eq. (1) as[13],

$$\phi_{l,m,n} = e^{i\Lambda t}\left[u_0 + \epsilon\left(a_{l,m,n}e^{-i\omega t} + b_{l,m,n}e^{i\omega^* t}\right)\right], \tag{3}$$

where u_0 is the unperturbed stationary solution, ϵ is an infinitesimal amplitude of the perturbation, and ω is its eigenfrequency. To order $\mathcal{O}(\epsilon)$, we thus obtain linear equations for the perturbation eigenmodes,

$$\omega\begin{pmatrix} a_k \\ b_k^* \end{pmatrix} = \begin{pmatrix} \dfrac{\partial F_k}{\partial u_j} & \dfrac{\partial F_k}{\partial u_j^*} \\ -\dfrac{\partial F_k^*}{\partial u_j} & -\dfrac{\partial F_k^*}{\partial u_j^*} \end{pmatrix}\begin{pmatrix} a_k \\ b_k^* \end{pmatrix},$$

where $F_k \equiv -C(u_{k+1}+u_{k-1}+u_{k+N}+u_{k-N}+u_{k+N^2}+u_{k-N^2}-6u_k)+\Lambda u_k - |u_k|^2 u_k$, and the use of the string index, $\{l,m,n\} \to l+(r-1)m+(s-1)n$, $r,s = 1,2,..,N$, maps the problem on the $N \times N \times N$ lattice into a vector problem of the size N^3. At the domain boundaries, Dirichlet boundary condition were used.

We examine solutions of Eq. (2) by fixing the frequency, $\Lambda = 2$, and varying the coupling C. The solutions with different values of the topological charge S (here, it ranges between 0 and 3) are generated by the iterative scheme with an initial ansatz which is motivated by known features of localized 3D vortices in the continuum limit[14],

$$\begin{aligned} u_{l,m,n}^{(\text{init})} &= A[(l-l_0)+i(m-m_0)]^S \exp\left(-|n-n_0|\right) \\ &\quad\times \operatorname{sech}\left(\eta\sqrt{(l-l_0)^2+(m-m_0)^2}\right), \end{aligned} \tag{4}$$

where (l_0, m_0, n_0) is the location of the vortex' center, and η is a scale parameter. The Newton algorithm was iterated until it converged to 1 part in 10^7. Our results are typically shown for $9\times9\times9$ and $11\times11\times11$ lattices, but larger sizes were also used.

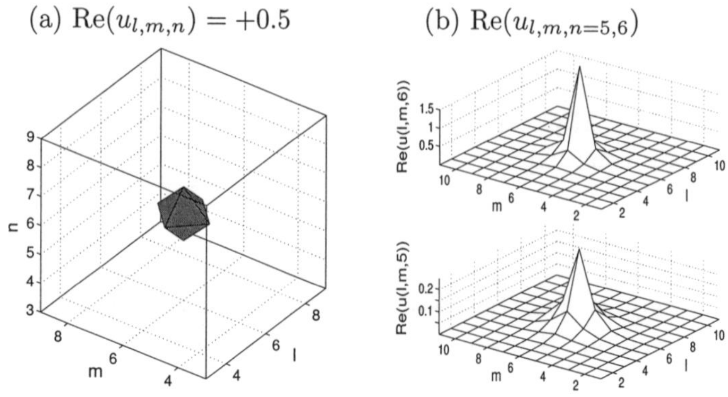

Figure 1. The ILM with $S = 0$ is shown for $C = 1$. The left panel shows a 3D plot of the contours where the real field of $u_{l,m,n}$ assumes the value 0.5. The right panels show 2D cross sections of the solution at $n = 6$ (top) and $n = 5$ (bottom) for the $11 \times 11 \times 11$ lattice used here.

3. Results

We now give a summary of basic results for the solutions with different topological charges. Ordinary ILMs with $S = 0$ are stable below a critical value of the coupling, as they are strongly unstable against collapse in the continuum limit of $C \to \infty$[17] (as this critical value is, in fact, quite large, it may depend on the size of the lattice used in the numerical computation). An example of a stable ordinary soliton is shown in Fig. 1. As the $S = 0$ ILMs have the largest stability interval $C < C_{\rm cr}^{(0)}$ in comparison with topologically charged ILMs (see below), they can be only destroyed by the instability, rather than transformed into ILMs of other types.

3D vortices with $S = 1$ have also been found. They are stable (see Fig. 2) below a critical value $C_{\rm cr}^{(1)} = 0.65$ (similarly to their 2D counterparts[15]). At the instability threshold, a quartet of complex eigenvalues emerges from collision of two imaginary eigenvalue pairs (for details, see, e.g.,[15,16]). The development of the instability is shown in Fig. 2 for a typical case with $C = 0.7 > C_{\rm cr}^{(1)}$. The perturbations destroy the vortex structure and, as a result, an ordinary ($S = 0$) ILM emerges; obviously, the change of the topological charge is possible in the lattice dynamics, in which the angular momentum is not a dynamical invariant.

An example of a double vortex (with $S = 2$) is displayed in Fig. 3 for $C = 0.01$. Similar to its 2D counterpart[15], this complex solution is unstable through a real eigenvalue pair at all values of C (notice, however, that

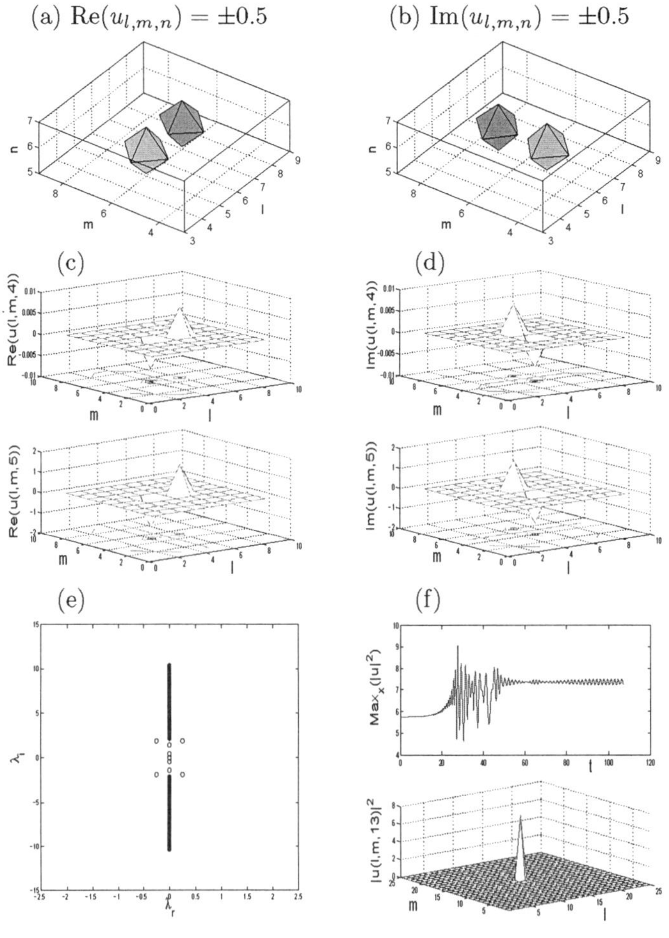

Figure 2. The top panels show contours at the level ± 0.5 for the fields $\mathrm{Re}(u_{l,m,n})$ [left] and $\mathrm{Im}(u_{l,m,n})$ [right] of the vortex ILM with $S = 1$. Red (color) / light gray (BW) surfaces correspond to levels for negative values and blue (color) / dark gray (BW) to levels with positive values. 2D cross sections of the vortex are shown in four middle panels ((c) and (d)). The bottom row displays the development of instability of the vortex for $C = 0.7$, through the time evolution of its amplitude ((f) top), and a 2D cut of the profile at $t = 100$ ((f) bottom). The unstable vortex transforms itself into an ordinary ILM with $S = 0$. The left bottom panel (e) shows the spectral plane (λ_r, λ_i) of the linear stability eigenvalues for the same unstable vortex.

purely real *stable* solutions to the 2D DNLS equation, that also may be identified as having the topological charge $S = 2$, have been very recently

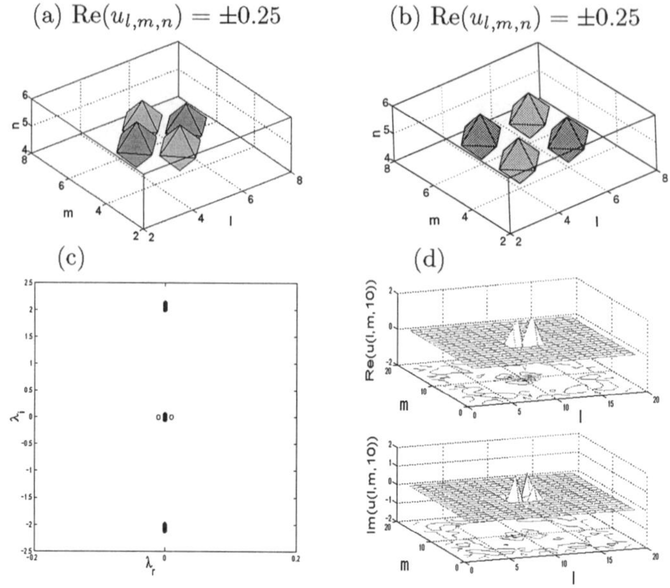

Figure 3. The ILM vortex with $S = 2$ for $C = 0.01$. Top panels have the same meaning as the top panels in Fig. 2. The bottom left panel (c) displays the linear stability eigenvalues, while (d) shows the result of long evolution of this unstable vortex. The eventual state, shown through its 2D cross-sections at $t = 1000$, is a vortex with $S = 3$, which is *stable* for this value of C.

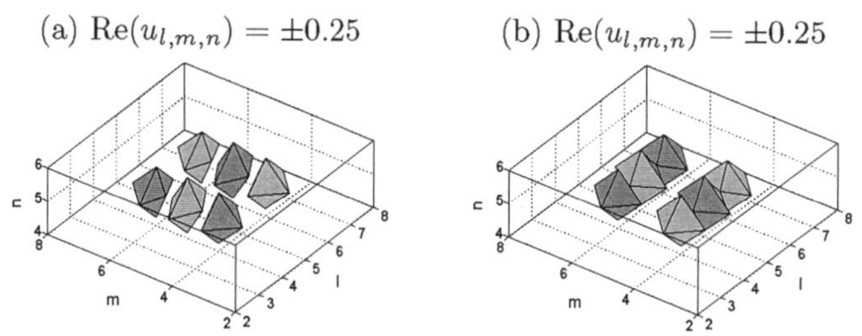

Figure 4. Stable stationary vortex ILM with $S = 3$ for $C = 0.01$. The panels have same meaning as top panels in Fig. 2.

found[16]; counterparts of such solutions exist in the 3D case as well). What is more interesting, however, is that this unstable ILM reshapes itself *not* into the one with $S = 0$, but rather towards a stable vortex with $S = 3$,

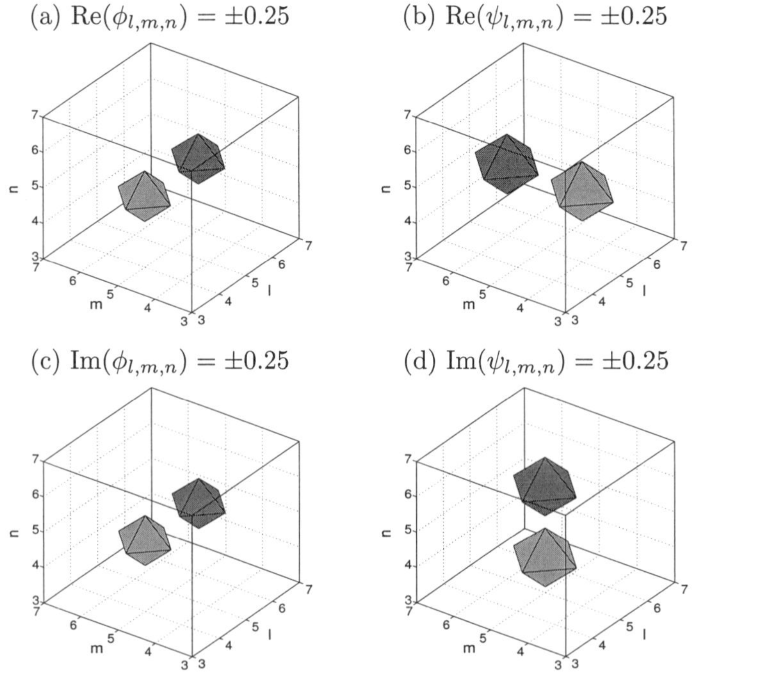

Figure 5. A two-component vortex ILM is shown for $C = 0.01$. Top and bottom panels correspond, respectively, to the first and second components. This state is a stable bound state of two $S = 1$ vortices with perpendicular orientations. The panels have same meaning as top panels in Fig. 2.

as is also shown in Fig. 3. The stabilization of the 3D vortex ILM through spontaneous *increase* of S is a striking result (again feasible only in the realm of dynamical lattices). To further illustrate it, in Fig. 4 we show the stable $S = 3$ vortex constructed for the same case, $C = 0.01$ (similar *stable* higher-order vortex ILMs were very recently found in the 2D DNLS model[16]).

Another striking feature, unique to the 3D case, is a possibility of the existence (and potential stability) of ILM vortex complexes in multi-component systems, with the vortices in different components being *orthogonal* to each other. We consider, in particular, two coupled DNLS equations,

$$\begin{aligned} &\left[i\tfrac{d}{dt} + C\Delta_2 + \left(|\phi_{l,m,n}|^2 + \beta|\psi_{l,m,n}|^2\right)\right]\phi_{l,m,n} = 0, \\ &\left[i\tfrac{d}{dt} + C\Delta_2 + \left(|\psi_{l,m,n}|^2 + \beta|\phi_{l,m,n}|^2\right)\right]\psi_{l,m,n} = 0. \end{aligned} \tag{5}$$

This model describes the array of BEC droplets composed of a mixture of two different species (as in the experiment of Ref.[18], but with attractive interactions, which may be induced via the Feshbach resonance[19]); in this case, β is the relative strength of the interaction between the species. We examine a complex of two orthogonal vortices, in which the one in the first component is oriented perpendicularly to the (l,m) plane, while in the second component, the vortex is perpendicular to the (l,n) plane. An example of such a stable complex is shown, for $\beta = 0.5$, in Fig. 5. The stability critically depends on β, as it was found that they may be stable for $\beta < 1$, and are unstable if $\beta > 1$.

4. Conclusions

The above results are a first step towards search for topologically nontrivial intrinsic localized modes in 3D nonlinear dynamical lattices and exploration of the rich complexity that the third dimension may add to what was previously known in 2D lattice dynamics. Besides identifying stable discrete solitons which would be subject to strong collapse in the continuum limit, without ($S = 0$) or with ($S = 1, 2, 3$) topological charge, the 3D lattice gives rise to a variety of novel dynamical effects, such as the reshaping of the $S = 2$ unstable vortex into a stable one with $S = 3$. Furthermore, the 3D dynamical lattice can sustain quite unusual but stable excitations, such as the two-component complex, in which the individual vortices in the components are orthogonal to each other. Studying more complex configurations in such a rich setup as well as examining their interactions, suggest challenging problems for future work.

The work of B.A.M. was supported, in a part, by a grant No. 8006/03 from the Israel Science Foundation. PGK gratefully acknowledges the support of NSF-DMS-0204585, NSF-CAREER and the Eppley Foundation for Research.

Bibliography

1. S. Aubry, Physica **103D**, 201 (1997); S. Flach and C.R. Willis, Phys. Rep. **295**, 181 (1998); see also focus issues of Physica **119D** (1999), and Chaos **13**, 586–799 (2003).
2. A.A. Sukhorukov *et al.*, IEEE J. Quantum Elect. **39**, 31 (2003); U. Peschel *et al.*, J. Opt. Soc. Am. B **19**, 2637 (2002).
3. A. Trombettoni and A. Smerzi, Phys. Rev. Lett. **86**, 2353 (2001); F.Kh. Abdullaev *et al.*, Phys. Rev. **A64**, 043606 (2001); F.S. Cataliotti *et al.*, Science **293**, 843 (2001); A. Smerzi *et al.*, Phys. Rev. Lett. **89**, 170402

(2002); G.L. Alfimov, P.G. Kevrekidis, V.V. Konotop, and M. Salerno, Phys. Rev. E **66**, 046608 (2002); R. Carretero-González and K. Promislow, Phys. Rev. A **66** 033610 (2002).

4. E.N. Pelinovsky and S.K. Shavratsky, Physica **3D**, 410 (1981).
5. P. Binder *et al.*, Phys. Rev. Lett. **84**, 745 (2000); E. Trías, J.J. Mazo, and T.P. Orlando, Phys. Rev. Lett. **84**, 741 (2000).
6. M. Peyrard, and A.R. Bishop, Phys. Rev. Lett., **62**, 2755 (1989); T. Dauxois, M. Peyrard and A.R. Bishop, Phys. Rev. E, **47**, R44 (1993); T. Dauxois, M. Peyrard and A.R. Bishop, Phys. Rev. E, **47**, 684 (1993); M. Peyrard *et al.*, Physica **68D**, 104 (1993); A. Campa, and A. Giansanti, Phys. Rev. E, **58**, 3585 (1998).
7. P.G. Kevrekidis, K.Ø. Rasmussen and A.R. Bishop, Int. J. Mod. Phys. B **15**, 2833 (2001).
8. D.N. Christodoulides and R.I. Joseph, Opt. Lett. **13**, 794 (1988); A. Aceves *et al.*, Opt. Lett. **19**, 1186 (1994).
9. H. Eisenberg *et al..*, Phys. Rev. Lett. **81**, 3383 (1998); R. Morandotti *et al..*, Phys. Rev. Lett. **83**, 2726 (1999); H. S. Eisenberg *et al.*, J. Opt. Soc. Am. B **19**, 2938 (2002); J. Meier *et al.*, Phys. Rev. Lett. **91**, 143907 (2003).
10. M. Greiner *et al.*, Nature (London) **415**, 39 (2002).
11. S. Komineas and N. Papanicolaou, Phys. Rev. Lett. **89**, 070402 (2002); Phys. Rev. A **67**, 023615 (2003).
12. B.B. Baizakov, V.V. Konotop, and M. Salerno, J. Phys. B **35**, 5105 (2002); B. Baizakov, B.A. Malomed, and M. Salerno, Europhys. Lett. **63**, 642 (2003).
13. M. Johansson and S. Aubry, Phys. Rev. E **61**, 5864 (2000).
14. D. Mihalache *et al.*, Phys. Rev. Lett. **88**, 073902 (2002).
15. B.A. Malomed and P.G. Kevrekidis Phys. Rev. E **64**, 026601 (2001).
16. P.G. Kevrekidis, B.A. Malomed, Z. Chen, and D.J. Frantzeskakis, *Stable Higher Charge Vortices in the Discrete Nonlinear Schrödinger Equation*, submitted to Phys. Rev. E.
17. C. Sulem and P.L. Sulem, *The Nonlinear Schrödinger Equation*, Springer-Verlag (New York, 1999).
18. C.J. Myatt *et al.*, Phys. Rev. Lett. **78**, 586 (1997).
19. S. Inouye *et al.*, Nature **392**, 151 (1998); E.A. Donley *et al.*, Nature **412**, 295 (2001).

ULTRAWIDE SPECTRAL BROADENING AND PULSE COMPRESSION BY HIGH-ORDER STIMULATED RAMAN SCATTERING

V. P. Kalosha and J. Herrmann

Max Born Institute for Nonlinear Optics and Short Pulse Spectroscopy, Max-Born-Str. 2a, 12489 Berlin, Germany

E-mail: kalosha@mbi-berlin.de

We present the results of analytical and numerical studies of ultrawide spectral broadening and compression of isolated extremely short visible and UV/VUV pulses by high-order stimulated Raman scattering (HSRS) in hollow waveguides. Spectral and temporal characteristics of the output pulses and the mechanism of pulse compression using dispersion of the gas filling and output glass window are investigated without the slowly varying envelope approximation. It is shown that phase-locked pulses as short as ~1 fs in the visible region can be generated by coherent scattering in impulsively excited Raman media without the necessity of external phase control.

1. Introduction

In the last decade ultrafast laser technology has been developed rapidly and successfully applied to a large variety of fields, especially for time-resolved studies of fast-evolving fundamental physical and chemical processes. Three main ingredients of this technological advance are connected with the development of Kerr-lens mode-locked solid state lasers, chirped pulse amplification and extra-cavity compression techniques[1,2]. The traditional method for pulse compression is based on the Kerr nonlinearity in waveguides to produce spectral broadening by self-phase modulation (SPM) and subsequent chirp compensation by anomalous dispersion in carefully designed linear optical elements like prism pairs, chirped mirrors or spatial light modulators. Pulses down to 5 fs were generated using single-mode fibers[3] and 4.5 fs by hollow fibers filled with a noble gas[4]. Though the theoretical limit of the bandwidth which can be achieved by this method allows still shorter pulses[5], the experimental results obtained to date by this method are difficult to improve practically due to problems in chirp compensation over an extremely broad spectrum and the onset of instabilities for too high input intensities due to plasma generation by ionization. Note that the method of pulse compression by SPM is only applicable in the near-IR and optical range.

There is also a great interest in an efficient pulse compression method in UV and VUV as well as in MIR for ultrafast time-resolved measurements in material science, chemistry and biology, where the methods known up to now suffer from low efficiency and too long pulse durations. Numerous applications demand methods for the generation of shorter pulses in optical as well in UV/VUV and MIR and also in the soft X-ray spectral range. Recently two complementary approaches were studied to push the duration of pulses to shorter limits in different spectral ranges and with different characteristics. On the one hand, using a new measurement technique recently attosecond trains[6] as well as single attosecond X-ray pulses[7] have been demonstrated. Based on this progress first time-resolved spectroscopic measurements in the attosecond domain have been reported in Ref.[8].

In a second approach high-order stimulated Raman scattering (HSRS) was studied with the aim to generate ultra-broadband spectra with high conversion efficiency sufficient for sub-femtosecond pulses. Two different schemes were considered for extremely short pulse generation. Ref.[9] theoretically predicted that high-repetition trains of sub-fs pulses can be obtained using adiabatic excitation of the medium by a properly detuned two-color cw field. Recently this technique was experimentally demonstrated in Ref.[10].

In another much promising scheme impulsive excitation by ultrashort pulses and a pump-probe arrangement is used to compress a probe pulse by the Raman medium excited by the pump pulse. Using such arrangement in Refs.[11,12], the first experimental demonstration of pulse compression by HSRS was reported with the generation of a train of 6-fs pulses. A mechanism for the generation of single compressed pulses by HSRS for probe pulses shorter than the period of Raman oscillations was first proposed and theoretically studied in Refs.[13,14] predicting compression of single pulses without additional chirp control caused by the material dispersion of the medium itself. This regime was used to produce the shortest single optical pulses with a duration of 3.8 fs in SF_6[15].

Optical excitation of coherent molecular vibrations or rotations via stimulated Raman scattering has mainly been carried out in one of three ways. The original method consists of focusing of an intense pump field with carrier frequency $\omega_{\rm pu}$ into a Raman medium where, due to the two-photon stimulated process, a Stokes wave with frequency $\omega_{\rm pu} - \Omega$ is generated growing out from spontaneous Raman emission where Ω is the Raman frequency shift (see, e.g., Ref.[16]). This process may be observed if the laser intensity exceeds a certain threshold. Anti-Stokes radiation often appears on a cone determined by the phase-matching condition. The second method which is possible in a collinear setup is to excite the Raman medium by two tunable laser outputs with frequencies $\omega_{\rm pu}$ and $\omega_{\rm pu} - \Omega$ (two-color excitation)[17], which are adjusted to match the Raman frequency Ω. Coherent Raman excitation can also be achieved using a third method, the impulsive technique[18], in which a single ultrashort pulse with duration shorter than the Raman oscillation period $T_{\rm R} = 2\pi/\Omega$ initiates the coherent vibrational or rotational motion. Then the spectral width exceeds the Raman shift and stimulated Raman scattering occurs through mixing among frequency components contained within the pulse bandwidth. In the impulsive regime the vibrational amplitudes of Raman excitation can be increased by using synchronized pulse sequences[19].

The last two methods do not involve a laser intensity threshold and for appropriate conditions they can be used for the aim of pulse compression to extremely short durations, because multiple Raman lines are generated in collinear direction after a certain propagation length by HSRS. An early theoretical study of HSRS in the framework of the slowly-varying envelope approximation (SVEA) is presented in Ref.[20] which shows a distribution of the spectral lines described by Bessel functions. In the case of two-color excitation with a pump and a first Stokes pulse of comparable intensities, a large number of Stokes and anti-Stokes lines are generated in collinear direction spanning the range from IR to UV[17,21]. Theoretical studies of this regime were published in Refs.[20,22]. In Refs.[13,14] we predicted that for probe pulses with a duration smaller than the period of Raman oscillations, the HSRS spectrum becomes continuous with ultra-broadband width.

The present chapter is based on our papers[13,14,23–25] and contains the investigation of pulse compression by HSRS. In section 2 the basic propagation equation is derived for the description of ultra-broad HSRS without standard approximations as the use of SVEA and a Taylor expansion for the dispersion. In section 3 the specific character of HSRS is investigated by analytical solutions of the basic equations. This allows to reveal the main physical specifics of phase-amplitude modulation and pulse compression by HSRS, as the behavior of the spectral phases, the accessible bandwidths, the possibility to compress pulses by normal dispersion simply by a piece of glass, but also by anomalous dispersion, and specific features for a probe-pulse delay near the out-of-phase point of Raman oscillations, where the Fourier-transformed field is real but changes the sign, i. e. the spectral phase jumps from zero to π inside the spectrum. In section 4 numerical solutions are presented for pulse compression in the optical range, which takes into account the full influence of dispersion and the change of the pump by HSRS during propagation in the Raman medium. In particular, the influence of dispersion is studied and optimum condition are found with compression up to 1.7 fs in SF_6.

2. Theoretical fundamentals

The standard theoretical method in nonlinear optics is the SVEA in which the rapidly varying part of the electric field is separated from the slowly varying envelope. It is valid only if the spectral width of a pulse $\Delta\omega$ is much smaller than the carrier frequency ω_0: $\Delta\omega \ll \omega_0$. Therefore, the SVEA in time can not be used for conditions studied here.

Nonlinear pulse propagation without special prerequisites of the SVEA can be studied by the numerical solution of Maxwell equations by the finite-difference time-domain method along with solutions for the material equations (see, e.g., Ref.[5] and references therein). However, the large numerical effort of this approach, which directly resolves high optical oscillations of the field both in time and space, limits possible propagation lengths to a few mm. In several papers various improved approximated equations have been derived that allow the theoretical description beyond the validity of the standard approach[2,27,28].

The propagation of the optical pulses in nonlinear media is described by the wave equation

$$\left(\frac{\partial^2}{\partial z^2} + \Delta_\perp\right)\mathbf{E} - \nabla(\nabla\cdot\mathbf{E}) - \frac{1}{c^2}\frac{\partial^2\mathbf{E}}{\partial t^2} = \mu_0\frac{\partial^2\mathbf{P}}{\partial t^2}, \qquad (1)$$

where $\Delta_\perp$ is the transverse Laplacian, $\mathbf{P} = \mathbf{P}_\mathrm{L} + \mathbf{P}_\mathrm{NL}$ is the medium polarization with linear nonresonant contributions $\mathbf{P}_\mathrm{L}$ and nonlinear or resonant (as, e.g., given by the Raman effect) contributions $\mathbf{P}_\mathrm{NL}$. The z direction is marked out here because we suppose a predominant propagation in this direction but without the assumption of the paraxial approximation. Substituting the Fourier-transformed field $\mathbf{E}(\mathbf{k}_\perp, z, \omega)$ and polarization $\mathbf{P}(\mathbf{k}_\perp, z, \omega)$ into Eq. (1), splitting the transverse component of the electric field $\mathbf{E}_\perp = \{E_x, E_y\}$ into a forward $\mathbf{E}^+$ and a backward $\mathbf{E}^-$ propagating parts along the z-axis: $\mathbf{E}(\mathbf{k}_\perp, z, \omega) = \mathbf{E}^+(\mathbf{k}_\perp, z, \omega) + \mathbf{E}^-(\mathbf{k}_\perp, z, \omega)$ and neglecting the waves propagating backward we arrive to the equations[25]

$$\frac{\partial E_j^+}{\partial z} = ih_j E_j^+ \tag{2}$$

with

$$h_j^2(\mathbf{k}_\perp, z, \omega) = \omega^2/c^2 - \mathbf{k}_\perp^2 + ik_j \tilde{\nabla} \cdot \mathbf{E} + \mu_0 \omega^2 P_j/E_j, \tag{3}$$

$\tilde{\nabla} \cdot \mathbf{E} = -i\mathbf{k}_\perp \mathbf{E}_\perp + \partial E_z/\partial z$ and $j = \{x, y\}$. Note that the right-hand side term $h_j E_j^+$ in Eq. (2) has no singularity for $E_j^+ \to 0$.

Equations (2) are the first-order evolution equations in 3D without SVEA, paraxial approximation or perturbative inclusion of vectorial effects and valid for isotropic and anisotropic media with an arbitrary functional form for the nonlinear polarization. They can be solved numerically as a boundary problem by the split-step operator method. They take into account a possible extremely broad bandwidth, sharp temporal features and non-axial effects in space-time coupling, higher-order nonlinear and dispersive and also vectorial and anisotropic effects. Eqs. (2) reduce the numerical effort in solving the exact 3D Maxwell equations. In general, the full 3D integration of the exact Maxwell equation for propagation lengths in the range of tens of cm is outside the capacity of computers available today. The only approximation used here is the assumption of weak backward waves and the relation $\partial \Delta n_{\mathrm{NL}}(E)/\partial z \ll k(\omega_0)$, where Δn_{NL} is defined by $P_{\mathrm{NL}} = \varepsilon_0 \Delta n_{\mathrm{NL}} E$. It is much weaker condition than the commonly used SVEA.

Previously derived evolution equations without SVEA as presented in Refs.[2,27,28] can be derived from Eqs. (2). Introducing the refractive index $n^2(\omega) = 1 + \chi(\omega)$ for an isotropic medium, $\nabla \cdot \mathbf{E} = 0$ and assuming small contribution of the beam's non-paraxiality as well as nonlinearity, we use the first-order expansion of the square root in Eq. (3) and then introduce the moving time coordinate $\eta = t - z/c$, make the substitution $\partial/\partial z \to \partial/\partial z - \partial/c\partial\eta$ and after back Fourier-transformation to the real spatial coordinates from Eqs. (2) obtain the following equation[28]:

$$\begin{aligned}\frac{\partial \mathbf{E}_\perp(\mathbf{r}, \omega)}{\partial z} &= \frac{i\omega}{c}\left[n(\omega) - 1\right]\mathbf{E}_\perp(\mathbf{r}, \omega) + \frac{i}{2k(\omega)}\Delta_\perp \mathbf{E}_\perp(\mathbf{r}, \omega) \\ &\quad + \frac{i\mu_0\omega^2}{2k(\omega)}\mathbf{P}_{\mathrm{NL}}(\mathbf{r}, \omega),\end{aligned} \tag{4}$$

where here and below the superscript '+' is omitted.

Equation (4) is a generalization of the so-called *reduced* Maxwell equation[27] which is valid only for a refractive index close to unity. The reduced Maxwell equation is obtained from Eq. (4) by the substitution $n(\omega) - 1 \approx [n^2(\omega) - 1]/2$ and back Fourier transformation into the time domain:

$$\frac{\partial E}{\partial z} = -\frac{\mu_0 c}{2} \frac{\partial P}{\partial \eta}. \tag{5}$$

In a bulk medium the nonuniform transverse intensity profile of the beam leads to nonuniform phase relations due to diffraction and nonlinear transverse reshaping effects. Pulse compression by SPM or HSRS in a nonlinear bulk medium is therefore substantially limited. A possible solution of this problem is using hollow fibers filled with a gas as nonlinear medium. This technique was first applied for optical pulse compression by self-phase modulation using hollow-core silica waveguides filled with a noble gas[2,4,29]. The same technique was also used for impulsive Raman scattering with a Raman-active gas as H_2 or SF_6[11,12].

The Fourier-transformed linear polarization in a hollow waveguide is

$$P_{\mathrm{L}}(z,\omega) = \epsilon_0[\chi_{\mathrm{b}}(\omega) + \chi_{\mathrm{g}}(\omega)]E(z,\omega), \tag{6}$$

where χ_{b} is the bulk susceptibility defined by the Sellmeyer formula $\chi_{\mathrm{b}}(\omega) = \sum_\ell S_\ell/(\omega_\ell^2 - \omega^2)$ with experimentally determined parameters S_ℓ, ω_ℓ. For the hollow waveguide with radius $R \gg \lambda$ the waveguide contribution for the fundamental EH_{11} mode is given by[30] $\chi_{\mathrm{g}}(\omega) = (2.405c/\omega R)^2[1 + i(1+\nu^2)/(2\pi R\sqrt{1-\nu^2})]$, where $\nu \simeq 1.45$ is the ratio between the refractive indices of fused silica and the internal gas.

For the description of the Raman effect we assume that only two rotational or vibrational levels are excited by the pump pulse. This assumption is justified if the Raman frequency is not too small, as, e.g., in H_2 or SF_6. Then the Raman polarization $P_{\mathrm{R}} = N\mathrm{Sp}(\hat{\alpha}\hat{\rho})E$ can be described by the two-photon two-level system equations (see, e.g.,[31]):

$$\left(\frac{\partial}{\partial \eta} + \frac{1}{T_2} - i\Omega\right)\rho_{12} = \frac{i}{2\hbar}\left[(\alpha_{11} - \alpha_{22})\rho_{12} + \alpha_{12}w\right]E^2, \tag{7a}$$

$$\frac{\partial w}{\partial \eta} + \frac{w+1}{T_1} = \frac{i\alpha_{12}}{2\hbar}\left(\rho_{12} - \rho_{12}^*\right)E^2, \tag{7b}$$

where $\rho_{12} = \frac{1}{2}(u + iv)$, $w = \rho_{22} - \rho_{11}$, $\hat{\rho}$ is the density matrix, $\hat{\alpha}$ is the two-photon polarizability matrix responsible for the Raman process and Stark shift, Ω is the Raman frequency shift, N is the density, T_1, T_2 are the relaxation times.

3. Analytical solution for the impulsive pump-probe regime

3.1. *General solution*

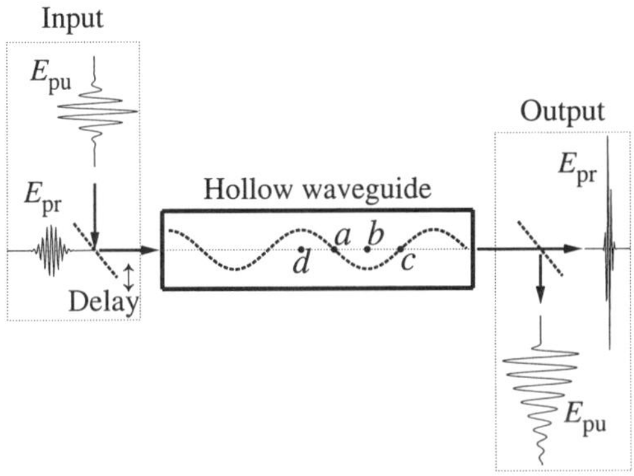

Figure 1. Schematic of an impulsive pump-probe regime for pulse compression using high-order stimulated Raman scattering in a gas-filled hollow waveguide. The periodic response of the molecular system in hollow waveguide is shown where the positions of the input probe pulse at the out-of-phase point, minimum, in-phase point and maximum are indicated by a, b, c and d, respectively.

First, we derive an analytical solution of Eqs. (5) and (7a) for the impulsive regime where a short pump pulse with duration

$$\tau_{pu} < T_R = 2\pi/\Omega \tag{8}$$

excites the Raman medium and a delayed weaker probe pulse is amplitude- and phase-modulated due to the long-living molecular excitation of the foregoing pump pulse (see Fig. 1). Then the probe-pulse propagation is in a linear regime with polarization

$$P_R = N\alpha_{12}u(z,\eta)E_{pr}(z,\eta), \tag{9}$$

where $u(z,\eta)$ depends only on the pump pulse. For rather general situations the medium response for the probe pulse can be presented approximately by the following general separation ansatz

$$u(z,\eta) = u_0(z)a(\eta), \tag{10}$$

where u_0 is the amplitude of Raman excitation.

In the case of a two-level Raman medium the functions $u_0(z)$ and $a(\eta)$ can be calculated in a simple way if relaxation, population changing and

Stark effect can be neglected. Then Eqs. (7) for the density matrix is simplified to the equation

$$\left(\frac{\partial}{\partial\eta} - i\Omega\right)\rho_{12} = -\frac{i\alpha_{12}}{2\hbar}E_{\mathrm{pu}}^2. \tag{11}$$

The solution of Eq. (11) for $u = 2\Re\rho_{12}$ is given by

$$u(z,\eta) = u_0(z,\eta)\sin\Omega\eta \tag{12}$$

with

$$u_0 = \frac{\alpha_{12}}{\hbar} == \int_{-\infty}^{\eta} E_{\mathrm{pu}}^2(z,\eta')\cos\Omega\eta'\mathrm{d}\eta',$$

where a small $z-$ and $\eta-$depending phase is neglected.

In a gaseous medium the main mechanism of the ultrawide spectral broadening and pulse compression can be understood with the neglect of linear dispersion. Then the propagation equation (5) for the probe pulse $E_{\mathrm{pr}}(z,\eta)$ is

$$\frac{\partial E_{\mathrm{pr}}}{\partial z} = -\gamma u_0(z)\frac{\partial}{\partial\eta}\left[a(\eta)E_{\mathrm{pr}}\right], \tag{13}$$

where $\gamma = \frac{1}{2}\mu_0 cN\alpha_{12}$. The first-order partial differential equation (13) can be solved using the equations of its characteristics given by

$$E_{\mathrm{pr}}(z,\eta) = E_{\mathrm{pr}}(s)\frac{a(s)}{a(\eta)}, \tag{14}$$

where $E_{\mathrm{pr}}(\eta)$ is the input probe pulse field at $z = 0$ and s is the parameter of the characteristic lines defined as follows[32]:

$$\int_{\eta}^{s}\frac{\mathrm{d}\eta'}{a(\eta')} = \gamma\int_0^z u_0(z')\mathrm{d}z', \tag{15}$$

From Eqs. (10) and (12) we have $a(\eta) = \sin\Omega\eta$ and the solution for the probe pulse is as follows[13,23]:

$$E_{\mathrm{pr}}(z,\eta) = E_{\mathrm{pr}}(s - t_{\mathrm{m}})\frac{\sin\Omega s}{\sin\Omega\eta}, \tag{16}$$

where $E_{\mathrm{pr}}(\eta) = A_{\mathrm{pr}}(\eta)\cos\omega_{\mathrm{pr}}\eta$ and the initial probe pulse envelope is chosen here to be the Gaussian: $A_{\mathrm{pr}}(\eta) = A_{\mathrm{pr}}\exp(-2\ln 2\eta^2/\tau_{\mathrm{pr}}^2)$, τ_{pr} and ω_{pr} are the duration (intensity FWHM) and the carrier frequency, respectively, t_{m} is the time of the peak position of the input pulse with respect to one of the out-of-phase point $\eta = 0$ (Fig. 1). From Eq. (15) we obtain

$$s(z,\eta) = \frac{2}{\Omega}\mathrm{arctg}\left[\Gamma\mathrm{tg}\left(\frac{\Omega\eta}{2}\right)\right], \tag{17}$$

where the amplification parameter $\Gamma(z) = \exp[\gamma\Omega\int_0^z u_0(z')\mathrm{d}z']$. This parameter increases exponentially with the Stokes shift Ω. On the other hand, for the impulsive regime the condition $\tau_{\mathrm{pu}} < T_{\mathrm{R}}$ has to be satisfied. Since u_0 increases with the pulse duration, a smaller value of Ω can be compensated by a larger duration of the pump pulse with the same intensity. $\Gamma(z)$ is determined by the change of the pump pulse during propagation which significantly influences the amplitude-phase modulation for the probe pulse (see Section 3.2). Eqs. (14) takes into account the decrease of the Raman amplitude $u_0(z)$ and generalize a solution given in Refs.[13] and[14] for the constant u_0. The decrease of u_0 limits the maximum possible value of $\Gamma(z)$ for a given pump-pulse energy. This leads to one of the main limitations of shortest accessible durations by HSRS. The specific dependence of $u_0(z)$ on distance can be calculated numerically, as it will be done in section 4.

Here we estimate the optimum pump parameters and maximum possible $\Gamma(z)$. The Raman amplitude u_0 at $z = 0$ for a Gaussian pump pulse can be found from Eq. (12) as

$$u_0 = \frac{\sqrt{\pi}\alpha_{12}\tau_{\mathrm{pu}}}{4\hbar\sqrt{\ln 2}} A_0^2 \exp(-\Omega^2\tau_{\mathrm{pu}}^2/16\ln 2) \tag{18}$$

where A_0 is the pump-pulse amplitude. Therefore u_0 has a maximum for the pump-pulse input duration $\tau_{\mathrm{opt}} = \sqrt{2\ln 2}T_{\mathrm{R}}/\pi$. Due to the interaction with the medium the pump pulse is depleted and the amplitude of the Raman excitation is decreased. With the help of Eqs. (5) and (12), the next-order approximation to Eq. (18) is found as follows:

$$u_0(z) = u_0(1 - \sigma z), \tag{19}$$

where $\sigma = \sqrt{\pi}\gamma\alpha_{12}\Omega\tau_{\mathrm{pu}}A_0^2/8\hbar\sqrt{\ln 2}$. This gives the amplification parameter

$$\Gamma(z) = \exp\left[\Omega\gamma u_0(z - \sigma z^2/2)\right] \tag{20}$$

which reaches the maximum possible value $\Gamma_{\mathrm{max}} \approx 2$ independent on pump and medium parameters at distance $L_{\mathrm{opt}} = 1/\sigma$. However, L_{opt} depends on the pump energy and the medium. Here Γ_{max} is slightly underestimated because Eq. (20) is valid only for small distances.

The fluence which provides the maximum possible Raman amplitude $u_0 = 1$ and the population difference $w = 0$ is

$$J_{\mathrm{max}} \approx 1.6\frac{c\varepsilon_0\hbar}{\alpha_{12}}. \tag{21}$$

For typical Raman media $J_{\mathrm{max}} \approx 4\ \mathrm{J/cm^2}$ which corresponds to a pump energy 0.6 mJ for a beam radius 70 μm and the typical Raman polarizability $\alpha_{12} \sim 10^{-41}\ \mathrm{cm^2/V}$.

3.2. *Amplitude-phase modulation and compression of a short probe pulse*

In the method for the compression of single pulses by HSRS[13–15], the temporally delayed probe pulse with duration

$$\tau_{\rm pr} < T_{\rm R} \tag{22}$$

coherently interacts with the impulsively excited Raman medium. The analytical solution (16-17) allows an understanding of the main physical features of the spectral broadening and phase-amplitude modulation of the probe pulse[23,25].

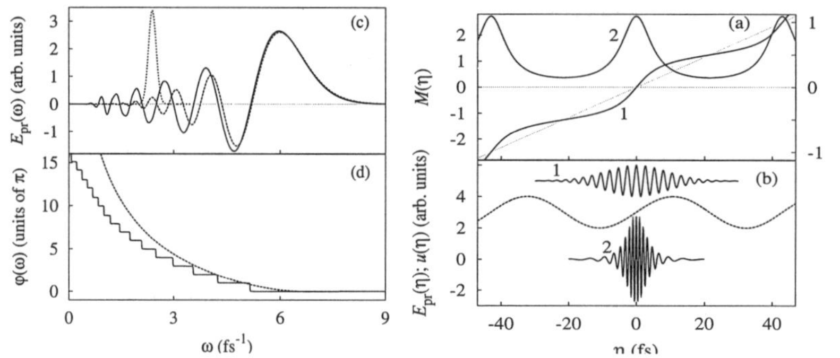

Figure 2. Temporal and spectral characteristics of the solution Eq. (16) for the out-of-phase position: (a) 1, function $s(\eta)$, 2, amplitude modulation factor $M(\eta) = \sin\Omega s/\sin\Omega\eta$; (b) 1, input probe pulse, 2, output pulse for $\Gamma = e^1$, dashed line, molecular oscillations; (c) solid line, spectrum of the probe pulse, dashed, approximation Eq. (29), dotted, input spectrum; (d) spectral phase of the output pulse, dashed, phase as defined by Eq. (28). Input pulse duration is 15 fs, wavelength is 790 nm, period of the molecular oscillation is 43 fs (SF_6).

The solution (16) is determined by the function $s(\eta)$ and the amplitude modulation factor $M(\eta) = \sin\Omega s/\sin\Omega\eta$. Both define the character of phase-amplitude modulation of the probe pulse by the Raman-active molecular modulator and are presented in Fig. 2 (a). The function $s(\eta)$ (curve 1) oscillates periodically around the straight line $s = \eta$ and has the steepest slope at the out-of-phase points $t_{\rm o} = lT_{\rm R}$ ($l = 0, \pm 1, \dots$), where the main contribution to the phase modulation arises. In the vicinity of $\eta = t_{\rm o}$ the expansion $s \approx t_{\rm o} + \Gamma(\eta - t_{\rm o}) - \Gamma\varepsilon(\eta - t_{\rm o})^3/3$ can be applied where $\varepsilon = (\Gamma^2 - 1)\Omega^2/4$. Near the in-phase points $t_{\rm i} = (l + \frac{1}{2})T_{\rm R}$ the function $s(\eta)$ has the smallest slope where $s \approx t_{\rm i} + (\eta - t_{\rm i})/\Gamma + \varepsilon(\eta - t_{\rm i})^3/3\Gamma^3$.

Then from Eq. (16) the main qualitative properties of the temporal pulse characteristics can be immediately seen, such as compression to a duration

$$\tau \approx \tau_{\rm pr}/\Gamma, \tag{23}$$

a change of the instantaneous frequency spanning from $\omega_{\rm min} \approx \omega_{\rm pr}/\Gamma$ to $\omega_{\rm max} \approx \Gamma\omega_{\rm pr}$. The instantaneous frequency nearby the peak of the probe pulse is

$$\omega(\eta) \approx \Gamma\omega_{\rm pr}[1 - \varepsilon(\eta - t_{\rm o} - t_{\rm m})^2] \tag{24}$$

and has a linear and a quadratic chirp depending on the position of the input pulse $t_{\rm m}$ in respect to the out-of-phase points $t_{\rm o}$ of the Raman oscillations. At the out-of-phase point this chirp is purely quadratic. The peak amplitude and the pulse energy are increased by the factor Γ and the pulse peak is shifted toward the out-of-phase point $t_{\rm o} = 0$ as $t_{\rm m}(z) \approx t_{\rm m}/\Gamma$ due to the Raman-induced refractive index modulation.

By the change of the variables and using the relation $a(s)/a(\eta) = \partial s/\partial\eta$, the Fourier-transformed field can be presented in the form

$$E_{\rm pr}(\omega) = \int_{-\infty}^{\infty} A_{\rm pr}(s - t_{\rm m}) \exp\left[i\omega_{\rm pr}(s - t_{\rm m}) - i\omega\eta\right] {\rm d}s \tag{25}$$

where $\eta = \eta(s)$ is the inverse function of Eq. (17). In the vicinity of the out-of-phase point, $\eta(s)$ can be approximated by $\eta \approx t_{\rm o} + (s - t_{\rm o})/\Gamma + \varepsilon(s - t_{\rm o})^3/3\Gamma^3$. Then to find the Fourier-transformed field (25) we use the stationary-point method and obtain

$$E_{\rm pr}(\omega) = \frac{C}{\sqrt[3]{\varepsilon\omega}} \left[A_{\rm pr}^*(s_0 + t_{\rm m}){\rm Ai}(x) + A_{\rm pr}(s_0 - t_{\rm m}){\rm Ai}^*(x)\right], \tag{26}$$

where ${\rm Ai}(x) = \sqrt{\pi}\int_0^\infty \exp(ixs + is^3/3){\rm d}s$, $s_0^2 = \Gamma^2(\Gamma\omega_{\rm pr} - \omega)/\varepsilon\omega$, $x = (\omega - \Gamma\omega_{\rm pr})/\sqrt[3]{\varepsilon\omega}$, $C = \Gamma\exp(-i\omega t_{\rm m})/\sqrt{\pi}$. The expression (26) can be simplified by the asymptotics of the Airy function:

$${\rm Ai}(x) = \begin{cases} \exp[-i\varphi_{\rm a}(\omega)]/(-x)^{1/4} & \text{for } \omega < \Gamma\omega_{\rm pr}, \\ \exp(-2x^{3/2}/3)/2x^{1/4} + i\sqrt{\pi}/x & \text{for } \omega > \Gamma\omega_{\rm pr}, \end{cases} \tag{27}$$

where

$$\varphi_{\rm a}(\omega) = \frac{2}{3}(\Gamma\omega_{\rm pr} - \omega)^{3/2}/\sqrt{\varepsilon\omega} - \frac{\pi}{4}. \tag{28}$$

Now we can consider the analytical solution (16) and the approximate expression (26) for different pump-probe delay. We discuss the properties of the output probe for a delay when the peak of the input probe pulse has

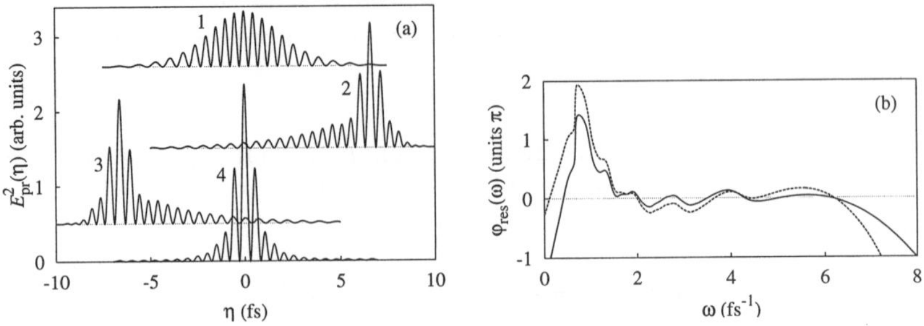

Figure 3. Compression of the probe pulse in the out-of-phase position: (a) 1, probe pulse propagated in the Raman excited medium described by Eq. (16); 2, pulse after the phase compensation with $\varphi_c(\omega) = -0.9\omega^2$; 3, pulse after the phase compensation with $\varphi_c(\omega) = 0.9\omega^2$; 4, pulse after ideal compression; (b) residual phase $\varphi_{res}(\omega)$ for $\varphi_c(\omega) = -0.9\omega^2$ (solid) and for phase compensation by a 18-μm thick SiO_2 plate (dashed). All parameters are as in Fig. 2.

a position near the out-of-phase point $t_m = 0$ (see Fig. 1). In Fig. 2 (b) the solution is presented for a 15-fs Gaussian input probe pulse (curve 1) together with the Raman oscillations $u(\eta)$ for $\Gamma = 2.7$ and $T_R = 43$ fs (SF_6). The output pulse (curve 2) is compressed to 4.4 fs, in agreement with the estimate (23). The pulse spectrum is significantly broadened and blue-shifted with sign changes inside the spectrum. This specific behavior can be understood from the asymptotic expressions for $E_{pr}(\omega)$. For $t_m = 0$ the field (26) is real and for $\omega < \Gamma\omega_{pr}$ can be written as

$$E_{pr}(\omega) = \frac{CA_{pr}(s_0)\cos\varphi_a(\omega)}{[\varepsilon\omega(\Gamma\omega_{pr} - \omega)]^{1/4}}. \tag{29}$$

Equation (29) rather precisely agrees with the numerically calculated value of the solution (16) [compare corresponding solid and dashed line in Fig. 2 (c)]. The spectral field $E_{pr}(\omega)$ changes its sign at frequencies where $\varphi_a(\omega) = (l + \frac{1}{2})\pi$ $(l = 0, 1, \dots)$ and the spectral phase presented in Fig. 2 (d) by solid line shows jumps to π. The dashed line in Fig. 2 (d) shows $\varphi_a(\omega)$ given by Eq. (28). The appearance of a jump-like phase modulation is rather unusual in ultrafast optics. As seen in Fig. 3, after phase compensation by normal group velocity dispersion (GVD) characterized by the spectral phase $\varphi_c(\omega) = -\delta\omega^2$ with $\delta = 0.9$ fs^2, the pulse is compressed to 1.3 fs (curve 2) and exhibits an asymmetric shape with a longer front. This unexpected behavior can be understood on the basis of Eq. (29) which can be expressed by two complex components with opposite spectral phases $\pm\varphi_a(\omega)$ and equal amplitudes. The compressor decomposes the pulse into two parts: a short bandwidth-limited pulse arising from the first component with the phase $\varphi_a(\omega)$ and a second long phase-modulated pulse arising from the second one with the phase $-\varphi_a(\omega)$. Pulse compression of the same pulse can also be achieved by a compressor with anomalous dispersion $\varphi_c(\omega) = \delta\omega^2$ and results also in a 1.3 fs pulse but with a longer edge at the front [Fig. 3 (a), curve 3]. From the representation of the cos-function in Eq. (29) by two exponentials this rather surprising behavior can be easily understood.

From the bandwidth of the spectrum given by the minimum and the maximum frequencies of the pulse, the shortest possible duration accessible after exact phase compensation can be estimated. Fitting of a shape-depending numerical factor in the relation $\tau_{\min} \sim \Delta\omega^{-1}$ results in the pulse duration after ideal compression

$$\tau_{\min} \approx 4\Gamma/\omega_{\text{pr}}(\Gamma^2 - 1). \tag{30}$$

An ideally compressed pulse is presented in Fig. 3 by curve 4. Eq. (30) rather precisely predicts its duration of 1.1 fs. The important conclusion followed from Eq. (30) is that the ultimately shortest duration of the probe pulse is inversely proportional to ω_{pr} and Γ and independent of the input probe-pulse duration. However, the largest pulse duration for single-pulse compression is limited by the condition $\tau_{\text{pr}} < T_{\text{R}}$.

Analogously, one can show that near the minimum of the Raman oscillations the probe pulse has a smooth spectral phase given by Eq. (28) which can be compensated only by normal dispersion. Near the maximum the pulse has the smooth phase $-\varphi_a(\omega)$ which can be compensated only by anomalous dispersion. Near the in-phase point two pulses with the same amplitude and opposite sign of chirp with phases $\pm\varphi_a(\omega)$ are generated in the Raman medium. By an optical element with either normal or anomalous dispersion one of the pulse is compressed while the other is suppressed.

3.3. *Comparison with the slowly varying envelope approximation for small Raman conversion*

Let us derive from Eqs. (16)–(17) approximate expressions for a probe pulse in the Raman-excited medium in the case of a small conversion efficiency $\ln\Gamma \ll 1$ and compare the above results with corresponding expressions with the use of SVEA. For this limiting case, from Eq. (17) we find $s \approx \eta + \tilde{\gamma}\sin\Omega\eta$, where $\tilde{\gamma}(z) = \gamma\int_0^z u_0(z)\mathrm{d}z$, and obtain

$$E_{\text{pr}}(z,\eta) = A_{\text{pr}}(\eta - t_{\text{m}})\cos\omega_{\text{pr}}\left(\eta - t_{\text{m}} + \tilde{\gamma}\sin\Omega\eta\right). \tag{31}$$

This is a frequency-modulated pulse without temporal amplitude modulation and this means that it does not become shorter during propagation. With the use of the relation $\exp(ix \sin \Omega\eta) = \sum_l J_l(x) \exp(il\Omega\eta)$, the Fourier-transformed field is given by

$$E_{\rm pr}(\omega) = \sum_{l=-\infty}^{\infty} J_l A_{\rm pr}(\omega_{\rm pr} + l\Omega - \omega) \exp[i(l\Omega - \omega)t_{\rm m}], \tag{32}$$

where $J_l = J_l(\tilde{\gamma}\omega_{\rm pr})$ are the Bessel functions, $A_{\rm pr}(\omega)$ is the Fourier-transformed pulse envelope. If the input pulse is long with duration $\tau_{\rm pr} \gg T_{\rm R}$, the different terms in Eq. (32) describe discrete high-order Raman components which are symmetric in respect to the input carrier frequency (it is supposed that $\tilde{\gamma}\omega_{\rm pr} < 2.405$). Independently from the delay $t_{\rm m}$ the components have the same phase in anti-Stokes side and change their sign periodically with l on the Stokes side owing to $J_{-l} = (-1)^l J_l$. Therefore, at the Stokes side the spectral phase jumps between 0 to π.

For a short probe pulse with $\tau_{\rm pr} < T_{\rm R}$, the Raman components given by Eq. (32) are broadened and merged into a continuous spectrum with a shape which is defined by their amplitudes and the phases. If the probe pulse is in the out-of-phase point, only the Stokes components show sign changes and therefore due to destructive interference the spectrum is shifted to the anti-Stokes side analogous as in Fig. 2 for a large Raman conversion. If the probe pulse is in-phase with molecular oscillations, only the spectral components on the anti-Stokes side change the sign and the spectral phases jump from 0 to π. Therefore, the spectrum is shifted to the Stokes side. However, as shown above with increasing $\tilde{\gamma}$ these simple relations between the Raman components and their phases are violated and the solution (16)–(17) should be used instead of Eq. (31)–(32) to characterize ultra-wide spectral broadening by HSRS.

The Bessel-like distribution of the spectral intensity followed from Eq. (32) for the case of a pump-probe regime with short input pump and long probe pulses was studied in Ref.[11]. However, it should be noted that the phases of the different Raman lines are not locked as assumed in this paper but have jumps of π as shown above. In the case of two-color excitation a similar Bessel-like disribution of the spectral components has been derived in Refs.[9,20].

4. Compression of visible and UV/VUV pulses: numerical results

In order to find analytical solutions in Section 3 the influence of linear dispersion was neglected. Whether dispersion plays a major role or can be neglected sensitively depends on the chosen spectral range and other parameters. In general, velocity walk-off causes physical limitations of the shortest pulse durations due to the shift of the position of pulse maximum to the in-phase point which typically leads to a pulse splitting after a certain propagation length. Besides, dispersion is the reason for an additional contribution to the spectral phases. Normal GVD of the gas filling leads to the compression of the pulse during propagation if the delay is in the range $T_R/20 \lesssim t_m < T_R/2$ in the same way as the glass window acts beyond the waveguide.

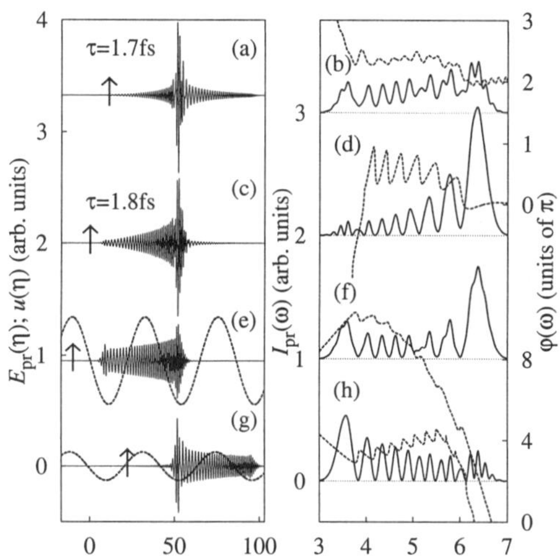

Figure 4. Calculated temporal shapes (left column) and spectral characteristics (right) for a 20-fs input probe pulse at at 395 μm in a hollow waveguide filled with SF_6. The spectra and spectral phases are plotted by solid and dashed lines, respectively. Input position shown by arrows is in minimum (a, b), the out-of-phase point (c,d), the maximum (e, f) and the in-phase point (g,h). Dashed lines present the Raman oscillations at the waveguide input (e) and the output (g). The pump intensity is 50 TW/cm^2, initial pump pulse duration τ_{pu} = 20 fs, wavelength λ_{pu} = 790 nm, hollow waveguide radius R = 70 μm.

To analyze the full influence of dispersion with phase mismatch, walk-off,

GVD as well as higher-order dispersion effects and the change of the pump pulse we have solved numerically the full basic equations given in Eqs. (5)–(7). We study pulse propagation in a waveguide filled with SF_6 with $\alpha_{12} = 1.3 \times 10^{-41}$ Cm^2/V, $C_1 = 0.7$ fs^{-2}, $\omega_1 = 31.9$ fs^{-1} [33]. In Fig. 4 the shape and the spectrum of an 20-fs, 395-nm input probe pulse is presented after propagation through SF_6 for a distance 1 m, a pressure 0.3 atm, a waveguide radius 70 μm and for four characteristic positions of the input probe pulse in respect to the Raman oscillations. The temporal shapes are presented in the left and the spectra in the right sections. For a probe-pulse position in the minimum of Raman oscillations with delay $t_m = T_R/4$, the spectrum is almost symmetrical and has the width $\Delta\omega \approx 3.5$ fs^{-1} [Fig. 4 (a, b)]. It shows a behavior analogous as described by the analytical formula in Eq. (26) with a quasi-oscillating behavior. In difference to the analytical results, now the spectral phase is compensated by normal dispersion of the medium itself and the changes of the output-pulse phase is much smaller than π in a rather large spectral interval [Fig. 4 (b), dashed curves]. Therefore, an extremely short pulse with duration of 1.7 fs is formed during propagation in the Raman medium itself, close to the bandwidth-limited value. For the out-of-phase delay $t_m = 0$ (c,d), the maximum of the spectrum is significantly enhanced on the blue side and the spectrum shows zeros located near the positions predicted by Eq. (29). In this case the temporal shape shows a short bandwidth-limited pulse with duration of 1.8 fs and a long phase-modulated part. This agrees with the scenario of pulse compression near the out-of-phase position as described in Section 3.2. The Raman medium now acts as phase modulator, which broadens the spectrum and causes phase jumps of π within the spectrum, and as compressor with normal GVD, which decomposes the pulse into a bandwidth-limited short and a phase-modulated long part. For an initial probe delay at the maximum of Raman oscillations, we find the opposite behavior [Fig. 4 (e,f)] compared with the position in the minimum, as the chirps both due to the medium dispersion and the Raman modulation have the same sign. This explains the temporal shape in (e) with a long front. For the case of the in-phase delay $t_m = T_R/2$ presented in Fig. 4 (g,h), the Raman amplitude modulation splits the pulse into two parts with approximately the same peaks and located at neighboring out-of-phase points. However, the additional normal GVD of the medium compresses the front part of the pulse and leads to a suppression of the tail maximum as seen in Fig. 4 (g).

Ultrashort UV/VUV pulses are also of great interest for ultrafast time-resolved measurements of atoms, molecules and solids. The analytical

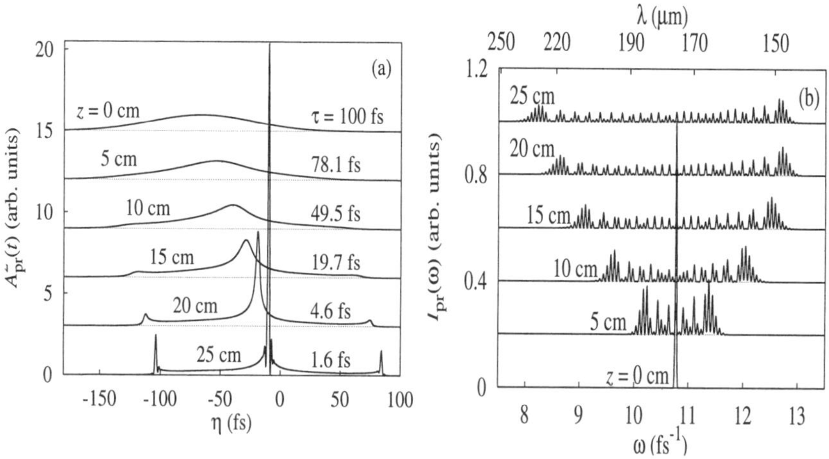

Figure 5. Evolution of temporal shape (a) and spectrum (b) of 100-fs input-probe pulse at 175 nm for pumping of *para*-H_2 at 0.3 atm in a waveguide of radius 70 μm by a train of four 20-fs pulses each of intensity 50 TW/cm^2 at 790 nm.

description of UV/VUV pulse compression is identical with that in the optical range given in Section 3. In difference to the optical region in UV/VUV dispersion has stronger influence and leads to pulse splitting at 1 atm already at short propagation lengths when the spectrum is still not broad enough. This requires to reduce phase mismatch and walk-off by dispersion control of the waveguide using an optimized pressure. As a result of such optimization we found that for too low a pressure the pulse compression and the spectral broadening is small because of the reduced Raman polarization, while with too high a pressure the bulk contribution to dispersion increases and the effective spectral width remains relatively small because after a certain propagation length pulse splitting occurs.

In Fig. 5 the evolution of the probe pulse shape (a) and spectrum (b) is shown for the pumping by a sequence of four synchronized 50-TW/cm^2 pulses. After 25 cm the spectrum is significantly broadened and is extended from 230 nm to 145 nm and the temporal shape shows a compression up to 1.6 fs duration. Note that the probe pulse at distance 25 cm is almost bandwidth-limited and the compression occurs due to the normal dispersion of the gas filling only. In Fig. 5 (a) two side peaks are formed, temporally shifted from the main pulse by $\pm T_R$, because the input pulse duration is already in the range of T_R. For a probe input with a duration shorter than about 70 fs such side peaks do not occur.

5. Conclusions

We have reported the results of the theoretical study of ultra-broadband phase-amplitude modulation and pulse compression by Raman-active molecular modulators for the impulsive pump-probe regime. Using the analytical solution (16)–(17) without the preposition of SVEA, the simple analytical expressions (23)–(30) were derived for the temporal and spectral characteristics of a probe pulse modulated by the impulsively excited Raman medium, which gives a clear physical understanding of the specific character of pulse compression by HSRS with remarkable different features as compared to SPM. As discussed in Section 3.2, in different regions of the pump-probe delay smooth spectral phases with positive or negative chirp arise, which can be compensated simply by dispersion of a glass plate with normal dispersion in the optical or UV/VUV or anomalous dispersion in the MIR for the generation of extremely short single pulses. Near the out-of-phase position phase jumps occur within the spectra and in this case a compressor with normal as well with anomalous dispersion decomposes the pulse into a short bandwidth-limited and a long weak phase-modulated part.

Acknowledgments

We acknowledge financial support by the Deutsche Forschungsgemeinschaft. We are thankful to Prof. I. D. Feranchuk for the invitation to contribute into this collection dedicated to the anniversary of the Department of Theoretical Physics, Belarus State University. One of the authors (VPK) is grateful to all his teachers at the Department of Theoretical Physics.

Bibliography

1. G. Steinmeyer, D. H. Sutter, L. Gallmann, N. Matuschek, and U. Keller, *Science*, **286**, 1507 (1999).
2. T. Brabec and F. Krausz, *Rev. Mod. Phys.* **72**, 545 (2000).
3. A. Baltuvska, Z. Wei, M. S. Pshenichnikov, and D. A. Wiersma, *Opt. Lett.* **22**, 102 (1997).
4. M. Nisoli, S. De Silvestri, O. Svelto, R. Szipcs, K. Ferencz, C. Spielmann, S. Sartania, and F. Krausz, *Opt. Lett.* **22**, 522 (1997).
5. V. P. Kalosha and J. Herrmann, *Phys. Rev.* **A62**, 011804 (2000).
6. P. M. Paul, E. S. Toma, P. Breger, G. Mullot, F. Augé, P. Balcou, H. G. Muller, and P. Agostini, *Science* **292** 1689 (2001).
7. M. Hentschel, R. Kienberger, C. Spielmann *et al.*, *Nature* **414**, 509 (2001).
8. M. Drescher, M. Hentschel, R. Kienberger *et al.*, *Nature* **419**, 803 (2002).

9. S. E. Harris and A. V. Sokolov, *Phys. Rev. Lett.* **81**, 2894 (1998).
10. A. V. Sokolov, D. R. Walker, D. D. Yavuz, G. Y. Yin, and S. E. Harris, *Phys. Rev. Lett.* **87**, 033402 (2001).
11. M. Wittmann, A. Nazarkin, and G. Korn, *Phys. Rev. Lett.* **84**, 5508 (2000).
12. M. Wittmann, A. Nazarkin, and G. Korn, *Opt. Lett.* **26**, 298 (2001).
13. V. P. Kalosha and J. Herrmann, *Phys. Rev. Lett.* **85**, 1226 (2000).
14. V. P. Kalosha and J. Herrmann, *Opt. Lett.* **26**, 456 (2001).
15. N. Zhavoronlov and G. Korn, Phys. *Rev. Lett.* **88**, 203901 (2002).
16. Y. R. Shen, *The Principles of Nonlinear Optics* (John Wiley, New York, 1984).
17. V. Schulz-von der Garten, T. Bornemann, V. Kornas, and H. F. Döbele, *IEEE J. Quant. Electr.* **26**, 739 (1990).
18. Y.-X. Yan, E. B. Gamble, and K. A. Nelson, *J. Chem. Phys.* **83**, 5391 (1985).
19. A. M. Weiner, D. E. Leaird, G. P. Wiederrecht, and K. A. Nelson, *J. Opt. Soc. Amer.* **B8**, 1264 (1991).
20. D. Eimerl, D. Milam, and J. Yu, *Phys. Rev. Lett.* **70**, 2738 (1993).
21. H. Kawano, Y. Hirakawa and T. Imasaka, *IEEE J. Quant. Electron.* **34**, 260 (1998).
22. A. P. Hickman and W. K. Bischel, Phys. Rev. A **37**, 2516 (1988).
23. V. P. Kalosha and J. Herrmann, *Phys. Rev.* **A67**, 031801 (2003).
24. V. P. Kalosha and J. Herrmann, *Opt. Lett.* **28**, 950 (2003).
25. V. P. Kalosha and J. Herrmann, *Phys. Rev.* **A68**, 023812 (2003).
26. K. S. Yee, *IEEE Trans. Antennas Propag.* **14**, 302 (1966).
27. R. K. Bullough, P. M. Jack, P. W. Kitchenside, R. Saunders, *Physica Scripta*, **20**, 364 (1979).
28. A. V. Husakou and J. Herrmann, *Phys. Rev. Lett.* **87**, 203901 (2001).
29. M. Nisoli, S. De Silvestri, and O. Svelto, *Appl. Phys. Lett.* **68**, 2793 (1996).
30. E. A. J. Marcatili and R. A. Schmeltzer, *Bell Syst. Techn. J.* **43**, 1783 (1964).
31. V. S. Butylkin, A. E. Kaplan, Y. G. Khronopulo, and E. I. Yakubovich, *Resonant Nonlinear Interactions of Light with Matter* (Springer-Verlag, New York, 1977).
32. V. Kalosha, M. Spanner, J. Herrmann, and M. Ivanov, *Phys. Rev. Lett.* **88**, 103901 (2002).
33. A. Weber (Ed.), *Raman Spectroscopy of Gases and Liquids* (Springer-Verlag, Berlin, 1979).
34. H. Kawano, T. Mori, Y. Hirakawa, and T. Imasaka, Phys. Rev. A **59**, 4703 (1999).

Part III

CONDENSED MATTER AND COMPLEX SYSTEMS

THE DEVELOPMENT OF NEW OPERATOR METHODS FOR DESCRIBING BODY AND SURFACE ELECTROMAGNETIC WAVES IN COMPLEX MEDIA

L.M. Barkovsky, G.N. Borzdov, A.N. Furs, A.V. Novitsky, V.M. Galynsky

Department of Theoretical Physics, Belarusian State University,
4 Fr. Skarina Ave., Minsk 220050, Belarus
Barkovsky@bsu.by

New applications of Fedorov covariant methods to the theory of light propagation in dielectric circular waveguides, to the theory of surface electromagnetic waves in complex anisotropic and gyrotropic media and to the inverse problems of optics of bianisotropic media are presented.

1. Introduction

Maxwell field theory and the electromagnetic theory of light are regarded today as the most fundamental physical theories. Physicists now recognize that Maxwell's contribution equals that of Newton and Einstein. Maxwell might say if he lived now that his field theory and statistical physics stand as progenitors of relativity and quantum mechanics[1]. For many years theoretical researches of optical properties of crystals or, more exactly, researches according to Maxwell's crystal optics have been carried out in Belarus under the supervision of the founder of the Theoretical Physics Department of the Belarusian State University, Hero of Socialist Labour, Academician F.I. Fedorov. In early years he recognized great importance of Maxwell's theory for world science. Fedorov's works on theoretical physics have one general feature in common. This is a covariant approach and all possible developments and applications of direct free-coordinate method of analysis which includes as many as possible dynamical problems related by one indication or an other. It is the question of using specially prepared and adapted algebraic-geometrical methods. It appeared possible to avoid using the explicit form of the matrix-bases representing dynamical equations, their solutions and symmetries by means of the properties of their minimal polynomials. This conception has evident advantages. It goes back to well-known works of Hamilton, Cayley, Grassman on noncommutative algebras, to Lie's algebraic-geometrical studies and to famous Erlangen program of F. Klein. Before Maxwell's epoch differential geometry was already able to do without Cartesian coordinates using tensors and vectors as varieties of generalized numbers. The advantages of such approach to physical problems were pointed out by Maxwell in the preliminary chapter of his treatise[2] :"...But for many purposes of physical reasoning, as distinguished from calculation, it is desirable to avoid explicitly introducing the Cartesian coordinates, and to fix the mind at once on a point of space instead of its three coordinates, and on the magnitude and direction of a force instead of its three components. This mode of contemplating geometrical and physical quantities is more primitive and more natural than the other, although the ideas connected with it did not receive their full development till Hamilton made the next great step in dealing with space, by the invention of his Calculus of Quaternions." Further he continued "I am convinced, however, that the introduction of the ideas, as distinguished from the operations and methods of Quaternions will be of great use to us in the study of all parts of our subject, and especially in electrodynamics where we have to deal with a number of physical quantities, the relations of which to each other can be expressed far more simply by a few expressions of Hamilton's, than by the ordinary equations."

At the Theoretical Physics Department F.I. Fedorov together with L.M. Barkovsky, G.N. Borzdov, V.V. Zhilko, A.V. Lavrinenko and A.N. Furs have worked out the direct coordinate-free method of evolution operators for describing body and surface electromagnetic and elastic waves in various layered media such as bianisotropic, bigyrotropic, absorbing, magnetic, dispersive, modulated and moving ones[3–6]. In the present paper we extend the formalism of evolution operators to the theory of light propagation in dielectric circular cylindrical waveguides, explain the operator approach in the theory of surface electromagnetic waves in complex anisotropic and gyrotropic media and give some general covariant solutions of an inverse problem of optics of bianisotropic media.

2. Evolution operators of electromagnetic waves in circular fibres

Circular waveguides are widely used in laser physics, in fiber and nonlinear optics. By selecting the core and cladding materials the necessary waveguide properties (new guiding channels, frequency and polarization-mode dispersion compensation, single mode frequency domain extension) can be obtained[7,8]. Recently the existence of composite materials with negative dielectric permittivity ε and magnetic permeability μ has been proved experimentally. Such materials represent a unique basis for constructing new devices with unconventional properties[9-11].

Let us apply operator methods for investigating mode, polarization and energy characteristics of radiation in circular fibers. We consider a circular waveguide with simultaneously negative dielectric permittivity and magnetic permeability of the core as an example.

Let the electromagnetic field has a harmonic time dependence. We use the circular cylindrical coordinates (r, φ, y) with basis vectors $\mathbf{e}_r(\varphi)$, $\mathbf{e}_\varphi(\varphi)$, $\mathbf{b}$, which possess the following properties:

$$\mathbf{e}_r \times \mathbf{e}_\varphi = \mathbf{b}, \qquad \frac{\partial \mathbf{e}_\varphi}{\partial \varphi} = -\mathbf{e}_r, \qquad \frac{\partial \mathbf{e}_r}{\partial \varphi} = \mathbf{e}_\varphi.$$

The vector $\mathbf{b}$ is directed along the waveguide axis, $\mathbf{e}_r(\varphi)$ and $\mathbf{e}_\varphi(\varphi)$ have radial and tangent directions. Let us consider an electromagnetic wave with y-translational and φ-circular symmetry:

$$\begin{pmatrix} \mathbf{H}(\mathbf{r}) \\ \mathbf{E}(\mathbf{r}) \end{pmatrix} = \exp\left(i\beta y + i\nu\varphi\right) \begin{pmatrix} \mathbf{H}(r,\varphi) \\ \mathbf{E}(r,\varphi) \end{pmatrix}, \tag{1}$$

where β is a mode propagation constant, ν is an integer number. Such waves propagate in a circular waveguide, if dielectric permittivity and magnetic permeability tensors have the form

$$\varepsilon(r,\varphi) = \sum_{i,j=1}^{3} \varepsilon_{ij}(r)\mathbf{e}_i \otimes \mathbf{e}_j, \quad \mu(r,\varphi) = \sum_{i,j=1}^{3} \mu_{ij}(r)\mathbf{e}_i \otimes \mathbf{e}_j, \tag{2}$$

where $\mathbf{e}_1 = \mathbf{e}_r$, $\mathbf{e}_2 = \mathbf{e}_\varphi$, $\mathbf{e}_3 = \mathbf{b}$, the components of tensors ε and μ depending only on the radial coordinate. Some particular cases of the media (2) are an isotropic medium $\varepsilon = \varepsilon(r)\,1$, an uniaxial crystal $\varepsilon = \varepsilon_1(r)(1 - \mathbf{b}\otimes\mathbf{b}) + \varepsilon_2(r)\mathbf{b}\otimes\mathbf{b}$, an optically active medium $\varepsilon = \varepsilon(r)1 + i\chi(r)\mathbf{b}^{\times}$. Here 1 is a three-dimensional unit operator, $\mathbf{b}^{\times}$ is the tensor dual to the vector $\mathbf{b}$[6,4]. Assuming the field components $\mathbf{H}(r,\varphi)$ and $\mathbf{E}(r,\varphi)$ to be φ-independent we derive from the Maxwell equations

$$\begin{aligned} \mathbf{e}_r^{\times}\frac{d\mathbf{H}}{dr} + \left(i\beta\mathbf{b}^{\times} + \frac{i\nu}{r}\mathbf{e}_\varphi^{\times} + \frac{1}{r}\mathbf{b}\otimes\mathbf{e}_\varphi\right)\mathbf{H} = -ik\varepsilon(r)\mathbf{E}, \\ \mathbf{e}_r^{\times}\frac{d\mathbf{E}}{dr} + \left(i\beta\mathbf{b}^{\times} + \frac{i\nu}{r}\mathbf{e}_\varphi^{\times} + \frac{1}{r}\mathbf{b}\otimes\mathbf{e}_\varphi\right)\mathbf{E} = ik\mu(r)\mathbf{H}. \end{aligned} \tag{3}$$

In Eqs. (3) $\mathbf{e}_r$, $\mathbf{e}_\varphi$, $\mathbf{b}$ define the vector structure of the fields $\mathbf{H}$ and $\mathbf{E}$. The φ-dependence of $\mathbf{e}_r$, $\mathbf{e}_\varphi$ has already been taken into account in (3), that is why these vectors can be considered constant in what follows.

Introducing tangential components of the electric and magnetic field strengths we can rewrite (3). The tangential components $\mathbf{H}_t = I\mathbf{H}$ and $\mathbf{E}_t = I\mathbf{E}$ are located in the plane tangent to the cylinder surface. Therefore they are continuous on the circular cylinder interface between the two media. Here I is the projection operator onto the interface normal to $\mathbf{e}_r$. From (3) we obtain the equation for $\mathbf{H}_t$ and $\mathbf{E}_t$:

$$\frac{d}{dr}\mathbf{W}(r) = ikM(r)\mathbf{W}(r), \tag{4}$$

where

$$\begin{aligned}
M &= \begin{pmatrix} A & B \\ C & D \end{pmatrix}, \quad \mathbf{W} = \begin{pmatrix} \mathbf{H}_t(r) \\ \mathbf{E}_t(r) \end{pmatrix}, \quad \mathbf{u} = (\beta/k)\mathbf{e}_\varphi - \nu/(kr)\mathbf{b}, \\
A &= \frac{i}{kr}\mathbf{e}_\varphi \otimes \mathbf{e}_\varphi + \frac{1}{\varepsilon_r}\mathbf{e}_r^\times \varepsilon \mathbf{e}_r \otimes \mathbf{u} - \frac{1}{\mu_r}\mathbf{e}_r^\times \mathbf{u} \otimes \mathbf{e}_r \mu I, \\
B &= -\frac{1}{\mu_r}\mathbf{e}_r^\times \mathbf{u} \otimes \mathbf{u} + \frac{1}{\varepsilon_r} I\tilde{\bar{\varepsilon}}\mathbf{e}_r^\times, \\
C &= \frac{1}{\varepsilon_r}\mathbf{e}_r^\times \mathbf{u} \otimes \mathbf{u} - \frac{1}{\mu_r} I\tilde{\bar{\mu}}\mathbf{e}_r^\times, \\
D &= \frac{i}{kr}\mathbf{e}_\varphi \otimes \mathbf{e}_\varphi + \frac{1}{\mu_r}\mathbf{e}_r^\times \mu \mathbf{e}_r \otimes \mathbf{u} - \frac{1}{\varepsilon_r}\mathbf{e}_r^\times \mathbf{u} \otimes \mathbf{e}_r \varepsilon I.
\end{aligned} \tag{5}$$

Using a restoration operator V the vectors $\mathbf{E}$ and $\mathbf{H}$ can be expressed by means of their tangential components as follows:

$$\begin{pmatrix} \mathbf{H}(r) \\ \mathbf{E}(r) \end{pmatrix} = V \begin{pmatrix} \mathbf{H}_t(r) \\ \mathbf{E}_t(r) \end{pmatrix}, \quad V = \begin{pmatrix} 1 - \mathbf{e}_r \otimes \mathbf{e}_r \mu/\mu_r & -\mathbf{e}_r \otimes \mathbf{u}/\mu_r \\ \mathbf{e}_r \otimes \mathbf{u}/\varepsilon_r & 1 - \mathbf{e}_r \otimes \mathbf{e}_r \varepsilon/\varepsilon_r \end{pmatrix}, \tag{6}$$

If $r \to \infty$, $\mathbf{e}_r \to \mathbf{q}$, $\mathbf{e}_\varphi \to \mathbf{a} = \mathbf{b} \times \mathbf{q}$ then expressions (4)–(6) transform into the similar ones for a plane-stratified anisotropic medium, where $\mathbf{q}$ is the unit normal to the plane layer interface[12,13].

The general solution of Eq. (4) is given by an evolution operator (the product integral) $\Omega_a^r[ikM(r)]$:

$$\begin{aligned}
\mathbf{W}(r) &= \Omega_a^r[ikM(r)]\mathbf{W}(a), \quad a \neq 0, \\
\Omega_a^r[ikM(r)] &= \int_a^r (E + ikM(r)dr), \quad E = \begin{pmatrix} I & 0 \\ 0 & I \end{pmatrix}.
\end{aligned} \tag{7}$$

Thus, the fields at an arbitrary point r can be obtained by multiplying the evolution operator by the initial field vector $\mathbf{W}(a)$.

In the case of a homogeneous isotropic medium with dielectric permittivity ε and magnetic permeability μ the tensors (5) take the form

$$A=\frac{i}{kr}\mathbf{e}_\varphi\otimes\mathbf{e}_\varphi,\quad B=\frac{\beta\nu}{k^2\mu r}(\mathbf{b}\otimes\mathbf{b}-\mathbf{e}_\varphi\otimes\mathbf{e}_\varphi)+\frac{u_1}{k^2\mu}\mathbf{b}\otimes\mathbf{e}_\varphi-\frac{u_2}{k^2\mu}\mathbf{e}_\varphi\otimes\mathbf{b},$$
$$D=\frac{i}{kr}\mathbf{e}_\varphi\otimes\mathbf{e}_\varphi,\quad C=-\frac{\beta\nu}{k^2\varepsilon r}(\mathbf{b}\otimes\mathbf{b}-\mathbf{e}_\varphi\otimes\mathbf{e}_\varphi)-\frac{u_1}{k^2\varepsilon}\mathbf{b}\otimes\mathbf{e}_\varphi+\frac{u_2}{k^2\varepsilon}\mathbf{e}_\varphi\otimes\mathbf{b},\quad (8)$$

where $u_1 = k^2\varepsilon\mu - \beta^2$, $u_2 = k^2\varepsilon\mu - \nu^2/r^2$. Then the solution of (4) can be written as follows

$$\mathbf{W}(r) = P(r)P^-(a)\mathbf{W}(a),\quad P(r) = \begin{pmatrix}\alpha_1 & \alpha_2\\ \beta_1 & \beta_2\end{pmatrix},\tag{9}$$

where P^- is a block matrix pseudoinverse to P. Planar tensors α_1, α_2, β_1, β_2 have the form

$$\alpha_m = F_{\nu\pm}^{(m)}\left(\mathbf{b}\mp\frac{\beta\nu a^2}{u_\pm^2 r}\mathbf{e}_\varphi\right)\otimes\mathbf{b}\pm\frac{ika\varepsilon}{u_\pm}F_{\nu\pm}^{(m)\prime}\mathbf{e}_\varphi\otimes\mathbf{e}_\varphi,$$
$$\beta_m = \mp\frac{ika\mu}{u_\pm}F_{\nu\pm}^{(m)\prime}\mathbf{e}_\varphi\otimes\mathbf{b}+F_{\nu\pm}^{(m)}\left(\mathbf{b}\mp\frac{\beta\nu a^2}{u_\pm^2 r}\mathbf{e}_\varphi\right)\otimes\mathbf{e}_\varphi,\ m=1,2;\quad (10)$$

where $u_\pm^2 = \pm k^2a^2(\varepsilon\mu-\beta^2/k^2)$, $F_{\nu+}^{(1)} = J_\nu(u_+r/a)$ is the Bessel function of the first kind, $F_{\nu+}^{(2)} = Y_\nu(u_+r/a)$ is the Bessel function of the second kind, $F_{\nu-}^{(1)} = I_\nu(u_-r/a)$ and $F_{\nu-}^{(2)} = K_\nu(u_-r/a)$ are the modified Bessel functions. The upper indices 1 and 2 number the independent solutions of the Bessel equation, the signs + and − correspond to the choice of u_+ or u_- in the function argument. From (9) we find the evolution operator of a circular homogeneous isotropic layer as follows:

$$\Omega_a^r = P(r)P^-(a).\tag{11}$$

On the other hand Eq. (9) can be reduced to the superposition of the two waves which correspond to the independent solutions of the Bessel equation:

$$\begin{pmatrix}\mathbf{H}_t(r)\\ \mathbf{E}_t(r)\end{pmatrix} = \begin{pmatrix}\mathbf{H}_{t1}(r)\\ \mathbf{E}_{t1}(r)\end{pmatrix}+\begin{pmatrix}\mathbf{H}_{t2}(r)\\ \mathbf{E}_{t2}(r)\end{pmatrix},\tag{12}$$

where $\mathbf{H}_{t1} = \alpha_1\mathbf{c}_1$, $\mathbf{H}_{t2} = \alpha_2\mathbf{c}_2$, $\mathbf{E}_{t1} = \beta_1\mathbf{c}_1$, $\mathbf{E}_{t2} = \beta_2\mathbf{c}_2$. The impedance tensor of each wave relates tangential components of the electric and magnetic field strengths $\mathbf{E}_{tm} = \Gamma_m\mathbf{H}_{tm}$ and can be described in terms of α_m, β_m:

$$\Gamma_m = \beta_m\alpha_m^-.\tag{13}$$

If we know the initial amplitudes of two independent waves $\mathbf{H}_{t1}(a)$ and $\mathbf{H}_{t2}(a)$, than the field evolution is reduced to the formulae

$$\mathbf{H}_{tm}(r) = \alpha_m(r)\alpha_m^{-}(a)\mathbf{H}_{tm}(a), \quad \mathbf{E}_{tm}(r) = \beta_m(r)\alpha_m^{-}(a)\mathbf{H}_{tm}(a). \tag{14}$$

In the waveguide core and cladding there exist only one of the wave. In the core the wave corresponding to the solution $F_{\nu+}^{(1)} = J_\nu(u_+ r/a)$ propagates (the second solution has a peculiarity in the waveguide center). In the cladding we have a single wave with the solution $F_{\nu-}^{(2)} = K_\nu(u_- r/a)$, which tends to zero while moving from the core.

Let us apply the obtained results for describing electromagnetic wave propagation in isotropic circular fibers. We consider a waveguide having a core, a cladding and n intermediate layers between them. Such a waveguide has the following dielectric permittivities and magnetic permeabilities

$$\varepsilon = \begin{cases} \varepsilon_{co},\ r < a_0 \\ \varepsilon_j,\ \ a_{j-1} \le r < a_j \\ \varepsilon_{cl},\ r \ge a_n \end{cases}, \quad \mu = \begin{cases} \mu_{co},\ r < a_0 \\ \mu_j,\ \ a_{j-1} \le r < a_j \\ \mu_{cl},\ r \ge a_n \end{cases}, \quad j = 1...n, \tag{15}$$

while the boundary conditions (the continuity of tangential field components) take the form

$$\begin{pmatrix} \mathbf{H}_t^{(cl)}(a_n) \\ \Gamma_{cl}\mathbf{H}_t^{(cl)}(a_n) \end{pmatrix} = \Omega_{a_0}^{a_n} \begin{pmatrix} \mathbf{H}_t^{(co)}(a_0) \\ \Gamma_{co}\mathbf{H}_t^{(co)}(a_0) \end{pmatrix}. \tag{16}$$

In formula (16) $\mathbf{H}_t^{(cl)}$, $\mathbf{H}_t^{(co)}$ are the tangential components of the magnetic field strength in the cladding and in the core, respectively; Γ_{cl} and Γ_{co} are the surface impedance tensors of the cladding and the core, $\Omega_{a_0}^{a_n}$ is the evolutional operator of n layers. One can observe the same structure of boundary conditions in the plane-stratified media[12,13]. Using (11) the evolution operator can be expressed by multiplying block matrices in the following order

$$\Omega_{a_0}^{a_n} = P_n(a_n)P_n^{-}(a_{n-1})P_{n-1}(a_{n-1})P_{n-1}^{-}(a_{n-2})\dots P_1(a_1)P_1^{-}(a_0). \tag{17}$$

Substituting Eqs. (10) in (13) we obtain the surface impedance tensors

$$\begin{aligned}\Gamma_{co} =& -\frac{iu}{ka_0\varepsilon_{co}}\frac{J_\nu(u)}{J'_\nu(u)}\left(\mathbf{b}-\frac{\beta\nu a_0}{u^2}\mathbf{e}_\varphi\right)\otimes\left(\mathbf{e}_\varphi+\frac{\beta\nu a_0}{u^2}\mathbf{b}\right)\\ &-\frac{ika_0\mu_{co}}{u}\frac{J'_\nu(u)}{J_\nu(u)}\mathbf{e}_\varphi\otimes\mathbf{b},\\ \Gamma_{cl} =& \frac{iw}{ka_0\varepsilon_{cl}}\frac{K_\nu(wa_n/a_0)}{K'_\nu(wa_n/a_0)}\left(\mathbf{b}+\frac{\beta\nu a_0^2}{w^2a_n}\mathbf{e}_\varphi\right)\otimes\left(\mathbf{e}_\varphi-\frac{\beta\nu a_0^2}{w^2a_n}\mathbf{b}\right)\\ &+\frac{ika_0\mu_{cl}}{w}\frac{K'_\nu(wa_n/a_0)}{K_\nu(wa_n/a_0)}\mathbf{e}_\varphi\otimes\mathbf{b},\end{aligned} \tag{18}$$

where $u^2 = k^2a_0^2\left(\varepsilon_{co}\mu_{co}-\beta^2/k^2\right)$, $w^2 = k^2a_0^2\left(\beta^2/k^2-\varepsilon_{cl}\mu_{cl}\right)$. Multiplying (16) by the block matrix $(\Gamma_{cl}, -I)$, we have

$$\Theta\mathbf{H}_t^{(co)}(a_0) = 0, \quad \Theta = \begin{pmatrix}\Gamma_{cl} & -I\end{pmatrix}\Omega_{a_0}^{a_n}\begin{pmatrix}I\\ \Gamma_{co}\end{pmatrix}. \tag{19}$$

From (19) we can derive the dispersion equation and determine the polarization structure of electromagnetic radiation in the waveguide. To obtain the dispersion equation it should be noted that the planar tensor Θ has two eigenvectors $\mathbf{e}_r$ and $\mathbf{H}_t^{(co)}(a_0)$ with zero eigenvalues. Then $\overline{\Theta}$ is a dyad and the dispersion equation is[13]

$$\overline{\Theta}_t = 0, \tag{20}$$

where $\overline{\Theta}_t$ is a trace of an adjoined tensor. The solution of the dispersion equation yields the dependence of the mode propagation constant on the frequency. The intermediate waveguide layers are taken into account by the evolution operator (17). Since the evolution operator and the impedance are of the matrix form we can easily apply (20) for the numerical calculation of modes in the waveguide with any number of layers. If there is no intermediate layers ($n = 0$), then the evolution operator is equal to the unit matrix E (see (7)), $\Theta = \Gamma_{cl} - \Gamma_{co}$ and the dispersion equation can be reduced to the well-known equation for a two-layer circular waveguide[14].

From (19) and the Cayley-Hamilton theorem the tangential component of the magnetic field strength at the interface is

$$\mathbf{H}_t^{(co)}(a) = (\Theta - \Theta_t)\mathbf{p}. \tag{21}$$

where $\mathbf{p}$ is an arbitrary vector satisfying the condition $\Theta\mathbf{p} \neq 0$. The electric and magnetic field vectors in the core have the form

$$\begin{pmatrix}\mathbf{H}^{(co)}(r)\\ \mathbf{E}^{(co)}(r)\end{pmatrix} = V\begin{pmatrix}\alpha_1(r)\\ \beta_1(r)\end{pmatrix}\alpha_1^-(a)(\Theta-\Theta_t)\mathbf{p}. \tag{22}$$

The field vector in the cladding and n intermediate layers can be found in the same way. The waveguide modes are determined by the dispersion equation (20). Each mode is characterized by a certain polarization type. TE- and TM-modes have no longitudinal components in the electric and magnetic field vectors, i.e. $\mathbf{bE}(r) = 0$ and $\mathbf{bH}(r) = 0$ for any r. In the multilayer isotropic dielectric waveguide TE- and TM-modes exist at $\nu = 0$. At other values of ν there are hybrid modes only. The type of a hybrid mode is determined by the ratio of the electric and magnetic field longitudinal components[14]

$$\delta = \frac{E_y}{iH_y} = \frac{\mathbf{b}\Gamma_{co}(a)(\Theta - \Theta_t)\mathbf{p}}{i\mathbf{b}(\Theta - \Theta_t)\mathbf{p}}. \tag{23}$$

The values $\delta > 0$ and $\delta < 0$ correspond to HE- and EH-modes, respectively.

Let us calculate the energy flux in the direction of wave guiding. The Pointing vector projection onto the y-axis is

$$S_y = \frac{c}{16\pi}\mathbf{W}^\dagger(r)V^\dagger \begin{pmatrix} 0 & \mathbf{b}^\times \\ -\mathbf{b}^\times & 0 \end{pmatrix} V\mathbf{W}(r), \tag{24}$$

where $\dagger$ means the Hermitian conjugation. Substituting the restoration operator (6) we obtain

$$S_y = \frac{c}{16\pi}\mathbf{W}^\dagger(r) \begin{pmatrix} \dfrac{\mathbf{e}_\varphi \otimes \mathbf{u}}{\varepsilon} + \dfrac{\mathbf{u} \otimes \mathbf{e}_\varphi}{\varepsilon^*} & 0 \\ 0 & \dfrac{\mathbf{e}_\varphi \otimes \mathbf{u}}{\mu} + \dfrac{\mathbf{u} \otimes \mathbf{e}_\varphi}{\mu^*} \end{pmatrix} \mathbf{W}(r). \tag{25}$$

The vector $\mathbf{W}(r)$ is given by the formulae (14) in the core and the cladding and (9) in the intermediate layers. Here we introduce one of the most important waveguide characteristics called a mode localization coefficient. It equals the power localized in the core divided by the total power[14]

$$\eta = \frac{P_{co}}{P}, \quad P_{co} = \int_0^{a_0} S_y(r) r dr, \qquad P = \int_0^{\infty} S_y(r) r dr. \tag{26}$$

3. Surface electromagnetic waves in complex anisotropic and gyrotropic media and the effects of unidirectional wave propagation

Surface electromagnetic waves (surface polaritons) are inhomogeneous waves whose intensity decreases exponentially with moving from the boundary. If $\mathbf{q}$ is a unit normal vector to the interface and the equation

$\mathbf{qr} = 0$ defines the interface, then the magnetic field strength of a surface wave in an isotropic medium is described by the formula

$$\mathbf{H}(\mathbf{r},\, t) \;=\; \mathbf{H}_0 \, \exp[i\, k\, (\mathbf{b} + \eta\, \mathbf{q})\, \mathbf{r} - i\, \omega\, t], \tag{27}$$

where a unit vector $\mathbf{b}$ determines the propagation direction along the interface, $\mathbf{H}_0$ is the field amplitude of wave at the interface, ω is a wave frequency, k is the wave number. The imaginary part of η characterizes the degree of the field decay when moving away from the interface (here $\operatorname{Im}\eta < 0$ for the medium in half-space $\mathbf{qr} < 0$ and $\operatorname{Im}\eta > 0$ for the second medium), its real part characterizing the form of the field decay (at $\operatorname{Re}\eta = 0$ it is an ordinary exponential decay, and at $\operatorname{Re}\eta \neq 0$ it is a convergent oscillation). The strength of the electric field is described by a formula similar to (27).

The interest to the surface polaritons is first of all due to the development of new information processing devices, new types of lasers with a higher amplification of light[15,16], to the construction of light-driving micromachines[17] and other causes.

Surface electromagnetic waves at the interface between isotropic dielectric media are well studied[18]. The solution of Maxwell's equations under boundary conditions in this case leads to well-known dispersion equations for TM and TE-wave, respectively

$$\begin{aligned} &\varepsilon_2 \sqrt{\frac{1}{\nu^2} - \varepsilon_1\, \mu_1} + \varepsilon_1 \sqrt{\frac{1}{\nu^2} - \varepsilon_2\, \mu_2} = 0, \\ &\mu_2 \sqrt{\frac{1}{\nu^2} - \varepsilon_1\, \mu_1} + \mu_1 \sqrt{\frac{1}{\nu^2} - \varepsilon_2\, \mu_2} = 0, \end{aligned} \tag{28}$$

where the *reduced frequency* $\nu = \omega/(c\, k)$ is a phase velocity of the surface wave in the units of speed of light in vacuum c. Equations (28) have solutions only if one of the dielectric permittivities for TM-waves or magnetic permeabilities for TE-waves is negative. Until the 90s years of the last century researchers focused their attention mainly on the surface polaritons at the boundary of the media with negative dielectric permittivity ε, such as metals, plasma, semiconductors excited near resonant frequencies. Realization of the composite materials, in which dielectric permittivity and magnetic permeability are either negative (left-handed (LH) materials[9,19]) or negative separately[20,21], stimulated the research of new types of surface polaritons and led to the concept of a perfect lens[10], which can focus not only body waves, but also evanescent waves. Such lenses can be used for creating a perfect image of the source.

In a series of papers the surface polaritons at the interface between nonlinear self-focusing media are investigated. In such media surface waves can exist only when the field strength exceeds some threshold value[22].

In this paper we are interested in the surface waves at the interface between complex nonmagnetic anisotropic and gyrotropic media, and in the existence conditions of such waves. It turns out that, if at least one of the boundary media is anisotropic, then solutions of Maxwell's equations for a surface wave can exist, and the condition of negativity of the principal values of the dielectric permittivity tensor is no longer required. Such a new type of surface polaritons (the polaritons due to the anisotropy of the boundary media or dispersionless polaritons) has been discovered quite recently by D'yakonov[23]. In paper[23] the interface between isotropic media and an uniaxial crystal was considered, with optical axis being located in the boundary plane. It was shown that if dielectric permittivities $\varepsilon_{\parallel}$, $\varepsilon_{\perp}$ for the uniaxial crystal and ε for an isotropic medium satisfy the condition $\varepsilon_{\perp} < \varepsilon < \varepsilon_{\parallel}$, then surface waves can propagate along certain directions, the set of which forms four sectors in the boundary plane. These sectors are characterized by the limiting angles α_1 and α_2 with respect to the direction of the optical axis ($\alpha_1 < \alpha_2$), so that

$$\begin{aligned} \sin^2\alpha_1 &= \frac{\xi}{2}\{1-\chi\xi+[(1-\chi\xi)^2+4\chi]^{1/2}\}, \\ \sin^2\alpha_2 &= \frac{(1+\chi)^3\xi}{(1+\chi)^2(1+\chi\xi)-\chi^2(1-\xi)^2}, \end{aligned} \tag{29}$$

where $\chi = (\varepsilon_{\parallel}/\varepsilon_{\perp}) - 1$ and $\xi = (\varepsilon - \varepsilon_{\perp})/(\varepsilon_{\parallel} - \varepsilon_{\perp})$. The angular width $\Delta\alpha = \alpha_2 - \alpha_1$ of the sectors of the allowed propagation directions is found to be the higher, the higher is the degree of the crystal anisotropy χ. For natural materials the values $\Delta\alpha$ are small and can reach several angular minutes only. Using artificial materials with strong anisotropy ($\chi \sim 10$) will certainly allow to observe and study dispersionless surface polaritons experimentally. It's notable that such polaritons can be excited even when the boundary is formed by different cut planes of the same crystal[24].

The theory of surface electromagnetic waves in anisotropic dielectric media is more complicated in comparison with isotropic ones. Surface wave field distribution in the anisotropic crystal characterized by the dielectric permittivity tensor ε is determined by the contribution of two inhomogenious eigenwaves. For instance, at $\mathbf{qr} < 0$

$$\mathbf{H}(\mathbf{r}, t) = \sum_{s=1}^{2} C_s \mathbf{H}_s^0 \exp[i\,k\,(\mathbf{b} + \eta_s\,\mathbf{q})\mathbf{r} - i\,\omega\,t], \tag{30}$$

where C_s — weight factors, and $\mathrm{Im}\,\eta_s < 0$. Decay coefficients η_s for eigenwaves as a function of the reduced frequency ν are found from Fresnel equation in Fedorov's form[4]

$$\det(\mathbf{m}^{\times}\varepsilon^{-1}\,\mathbf{m}^{\times} + 1) = 0 \tag{31}$$

where the refraction vector $\mathbf{m}$ is represented as $\mathbf{m} = (\mathbf{b} + \eta\,\mathbf{q})/\nu$ (for any vector $\mathbf{m}$ the dual tensor $\mathbf{m}^{\times}$ equals $(\mathbf{m}^{\times})_{ik} = e_{ijk}\,m_j$[3,4], where e_{ijk} is Levi-Civita pseudotensor). This algebraic quartic equation for η has real coefficients if the medium under consideration is a nonabsorbing anisotropic and gyrotropic one ($\varepsilon^{+} = \varepsilon$). If the reduced frequency ν does not exceed some limiting value ν_{L}[25], then solutions of equation (31) are complex-conjugate pairs of roots $\eta_1(\nu) = \eta_4^*(\nu)$, $\eta_2(\nu) = \eta_3^*(\nu)$. The physical meaning of the limitation $0 \leqslant \nu < \nu_{\mathrm{L}}$ is that phase velocity of surface waves is smaller than phase velocities of boby eigenwaves in this medium. Note that only roots with negative imaginary part must be included in equation (30). Then the amplitudes of inhomogenous waves are determined from the equations

$$\begin{gathered}(\mathbf{m}_s^{\times}\varepsilon^{-1}\,\mathbf{m}_s^{\times} + 1)\mathbf{H}_s^0 = 0, \quad \mathbf{E}_s^0 = -\varepsilon^{-1}\,\mathbf{m}_s^{\times}\mathbf{H}_s^0, \\ \mathbf{m}_s = \tfrac{1}{\nu}(\mathbf{b} + \eta_s\,\mathbf{q}), \quad s = 1,\, 2.\end{gathered} \tag{32}$$

Field distribution in the medium in half-space $\mathbf{qr} > 0$ which is characterized by the dielectric permittivity tensor ε' is found from the equations similar to (30)–(32), replacing $\varepsilon \to \varepsilon'$, $\eta_s \to \eta'_s$, $C_s \to C'_s$, $\mathbf{m}_s \to \mathbf{m}'_s$, $\mathbf{H}_s^0 \to \mathbf{H}_s^{\prime 0}$, $\mathbf{E}_s^0 \to \mathbf{E}_s^{\prime 0}$, and $\mathrm{Im}\,\eta'_s > 0$. Boundary conditions for magnetic and electric fields take the form

$$\begin{gathered}\mathbf{H}_\tau^0 = \textstyle\sum_{s=1}^2 C_s\mathbf{H}_{\tau s}^0 = \sum_{s=1}^2 C'_s\mathbf{H}_{\tau s}^{\prime 0} = \mathbf{H}_\tau^{\prime 0}, \\ [\mathbf{qE}^0] = \textstyle\sum_{s=1}^2 C_s[\mathbf{qE}_s^0] = \sum_{s=1}^2 C'_s[\mathbf{qE}_s^{\prime 0}] = [\mathbf{qE}^{\prime 0}],\end{gathered} \tag{33}$$

where index τ is used for denoting vectors projections onto the interface plane. Dispersion equation is derived from formulae (33) when coefficients C_s, C'_s are eliminated. If the solution $\nu = \nu_{\mathrm{S}}$ of this equation exists and $0 < \nu_{\mathrm{S}} < \min(\nu_{\mathrm{L}}, \nu'_{\mathrm{L}})$, then it corresponds to the surface wave propagating in **b**-direction along the interface between anisotropic media with dielectric permittivity tensors ε and ε'. In papers[25,26] we proposed a new approach for deriving the dispersion equations for surface polaritons in anisotropic dielectric media based on the integral representation of the surface impedance tensors. Surface impedance tensors γ and γ' relate the tangential components of the electric and magnetic field strength at the interface: $[\mathbf{qE}^0] = \gamma\mathbf{H}_\tau^0$, $[\mathbf{qE}^{\prime 0}] = \gamma'\mathbf{H}_\tau^{\prime 0}$. It is obvious, that boundary conditions (33) can be presented in the form $(\gamma - \gamma')\mathbf{H}_\tau^0 = 0$, then the dispersion equation is equal to

$$\overline{(\gamma - \gamma')}_t = 0. \tag{34}$$

In papers[25,26] it was established that tensors γ and γ' are calculated by the formulae

$$\gamma = \frac{1}{\nu} Q^-(-iI - S), \quad \gamma' = \frac{1}{\nu} Q'^-(iI - S'), \tag{35}$$

where $I = 1 - \mathbf{q} \otimes \mathbf{q}$ is a projective tensor onto the interface, and the sign $()^-$ denotes the operation of pseudoinversion[4]. The tensors Q and S are equal to

$$Q = -\frac{1}{\pi} \int_0^\pi (\mathbf{e}_2\mathbf{e}_2)^- d\phi, \quad S = -\frac{1}{\pi} \int_0^\pi (\mathbf{e}_2\mathbf{e}_2)^-(\mathbf{e}_2\mathbf{e}_1) d\phi. \tag{36}$$

Integration elements in (36) involve the bilinear forms defined for any vector arguments $\mathbf{u}$ and $\mathbf{v}$ as follows

$$(\mathbf{uv}) = I\mathbf{u}^\times \varepsilon^{-1} \mathbf{v}^\times I - \frac{I\mathbf{u}^\times \varepsilon^{-1} \mathbf{a} \otimes \mathbf{a} \varepsilon^{-1} \mathbf{v}^\times I}{\mathbf{a}\varepsilon^{-1}\mathbf{a} - \nu^2} + \nu^2 \mathbf{bu} \cdot \mathbf{bv} I, \tag{37}$$

vectors $\mathbf{e}_1$ and $\mathbf{e}_2$ being expressed in terms of $\mathbf{b}$ and $\mathbf{q}$

$$\mathbf{e}_1 = \mathbf{b} \cos\phi + \mathbf{q} \sin\phi, \quad \mathbf{e}_2 = -\mathbf{b} \sin\phi + \mathbf{q} \cos\phi, \quad \mathbf{a} = [\mathbf{e}_1\mathbf{e}_2] = [\mathbf{bq}]. \tag{38}$$

Tensors Q' and S' in (35) for the second medium $\mathbf{qr} > 0$ can be calculated by the same formulae (36)–(38) replacing $\varepsilon \to \varepsilon'$.

Dispersion equation (34) with the surface impedance tensors of the form (35) turns out to be convenient for studying surface polaritons, inasmuch as γ and γ' can be calculated uniformly by covariant formulae (36)–(38) at arbitrary orientations of vectors $\mathbf{b}$, $\mathbf{q}$ relative to the principal axes of tensors ε and ε'. Moreover, the left hand side of equation (34) monotonically depends on ν and this fact considerably simplifies the analysis of solutions of the dispersion equation existence.

As mentioned above a specific feature of dispersionless surface polaritons is their ability to get excited only in some certain directions along the interface, the path of a surface wave in anisotropic non-gyrotropic media being reversible. If the wave can propagate in some direction $\mathbf{b}$, then it can propagate in the opposite direction $-\mathbf{b}$. However, for gyrotropic media this property vanishes and the effect of unidirectional propagation of surface electromagnetic waves appears[27]. A non-absorbing gyrotropic medium is described by the inverse dielectric permittivity tensor ε^{-1} which is presented as decomposed into the symmetric and antisymmetric parts $\varepsilon^{-1} = \varepsilon_0^{-1} + i\mathbf{h}^\times$, where ε_0^{-1} is a real symmetric tensor, $\mathbf{h}$ is a real vector. Consider the simplest case of surface waves at the interface between an optically active Faraday medium ($\mathbf{qr} < 0$) and an isotropic optically inactive medium ($\mathbf{qr} > 0$) provided that on external magnetic field $\mathbf{H}^{\text{extr}}$ is applied parallel to the interface. Let an isotropic medium be characterized by the inverse permittivity a'. Then an inverse permittivity tensor for Faraday medium takes the form[27]

$$\varepsilon^{-1} = a + ib\mathbf{c}^{\times}. \tag{39}$$

We consider the unit vector $\mathbf{c}$ involved in (39) to be codirected with the external field vector $\mathbf{H}^{\text{extr}}$ for Faraday media with a positive rotation (Verdet constant $V > 0$) and counterdirected with it for media with a negative rotation ($V < 0$). Then parameters a and b are positive and expressed in terms of the refractive index n_0 of a linearly polarized wave propagating along $\mathbf{c}$-direction, its wavelength $\lambda_0 = 2\pi c/\omega$ and Verdet constant V as

$$a = \frac{1}{n_0^2}, \quad b = \frac{|V|\lambda_0\mathcal{H}^{\text{extr}}}{\pi n_0^3}. \tag{40}$$

Let us denote the angle between the surface wave propagation direction $\mathbf{b}$ and vector $\mathbf{c}$ as α: $\mathbf{c} = \mathbf{b}\cos\alpha + \mathbf{a}\sin\alpha$, $\mathbf{a} = [\mathbf{bq}]$. Using formulae (34)–(37) we can obtain the expressions for the surface impedance tensors of both media

$$\begin{aligned}\gamma = {} & \frac{i}{a - \nu^2 - a\eta_1\eta_2}\{i\nu a(\eta_1 + \eta_2)\mathbf{b}\otimes\mathbf{b} - \nu b\cos\alpha(\mathbf{b}\otimes\mathbf{a} + \mathbf{a}\otimes\mathbf{b}) \\ & + \frac{1}{\nu}[b\sin\alpha(a - \nu^2 - a\eta_1\eta_2) + ia^2\eta_1\eta_2(\eta_1 + \eta_2)]\mathbf{a}\otimes\mathbf{a}\}, \\ \gamma' = {} & -i\nu\sqrt{\frac{a'}{a' - \nu^2}}\mathbf{b}\otimes\mathbf{b} + \frac{i}{\nu}\sqrt{a'(a' - \nu^2)}\mathbf{a}\otimes\mathbf{a}\end{aligned} \tag{41}$$

and the dispersion equation

$$\begin{aligned}& \frac{i(\eta_1 + \eta_2)ab\sin\alpha}{a - \nu^2 - a\eta_1\eta_2} - \frac{a^3\eta_1\eta_2(\eta_1 + \eta_2)^2 + \nu^2b^2\cos^2\alpha}{(a - \nu^2 - a\eta_1\eta_2)^2} \\ & + \left[b\sin\alpha + \frac{ia^2\eta_1\eta_2(\eta_1 + \eta_2)}{a - \nu^2 - a\eta_1\eta_2}\right]\sqrt{\frac{a'}{a' - \nu^2}} \\ & - \frac{ia(\eta_1 + \eta_2)}{a - \nu^2 - a\eta_1\eta_2}\sqrt{a'(a' - \nu^2)} - a' = 0,\end{aligned} \tag{42}$$

where

$$\eta_{1,2} = -\frac{i}{a}\left[a(a - \nu^2) - \frac{1}{2}b^2\cos^2\alpha \pm \frac{1}{2}b\cos\alpha\sqrt{4a\nu^2 + b^2\cos^2\alpha}\right]^{1/2}, \tag{43}$$

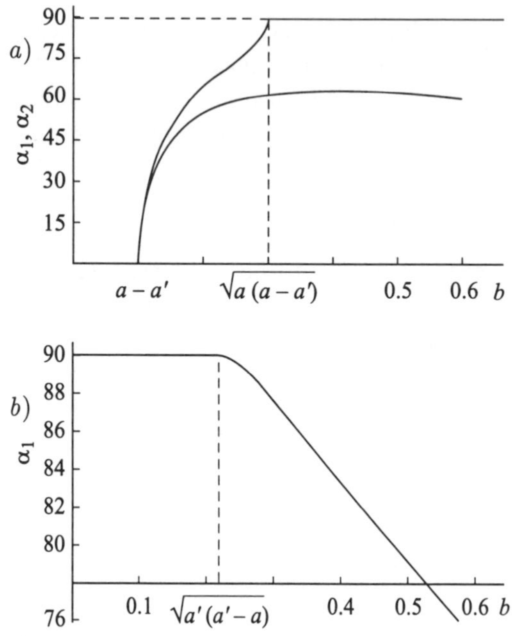

Figure 1. Limiting values of the angles α_1, α_2 (in degree) versus parameter b at fixed a, a': a) $a = 0.9$, $a' = 0.8$; b) $a = 0.9$, $a' = 0.95$

η_1, η_2 being the coefficients included in (30).

Dispersion equation (42) preserves its form substituting $\nu \to -\nu$ (or $k \to -k$). This fact might seem to reversibility of the surface waves at the interface between media under consideration. However, from Eq. (30) it is obvious, that the substitution leads not only to the inversion of the wave propagation direction $\mathbf{b} \to -\mathbf{b}$, but to an exponential increase of the field amplitude with its distance from the interface. From a physical point of view such a wave is unrealistic. Actually, the inversion of the propagation direction corresponds to the increase (or decrease) of the angle α of π: $\alpha \to \alpha \pm \pi$. Obviously that Eq. (42) isn't invariant with respect to such operation. One can show[27] that Eq. (42) has some solution if parameter b which linearly depends on external field strength $\mathcal{H}^{\text{extr}}$ exceeds some threshold value

$$b > a - a' \quad \text{at} \quad a > a', \qquad b > \sqrt{a'(a' - a)} \quad \text{at} \quad a < a'. \tag{44}$$

In paper[27] it was found that all vectors $\mathbf{b}$ corresponding to the allowed directions of surface wave propagation lie in the interface plane on one side of the vector $\mathbf{c}$, so that $[\mathbf{bq}]\mathbf{c} > 0$. Just this fact reveals the effect of unidirectional propagation of surface electromagnetic waves at the interface between Faraday and isotropic inactive media unlike the body waves, which can always propagate in such media directly and backwards. The set of the allowed directions $\mathbf{b}$ forms either two sectors at $a > a'$, $a - a' < b < \sqrt{a(a-a')}$ or one sector at $b > \sqrt{\max(a, a')|a - a'|}$. In the first case the positions of vectors $\mathbf{b}$ with respect to $\mathbf{c}$ are characterized by the angles α so that $\alpha \in (\alpha_1, \alpha_2) \cup (\pi - \alpha_2, \pi - \alpha_1)$, in the second by the angles $\alpha \in (\alpha_1, \pi - \alpha_1)$, where limiting angles α_1, α_2 are expressed in a complicated way in terms of parameters a, b and a'. In Fig. 1 typical dependence of angles α_1 and α_2 on the parameter b at fixed a and a' is shown.

Thus, the interface between Faraday medium and isotropic optically inactive medium behaves as an "optical diode" for surface waves, and this property can be used for constructing optical isolators and shutters for surface polaritons.

4. Covariant impedance techniques to characterize complex media

In the last three decades, considerable attention has been focused on complex electromagnetic media (chiral, anisotropic, bianisotropic). To characterize these media, four experimental approaches — free-space techniques, waveguide-based approaches, resonator measurements and time-domain methods — are mainly used. Free space measurement techniques are particularly attractive because they can be used to characterize complex media over a wide frequency range. They are developed in the frame of the plane-wave approximation of the incident beam. A non-chiral non-magnetic anisotropic medium has 9 independent constitutive parameters (components ϵ_{ij} of the permittivity tensor ϵ). To find them, it is sufficient to measure the reflection and transmission operators of three differently cut slabs for the case of normal incidence[28,29]. This approach is based on the use of the three-dimensional impedance operators which uniquely define the reflection and transmission operators[6,12,13].

A general bianisotropic medium is described by the constitutive equations

$$\mathbf{D} = \epsilon\mathbf{E} + \alpha\mathbf{B}, \quad \mathbf{H} = \beta\mathbf{E} + \mu^{-1}\mathbf{B}. \tag{45}$$

The techniques that have been tested in practice so far give no way of finding all 36 constitutive parameters — complex components ϵ_{ij}, α_{ij}, β_{ij}, μ_{ij} $(i,j = 1,2,3)$ of the permittivity tensor ϵ, the permeability tensor μ, and the magnetoelectric pseudotensors α and β.

In[30–33], the covariant impedance methods in electrodynamics of motionless and uniformly moving linear media are developed, and the exact solutions of the inverse scattering problems for such media are found, which can form a basis for the development of new free space techniques. Two such techniques are briefly outlined in this section. The both techniques are formulated in the Lorentz covariant form because this enables one to use the exterior algebra which has a much wider spectrum of mathematical operations in comparison to the component technique and the vector and matrix algebras. In particular, the application of pseudoinversion[30,31] has made it possible to compose a four–dimensional tensor with linear dependence on constitutive parameters from three operators which have two–dimensional kernels and a very involved dependence on constitutive parameters but can be readily calculated from the measured reflection and transmission coefficients. Naturally, the Lorentz covariant methods, developed in[30–33], can be effectively applied in electrodynamics of complex media at rest.

The obtained covariant solutions, as applied to a general bianisotropic medium, provide two different ways to extract all constitutive parameters. As intermediate results in calculating these parameters, one can extract either the coefficients of the four–dimensional tensor impedance equation[30,31] (technique IMPEDANCE), or the polarization parameters of eigenwaves in a slab[32,33] (technique POLARIZATION).

4.1. *Four–dimensional constitutive tensor*

The key problem under consideration is the reflection and transmission of a plane harmonic electromagnetic wave at oblique incidence onto a homogeneous bianisotropic slab. Using the four–dimensional tensors

$$\begin{aligned} F &= \sum_{1\le i<j\le 4} F_{ij}\theta^i\wedge\theta^j \\ &= B_3\theta^1\wedge\theta^2 - B_2\theta^1\wedge\theta^3 + B_1\theta^2\wedge\theta^3 + (E_1\theta^1 + E_2\theta^2 + E_3\theta^3)\wedge\theta^4, \end{aligned} \tag{46}$$

$$\begin{aligned} G &= \sum_{1\le i<j\le 4} G_{ij}\theta^i\wedge\theta^j \\ &= H_3\theta^1\wedge\theta^2 - H_2\theta^1\wedge\theta^3 + H_1\theta^2\wedge\theta^3 + (D_1\theta^1 + D_2\theta^2 + D_3\theta^3)\wedge\theta^4, \end{aligned} \tag{47}$$

one can replace Eqs. (45) by the Lorentz covariant constitutive equation

$$G = MF = \sum_{1 \le i < j \le 4,\, 1 \le k < l \le 4} M_{ij}{}^{kl} F_{kl} \theta^i \wedge \theta^j, \tag{48}$$

where

$$M = \sum_{1 \le i < j \le 4,\, 1 \le k < l \le 4} M_{ij}{}^{kl} \theta^i \wedge \theta^j \otimes \mathbf{e}_k \wedge \mathbf{e}_l \tag{49}$$

is the four–dimensional constitutive tensor. Here, $(\mathbf{e}_i)$ and (θ^i) are the dual orthonormal bases in the Minkowski vector space V and its dual V^* (the space of covariant vectors); $\mathbf{e}_i \rfloor \theta^j = \mathbf{e}_i \lfloor \theta^j = \delta_i^j$, $i, j = 1, 2, 3, 4$, δ_i^j is the Kronecker delta function; $\otimes$, $\wedge$, $\rfloor$, $\lfloor$ are the tensor, exterior and interior products; $\mathbf{e}_i \wedge \mathbf{e}_j = \mathbf{e}_i \otimes \mathbf{e}_j - \mathbf{e}_j \otimes \mathbf{e}_i$, $\mathbf{e}_1 \rfloor (\theta^1 \wedge \theta^2) = \theta^2$, $(\mathbf{e}_1 \wedge \mathbf{e}_2) \lfloor \theta^1 = \mathbf{e}_2$.

To solve the direct scattering problem for a bianisotropic slab with the constitutive tensor M, a numerical code (code DIR) was produced. It is based on the use of the Lorentz covariant impedance and characteristic matrix methods[30,31,32,33]. The code consists of the following main steps:

- calculation of the coefficients of the four–dimensional tensor impedance equation;
- solving the dispersion equation for the unknown normal components of wave vectors of eigenwaves excited in the slab;
- calculation of the four–dimensional tensors of normal refraction and surface impedance, and then the characteristic matrix of the slab[30,31,32,33];
- calculation of the reflection and transmission coefficients (in terms of tangential components of **E** fields).

Code DIR was used to provide the input data for codes solving the inverse scattering problem by making use the techniques IMPEDANCE and POLARIZATION.

4.2. *Extracting the characteristic matrix*

Extracting the characteristic matrix C from the reflection and transmission coefficients is the first step in solving the inverse scattering problem for both techniques IMPEDANCE and POLARIZATION. It is based on the covariant relations[33]

$$\mathcal{C} = \begin{pmatrix} C_1 & C_2 \\ C_3 & C_4 \end{pmatrix}, \tag{50}$$

where

$$\begin{aligned}
C_1 &= \left[d_1 - (r_2 + I)d_2^-(r_1 + \gamma_3^-\gamma_1)\right](I - \gamma_3^-\gamma_1)^-,\\
C_2 &= \left[d_1 - (r_2 + I)d_2^-(r_1 + I)\right](\gamma_1 - \gamma_3)^-,\\
C_3 &= \left[\gamma_4 d_1 - (\gamma_4 r_2 + \gamma_2)d_2^-(r_1 + \gamma_3^-\gamma_1)\right](I - \gamma_3^-\gamma_1)^-,\\
C_4 &= \left[\gamma_4 d_1 - (\gamma_4 r_2 + \gamma_2)d_2^-(r_1 + I)\right](\gamma_1 - \gamma_3)^-.
\end{aligned} \tag{51}$$

Here, γ_1, r_1 and d_1 (γ_2, r_2 and d_2) are the surface impedance, reflection and transmission tensors of the wave incident onto the first (second) boundary; γ_3 and γ_4 are the impedance tensors of the waves leaving the slab from the first and second boundaries, respectively; I is the projection operator of the interface; γ^- is the pseudoinverse tensor[30–33] ($\gamma^-\gamma = I$).

4.3. *To the constitutive parameters via the coefficients of the four–dimensional tensor impedance equation (technique IMPEDANCE)*

The coefficients of the impedance equation are related with the four–dimensional constitutive tensor M(49) as follows[30,31]

$$\begin{aligned}
A' &= A - A\tau\otimes\boldsymbol{\tau}A/\lambda, & A &= \mathbf{Q}\rfloor M\lfloor Q,\\
B_1' &= B_1 + B_1\tau\otimes\mathbf{Q}B_1/\lambda, & B_1 &= \boldsymbol{\tau}\rfloor M\lfloor Q,\\
B_2' &= B_2 + B_2 Q\otimes\boldsymbol{\tau}B_2/\lambda, & B_2 &= \mathbf{Q}\rfloor M\lfloor \tau,\\
C' &= C - CQ\otimes\mathbf{Q}C/\lambda, & C &= \boldsymbol{\tau}\rfloor M\lfloor \tau,
\end{aligned} \tag{52}$$

where $\mathbf{Q}\in V$ and $Q\in V^*$ are the contravariant and covariant four–dimensional vectors normal to the interface; $\boldsymbol{\tau}\in V$ and $\tau\in V^*$ are the contravariant and covariant vector parameters specifying the properties of the incident wave; $\lambda = \boldsymbol{\tau}A\tau\equiv\mathbf{Q}CQ\equiv Q^iC_i{}^jQ_j$; and the summation convention is applied to the repeated indices ($i, j = 1, 2, 3, 4$).

Technique IMPEDANCE is based on the linear dependence of A, B_1, B_2 and C on the constitutive tensor M. The covariant solutions, obtained in papers[30,31], make it possible to calculate A', B_1', B_2' and C' from the measured reflection and transmission coefficients of two similarly cut slabs of different thickness. Two samples are used here to remove the modulo 2π phase ambiguity. Extracting the coefficients A', B_1', B_2' and C' includes the following intermediate steps: calculation of the characteristic matrices and their eigenvalues and eigenvectors; calculation of tensors of normal refraction and surface impedance.

To find all components of M, it is necessary to extract first tensors A', B_1', B_2' and C' for a properly chosen set of the vector parameters $\mathbf{Q}$ and

$\boldsymbol{\tau}$, namely, $\mathbf{Q}_1$, $\mathbf{Q}_2$, $\mathbf{Q}_3$, $\boldsymbol{\tau}_1$, $\boldsymbol{\tau}_2$, $\boldsymbol{\tau}_3$, such that ($\mathbf{Q}_1$, $\mathbf{Q}_2$, $\mathbf{Q}_3$, $\mathbf{Q}_4 = \boldsymbol{\tau}_i$) and ($\boldsymbol{\tau}_1$, $\boldsymbol{\tau}_2$, $\boldsymbol{\tau}_3$, $\boldsymbol{\tau}_4 = \mathbf{Q}_i$) form bases in the Minkowski space V at any $i = 1, 2, 3$. This means that the measurements should be carried out for three differently cut slabs for three different orientations of the incidence plane.

The tensors A', B_1', B_2' and C' having been extracted, one can calculate

$$A_i = \mathbf{Q}_i \rfloor M \lfloor Q^i, \quad C_i = \boldsymbol{\tau}_i \rfloor M \lfloor \tau^i, \quad i = 1, 2, 3 \tag{53}$$

by making use of the pseudoinversion, and then to find components ${M'_{ij}}^{kl}$ of M in a basis ($\mathbf{Q}_i$, $i = 1, 2, 3, 4$) as follows.

From A_1, A_2 and A_3, one can directly find 24 tensor components. In particular, if ($\mathbf{Q}_i = \mathbf{e}_i$, $i = 1, 2, 3, 4$) is an orthonormal basis ($\mathbf{e}_i^2 = 1$, $i = 1, 2, 3$; $\mathbf{e}_4^2 = -1$) in the Minkowski space, we immediately obtain

$$\epsilon_{11} = {M'_{14}}^{14} = (A_1)_4{}^4, \qquad \left(\mu^{-1}\right)_{11} = {M'_{23}}^{23} = (A_2)_3{}^3 = (A_3)_2{}^2,$$

and so on.

When the first 24 components ${M'_{ij}}^{kl}$ are found, the other 12 components can be calculated from the expressions of the form[30–33]:

$$\begin{aligned} {M'_{14}}^{24} &= \left(\tau_4^3\right)^{-2}\left(\mathbf{Q}_1 C_3 Q^2 + \tau_1^3 \tau_2^3 {M'_{12}}^{12} + \tau_1^3 \tau_4^3 {M'_{14}}^{12} - \tau_2^3 \tau_4^3 {M'_{12}}^{24}\right), \\ {M'_{12}}^{34} &= \left[\tau_2^1 {M'_{12}}^{23} + \mathbf{Q}_1 {B'_1}^{(2,1)} Q^3 - \left(\mathbf{Q}_1 A_2 \tau^1\right)\left(\mathbf{Q}_2 C_1 Q^3\right)/\lambda_{21}\right]/\tau_4^1, \end{aligned}$$

and so on. Here, $\tau^i = \tau_j^i Q^j$, $j \neq i$, $\lambda_{ij} = \boldsymbol{\tau}_j A_i \tau^j$, ${B'_1}^{(i,j)}$ and ${B'_2}^{(i,j)}$ are the values of B_1' and B_2' at $\mathbf{Q} = \mathbf{Q}_i$ and $\boldsymbol{\tau} = \boldsymbol{\tau}_j$. The components ${M'_{ij}}^{kl}$ have being found, to find the constitutive parameters ϵ_{ij}, α_{ij}, β_{ij}, $\left(\mu^{-1}\right)_{ij}$, $i, j = 1, 2, 3$, it is sufficient to make a change from the basis ($\mathbf{Q}_i$, $i = 1, 2, 3, 4$) to the basis ($\mathbf{e}_i$, $i = 1, 2, 3, 4$).

4.4. *To the constitutive parameters via the polarization parameters of eigenwaves (technique POLARIZATION)*

Technique POLARIZATION is based on the fact that the constitutive tensor M can be treated as a linear operator acting in the six-dimensional space of two-forms (antisymmetric 2-covariant tensors) and relating the field F and induction G tensors ($G = MF$). Hence, M is uniquely defined by its action on any basis of this space. This basis can be composed from any six linearly independent amplitudes of eigenwaves propagating in the medium under study.

The necessary information on wave vectors and polarization parameters of eigenwaves can be obtained by calculating eigenvalues and eigenvectors of the characteristic matrix which in its turn can be calculated from the measured reflection and transmission coefficients of a bianisotropic slab. Since the eigenvectors of this matrix specify only tangential components of eigenwave amplitudes, it is necessary to carry out such measurements for two differently cut off slabs, exciting in each of them the same eigenwave, and then to reconstruct its full vector amplitudes. The amplitudes of any six eigenwaves have being found, the whole set of constitutive parameters can readily be calculated[32,33].

4.5. *Computer modeling*

For computer modeling of the suggested free space techniques IMPEDANCE and POLARIZATION two different numerical codes (INV1, INV2) were produced. To simulate measurements errors, the reflection and transmission coefficients calculated by code DIR, were first multiplied by coefficients of the form

$$C_{err} = (1 + h_1)\exp(ih_2), \tag{54}$$

where h_1 and h_2 were generated by a random numbers generator, and then were used as the input data for codes INV1 and INV2. For the simplicity sake, the rectangular distributions for random numbers h_1 and h_2 were used. The random numbers h_1 and h_2 simulate errors in amplitude and phase measurements, respectively.

The calculated constitutive parameters were compared with the given ones. The accuracy of restoration of constitutive parameters at different values of maximum errors in amplitude $(max(h_1))$ and phase $(max(h_2))$ measurements was evaluated. The numerical results for several complex media are presented in paper[33]. They show that the both suggested techniques make it possible to calculate all constitutive parameters of an anisotropic, chiral, or general bianisotropic medium, provided that its reflection and transmission coefficients are measured with sufficient accuracy. The validity of these techniques is supported over a wide range of constitutive parameters values at both microwave and optical frequencies.

Novel free–space techniques[30–33], based on the covariant impedance methods, make it possible to extract all constitutive parameters of an anisotropic, chiral, or general bianisotropic medium, provided that the reflection and transmission coefficients of planar samples under normal and oblique incidence of plane harmonic waves are measured with sufficient accuracy. The developed techniques do not use any speculations regarding the number of nonzero constitutive parameters or the number of linearly independent parameters. They based only on the most general assumptions such as linearity and macroscopic homogeneity of the medium.

Bibliography

1. P. Harman, *Physics World* **16**, No. 11, p. 50 (2003).
2. J. C. Maxwell, *Treatise on Electricity and Magnetism* (Oxford UP, Oxford, 1873).
3. F. I. Fedorov, *Optics of Anisotropic Media* (Acad. Nauk BSSR, Minsk, 1958).
4. F. I. Fedorov, *Theory of Gyrotropy* (Nauka i Teknika, Minsk, 1976).
5. L. M. Barkovsky and G. N. Borzdov, in *Advances in Complex Electromagnetic Materials*, Ed. A. Priou (Kluwer, Dordrecht, The Netherlands, 1997), p. 3.
6. A. N. Furs and L. M. Barkovsky, *Operator methods for description of optical fields in complex media* (Belaruskaya Navuka, Minsk, 2003).
7. J. Zweck, Jr. I. T. Lima, Sun Yu and et al, *Opt. Phot. News* **14**, No. 11, p. 30 (2003).
8. L. A. Zenteno and D. T. Walton, *Opt. Phot. News* **14**, No. 3, p. 39 (2003).
9. V. G. Veselago, *Sov. Phys. Uspekhi* **10**, p. 509 (1968).
10. J. B. Pendry, *Phys. Rev. Lett.* **85**, 3966 (2000).
11. I. V. Shadrivov, A. A. Sukhorukov and Yu. S. Kivshar, *Phys. Rev.* **E67**, 57602 (2003).
12. L. M. Barkovskii, G. N. Borzdov and A. V. Lavrinenko, *J. Phys.* **A20**, 1095 (1987).
13. G. N. Borzdov, *J. Math. Phys.* **38**, 6326 (1997).
14. M. J. Adams, *An introduction to optical waveguides* (John Wiley and Sons, Chichester, 1981)
15. J. J. Baumberg, *Phys. World* **15**, No. 3, p. 37 (2002).
16. G. S. McDonald, G. H. C. New and J. P. Woerdman, *Opt. Photon. News* **11**, No. 11, p. 22 (2000).
17. H. Rubinsztein-Dunlop and M. E. J. Friese, *Opt. Photon. News* **13**, No. 4, p. 22 (2002).
18. V. M. Agranovich, *Surface Polaritons*, Ed. V. M. Agranovich and D. C. Mills, (North-Holland, Amsterdam, 1982).
19. D. R. Smith, W. J. Padilla, D. C. Vier, S. C. Nemat-Nasser and S. Schultz, *Phys. Rev. Lett.* **84**, 4184 (2000);
    R. A. Shelby, D. R. Smith and S. Schultz, *Science* **292**, 77 (2001);
    A. Grbic and G. V. Eleftheriades, *J. Appl. Phys.* **92**, 5930 (2002).
20. J. B. Pendry, A. J. Holden, D. J. Robbins and W. J. Steward, *IEEE trans. MTT* **47**, 2075 (1999).
21. J. B. Pendry, A. J. Holden, W. J. Stewart and I. Youngs, *Phys. Rev. Lett.* **76**, 4773 (1996);
    J. B. Pendry, A. J. Holden, D. J. Robbins and W. J. Steward, *J. Phys.: Condens. Matter* **10**, 4785 (1998).
22. A. G. Litvak and V. A. Mironov, *Izv. Vuzov: Radiophysica* **11**, p. 1911 (1968).
23. M. I. D'yakonov, *Sov. Phys.–JETP* **67** , p. 714 (1988).
24. N. S. Averkiev and M. I. D'yakonov, *Opt. Spectrosk.* **68**, No. 5, p. 1118 (1990).

25. A. N. Furs and L. M. Barkovsky, *J. Opt. A: Pure Appl. Opt.* **1**, p. 109 (1999).
26. A. N. Furs and L. M. Barkovsky, *Microwave and Opt. Technol. Lett.* **14**, p. 301 (1997).
27. A. N. Furs and L. M. Barkovsky, *Technical Physics* **48**, p. 385 (2003).
28. V. V. Zhilko, *Vesti AN BSSR: Phys. Math. Nauki*, No. 4, p. 82 (1984).
29. F. I. Fedorov, L. M. Barkovsky, G. N. Borzdov, and V. V. Zilko, *Kristallografiya* **30**, p. 629 (1985).
30. G. N. Borzdov, *Opt. Commun.* **94**, p. 159 (1992).
31. G. N. Borzdov, *J. Math. Phys.* **34**, 3162 (1993).
32. G. N. Borzdov, in *Advances in Complex Electromagnetic Materials*, Ed. A. Priou, (Kluwer, Dordrecht, The Netherlands, 1997), p. 71.
33. G. N. Borzdov, in *Electromagnetic fields in unconventional materials and structures*, Eds. O. N. Singh and A. Lakhtakia, (Wiley Interscience, New York, 2000), p. 83.

OPTICAL WAVES IN ONE-DIMENSIONAL FRACTAL INHOMOGENEOUS MEDIA

Sergei V. Zhukovsky, Andrei V. Lavrinenko and Sergey V. Gaponenko

Department of Theoretical Physics, Belarusian State University,
4, Fr. Skorina Ave., 220050, Minsk, Belarus
Laboratory of Nanooptics, Institute of Molecular and Atomic Physics,
National Academy of Sciences of Belarus,
70, Fr. Skorina Ave., 220072, Minsk, Belarus
zhukovsky@imaph.bas-net.by

Fractal multilayer structures as a class of deterministic non-periodic discrete inhomogeneous media are investigated on the basis of classical wave propagation. This includes optical properties of dielectric layered media and quantum particle behaviour in a fractal stepwise potential. The primary focus of the present review is drawing a correlation between topological properties and eigenvalue spectra. Namely, it is shown both numerically and analytically that geometrical self-similarity of fractal structures directly results in spectral scalability. Some other spectral properties resulting from minor topological details of fractal multilayers (such as resonance peak splitting and spectral self-similarity) are considered.

1. Introduction

Concerning the state of research on wave propagation in complex discrete inhomogeneous media (of which one-dimensional (1D) multilayers are the simplest particular case), it can be said that the results can be separated into two large groups. On the one hand, there is a variety of numerical methods for modelling wave propagation process in arbitrary nanostructures[1,2,3,4,5]. These methods are universal and have a great practical value in optoelectronics and integrated optics.

On the other hand, there are analytical or semianalytical methods, which are much less universal and generally based on certain spatial relationships of the structure. However, unlike the above mentioned "microscopic" methods where the only way of drawing conclusions is generalization of many individual results, this "macroscopic" approach allows to consider the effects of a certain topological features of the structure on its wave propagation properties and thus to gain better theoretical understanding of wave propagation in complex media.

One prominent example of this approach is the study of periodic nanostructures which has lasted for a couple of decades[1,2,6] and resulted in a conclusion that periodicity of the structures leads to band structure of eigenvalue spectra, in the optical case known as photonic band gaps (PBGs). Another, opposite example is what is known of random, or disordered, structures where disorder has been shown to result in light trapping, localization, and coherent backscattrting[7,8].

The intermediate case between these two extremes, namely *non-periodic deterministic structures*, encompasses a large variety of structures too and thus has been subject to extensive scientific research during the past few years. However, most investigations so far were either numerical[9–16] or experimental[17,18]. A notable exception is the work of M. Kohmoto with colleagues[9] who analytically showed that quasiperiodicity of Fibonacci multilayers gives rise to self-similarity of spectral transmission bands and a critical (intermediate between ordered and chaotic) behavior of the wave function. Thus, the dependence of characteristic wave propagation properties on particular topological features of the structure can be summarized in Table 1.

Topology	**Spectral features**
Periodic	Photonic band gaps (PBGs)
Disordered	Light trapping and localization
Quasiperiodic	Self-similar transmission bands

This review has been created with an intent to further fill up Table 1 and investigate the spectral effects of another type of deterministic non-periodic media, namely *fractal multilayers*, which we describe in detail in Section 2. Earlier[3] we showed numerically that these structures possess *spectral scalability*. In this paper we confirm this conclusion analytically and show that this spectral scalability is a result of spatial self-similarity and thus a characteristic spectral feature of fractal multilayers. This is done in Section 3. Section 4 is devoted to other spectral properties of fractal multilayer structures. There we describe sequential peak splitting, as well as report on a very recently discovered *spectral self-similarity* and compare this effect with self-similarity of transmission bands in quasiperiodic structures.

2. Fractal Multilayers

2.1. *Construction Procedure*

A textbook example of fractal multilayers is a well-known Cantor stack generated using the "middle third removal" procedure[19] (see Fig. 1). However, this procedure can be generalized. The most straightforward way to do so is to complicate the removal routine, applying it not only to the middle third, but to arbitrary (yet similar from generation to generation) regions of the structure. Some variations have been described in[19] and investigated in[3,20–23].

Here we introduce a more general procedure, which encompasses most of 1D fractals that have prefractals and hence can be used in multilayer design. The algorithm starts with an *initiator*, a single dielectric layer (label it A) with the index of refraction n_A and the thickness d_A. The initiator is stacked together G times, and the layers are numbered in base G (starting with zero). Then, those parts whose numbers belong to a given subset of digits $\mathbf{C} \subset \{0, 1, \ldots, G-1\}$ are replaced with another dielectric layer (this time, labelled B), with the index of refraction $n_B \neq n_A$ and thickness d_B. This replication-replacement (RR) procedure is then repeated for the resulting structure (which now consists of G layers), with the only difference that a group of G B-type layers is now used to replace the appropriate fragments. Repeating this RR procedure multiple times yields the desired fractal multilayer.

Here, an arbitrary integer $G > 2$ together with the subset $\mathbf{C}$ form the *generator* of the structure, while the number N of RR procedures applied is called the *number of generations*. The whole structure can be referred to as a $(G, \mathbf{C}, N)$ structure. One can see that the usual N-stage middle third Cantor stack is nothing other than a particular case of $(3, \{1\}, N)$. Other particular cases include higher-G Cantor structures $(G = 3, 5, 7, \ldots;\ \mathbf{C} = \{1, 3, \ldots, G-2\}, N)$[3], non-symmetric stacks[19], and generalized Cantor bars[21].

Note that the above mentioned generation procedure can come in two different variations. In one variation (see the right side of Fig. 1), the *total* structure thickness remains the same regardless of the number of generations. In this case a 0-generation structure represents a large uniform slab of homogeneous material, while the structure becomes finer as N increases. The other variation (the left side of Fig. 1) is based on the constant thickness of the "elementary layers" (building blocks used to construct all subsequent generations). In this variation the 0-generation structure is a thin single layer, and the structure thickness increases with N. The *detail* of the structure, however, does not change from generation to generation.

Some sample stacks for these two variations are shown in Fig. 1, and the construction details can be inferred therefrom.

2.2. *Quarter-Wave Condition and Spectral Symmetries*

Prior to proceeding with further analysis of fractal multilayers, we note that it is fruitful to assume the structures to consist of two kinds of isotropic dielectric layers (A and B) with an additional restriction of their optical thicknesses (the *quarter wave condition*):

$$n_A d_A = n_B d_B = d^* \equiv \lambda_0/4 = \pi c/2\omega_0. \tag{1}$$

This condition has two outcomes regardless of the structure topology. First, the optical spectra of such quarter-wave structures appear to be periodic with respect to frequency, the period equal to $2\omega_0$. Secondly, within each period there is a mirror symmetry around the odd multiples of ω_0, i.e.,

$$\begin{gathered} T(\omega + 2k\omega_0) = T(\omega), \\ T[(2n+1)\omega_0 - \eta\omega_0] = T[(2n+1)\omega_0 + \eta\omega_0], \\ n \in \mathbb{N}, 0 \leqslant \eta \leqslant 1. \end{gathered} \tag{2}$$

This way, we not only have ω_0 as a natural frequency normalization factor, but it is also enough to consider just one period of spectra to be

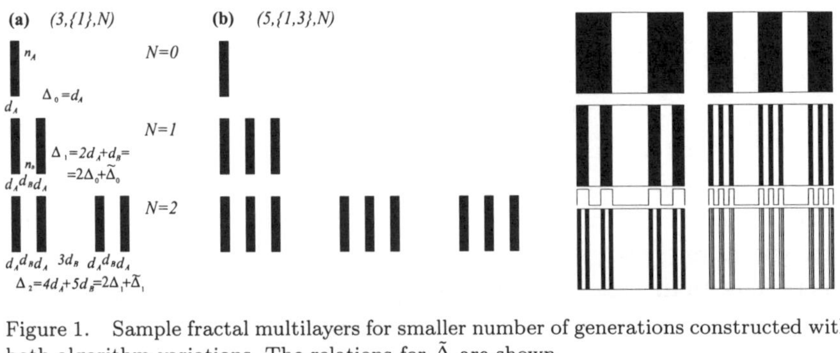

Figure 1. Sample fractal multilayers for smaller number of generations constructed with both algorithm variations. The relations for $\tilde{\Delta}$ are shown.

It can easily be seen from Eq.(1) that the quarter-wave condition is more compatible with the second variation of the construction algorithm. As such, it is this variation that was used throughout this review unless otherwise indicated.

2.3. *Some geometrical relations for fractal multilayers.*

To conclude this section, we list some simple but important geometrical relations concerning fractal multilayers. First of all, the total number of layers in such a structure is G^N (here and further, several adjacent layers of the same material count as separate layers). Among these layers, $(G-C)^N$ are A-type and the rest are B-type, C being the number of members in $\mathbf{C}$. Then, the total thickness of a $(G, \mathbf{C}, N)$ structure equals

$$\Delta_N = (G-C)^N d_A + \left(G^N - (G-C)^N\right) d_B \tag{3}$$

and can be written as a recurrent relation (see Fig. 1)

$$\Delta_N \equiv (G-C)\,\Delta_{N-1} + C\tilde{\Delta}_{N-1}. \tag{4}$$

Comparing Eqs. (3) and (4), one can obtain a scaling relation for $\tilde{\Delta}_N = G^N d_B$, e.g., for a middle third structure $(3, \{1\}, N)$

$$\tilde{\Delta}_N = 3^N d_B, \tag{5}$$

and in all cases $\Delta_0 = d_A$. Finally, the ratio between optical lengths of $(G, \mathbf{C}, N_2 > N_1)$ and $(G, \mathbf{C}, N_1)$ structures (essentially the *geometrical scaling factor*) is equal to

$$S_{geom} = G^{N_2 - N_1}. \tag{6}$$

3. Spectral Scalability

3.1. *Numerical Observation*

It has been earlier reported by us[3,24,25] that fractal multilayer structures possess *spectral scalability*, which can be defined as follows. First for Cantor structures[3,24] and then for all $(G, \mathbf{C}, N)$ structures[25] it has been found that the *whole* period of a $(G, \mathbf{C}, N_1)$ structure spectrum is similar in shape to the central part of a period of a $(G, \mathbf{C}, N_2 > N_1)$ structure spectrum magnified by a certain number S_{sp}, as can be seen in Fig. 2. This number, called the *spectral scaling factor*, equals

$$S_{sp} = G^{N_2 - N_1}. \tag{7}$$

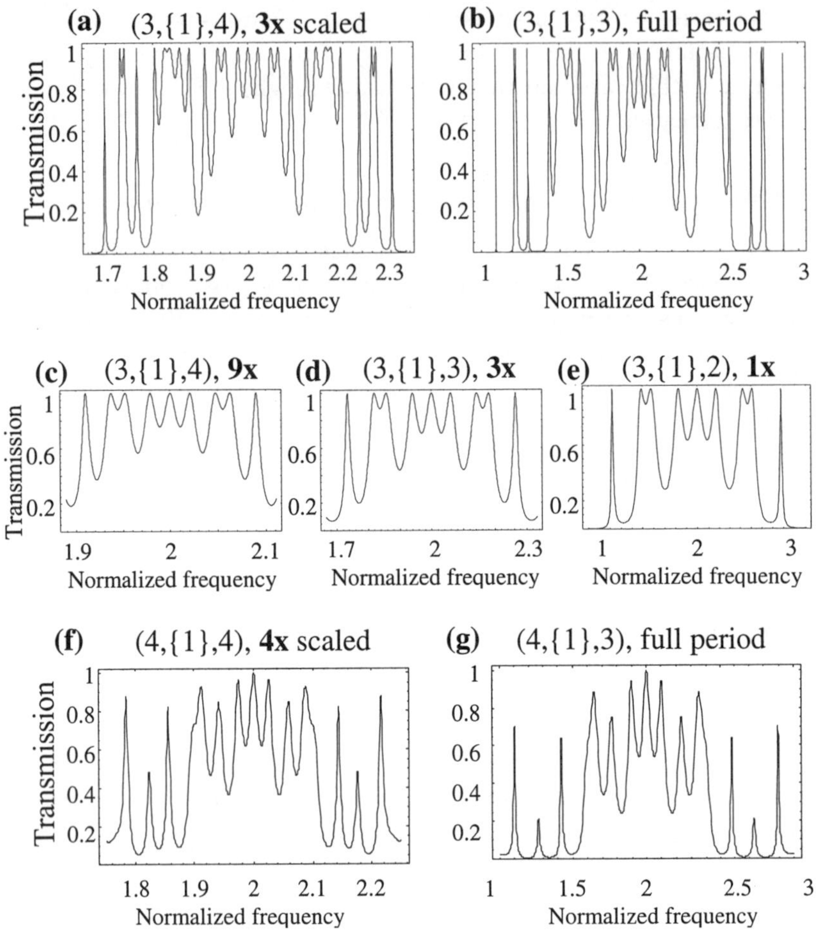

Figure 2. Spectral scalability for Cantor structures $(3, \{1\}, N)$ (a-e) and non-symmetric fractal structures $(4, \{1\}, N)$ (f,g). Compare the shapes of (a) and (b); (c-e); (f) and (g).

3.2. *Phase Terms Matching*

The equality between S_{geom} and S_{sp} hints at the idea that spectral scalability is in some way related to spatial self-similarity. However, to state that the former is the result of the latter, such qualitative speculations are clearly not enough and more rigorous reasoning is required. Moreover, close inspection of spectra in Fig. 2 reveals that there is no exact coincidence either in the value of transmission coefficient or in the frequencies of resonance transmission peaks (see Fig. 3a). Nevertheless, these discrepancies do not alter the shapes of the spectra and will be commented on later.

Consider first the simplest case of middle third Cantor structures. It is possible to use the recurrent relations for transmission and reflection derived by Sun and Jaggard[20,21], which directly employs the self-similarity of the structures. According to this method,

$$\begin{aligned} R_{N+1}(\omega) &= g_r\left[R_N(\omega), T_N(\omega), \tilde{\Delta}_N, \omega\right], \\ T_{N+1}(\omega) &= g_t\left[R_N(\omega), T_N(\omega), \tilde{\Delta}_N, \omega\right], \end{aligned} \tag{8}$$

where Δ_N is the overall thickness of the structure, and $\tilde{\Delta}_N$ is as defined in Eq. (4). The functions

$$\begin{aligned} g_r(x, y, d, \omega) &= x + \frac{xy^2 \exp\left[\frac{2i}{c}\omega n_B d\right]}{1 - x^2 \exp\left[\frac{2i}{c}\omega n_B d\right]}, \\ g_t(x, y, d, \omega) &= \frac{y^2 \exp\left[\frac{i}{c}\omega n_B d\right]}{1 - x^2 \exp\left[\frac{2i}{c}\omega n_B d\right]}, \end{aligned} \tag{9}$$

are obtained using effective medium formalism in[20].

Initial conditions (expressions for R_0 and T_0) for these recurrent relations can be found at once if one looks at Fig. 1. Since the $N = 0$ structure is in fact a single layer, according to the Airy formulas

$$\begin{aligned} R_0(\omega) &= -r + \frac{rtt' \exp\left[\frac{2i}{c}\omega n_A \Delta_0\right]}{1 - r^2 \exp\left[\frac{2i}{c}\omega n_A \Delta_0\right]}, \\ T_0(\omega) &= \frac{tt' \exp\left[\frac{i}{c}\omega n_A \Delta_0\right]}{1 - r^2 \exp\left[\frac{2i}{c}\omega n_A \Delta_0\right]}. \end{aligned} \tag{10}$$

Here r, t, t' are the normal-incidence Fresnel's coefficients[26].

Now, to proceed with the analysis of scalability, we need to compare the following quantities

$$T_{N+1}\left(\tfrac{\omega}{3}\right) \text{ and } T_N(\omega). \tag{11}$$

Using the relation (5) for $\tilde{\Delta}_N$ and substituting (8) and (9) into (11), one can obtain

$$T_{N+1}\left(\tfrac{\omega}{3}\right) = \frac{T_N^2\left(\frac{\omega}{3}\right) \exp\left[\frac{i}{c}\frac{\omega}{3} n_B \cdot 3^N d_B\right]}{1 - R_N^2\left(\frac{\omega}{3}\right) \exp\left[\frac{2i}{c}\frac{\omega}{3} n_B \cdot 3^N d_B\right]}, \tag{12}$$

$$T_N(\omega) = \frac{T_{N-1}^2(\omega) \exp\left[\frac{i}{c}\omega n_B \cdot 3^{N-1} d_B\right]}{1 - R_{N-1}^2(\omega) \exp\left[\frac{2i}{c}\omega n_B \cdot 3^{N-1} d_B\right]}. \tag{13}$$

We see that the phase exponents in (12) and (13) are exactly equal, and the sole difference between $T_{N+1}\left(\frac{\omega}{3}\right)$ and $T_N(\omega)$ lies in the coefficients, $T_N\left(\frac{\omega}{3}\right)$, $R_N\left(\frac{\omega}{3}\right)$ and $T_{N-1}(\omega)$, $R_{N-1}(\omega)$, respectively. But if one expands these coefficients in the same way, using (8) and (9), one can see that the difference will again manifest itself only in the coefficients, this time, $T_{N-1}\left(\frac{\omega}{3}\right)$, $R_{N-1}\left(\frac{\omega}{3}\right)$ and $T_{N-2}(\omega)$, $R_{N-2}(\omega)$, respectively.

Tracing this procedure down to $N = 0$ and seeing that all frequency-dependent exponents that appear along the way are equal for both terms in (11), we finally reach the point where subsequent substitution of Eqs. (8) is no longer possible. At this point the factors to be compared are $T_1\left(\frac{\omega}{3}\right)$, $R_1\left(\frac{\omega}{3}\right)$ and $T_0(\omega)$, $R_0(\omega)$. The corresponding phase terms are $\exp\left[\frac{i}{c}\omega n_B\tilde{\Delta}_0\right]$ and $\exp\left[\frac{i}{c}\omega n_A\Delta_0\right]$ and they are also equal if the condition (1) is met. The further difference in coefficients

$$\begin{aligned}
&\frac{tt'\varepsilon'^3}{1-r^2\varepsilon'^6} \text{ vs. } \frac{\left(\frac{tt'\varepsilon'}{1-r^2\varepsilon'^2}\right)^2\varepsilon}{1-\left(\frac{rtt'\varepsilon'^2}{1-r^2\varepsilon'^2}-r\right)^2\varepsilon^2},\\
&\frac{rtt'\varepsilon'^6}{1-r^2\varepsilon'^6} \text{ vs. } \frac{rtt'\varepsilon'^2}{1-r^2\varepsilon'^2}+\frac{\left(\frac{rtt'\varepsilon'^2}{1-r^2\varepsilon'^2}-r\right)\left(\frac{tt'\varepsilon'}{1-r^2\varepsilon'^2}\right)^2\varepsilon^2}{1-\left(\frac{rtt'\varepsilon'^2}{1-r^2\varepsilon'^2}-r\right)^2\varepsilon^2}
\end{aligned} \tag{14}$$

is smaller as r decreases, and the agreement is total if $r^2 \approx 0$, which is not too far from the truth since for real materials r rarely exceeds 0.3.

As we have seen, all frequency-dependent exponents in the expressions (11) are equal at any stage of decomposition. So it can be concluded that the quantities in Eq. (11) have *identical phase structure*, with a minor difference in the coefficients. Since the characteristic spectral features (transmission resonances and local band gaps) are essentially phase phenomena (resulting from constructive and destructive interference, respectively), similar phase structure results in similar appearance of spectral portraits as confirmed by Fig. 2. Finally, since the phase structure coincidence has been derived on the basis of geometrical self-similarity, it follows that this self-similarity is responsible for the coincidence of the spectral shapes in Fig. 2.

Rigorous generalization of these results to all fractal multilayers is rather straightforward and can be achieved by further generalizing the Sun-Jaggard computation procedure, e.g., according to the multiple-reflection effective-medium formalism presented in our joint paper in[27].

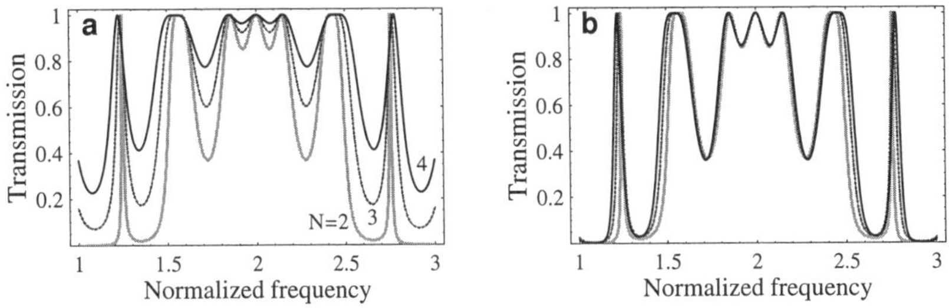

Figure 3. A closer look at the spectra in Fig. 2c,d,e: (a) as is; (b) modified by the power law (16).

3.3. *Amplitude Mismatch and the Power Law*

As have been mentioned earlier, a closer look at the spectra in Fig. 2 reveals that while the shapes of the spectra do coincide, there are still considerable differences, as can be seen in Fig. 3a. The most notable is the mismatch in the value of transmission coefficient in the areas *between* characteristic spectral features, i.e., transmission peaks and band gaps.

We have found that these discrepancies can be eliminated simply by introducing a *power law* in addition to frequency scaling in Eq. (7):

$$|T_{N+1}(\omega/G)|^{\gamma} = |T_N(\omega)|, \tag{15}$$

which has the meaning of a *scalability equation* for the quantities in Eq. (11). The confirmation is presented in Fig. 3.

The power γ was numerically found to vary slightly with frequency. However, in the vicinity of even multiples of ω_0, where scalability is most often observed, we can treat γ as approximately constant. Further investigations have shown that γ depends on the structural parameters, and this dependence takes the form

$$\gamma = \alpha + \tfrac{1-\alpha}{f^2}, \ f \equiv \tfrac{G-C}{G}, \tag{16}$$

Here, f is the dielectric filling fraction in the transition to the next generation ($N \to N+1$). Small parameter α has been found to be around 0.1. The numerical data for γ along with the approximation (16) is shown in Fig. 4).

It is an important result that γ depends *only* on the ratio between $G - C$ and G, (the dielectric filling fraction), so such structures as $(6, \{2,3\}, N)$, $(6, \{3,4\}, N)$, $(6, \{1,4\}, N)$ have the same γ. The fact that it does not depend on the position of "removed" layers, nor on the structure's

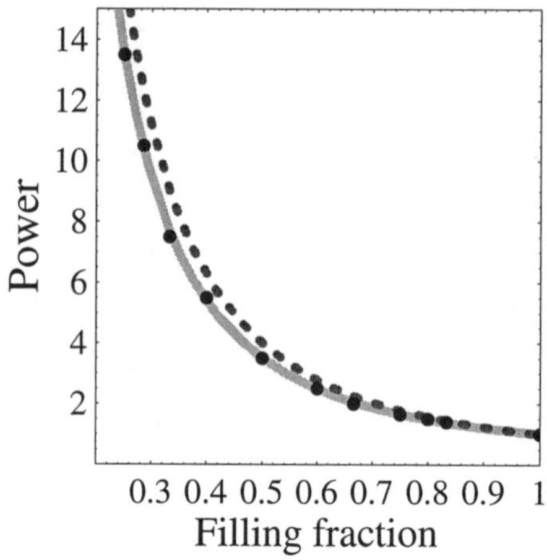

Figure 4. The dependence $\gamma(f)$: dots are numerical data, solid line is the numerical fitting (16), and the dashed line is the analytically obtained relation (18). dashed

lacunarity, nor on the value of n_B/n_A, suggests that *only the amount of dielectric "removed" during the transition* $N \to N+1$ *is important for* γ.

This invariance can also be explained physically if we mention that in the frequency region of interest the propagating wave exhibits little phase change and hence little internal reflection. Therefore, the influence of multiple reflections in negligible and thus each layer acts as an attenuator of the wave amplitude, and each layer has equal contribution, and hence the dependence on the *number* of substructures $G - C$. The exception to this reasoning is the case when there is a resonant, coherent interaction giving rise to a transmission peak or a local band gap. In this case multiple reflections are no longer neglected, the structure topology (and hence the position of digits in **C**) becomes important, and the power law (15) becomes irrelevant since in the gap $0^\gamma = 0$ and at the peak $1^\gamma = 1$ for any γ. Therefore we can state that the scalability effect can be separated into a *phase* (coherent) scaling where topology and far order becomes paramount, and an *amplitude* (incoherent) interaction between characteristic spectral features where the power law takes place and statistical parameters such as f are important.

So we can conclude that the only important parameter in f [see Eq. (15)] is its numerator $G - C$, while the sole part of the denominator G is to allow the frequency scaling (7). That's why it is enough to investigate the simplest configuration to establish analytically the function $\gamma(f)$. Instead of an arbitrary $(G, \mathbf{C}, N)$ structure, consider an equivalent one $(G, \mathbf{C}' = \{G - C, \ldots, G - 1\}, N)$, which is effectively a single layer whose thickness scales as

$$d_N = (G - C)^N d_A. \tag{17}$$

Using the Airy formulas for the transmission we can rewrite the scalability equation (15) in the form

$$\frac{1 + r^4 - 2r^2 \cos\delta}{1 + r^4 - 2r^2} = \left(\frac{1 + r^4 - 2r^2 \cos f\delta}{1 + r^4 - 2r^2}\right)^\gamma$$

where $\varepsilon' \equiv e^{i\delta} = \exp[(G - C)^N \frac{\omega}{\omega_0}\pi/2]$.

Since we are close to $\eta = 0$, we can assume $\delta \ll 1$. At the $2nd$ order of Taylor series of $\cos\delta$ (the first order obviously leads to an identity $1 = 1^\gamma$), we finally get

$$1 + \frac{2r}{(1 - r)^2}\delta^2 = 1 + \gamma\frac{2r}{(1 - r)^2}(f\delta)^2 \Rightarrow \gamma = \frac{1}{f^2}, \tag{18}$$

which has a good agreement with numerically obtained Eq. (16) (see Fig. 4). An even better analytical agreement can be achieved using a finer approximation for $\cos\delta$. This also results in a weak dependence $\gamma(\eta)$, as was numerically confirmed and found not significant for the observation of scalability.

3.4. *Perturbations Due to Finite Size*

So far we have shown that both phase structure matching and amplitude matching can be derived analytically, and in both cases the primary reason is the self-similarity of the structure. However, in Fig. 3b one can still observe small mismatches in the resonance peak locations for the spectral curves in (15). This agrees with the difference in the coefficients for the quantities (14) and shows that spectral scalability is only *approximate* in real multilayers.

However, we state that this perturbation results from the *finite size* of the structures under study. So, they are in fact prefractals rather than true fractals, so spatial self-similarity in them is not exact. It is a reasonable outcome that spectral scalability becomes not exact in finite-size multilayers. This perturbation of scalability is much the same as in other types of media. For example, finite periodic structures cannot exhibit completely zero transmission in the band gaps, and in finite disordered media light cannot be completely trapped, and self-similarity in finite

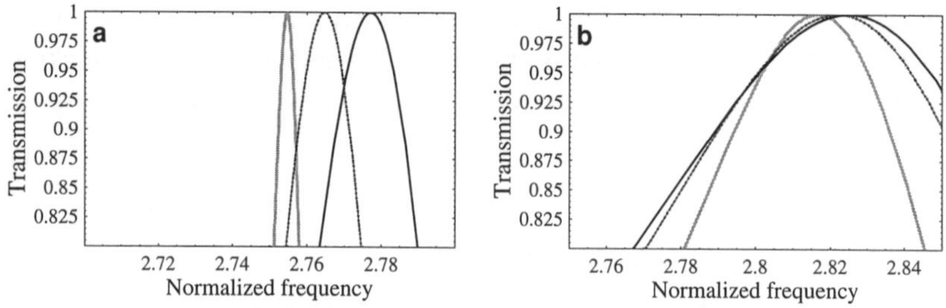

Figure 5. Enlarged part of a spectrum similar to that in Fig. 3b: (a) $n_A/n_B = 3$; (b) $n_A/n_B = 1.5$. It can be seen that in the latter case the peak mismatch is smaller

quasiperiodic (QP) structures is also just approximate. In this manner, N*th* generation Cantor multilayers can be compared to N-period 1D photonic crystals, while it is commonly known that band gaps are prominent with much larger N than were used for the plots in Figs. 2 and 3.

However, under certain conditions, one can observe decent band gaps even in periodic multilayers with as many as four periods. An analogous statement is true for scalability in fractal multilayers. But the condition to be desired is opposite. As was noted earlier and confirmed in numerical calculations, the peak mismatch decreases if the refractive index contract is smaller (see Fig. 5), while band structure is more pronounced if the contrast is large enough[1,2].

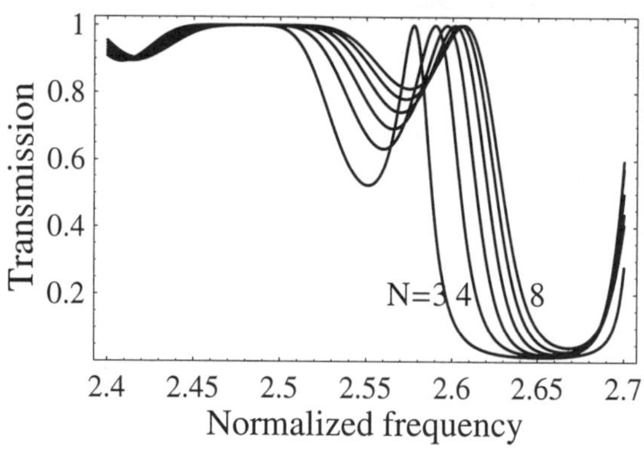

Figure 6. An enlargement of a small part of the $(3, \{1\}, N)$ spectra. N ranges from 3 to 8. The peak mismatch is smaller as N is larger.

Violating the quarter wave condition (1) destroys the similarity of phase structures, so it plays an important part in the manifestation of characteristic spectral properties in both fractal and periodic multilayers.

From these considerations it is likely that if N approaches infinity, spectral scalability will be exact. This can be indirectly confirmed by plotting the scaled spectra for several generations. As seen in Fig. 6, the mismatch goes smaller as N grows larger.

4. Other spectral properties of fractal multilayers

4.1. *Resonance peak splitting*

Note (see Fig. 2) that some resonances in the Cantor stack spectra turn out to be single peaks, while other are multiplets (e.g., doublets). This is a manifestation of another property of fractal multilayers, which is worth considering.

We have found a $(G,\ \mathbf{C},\ N+1)$ stack spectrum has multiplets at the frequencies where the $(G,\ \mathbf{C},\ N)$ stack spectrum has single peaks, while new peaks (singlets or multiplets) in a $(G,\ \mathbf{C},\ N+1)$ correspond to no peaks in a $(G,\ \mathbf{C},\ N)$ and "appear" in the forbidden gaps of the latter structure as N is increased. The number of components in the multiplets equals $(G+1)/2$, and, of course, the total number of peaks in one period equals the number of layers, i.e., G^N. As we see in Fig. 7 for $(3,\ \{1\},\ N)$ and $(5,\ \{1,3\},\ N)$, splitting occurs each time we increase the number of generations. That is why we have termed this property *sequential splitting.*

Numerical investigations show that sequential splitting is apparently observes in Cantor-like multilayers $(3,\ \{1\},\ N)$, $(5,\ \{1,3\},\ N)$, $(7,\ \{1,3,5\},\ N)$ and similar structures where previous-generation substructures and largest-size layers alternate with each other. As regards other fractal multilayer topologies, splitting is still pronounced but there is no exact distinction between split and new resonances, neither are there strict relations governing their multiplicity. Therefore we confine ourselves to Cantor-like structures and so we can omit $\mathbf{C}$ in the structure signature in this section.

The reason of such splitting is understood from the point of view of self-similarity of Cantor structures regarded in terms of coupled cavities. As the Cantor algorithm generates a fractal structure, an N_0^{th} generation stack contains all the structures with $N < N_0$ and the same G. For example, a $(G=3,\ N=4)$ stack contains two inclusions of $(G=3,\ N=3)$ stacks. Then we can make a statement that each single peak in the spectrum of a $(3,\ 4)$ stack, being a resonant mode introduced by the central "cavity"

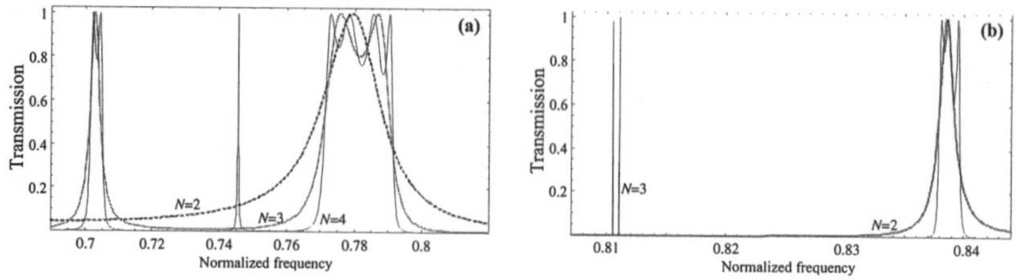

Figure 7. Sequential splitting for $G = 3$ **(a)** and $G = 5$ **(b)**. Note the number of components in each multiplet is equal to that given by (19). Also note that the half-width of a resonance is almost the same before and after splitting.

(or layer), splits into a doublet because there are two such cavities in a (3, 5) stack. Two new peaks have to appear in the (3, 5) spectrum, corresponding to the central cavity in this structure, and those peaks must be single, as there is only one such cavity present. It is obvious that the same reasoning applies to any generation, so all single peaks in the spectra of $(3,\ N)$ structures will split into doublets in the $(3,\ N+1)$ structures' spectra. Analogous considerations lead to the conclusion that in the case of $G = 5$ single peaks will split into triplets and new resonances will appear as doublets, in full accordance with Fig. 7b. Generalization of this statement results in a formula for the number of components S_{split} in multiplets that have just split and for the number of components S_{new} in newly appearing multiplets, respectively

$$S_{split} = \frac{G+1}{2} \quad \text{and} \quad S_{new} = \frac{G-1}{2}. \tag{19}$$

These relations state another connection between the geometrical and spectral properties of Cantor multilayers. Besides, note the half-width of a peak is almost unchanged during splitting, while the external "slopes" of the whole multiplet become steeper (see Fig. 7a), and its envelope can be roughly characterized as "rectangular". This effect can be made use of in optical band pass filters.

As an additional confirmation of the above mentioned correlation, we have plotted the field intensity profiles ($|\mathbf{E}_\tau|^2$, since it is this field component that undergoes no jumps at layer interfaces) for the frequencies that correspond to the resonance peaks (components of both split multiplets and new ones). The field intensity for a multiplet is shown to be actually localized in the regions (cavities) that have been claimed responsible for the resonance peak at the frequency in question (see Fig. 8). Thus, for the newly appeared multiplets the energy is strongly localized in newly appearing

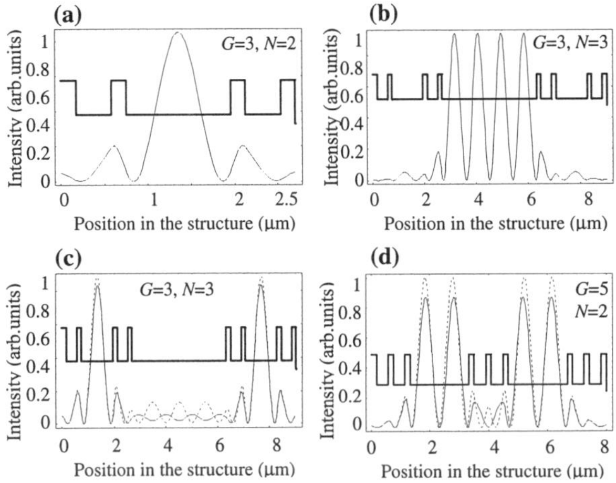

Figure 8. The field intensity profiles for a newly-appeared single peak in a $(3, \{1\}, 2)$ **(a)** and $(3, \{1\}, 3)$ **(b)**, for a split doublet in a $(3, \{1\}, 3)$ **(c)**, and for a newly-appeared doublet in a $(5, \{1, 3\}, 2)$ **(d)**.

layers (binding areas). In split multiplets there is a pronounced localization in the previous generation substructures. As regards the difference between the field profiles for the individual multiplet components, it is only observed *outside* (and between) localization areas. This indicates that, for instance, in split multiplets the substructures are responsible for the peak itself while the binding causes it to split. Note that this splitting property is essentially the same as energy level splitting in adjacent quantum wells and bound pendulums in mechanics, as well as resonance splitting in connected electrical LC-circuits in electricity.

4.2. *Spectral Self-Similarity*

We have shown earlier[3] that as N increases, new resonance peaks appear between every two existing peaks. This can be seen most clearly on the *resonance map* of the structure, i.e., a distribution of resonance peaks along the spectrum in dependence on the generation number (see Fig. 9). This means that transmission bands can never constitute a Cantor set the way it takes place in QP structures[10,12]. However, we also noted[28] that the spectra themselves do possess fractal properties. In this paper, we report that fractal multilayers exhibit spectral self-similarity, as can be seen in Fig. 10.

Spectral self-similarity present in fractal structures puts them in a certain relation with QP multilayers. However, there are a number of important differences between self-similarity in these two kinds of media. The first and the most important, in QP structures the transmission bands (i.e., the pattern of transmission *peaks*) are self-similar, while in fractal structures self-similarity is in the envelope, i.e., in the pattern of transmission *valleys*. Secondly, self-similarity here is around even multiples of ω_0 rather than around odd ones as it takes place in QP structures. Thirdly, the scaling factor is rational (and equal to G), again unlike QP multilayers. Finally, for spectral self-similarity in fractal structures to better manifest itself, it is necessary to introduce the *power law* similar to Eq.(15). Empirically the best value for the power was found to equal 2γ [see Eq. (16)].

Conclusions

To summarize, we have used the method that inherently contains spatial self-similarity[20] along with scaling relations (7) and (17), to find that fractal multilayers exhibit scalability both in the phase and in the value of transmittance according to the Eq. (15). So, it can be concluded that *spectral scalability actually results from spatial self-similarity*, and moreover, it is a *characteristic relation* between a topological property of a multilayer structure and a spectral property of wave propagation.

Besides, we have investigated that *fractal multilayer structures exhibit sequential transmission peak splitting* and reported that *spectral self-similarity is present in fractal multilayers.* We have also outlined similarities and differences with the same spectral effect in QP multilayers. However, the precise relations for this self-similarity, along with its connection with geometrical fractality, is yet to be understood and requires further investigation.

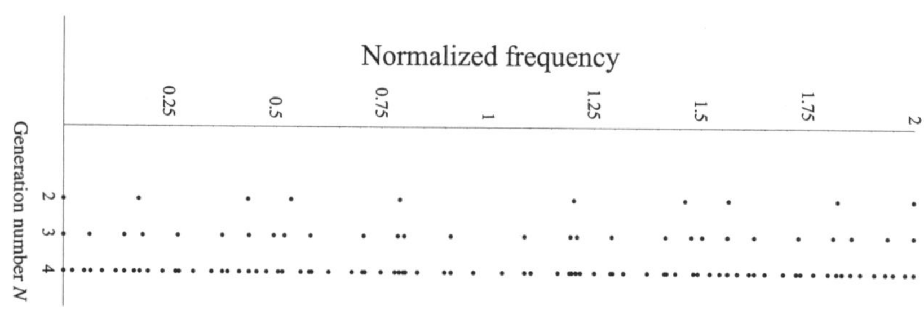

Figure 9. The resonance map (frequencies of resonance peaks versus generation number) for a Cantor structure with $G = 3$. One can see that new dots appear in a definite location, but seem to cover the whole frequency range. Thus, the problem of self-similarity of the transmission band spectra for Cantor structures requires further investigations.

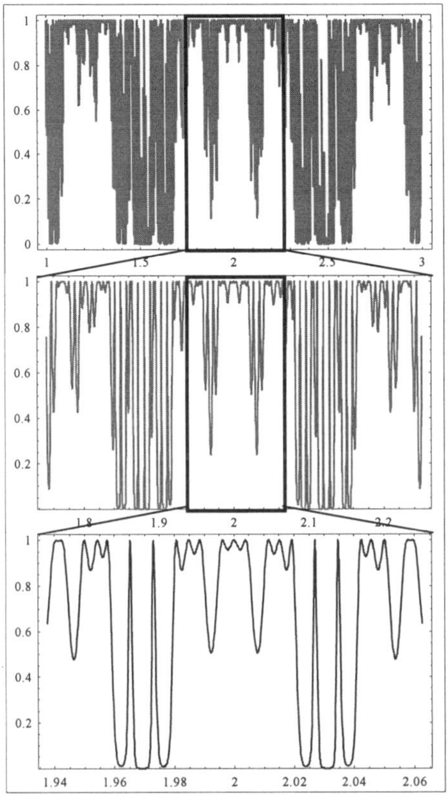

Figure 10. Self-similarity in fractal multilayers for structures $(4, \{1,2\}, 5)$. The central parts are scaled by 4 times along the frequency axis and raised to a power $2\gamma = 7$.

Although this review has been written with optical properties in mind, all the results are equally applicable for the electronic spectra in a fractal potential, in accordance to the analogy between electromagnetic wave properties and quantum particle behavior, which dates back to Feynman[29]. The results obtained are analytical; however, earlier report[3] demonstrated a good agreement between our numerical results and experiment. However, the experimental fabrication of fractal multilayers with material properties adequate for the observation of the effects described in this review still remains a challenge.

The authors acknowledge I. S. Nefedov, who first introduced to us the ideas of making use of spatial self-similarity in calculations, as well as helpful discussions with A. Smirnov and G. Krylov. This work was supported in part by INTAS (2001-0642), the NATO for Peace Project, and the Basic Research Foundation of Belarus (grant no. F03M-097).

Bibliography

1. A. Yariv and P. Yeh, *Optical waves in crystals* (Wiley, New York, 1984).
2. P. Yeh, *Optical waves in layered media* (Wiley, New York, 1988).
3. A. V. Lavrinenko, S. V. Zhukovsky and et al, *Phys. Rev.* **E65**, 036621 (2002).
4. A. Lavrinenko, P. I. Borel, L. H. Fradsen, M. Thorhauge, A. Harpoth, M. Kristensen, T. Niemi and H. M. H. Chong, *Opt. Express* **12**, p. 234 (2004), http://www.opticsexpress.org/abstract.cfm?URI=OPEX-12-2-234
5. K. Busch, S. F. Mingaleev and et al, *J. Phys.: Condens. Matter* **15**, R1233-R1256 (2003).
6. J. D. Joannopoulos, R. D. Meade, J. N. Winn, *Photonic Crystals: Molding the Flow of Light* (Princeton Univ. Press, Princeton, 1995).
7. P. Anderson, The quastion of classical localization: A theory of white paint?, *Philos. Mag.* **B52**, No. 3, p. 505 (1985).
8. D. S. Wiersma, P. Bartolini and at al, *Nature* **390**, 671 (1997).
9. M. Kohmoto, B. Sutherland and C. Tang, *Phys. Rev.* **B35**, 1020 (1987).
10. M. Kohmoto, B. Sutherland and K. Iguchi, *Phys. Rev. Lett.* **58**, 2436 (1987).
11. E. Maciá and F. Domìniguez-Adame, *Phys. Rev.* **B50**, 16856 (1994).
12. M. S. Vasconcelos and E. L. Albuquerque, *Phys. Rev.* **B57**, 2826 (1998).
13. M. S. Vasconcelos, E. L. Albuquerque and A. M. Mariz, *J. Phys.: Condens. Matter* **10**, 5839 (1999).
14. R. W. Peng, Mu Wang and at al, *Phys. Rev.* **B57**, 1544 (1998).
15. X. Huang, Y. Wang and C. Gong, *J. Phys.: Condens. Matter* **11**, 7645 (1999).
16. E. Maciá, *Phys. Rev.* **B60**, 10032 (1999).
17. R. Merlin, K. Bajema and et al, *Phys Rev. Lett.* **55**, 1768 (1985).
18. J. Todd, R. Merlin and et al, *Phys Rev. Lett.* **57**, 1157 (1986).
19. J. Feder, *Fractals* (Plenum Press, New York, 1988).
20. D. L. Jaggard and X. Sun, *Opt. Lett.* **15**, 1428 (1990).
21. X. Sun and D. L. Jaggard, *J. Appl. Phys.* **70**, 2500 (1991).
22. D. L. Jaggard and A. D. Jaggard, *Opt. Lett.* **22**, 145 (1997).
23. C. Sibilia, I. S. Nefedov and et al, *J. Opt. Soc. Am.* **B15**, 1947 (1998).
24. K. S. Sandomirskii, S. V. Gaponenko and et al, *Proc. SPIE* **4748**, p. 443 (2002)
25. S. V. Zhukovsky, A. V. Lavrinenko and S. V. Gaponenko, *Proc. SPIE* **4705**, p. 121 (2002).
26. S. V. Zhukovsky, A. V. Lavrinenko and S. V. Gaponenko, in *Proc. 4th INTON and ESPC, Vol. 2*, (Warsaw, 2002), p. 74.
27. A. Mandatori, C. Sibilia and et al, *J. Opt. Soc. Am.* (2003, in press).
28. S. V. Zhukovsky, S. V. Gaponenko and A. V. Lavrinenko, *Nonlinear Phenomena in Complex Systems* **4**, 383 (2001).
29. R. P. Feynman, *The Feynman Lectures on Physics*, Vol. 3, Ch. 14 (London, 1963).

PROPAGATION OF LIGHT IN PERIODIC LAYERED STRUCTURES AND WAVEGUIDES BASED ON PHOTONIC CRYSTALS

D. N. Chigrin†, A. V. Lavrinenko†‡¶, D. N. Shyroky‡, V. V. Zhilko‡

†*Institute of Material Science, University of Wuppertal, Germany,*
‡*Belarusian State University, Minsk, Belarus,*
¶*Institute of Molecular and Atomic Physics, National Academy of Science, Minsk, Belarus*

A range of optical phenomena occurring when light is propagating through periodically organized dielectric media is presented. The total omnidirectional reflection from 1D periodic media is explained and characterized. The splitting of photonic band gaps stemming from natural or induced anisotropy of a dielectric is shown to have perspective applications in controlling polarized light. The local density of modes is introduced for the case of anisotropic periodic media. While treating periodic layered media in a long wavelength approximation effective material tensors are introduced by operator methods. The waveguiding effect in 2D photonic crystals is numerically analyzed by full-vectorial FDTD technique.

1. Introduction

Periodic media are well acknowledged for their capability to control the propagation and emission of electromagnetic waves and have gained a substantial attention as *photonic crystals* (PhC) or *photonic bandgap structures* (PBG).

The first self-consistent treatment of light emission in a periodic medium with a strong modulation of the dielectric function was made by Bykov in 1972[1]. Bykov pointed out the possibilities to realize a complete photonic bandgap and the inhibition of the spontaneous emission of atoms embedded in a periodic medium. The first self-consistent treatment of electromagnetic eigenmodes and photonic band structure in a three-dimensional periodic medium with a large dielectric function modulation was given by Ohtaka in 1979[2]. These two pioneering works did not gain the attention they deserved at the time. It was only after the appearance of the papers of Yablonovitch[3] and John[4] in 1987, than the idea to control the flow of light by means of a periodic medium become popular. Yablonovitch[3] proposed to use a three-dimensional periodic medium, which he called a *photonic crystal*, to inhibit the spontaneous emission of atoms and to realize localized defect modes and consequently to enhance the spontaneous emission. In the same year, John[4] proposed the use of a disordered three-dimensional periodic medium to localize electromagnetic waves. These ideas stimulated research area, which leads both to various unexpected results in the fundamental understanding of light–matter interaction and to various new optoelectronics and photonics applications. Summaries of early studies can be found in[5].

To mention a potential PhCs applications an integrated optical circuit can be considered. A typical passive integrated optical circuit incorporating waveguides, bends, junctions and couplers extends over several centimeters rather than micrometers in size. Wave guided light does not readily negotiate sharp bends, so transitions need to be gradual. Couplers typically have long interaction lengths with respect to the wavelength of light. Photonic crystals possessing a complete PBG offer one of the possible solutions to miniaturization of integrated optics. Optical waveguides, sharp bends and large-angle Y-junctions with virtually no excess loss have been designed in the micrometer and sub-micrometer length scale using PhCs[5]. Another important issue for integrated optics is filtering. Photonic crystals are proven to be good materials for realization of high quality resonators with very sharp filtering characteristics[5]. A compact PhC based add/drop filters suitable for wavelength division multiplexing applications have been proposed based on photonic crystal waveguides (PCW) and resonators.

Although the main expectations associated with PhCs rely on the existence of a complete photonic bandgap, photonic crystals with an incomplete photonic bandgap can also influence dramatically the electromagnetic wave propagation. For example, in Sec. 2 we consider a 1D periodical multilayer dielectric stacks exhibiting remarkable properties of omnidirectional reflection for arbitrary polarized electromagnetic waves. In Sec. 3–4 we consider 1D periodic media based on anisotropic constituent materials. As an analytical approach the free-coordinate method manifested in optics by Academician F.I.Fedorov[6] has been employed. Another example of the system with incomplete 3D PBG is considered in Sec. 5. In this section a systematic FDTD study of 2D PhC waveguides is presented.

2. One-dimensional photonic crystals. Isotropic media

Probably one of the most widespread application of one-dimensional periodic media is a Bragg mirror. Mirrors come in two basic varieties: a metallic mirror with dissipative losses of few-percent, being in fact an omnidirectional reflector, and a dielectric Bragg mirror. A dielectric Bragg mirror can be made nearly lossless, but it is typically highly reflecting only within a limited angular range. A structure combining the properties of both mirror types, i.e., being omnidirectional and lossless, is of strong interest as it is likely to find many applications in optoelectronics and all-optical systems. Until recently, the possibility to design such a "perfect mirror" was exclusively associated with 3D PhCs having a complete PBG. Recently has been reported, that a simple-to-fabricate Bragg mirror suffices to design a low-loss omnidirectional reflector[7–11].

The projected band structure of an infinite periodic layered medium is depicted in Fig. 1. The refractive indices, $n_1 = 1.4$ and $n_2 = 3.4$, are chosen close to those of SiO_2 and Si in the near IR region. The layers have equal thickness ($d_1 = d_2$), which corresponds to the filling fraction $\eta = 0.5$, where the filling fraction is defined via $\eta = d_2/(d_1 + d_2)$. An infinite periodic structure can support both propagating and evanescent Bloch waves. In Fig. 1, gray areas correspond to the propagating states, whereas white areas contain the evanescent states only and are referred to as photonic bandgaps.

When the frequency and the wave vector of a wave, impinging externally at an angle, $\alpha_{\rm inc}$, from a homogeneous medium of refractive index n onto a one-dimensional periodic medium lies within the bandgaps, the incident wave undergoes strong reflection. Pronounced high reflection bands depend strongly on frequency and angle of incidence.

These can be easily understood from Fig. 1. PBGs *(i)* rapidly move to higher frequencies with increasing incident angle, denoted by the increase of the tangential component of the wave vector, and *(ii)* the TM band gaps tends to zero when approaching the Brewster light-line, where $\omega = c\left|\mathbf{k}_{\|}\right|/n_1 \sin\alpha_B$ (Fig. 1), and $\alpha_B = \arctan n_2/n_1$ is the Brewster angle at $n_1 - n_2$ interface. These properties of the band structure restrict the angular aperture of a polarization insensitive range of high reflectance.

In essence, omnidirectional reflectance can be achieved due to the limitation of the number of modes that can be excited by externally incident waves inside the Bragg mirror. Light coming from the low-index ambient medium ($n < n_1$, n_2) is tunneled into the internal cone narrowed by Snell's

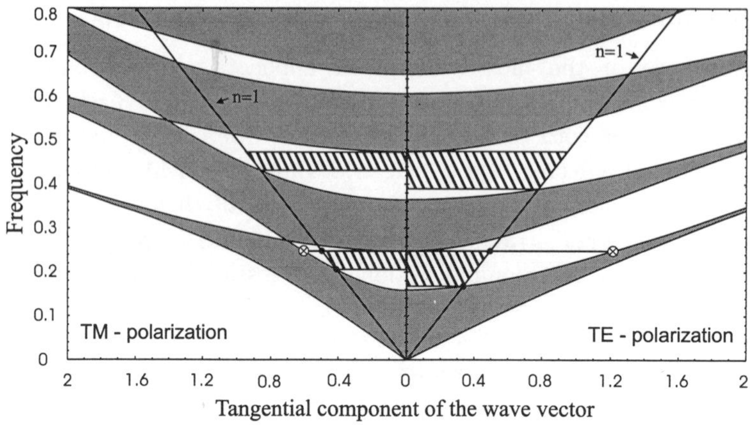

Figure 1. Projected band structure of a typical 1D photonic crystal for TE (right panel) and TM (left panel) polarizations. The frequency and the tangential component of the wave vector are normalized as $\omega\Lambda/2\pi c$ and $\left|\mathbf{k}_{\|}\right|\Lambda/\pi$, respectively. The hatched areas correspond to omnidirectional reflection bands. The solid lines are the ambient-medium light-lines

law. Then, *(i)* a sufficiently large contrast between refractive indices of layers constituting the periodic medium with respect to a refractive index of an ambient medium ensures that light coming from the outside will never go below the Brewster's angle inside the crystal and *(ii)* a sufficiently large refractive index contrast of the layers themselves can keep the bandgaps open up to grazing angles[7–11].

A reduced region of **k**-space, where propagating Bloch modes of the photonic crystals can be excited by externally incident wave, lies above the light-line (Fig. 1), which is a projection of the band structure of an ambient medium on the $\omega\left|\mathbf{k}_{\|}\right|$ plane. For example, in the case of the Si/SiO_2 periodic medium situated in air the first two bandgaps are open for all external angles of incidence (hatched areas in Fig. 1). No propagating mode is allowed in the stack for any propagating mode in the ambient medium for either polarizations. This is how total omnidirectional reflection arises.

To design an omnidirectional mirror, one needs to have total reflection for all incident angles and all polarization states. The TM photonic band gap is narrower than the TE one and this determines the bandwidth of an omnidirectional reflection band. The upper edge of the reflection band corresponds to the upper edge of the forbidden gap at normal incidence. The lower edge is defined by the intersection of the ambient-medium light-line with the upper edge of the corresponding TM band (Fig. 1)).

For given parameters of the periodic structure, the refractive index of an ambient medium may be used to control the bandwidth. By increasing the refractive index of an ambient medium from $n = 1$ to some $n = n_{\max}$ the bandwidth decreases till an omnidirectional reflection band is closed. The filling fraction $\eta = d_2/\Lambda$ optimizes the relative reflection bandwidth $\Delta\omega/\omega_0$ of an omnidirectional reflection band with respect to the given refractive indices of the layers constituting one-dimensional periodic medium and the index of an ambient medium. We found that a quarter-wave stack in not the optimal configuration to achieve the maximal bandwidth. Contour plots resembling different dependencies of the relative bandwidth on the parameters of the stack have been presented in paper[12].

For the SiO_2/Si ($n_1 = 1.4$, $n_2 = 3.4$) structure in air ($n = 1.0$) the omnidirectional reflection band is centered at the normalized frequency $\omega\Lambda/2\pi c = 0.275$ with the optimal filling fraction $\eta_{opt} = 0.324$. The relative bandwidth is about 25%. For example, to obtain an omnidirectional reflection centered at the wavelength $\lambda = 1.5\,\mu m$, the period of the structure should be about $0.412\,\mu m$.

Theoretical predictions were checked at the Institute of Molecular and Atomic Physics, Minsk, Belarus[10,12,13]. A Bragg mirror in the optical wavelength range was chosen for testing. A lattice consisting of 19 layers of Na_3AlF_6 and ZnSe ($n_1 = 1.34$ and $n_2 \in [2.5, 2.8]$ correspondingly) was fabricated by standard optical technology using layer by layer deposition of the materials on a glass substrate. The multilayer stack was terminated at both ends with ZnSe layer. The thickness of each layer was $d_1 = d_2 = 90$ nm. An omnidirectional total reflection was expected within the spectral band $\Delta\lambda \in [604.3, 638.4]$ nm with a relative bandwidth of 5.3%.

The dependence of the transmission coefficients for TE- and TM-polarized incident radiation of a He-Ne laser at 632.8 nm on angle of incidence is presented in Fig. 2. For TM-polarization circles depict the directly measured transmission coefficient, and squares depict data obtained from reflection measurements. Mismatch between them can be explained by additional reflection from air–ZnSe, ZnSe–substrate and substrate–air interfaces. The solid (dashed) curve gives theoretically calculated transmission coefficients for TE-(TM-) polarized light. For TE-polarization (triangles) the signal at angles larger than 60° is out of the sensitivity range of the experimental set-up. The dotted curve corresponds to the transmission coefficient of the structure terminated at both ends with a low index layer, Na_3AlF_6, for TM-polarized light.

As can be seen in Fig. 2, the transmission coefficient of TM-

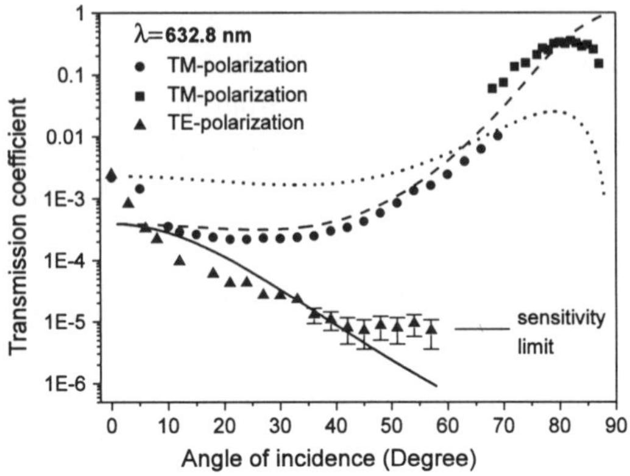

Figure 2. Dependence of the transmission coefficient of a Na_3AlF_6/ZnSe structure on angles of incidence at 632.8 nm (HeNe laser) for two polarizations of incident light.

polarized radiation remains below 10^{-3} over a wide angular range. Due to the Brewster effect at the air–ZnSe, ZnSe–substrate and substrate–air interfaces at large angles, it increases to 0.33 at 80° and then decreases again. In contrast, the transmission of TE-polarized radiation decreases monotonically with growing angle of incidence being less than 10^{-5} for angles larger than 40°. Transmission coefficients of less than 10^{-5} are beyond the capabilities of the experimental set-up used. For this reason, the transmitted signal at angles of incidence larger than 60° cannot be detected and therefore no data points for TE-polarization at these angles are presented in Fig. 2.

The reflectivity of TM-polarized radiation at large angles can be enhanced in structures terminated at both ends with a low index layer, Na_3AlF_6, (Fig. 2). In this case, the transmission at 80° is as small as 0.03, corresponding to a reflection coefficient of 97%. The overall reflectivity can be enhanced in structures with larger number of layers.

Since the first prediction and demonstration of the omnidirectional reflection from a one-dimensional periodic medium[7–11], several further developments have been made. The idea of omnidirectional mirror has also obtained a favorable press coverage over last years. The total number of publication devoted to this subject exceeds one hundred and many new publications appear every year. Some examples of the state-of-the-art applications of this idea comprise an improved planar microcavity, optical fibers and Fabry-Perot resonator.

3. One-dimensional photonic crystals. Anisotropic media

The usage of anisotropic media in structure composition of PhC leads to the splitting of a band structure. These band structures correspond to the eigen polarizations of the whole period. To simplify the consequent analysis we chose the system with only one anisotropic layer. Thus eigenmodes of this layer coincide with the eigenmodes of the full period.

Let us consider the 1D periodic medium, consisting of two kinds of layers: crystal d thick plates, described by tensor of dielectric permittivity ε, separated by b-thick isotropic layers ($p = b+d$ — is a period of considered 1D medium). By using of transfer matrix method we managed to obtain dispersion equation[14,15]

$$[|TL - \exp(i\mu p)E| = 0] \tag{1}$$

where

$$T = \begin{pmatrix} \exp(ikp)I & 0 \\ 0 & \exp(-ikp) \end{pmatrix}, \quad E = \begin{pmatrix} I & 0 \\ 0 & I \end{pmatrix} \tag{2}$$

are 4×4 block operator matrices. This equation is valid for any kind of natural or forced anisotropy of crystal.

For example, let's consider bilayered periodic structure, one layer of which is uniaxial. After some cumbersome calculations we derived the dispersion equation consisting of two branches, corresponding to separated o- and e-waves. These equations finally are transformed to standard form

$$\cos(\mu p) = \cos(\phi_\pm)\cos(kb) - 1/2\sin(\phi_\pm)\sin(kb)(n/n_\pm + n_\pm/n) \tag{3}$$

where $\phi_\pm = kn_\pm d$, $n_- = \sqrt{\varepsilon}$, $n_+ = \sqrt{\varepsilon_o\varepsilon_e/(\varepsilon_e - \Delta\varepsilon\sin^2\theta)}$, $\Delta\varepsilon = \varepsilon_e - \varepsilon_o$, ε_o, ε_e are the principal values of tensor ε in common notation, angle θ is an angle between vector of wave normal and direction of optical axes, μ is length of Bloch's vector. When right-hand side of the dispersion equations is greater than 1 or less than -1, wave number becomes complex. The waves are evanescent and cannot propagate in such periodic structure. So the photonic band structure is formed. A plot of the o- and e- first band gaps for a filling factor $d/p = 0.5$ versus principle values of permittivity is shown on Fig. 3.

The very intriguing are the places, where zones of nearly total transmission for one polarization superimpose with PBG for the other, e.g. see their disposition on line B. For confirmation of our guess we calculated transmission spectra for o- and e- waves (Fig. 4), using Fresnel's operator technique[14,15]. These spectra showed the shift of transmission

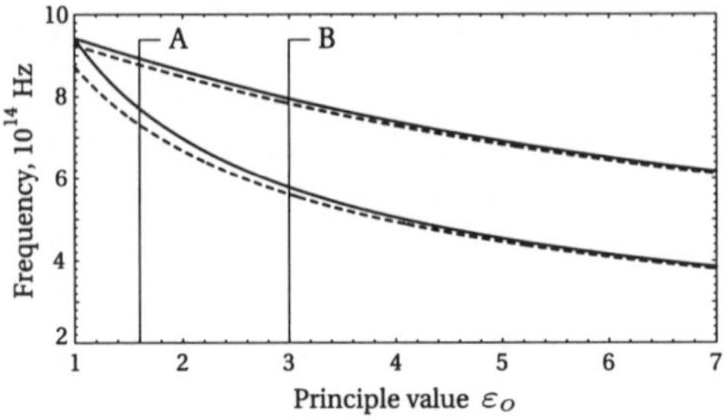

Figure 3. The position of the first PBG for 1D anisotropic PhC as function of the ordinary principal value of the dielectric permittivity tensor. The difference between the principal values is fixed and equals 0.2. All layers have equal thicknesses. Solid line — ordinary wave PBG, dashed line — extraordinary wave PBG. Optical axis lies in the plane interface.

bands. Moreover, one can find the frequencies on which the transmission maximum for e-wave coincides with the transmission minimum of o-wave. This fact may be exploited for constructing efficient switches and filters for electromagnetic radiation based on element, which is capable to change the state of polarization.

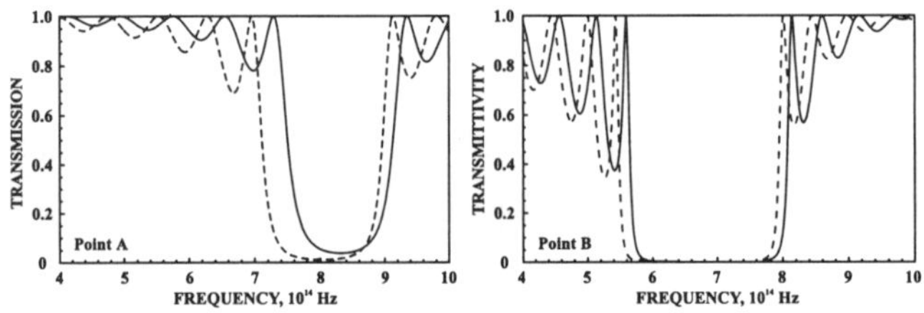

Figure 4. Transmission spectra for the 1D anisotropic PhC. Left panel — for dielectric constant 1.6 (line A); right panel — for dielectric constant 3.0 (line B). Solid line denotes o-wave spectrum, dashed line — e-wave spectrum.

Another interesting region is lying on the line A on the Fig. 3. Here the possibility exists of polarization control of radiation by changing of dielectric permittivity. The electrooptic effect (both linear and quadratic) or Frederics transition in nematic liquid crystals are apt for such processes. We investigated the system consisted of 10 periodically arranged nematic liquid crystal cell (Fig. 5, left). Liquid crystal is PCB, all layers have equal thickness 0.5 microns. Transmission spectra are on the right side of Fig. 5, the initial one marked by solid line and the one after Frederics transition — by dashed line[16]. On the certain frequencies transmission coefficient has been drastically changed from 0.99993 to less -60 dB after the changing in the optical axis orientation. So this effect can be applied in polarization switches devices.

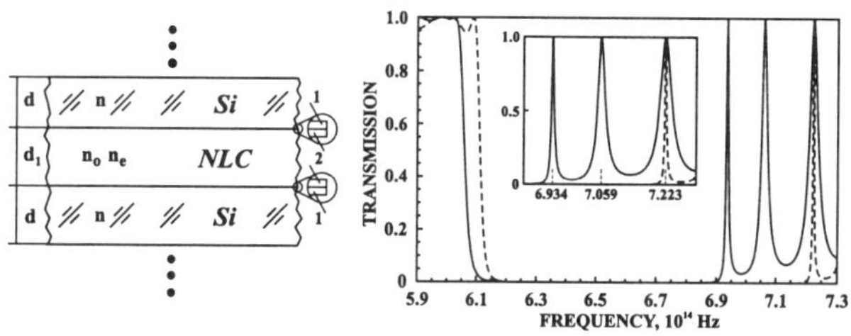

Figure 5. Left panel. One period of 1D PhC with a nematic liquid crystal cell. Right panel transmission spectrum for the 10-period structure. Solid line — before and dashed line — after the Frederics transition.

One of the principal characteristics complimentary to the dispersion equation in optics of PhCs is photonic density of modes (DOM). In 2D and 3D cases it is computed numerically, however in 1D case DOM has its analytical representation. The expression derived in Ref.[17] for arbitrary profile of scalar index of refraction in the limits of one period of PhC was related to the case of isotropic periodic media.

The direct application of wave operator approach in crystal optics comes out with a tensorial generalization of refractive index concept[18]. So it is naturally to expect the tensorial generalization of DOM in anisotropic media[19]. We obtained though the cumbersome, but explicit operator expression for the tensor of DOM via decomposition on the eigenvalues parts by polarization projective operators[18]. As an application of the tensor of DOM technique we made the direct comparison of transmission spectra with DOM ones for the binary superlattices $(Hg_2Cl_2/Si)^5Hg_2Cl_2$ ($n_1^+ = 2.65$, $n_1^- = 1.97$, $n_2 = 3.55$) and $(TiO_2/Si)^5TiO_2$ ($n_1^+ = 2.90$, $n_1^- = 2.60$, $n_2 = 3.55$) with quarterwave layers[19]. The plots are shown in Fig. 6. It is evident that information about the depth and spectral positions of the PBGs can be gained either by transmission or by DOM calculations. However, the very edges of the gaps are more pronouncedly expressed by the DOM curves due to the strong localization of light on them.

4. Covariant approach in effective medium theory

Homogenization is rather efficient approximation in the theory of periodic structures. The reduction of dimension is somehow compensated by the

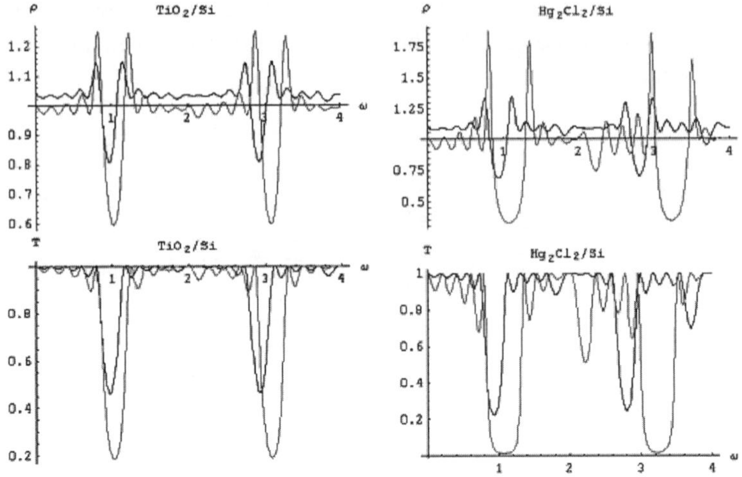

Figure 6. Spectral dependencies of DOM ρ and transmission coefficients T versus normalized frequency for two binary 1D PhC (marked on the top). All layers are of quarter-wave thickness. The smaller gaps on every graph belong to modes with +polarization, the dipper ones belong to modes with −polarization.

breaking of symmetry properties and as a results appearing anisotropy of material tensors.

It is well-known, that even a stack of isotropic layers produces an efficient uniaxial medium[20], described by the permittivity $\underline{\underline{\varepsilon}}^{eff}$ and permeability $\underline{\underline{\mu}}^{eff}$ tensors. Therefore, the analytics of the problem should be orientated on the tensorial calculus. But traditionally, only the component or principal values representation of tensors $\underline{\underline{\varepsilon}}^{eff}$ and $\underline{\underline{\mu}}^{eff}$ is used in literature. This fact significantly decreases the universality of obtained results. Moreover, coordinate form of material tensors might lead to a certain mismatch with the free-coordinate operator formalism (see, for example[18]).

Let us consider one period of planar periodic stratified medium, consisting of K planar infinite layers with the thicknesses d_j, $j = 1, 2, \ldots, K$. Electric and magnetic properties of each layer are describing by the permittivity $\underline{\underline{\varepsilon}}^j$ and permeability $\underline{\underline{\mu}}^j$ dyadics, which are in general case complex and non-symmetrical. The explicit expression for the effective parameters were derived in papers[21–23]: $\underline{\underline{\varepsilon}}^{eff}$:

$$\varepsilon^{eff} = \underline{\underline{\varepsilon}}_{av} + \frac{1}{W_\varepsilon} \sum_{j=1,l=1}^{K} \frac{f_j f_l}{\varepsilon_q^j \varepsilon_q^l} \underline{\underline{\varepsilon}}^{jl} \cdot \underline{\underline{\rho}}_q \cdot \underline{\underline{\varepsilon}}^l, \tag{4}$$

where

$$\underline{\underline{\varepsilon}}_{av} = \sum_{j=1}^{K} f_j \underline{\underline{\varepsilon}}^j, \quad W_\varepsilon = \sum_{j=1}^{K} \frac{f_j}{\varepsilon_q^j}, \quad \underline{\underline{\varepsilon}}^{jl} = \underline{\underline{\varepsilon}}^j - \underline{\underline{\varepsilon}}^l. \tag{5}$$

This expression gives us the effective dielectric tensor for one period of multilayer system. It is written in free-coordinate form, so it is ready for subsequent covariant operations. In principle it is easy to determine its components in any particular coordinate system. The analog expression exists and for effective magnetic constant of the superlattice.

Some examples of application of free-coordinate expression in waveguide calculations as well as comparison with exact formulas are given in Ref.[24,25].

5. Waveguides in photonic crystals

Waveguides in PhC are nothing else than linear defects, created by removing rows of holes, rods or other elements of a photonic crystal in specific direction of 2D periodic lattice. The most common is the triangular lattice of holes in SOI or GaAs-AlGaAs layered platforms. Such PhCs have a PBG for TE polarization[5]. Confinement of light in the third dimension (perpendicular to the plane of the lattice) is due to the contrast of indices of refraction.

The main characteristics of the PCW are dispersion dependencies and transmission spectra (or losses). We mostly calculated the latter one by the use of 3D FDTD numerical method. Our version of the FDTD technique together with the analytical transformation of formulas for uniaxial perfect matched layers (PML) used as absorbing boundary conditions were described in paper[26].

2D calculations for PCW with 1-row-thick channel show that in PBG zone transmission does not depend on the length of the waveguide. It means that there are no any significant scattering in-plane losses for defect modes excited in the channel of such waveguide in 2D model. All modes in 2D case can be classified due to their polarization and parity relatively to the axe of the channel. Even and odd modes can be excited by application of the initial field with the same kind of parity.

In ideal case interaction of modes with different parities is prohibited, so the crossing of dispersion curves for such modes does not lead to any interaction of them, and no any anticrossing effects on dispersion diagram are visible. However, our nonsymmetrical numerical scheme breaks the prohibition, and corresponding dips on transmission spectra show that interaction takes place. This numerical artifact as we show in[26] can be overwhelmed by increasing of spatial resolution (the number of mesh points per lattice constant Λ).

In 3D case transmission spectra depend on the length of a waveguide. To have a realistic modeling we took a three-layered SOI structure with Si core of 250 nm thickness and 100 nm SiO_2 claddings. Waveguide has one row of atoms removed in Γ-J direction of triangular lattice with filling factor about 50% (radius of the holes $r = 0.375\Lambda$). Analyzing data for transmission of waveguides with different lengths we were able to estimate losses of PCW. Increasing the size of the modeled system in vertical direction can drastically reduce calculated losses. When we inserted an air layer above PCW and thicker layer of SiO_2 as a substrate below the core we managed to reduce the minimal losses from 107 dB/mm to 35 dB/mm[26]. We attributed the fact of losses decreasing to the contribution of the leaky modes, whose dispersion curves are lying above the light cone on the dispersion diagram. They can couple to the air modes thus radiating the energy out from the waveguide. Additional notification of leaky mode contribution to the total transmission is the width of the transmission window on the spectrum, which is much broader than predicted by the dispersion curves of diffracted guided modes.

Direct comparison with the experiment made at COM center of Danish Technical University, Lyngby, Denmark is presented in Fig. 7, where experimental losses versus calculated ones are plotted for the telecommunication band round 1.33 microns[26]. Correlation between curves is obvious.

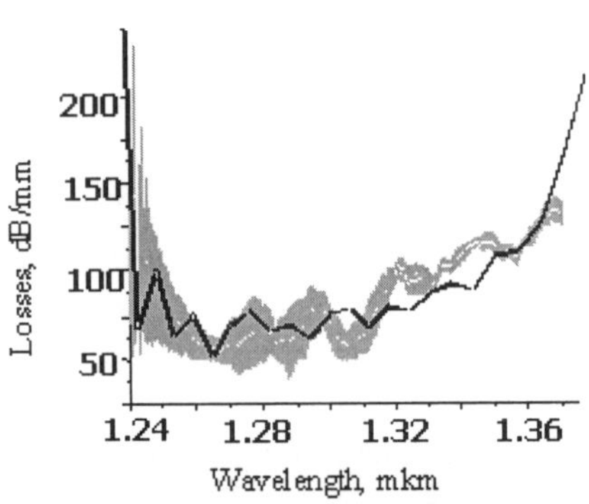

Figure 7. Losses of the straight PCW for TE-polarization. Experimental values — gray lines, 3D FDTD — black solid line.

We found by modeling that in the case of TM polarization of waves losses might be even smaller than for TE modes, up to 5 dB/mm in spite of the absence of PBG. The fundamental mode is responsible in that case for the high transmission. The result has been confirmed experimentally[27].

Losses per unit length for both polarizations are rather high while comparing with the ones for standard fibers or ridge waveguides. Nevertheless, the absolute values will be acceptable if the lengths of waveguides are a couple of dozens of microns only.

In more sophisticated optoelectronic device like Y-splitter 3D transmission is much lower, e.g. 23%. The device consists of a main channel, a joint, where two new output arms start, two 60°-degree bends in each of the arms. Certainly such a low transmission hampers direct implementation of the device into integrated circuits. However, if realization of low-loss Y-splitter can be fulfilled, then it will cause a lot of advantages for possible shrinkage of monoblock structures sizes.

For optimization we chose the following layout. First, a set of parameters was defined with the most crucial influence on the transmission properties. Then global maximums for transmission were determined by the use of Kallistos commercial software (Photon Design Ltd, Oxford, UK)[28]. Transmission was calculated in 2D frequency domain by eigenmode expansion technique. We obtained several sets of parameters, each yielding the 2D transmission raised up to 99.9%. To choose the best one the 3D FDTD modeling was used, which showed unambiguously that the so-called steering design is preferable to the resonant one. 3D transmission spectra for the non-optimized design and the best performing steering one are shown in Fig. 8. The effective optimization is evident and what is also very important — it is completed with broadening in spectral range[28].

Another semiconductor platform, which has been using in photonic crystals production, is a GaAs-AlGaAs layered stack. It has low contrast of indices of refraction what influenced a lot its possibilities to confine light in the vertical direction. Resembling this fact such systems are called weak-guiding. To prevent from scattering of light in the buffer layer the holes have to be deep enough. We investigated the behavior of transmission spectra with the depth of the holes, taking as an example the coupled resonator PCW. When the holes penetrated through the core layer and claddings only (total thickness 600 nm) transmission is about 1%. But when the depth was doubled with holes piercing the buffer layer transmission went up to 10%. Further increasing of depth changed the transmission weakly showing the saturation effect. The SOI platform having stronger confinement of light by

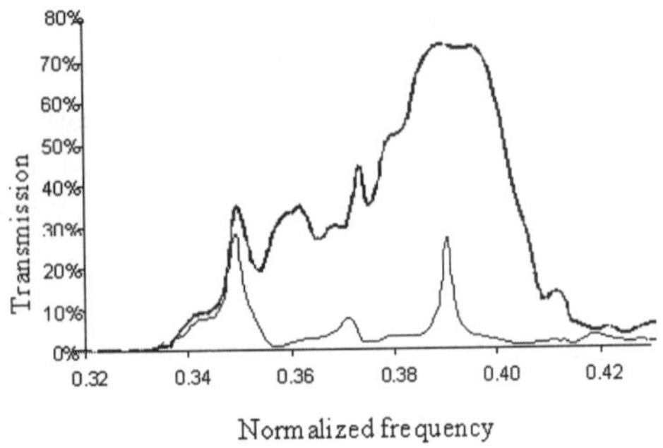

Figure 8. Transmission 3D FDTD spectra for the Y-splitter on PCW with 60°-bend in both output arms. Y-splitter is made in suspended in air membrane with dielectric constant 6.25. Thin line is a spectrum for the non-optimized device, thick line — for the device with the steering optimization.

reflection showed good tolerance for the depth of the holes.

Among others optoelectronic devices coupled waveguides on the base of PhC were treated. It was shown that they could find possible application as a PCW analog of devices known in integrated optics[29]. The full coupling length for excitation of modes in the neighboring channel is nearly 20 lattice constants, what is small enough comparatively with traditional systems[29]. TM-polarized modes resembled high coupling properties too[30]. Both theoretical and experimental TE-transmission spectra for the main channel were excitation occurred are shown in Fig. 9.

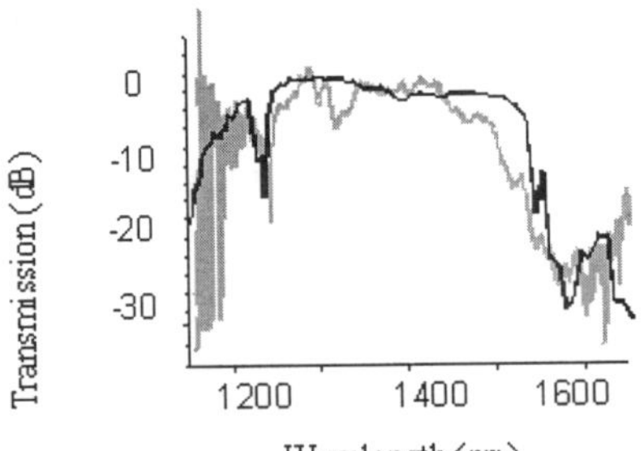

Figure 9. Transmission spectra of TE-light in the main channel of coupled photonic crystal waveguides system. Gray lines — experimental results, solid black line — 3D FDTD.

Another optimization approach, which has been addressed in our works, is the enhancement of the steepness of dispersion curves. Usually diffracted-guided modes have very flat dispersion curves. So the group velocity is very small. We made successful attempt to elaborate the comprehensive principle of the SOI PCW steepness enlarging[31]. Summarizing the results we can conclude that it is necessary to reduce the effective index of the guiding channel itself. Using light dielectric like glass instead of Si, or etching holes with the radius much bigger than that of the PhC can achieve the necessary result.

Bibliography

1. V. P. Bykov, *Soviet Physics - JETP* **35**, 269 (1972).
2. K. Ohtaka, *Phys. Rev. B* **19**, 5057 (1979).
3. E. Yablonovitch, *Phys. Rev. Lett.* **58**, 2059 (1987).
4. S. John, *Phys. Rev. Lett.* **58**, 2486 (1987).
5. J. D. Joannopoulos, R. D. Meade and J. N. Winn, *Photonic Crystals* (Univ. Press, Princeton, 1995).
6. F.I.Fedorov, *Theory of gyrotropy* (Nauka i Tekhnika, Minsk, 1976).
7. D. N. Chigrin and A. V. Lavrinenko, *APS e-print*, No. aps1998may30_001.
8. J. N. Winn, et al, *Opt. Lett.* **23**, 1573 (1998).
9. Y. Fink, et al, *Science* **282**, 282 (1998).
10. D. N. Chigrin, et al, *Appl. Phys. A.* **68**, 25 (1999).
11. P. S. J. Russell, S. Redwell and P. J. Roberts, *Opt. Commun.* **160**, 66 (1999).
12. D. N. Chigrin, et al, *J. of Light. Techn.* **17**, 2018 (1999).
13. D. N. Chigrin, et al, *Optics & Photonic News* **10**, 33 (1999).
14. A. V. Lavrinenko, et al, in *Proc. of the Intern. Workshop on Quantum Systems.* Minsk, 3-7 June 1996, (World Scientific, Singapore, 1997), p. 184.
15. A. V. Lavrinenko, et al, *Proc. SPIE* **3580**, 2 (1998).
16. D. N. Chigrin and A. V. Lavrinenko, *Proc. 19th Topic. Meeting Europ. Opt. Society "Electromagnetic Optics"*, 7-9 Sept. 1998, (Hyeres, France, 1998), p. 33.
17. J. M. Bendickson, J. P. Dowling and M. Scalora, *Phys. Rev. E.* **53** 4107 (1996).
18. A. N. Furs and L. M. Barkovsky, *Operator methods for description of optical fields in complex media* (Belaruskaya Navuka, Minsk, 2003).
19. A. V. Lavrinenko and D. N. Shyroki, *SPIE Proc.* **4517**, 183 (2001).
20. M. Born and E. Wolf, *Principles of optics*, (Pergamon press, Oxford, 1968).
21. V. V. Zhilko and A. V. Lavrinenko, *Vesti AN BSSR. Serija fis.-mat. navuk* No. 4, p. 29 (1995).
22. V. V. Zhilko and A.V. Lavrinenko, *Vestnik BGU. Serija 1* No. 3, p. 17 (1995).
23. A. V. Lavrinenko and V. V. Zhilko, *Micr. and Opt. Techn. Lett.* **15** 54 (1997).
24. A. V. Lavrinenko, D. N. Shyroki and V. V. Zhilko, *Nonlinear Phenomena in Complex System* **5**, No. 1, pp. 84-89 (2002).
25. A. V. Lavrinenko, D. N. Shyroki and V. V. Zhilko, *J. Magnetism and Magnetic Materials* **247**, No. 2, pp. 171-177 (2002).

26. A. V. Lavrinenko, et al, *Opt. Express* **12** 234 (2004), http://www.opticsexpress.org/abstract.cfm?URI=OPEX-12-2-234.
27. P. I. Borel, et al, *Conference on Lasers and Electro - Optics CLEO/QELS - USA 2003* (Paper JWC3, Baltimore, 2003).
28. T. Felici, et al, *29th Europian Conference on Optical Communication*, ECOC-03, September 21-25 (paper Tu 4.2.1. Riminy, Italy, 2003)
29. P. I. Borel, et al, *Conference on Lasers and Electro-Optics CLEO-Europe 2003* (Paper CM4-2-FRI, Munich, Germany, 2003).
30. M. Thorhauge, et al, *Optical Fiber Communications*, OFC-2003: Technical Digest: in 2 vol. (Atlanta, USA, 2003), V.2. pp.494-495.
31. T. Sondergaard and A. Lavrinenko, *Opt.Commun.* **203**, 263 (2002).

RADIATION FORCES IN CLASSICAL AND QUANTUM PHYSICS

Serguei T. Zavtrak
AgResearch Limited, Ruakura MIRINZ Centre, Private Bag 3123, Hamilton, New Zealand
E-mail: serguei.zavtrak@agresearch.co.nz

This article presents theoretical and experimental results of the research of radiation forces, which includes acoustics and both classical and quantum electrodynamics. These forces are generally caused by external radiation acting on small particles and can be represented as the sum of the so-called primary and secondary forces. The first type can be treated as an integral response of a single particle to the initial external radiation, which is the combination of re-emission, thermal losses and viscous currents inside and outside the particles (liquid drops or gas bubbles in a liquid, for example). The second type is caused by re-emission from the particles, which is induced by the incident external waves. The radiation forces are known for their numerous effects and are useful for many practical applications. The author focuses mainly on secondary radiation forces. It is shown that these secondary forces can produce quite unusual long-range effects such as the formation of clusters from dispersed particles in a liquid (gas) in the presence of acoustic waves and the appearance of attraction zones for electrons in the presence of electromagnetic waves. The article is written in a very descriptive way in order to eliminate rather cumbersome mathematical calculations (all the details may be found in Refs), to explain the main results as simple as possible and to encourage further research in those who might be interested.

1. Introduction

It is well known that when small particle like gas bubbles, solid particles, liquid drops are affected by sound waves, this leads to time-averaged hydrodynamic forces. These forces are called acoustic radiation forces. Herewith, they formally may be divided in two types. The first one is related to direct action of the sound wave, which takes place for a lonely particle.

This type is called primary radiation force. Acoustic levitation is a good example of primary radiation forces and known by its technical applications (a lot of information on this topic may be found in the Internet by just typing "Acoustic levitation" and using Google or other search engines). Usually, this type makes the particles drift to certain points in liquid (gas). Positions of those points, direction and speed with which the particle move are function of their dimensions, frequency of sound, type of materials from which the particles are made, type of liquid or gas in which there are placed and so on. The second type may be associated with interaction between the particles, which may be either attraction or repulsion. This type of forces is called secondary radiation forces because its caused by secondary radiation from the particles induced by the primary fields. Coagulation of aerosols[1,2] and cluster formation[3] is an example of the secondary radiation forces. Besides, both primary and secondary forces play an important role in acoustic cavitation and sonoluminescence[4–7]. Similar effects are known in optics, when small particles being electrically or magnetically responsive are excited by electromagnetic waves[8,9]. But, in acoustics such the effects are significantly more powerful. So, we start our consideration from acoustic radiation forces. The first who reported on the radiation forces were C.A. Bjerknes and his son V.F.K. Bjerknes. They observed drift and interaction between two gas bubbles placed in liquid and excited by acoustic waves[10]. After that, interaction between gas bubbles was always associated with their names. Radiation forces between solid particles is typically associated with W. Konig, who first studied interaction between two rigid spheres in a fluid flow and explained their coagulation in a sound field[11]. Since that, acoustic radiation forces were subjects of investigation by many scientists. At the early stage, theory was based on quite simple approximations, which however could not explain new effects such as stable bubble clusters (bubble grapes)[3], unusual bunching of solid particles in highly viscous liquid near velocity nodes instead of antinodes[12], acoustic streamer formation[4,13] and so on. Recent thorough investigations took into account viscous currents outside and inside the particles, thermal and radiation losses and non-linear contributions. This allowed not only to explain many of those effects, but to predict new ones.

2. Primary Radiation Forces

King[14] was the first, who did thorough theoretical research on primary radiation forces. He obtained a general formula for the radiation force that is produced by a sound field on a rigid sphere of arbitrary radius freely suspended in an ideal fluid. As an example, we briefly derive the radiation force on a gas bubble under the action of a travelling sound wave. In the first approximation[15], when both liquid and gas in the bubble are practically ideal and wavelength in liquid is much bigger than the bubble, this force may be written as average value

$$\mathbf{F} = -\langle V(t)\nabla P\rangle \tag{1}$$

Here, $V(t) \approx V_0 + 4\pi R_0^2 R_1(t)$ is the bubble's volume, which is a sum of static and oscillating contributions, R_0 is its static radius, $R_1(t)$ is radius shift and $P(t) = A\cos(\omega t - \mathbf{k}\cdot\mathbf{r})$ is pressure wave in liquid. Equation for bubbles oscillations may be written as[16]

$$\ddot{R}_1 + \omega_r^2 R_1 + \omega\delta\,\dot{R}_1 = -\frac{A\cos(\omega t - \mathbf{k}\cdot\mathbf{r})}{\rho_l R_0} \tag{2}$$

Here,

$$\omega \approx \sqrt{\frac{3\gamma P_0}{\rho_l R_0^2}}$$

is the bubble's resonant frequency, P_0 is static pressure, γ is adiabatic constant of gas inside the bubble, ρ_l is density of liquid and $\delta = \delta_{th} + \delta_\nu + \delta_r$ is an absorption constant, which is a result of thermal, viscous and radiation losses. As a result, one can obtain

$$\mathbf{F} \sim \frac{\mathbf{k}\,|A|^2\,\delta}{\left(\omega_r^2/\omega^2 - 1\right)^2 + \delta^2}$$

It can be seen that the radiation forces is directed along the wave vector of the travelling wave, it is proportional to the square of pressure amplitude and riches a maximum at the resonance frequency. Radiation force for the standing wave is a bit more complicated. It is still directed along the wave vector, but its value and sign depends on the bubble's position and ratio ω_r/ω[16]. Note that King obtained his equations[14] for rigid spheres only. In spite of quite good agreement with many tests, early theory assumed the host liquid to be ideal. Yosioka and Kawasima[17] applied the equations of motion of an ideal compressible liquid to both the host liquid and the medium inside the particles and kept their spherical shape only at rest. They also introduced thermal δ_{th} and radiation δ_r absorption constants. Their result for the gas bubbles was improved by Crum and Eller[18], who introduced constants δ_{th} and δ_ν. However, the theory outlined up to here was based on assumption that the host liquid is ideal, i.e., non-viscous and non-heat-conducting. In reality, the liquid should be described by the Navier-Stokes equation instead of Euler equiation. Equations of liquid motion should be supplemented with the equation of heat transfer. Viscosity results in vortical components in the liquid motion, which should be taken into account. The boundary conditions on the particle's surface become significantly more complicated. Not only continuity of the normal component of the fluid velocity and stress, but also continuity of tangential components of the fluid velocity and continuity of temperature and heat flux is required. Calculation of radiation forces in a real fluid requires solving not only linearized equations of fluid motion, but the so-called equations of acoustic streaming as well. All these circumstances were taken into account by Doinikov[19–24], who built the most accurate theory of primary radiation forces.

3. Secondary Radiation Forces in Acoustics

What will happen to equation (4), if one puts two gas bubbles in the sound wave? The result will be (bubbles are distinguished by their bottom indexes, l is distance between the bubbles and c is sound speed in the host liquid)

$$\ddot{R}_1 + \omega_r^2 R_{r1} + \omega\delta_1 \dot{R}_1 = -\frac{A\cos(\omega t - \mathbf{k}\cdot\mathbf{r}_1)}{\rho_l R_{10}} + \frac{R_{20}^2}{R_{10} l}\ddot{R}_2(t - l/c)$$
$$\ddot{R}_2 + \omega_r^2 R_{r2} + \omega\delta_2 \dot{R}_2 = -\frac{A\cos(\omega t - \mathbf{k}\cdot\mathbf{r}_2)}{\rho_l R_{20}} + \frac{R_{10}^2}{R_{20} l}\ddot{R}_1(t - l/c)$$

Each of these equations takes into account not only the initial incident wave, but induced radiation from each neighbour, which is delayed in time, of course. Solving this system and calculating the radiation force according to equation (3), one can find[25]

$$\mathbf{F}_1 = \mathbf{F}_{1l} + \mathbf{F}_{1k},\ \mathbf{F}_2 = \mathbf{F}_{2l} + \mathbf{F}_{2k}$$

Here, the forces of the first type (in the direction of $\mathbf{l}$) are

$$\mathbf{F}_{1l} = \frac{2\pi R_{10} R_{20} A^2}{\rho\omega^2\left[(\omega_{r1}^2/\omega^2 - 1) + \delta_1^2\right]^{1/2}\left[(\omega_{r2}^2/\omega^2 - 1) + \delta_2^2\right]^{1/2}} \cdot \frac{\mathbf{l}}{l^3} \cdot \cos\Psi_1 \quad (3)$$

$$\mathbf{F}_{1l} = -\frac{2\pi R_{10} R_{20} A^2}{\rho\omega^2\left[(\omega_{r1}^2/\omega^2 - 1) + \delta_1^2\right]^{1/2}\left[(\omega_{r2}^2/\omega^2 - 1) + \delta_2^2\right]^{1/2}} \cdot \frac{\mathbf{l}}{l^3} \cdot \cos\Psi_2 \quad (4)$$

The forces of the second type are directed along wave vector of the incident wave

$$
\begin{aligned}
\mathbf{F}_{1k} =& \frac{2\pi R_{10} A^2 \delta_1 \mathbf{k}}{\rho\omega^2 \left[(\omega_{r1}^2/\omega^2 - 1) + \delta_1^2\right]} \\
&+ \frac{2\pi R_{10} R_{20} A^2 \cos\chi_1}{\rho\omega^2 \left[(\omega_{r1}^2/\omega^2 - 1) + \delta_1^2\right]^{1/2} \left[(\omega_{r2}^2/\omega^2 - 1) + \delta_2^2\right]^{1/2}} \cdot \frac{\mathbf{k}}{l}
\end{aligned} \tag{5}
$$

$$
\begin{aligned}
\mathbf{F}_{2k} =& \frac{2\pi R_{20} A^2 \delta_2 \mathbf{k}}{\rho\omega^2 \left[(\omega_{r2}^2/\omega^2 - 1) + \delta_2^2\right]} \\
&+ \frac{2\pi R_{10} R_{20} A^2 \cos\chi_2}{\rho\omega^2 \left[(\omega_{r1}^2/\omega^2 - 1) + \delta_1^2\right]^{1/2} \left[(\omega_{r2}^2/\omega^2 - 1) + \delta_2^2\right]^{1/2}} \cdot \frac{\mathbf{k}}{l}
\end{aligned} \tag{6}
$$

Ψ_1, Ψ_2, χ_1 and χ_2 are some phases, which linearly depend on spatial positions of the bubbles[25]. The first terms in equations (5) and (6) are usual short-range forces of interaction between the bubbles, which transform to well-known Bjerknes forces at $kl << 1$ (incompressible liquid). They decrease inversely proportional to the square of the distance between the particles, i.e. look like Coulomb forces. The first contributions in equations (7) and (8) are recognised as usual radiation pressure forces acting at lonely particles. The second contributions there are the most interesting because they are of the long-range type and decrease much slower, i.e. inversely proportional to the distance between the bubbles. We call this type as long-rage radiation forces. To our knowledge, they were discovered for the first time by Nemtsov[26] for the gas bubbles and then developed by the author and Doinikov[27–30] for any particles. In classical electrodynamics those forces were described by the author[25]. Quantum theory of the radiation forces was built by the author in[31]. Note, that at large distances ($kl >> 1$) between the bubbles the long-range radiation forces dominate over Bjerknes forces. This may lead to appearance of stable coupled pairs of the bubbles and even bubble grapes observed in the[3]. Situation is getting much more complicated, if viscous currents and heat transfer are taken into account, which is done by Doinikov[21–24], who thoroughly investigated interaction between the bubbles, liquid drops, rigid particles and so on and explained many experimental results. As shown in[25], sum of all radiation forces is equal to impulse taken away by the sound waves scattered by both bubbles.

4. Secondary Radiation Forces in Classical Electrodynamics

Secondary radiation forces between two charges in presence of an electromagnetic wave were described by the author in Ref.[25]. Lets put two charges in plane electromagnetic wave with electric and magnetic amplitudes $\mathbf{E}_{ext} = \mathbf{E}_0 \cos(\omega t - \mathbf{k}\cdot\mathbf{r})$ and $\mathbf{H}_{ext} = \mathbf{H}_0 \cos(\omega t - \mathbf{k}\cdot\mathbf{r})$ ($\mathbf{H}_0 = [\mathbf{k}/k, \mathbf{E}_0]$). Let present position of the charges as $\mathbf{r}_1 = \mathbf{r}_{10} + \xi_1$, $\mathbf{r}_2 = \mathbf{r}_{20} + \xi_2$, where $\mathbf{r}_{10}$, $\mathbf{r}_{20}$ are their mean positions and ξ_1, ξ_2 are their displacement caused by the incident wave (its supposed that $|\xi_{1,2}| << l, \lambda$, where $l = |\mathbf{r}_{10} - \mathbf{r}_{20}|$ and λ is wavelength). Taking into account electromagnetic fields scattering by neighbouring particles, one obtains the following ponderomotive equations

$$m_1 \ddot{\mathbf{r}}_1 = e_1 \left(\mathbf{E}_{ext}(\mathbf{r}_1, t) + \mathbf{E}_{r2}(\mathbf{r}_1, t)\right) + \frac{e_1}{c}\left[\dot{r}_1, \mathbf{H}_{ext}(\mathbf{r}_1, t) + \mathbf{H}_{r2}(\mathbf{r}_1, t)\right] \quad (7)$$

$$m_2 \ddot{\mathbf{r}}_2 = e_2 \left(\mathbf{E}_{ext}(\mathbf{r}_2, t) + \mathbf{E}_{r1}(\mathbf{r}_2, t)\right) + \frac{e_2}{c}\left[\dot{r}_2, \mathbf{H}_{ext}(\mathbf{r}_2, t) + \mathbf{H}_{r1}(\mathbf{r}_2, t)\right] \quad (8)$$

Here, scattering waves are taken into account, which may be described as delayed potentials[32]

$$\mathbf{E}_{r1,2} = -\frac{e_{1,2}}{c^2 R_{1,2}}\left(\ddot{\xi}_1\left(t'_{1,2}\right) - \mathbf{n}_{1,2}\left(\mathbf{n}_{1,2}\cdot \ddot{\xi}_1\left(t'_{1,2}\right)\right)\right) \quad (9)$$

$$\mathbf{H}_{r1,2} = -\frac{e_{1,2}}{c^2 R_{1,2}}\left[\mathbf{n}_{1,2}, \ddot{\xi}_1\left(t'_{1,2}\right)\right] \quad (10)$$

Here, $R_{1,2}$ are propagation distances of the radiation from the particles, unit vectors $\mathbf{n}_{1,2}$ give the direction of radiation. We shall examine only long-range limit $kl >> 1$ because the most interested results obtained there. Potentials (9) – (11) should be calculated at delayed moments[32]

$$t'_{1,2} = t - \left|r - r_{1,2}(t'_{1,2})\right|/c$$

Omitting usual Coulomb forces, with accuracy of the first order of the incident wave, equations (9) and (10) may be written as

$$m_1 \ddot{\xi}_1 = e_1 \mathbf{E}_0 \cos(\omega t - \mathbf{k}\cdot\mathbf{r}_{10}) - \frac{e_1 e_2}{c^2 l}\left(\ddot{\xi}_2(t - l/c) - \mathbf{n}\left(\mathbf{n}\cdot \ddot{\xi}_2(t - l/c)\right)\right)$$

$$m_2 \ddot{\xi}_2 = e_2 \mathbf{E}_0 \cos(\omega t - \mathbf{k}\cdot\mathbf{r}_{20}) - \frac{e_1 e_2}{c^2 l}\left(\ddot{\xi}_1(t - l/c) - \mathbf{n}\left(\mathbf{n}\cdot \ddot{\xi}_1(t - l/c)\right)\right)$$

Solutions of these calculations may be found in[25]. Here, we omit all the details and give expressions for long-range radiation forces at far zone ($kl >> 1$) only:

$$\mathbf{F}_1 = \partial U_1/\partial \mathrm{l}, \ \mathbf{F}_2 = -\partial U_2/\partial \mathrm{l}$$

$$U_{1,2} = -\frac{e_1^2 e_2^2\left(E_0^2 - (\mathbf{E}_0\cdot\mathbf{n})^2\right)}{2 m_1 m_2 \omega^2 c^2 l}\cos(kl - \mathbf{k}\cdot\mathbf{l})$$

Value $U_1(U_2)$ may be treated as energy of the first (second) particlee in the field scattered by the second (first) particle. This energy is inversely proportional to the distance between the charges, i.e. behaves like Coulomb potential, but has an oscillating factor, which getting crucial at the far zone. However, for the comparison of the forces, the gradients of the energies are important, not the energies themselves. Therefore, at large distances the radiation forces may dominate over Coulomb forces and lead to principle possibility of forming bound states even between two equal charges. Comparison between the radiation force and Coulomb force gives the value of electric intensity, for which those forces are getting compatible, namely $E_0 \sim m\omega ce^{-1}/\sqrt{kl}$. At wavelength $\lambda = 1\mu u$ and for electrons this gives energy current $W = cE_0^2/4\pi = 0.23 \cdot 10^{18} W/cm^2$. Let us compare the secondary radiation force with primary average force acting at the particle[32]

$$f_{1,2} = \frac{8\pi}{3} \left(\frac{e_{1,2}^2}{m_{1,2}c^2} \right) \langle \mathbf{E}_{ext}^2 \rangle \frac{\mathbf{k}}{4\pi k}$$

For equal particles one can find $F/f \sim (kl)^{-1}$. This means, under the action of primary radiation forces thay are moving together as the whole but they can form a bound state due to the secondary forces, which may be either attractive or repulsive.

Thus, there is a lot of similarities between radiation interaction between small particles in acoustics and charges in classical electrodynamics.

5. Quantum Theory of the Radiative Interaction of Charged Particles

This theory was built by the author and Komarov[31]. The simplest case is two electrons placed in a plane electromagnetic wave. The first step is to use Volkov's analytical solution[33] for electron in the plane electromagnetic wave instead of a wave function of free relativistic electron ($\hbar = c = 1$):

$$\Psi_p = \left(1 + \frac{e}{2(kp)} \hat{k}\hat{A} \cos(kx) \right) \frac{U(p)}{\sqrt{2p_0}} \exp iS(x) \tag{11}$$

where

$$S(x) = -(px) - \frac{e(pA)}{(kp)} \sin(kx) + \frac{e^2A^2}{8(kp)} \sin(2kx) \tag{12}$$

Here, $k = (\omega, \mathbf{k})$ and $A = (0, \mathbf{A}\cos(kx))$ are the wave 4-vector and 4-vector of the external electromagnetic wave, respectively (note that $(kA) = -(\mathbf{k} \cdot \mathbf{A}) = 0$); $x = (t, \mathbf{r})$ is 4-vector of electron; scalar product of vectors a

and calculating matrix element S with modified function K_0, one obtains

$$S = -i\,(2\pi)^4\,\delta^4\,(p_1 + p_2 - p_1' - p_2')\,V(\mathbf{p}) \tag{25}$$

where

$$V(\mathbf{p}) = \frac{4\pi e_1 e_2}{\mathbf{p}^2}\left[1 + \frac{1}{4}\left(\frac{e_1^2}{m_1^2} + \frac{e_2^2}{m_2^2}\right)\left(\mathbf{A}^2 - \frac{(\mathbf{A}\cdot\mathbf{p})^2}{\omega^2}\right)\right]$$
$$+ \frac{\pi e_1^2 e_2^2}{m_1 m_2}\left(\mathbf{A}^2 - \frac{(\mathbf{A}\cdot\mathbf{p})^2}{\omega^2}\right)\left(\frac{1}{\omega^2 - (\mathbf{p}-\mathbf{k})^2} + \frac{1}{\omega^2 - (\mathbf{p}+\mathbf{k})^2}\right) \tag{26}$$

Function $V(\mathbf{p})$ may be treated as the Fourier image of the static potential of interaction between two charges[33]. If one returns to usual spacial coordinates, then

$$V(\mathbf{r}) = \frac{e_1 e_2}{r} + \frac{e_1 e_2}{4}\left(\frac{e_1^2}{m_1^2} + \frac{e_2^2}{m_2^2}\right) A_\alpha A_\beta \left(\delta_{\alpha\beta} + \frac{1}{\omega^2}\frac{\partial^2}{\partial x_\alpha \partial x_\beta}\right)\frac{1}{r}$$
$$+ \frac{e_1^2 e_2^2}{2 m_1 m_2}\cos(\mathbf{k}\cdot\mathbf{r}) A_\alpha A_\beta \left(\delta_{\alpha\beta} + \frac{1}{\omega^2}\frac{\partial^2}{\partial x_\alpha \partial x_\beta}\right)\frac{\cos(kr)}{r} \tag{27}$$

In can be seen that presence of the electromagnetic waves results in an important change in the Coulomb interaction between the charged particles, which is an appearance of the long-range force. This force decreases inversely proportional to the distance between the particles. One can see that expression for $V(\mathbf{r})$ doesn't depend on quantum constant $\hbar$ at all. Hence, such the corrections to the Coulomb interaction are of the classical nature.

These results can be produced by other means. Let us consider usual quantum Lagrangian[33], in which the external electromagnetic wave is added

$$L = e\int \tilde{\Psi}\,(x)\gamma^\mu \Psi(x)\left(A_\mu(x) + A_\mu^{(cl)}(x)\right) d^3\mathbf{r} \tag{28}$$

Again, we restrict our consideration by accuracy up to e^4. One can check that Feynman diagrams, which are shown in Fig. 2 result in the same corrections of the Coulomb potential as expression (25).

These diagrams correspond to successive absorption and emission (or otherwise) of external quanta (elastic scattering) by system of two particles. As a result, combined energy-momentum vector remains the same. Those processes are presented by factor $\delta^4\,(p_1 + p_2 - p_1' - p_2')$ in matrix element S. As far as other contributions in the S-matrix are concerned, they lead

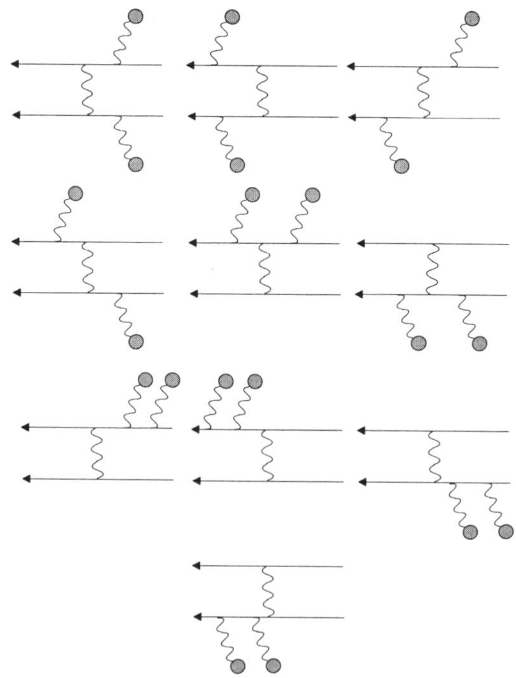

Figure 2. Feynman diagrams of the order of e^4, which result in modification of Coulomb potential

to appearance of terms with factors like $\delta^4(p_1 + p_2 - p'_1 - p'_2 \pm k)$ or $\delta^4(p_1 + p_2 - p'_1 - p'_2 \pm 2k)$, for which the combined energy-momentum is not preserved (inelastic scattering).

After simple, but unwieldy transformations the accurate expression for two-particle potential can be written as follows[31]

$$V(\mathbf{r}) = \frac{e_1 e_2}{2} \int_0^{2\pi} \frac{d\varphi}{2\pi} \int_{-\infty}^{\infty} d\tau \frac{\delta(\tau - |\rho(\tau, \mathbf{r}, \varphi)|) + \delta(\tau + |\rho(\tau, \mathbf{r}, \varphi)|)}{|\rho(\tau, \mathbf{r}, \varphi)|} \times \left(1 - \frac{e_1^2 \mathbf{A}^2}{4m_{1eff}^2} - \frac{e_2^2 \mathbf{A}^2}{4m_{2eff}^2} + \frac{R^2(\tau, \mathbf{r})\, \mathbf{A}^2}{2} \cos^2\varphi\right) \tag{29}$$

where

$$\rho(\tau, \mathbf{r}, \varphi) = \mathbf{r} - \frac{\mathbf{A}}{\omega} R(\tau, \mathbf{r}) \sin\varphi \tag{30}$$

$$R(\tau, \mathbf{r}) = \left(\frac{e_1^2}{m_{1eff}^2} + \frac{e_2^2}{m_{2eff}^2} - \frac{2e_1 e_2}{m_{1eff}\, m_{2eff}} \cos(\omega t - \mathbf{k} \cdot \mathbf{r})\right)^{1/2} \tag{31}$$

This expression is very complicated. But, it can be easily checked by considering extreme $m_2 \to \infty$. This case corresponds to hydrogen-like atom in a plane electromagnetic wave. At this limit

$$R(\tau, \mathbf{r}) \approx e_1/m_{1eff} \tag{32}$$

Assuming that $e_1 |\mathbf{A}| / m_1 << 1$, potential (28) takes the form

$$V(\mathbf{r}) = \frac{e_1 e_2}{2\pi} \int_0^{2\pi} \frac{d\varphi}{\left| \mathbf{r} - \frac{e_1 \mathbf{A}}{m_1 \omega} \sin\varphi \right|} \tag{33}$$

i.e. transforms to the well-known Kramers-Henneberger potential (see Ref.[34] and many references there). Such the potential is valid for high orbit electrons, when their motion around the nucleus is slow as compared with fast oscillations caused by the plane electromagnetic wave. Then, position of the electron can be presented is a sum of two parts $\mathbf{r} \to \mathbf{r} + \xi$, where ξ is the solution of equation

$$m_1 \ddot{\xi} = e_1 E \sin \omega t$$

Averaging the Coulomb potential with shifted electron's coordinates over fast oscillations results in expression (32).

6. Conclusions

1. Radiation pressure forces have the same nature in classical acoustics and both classical and quantum electrodynamics.
2. That is pure classical effect caused by scaterring of the incident wave by the particles
3. The are two types of radiation forces: primary and secondary. The first type is valid for lonely particles, whereas the second type is caused by secondary emission from the particles.
4. Roughly speaking, potential of the secondary emission is proportional to $\Phi\tilde{\ } A \exp(kr)/r$ ($\mathbf{r}$ is distance between the particle, A is amplitude of the incident wave). The radiation force is a product of gradient $\nabla\Phi$ and A. At far zone ($kr >> 1$) derivative of the exponent gives the main contribution, whereas derivative $\nabla(\mathbf{1}/\mathbf{r})$ may be neglicted. This means, that at far zone the secondary force decreases inversely proportional to r and begins to dominate over classical Bjerkness or Coulomb forces $(\tilde{\ } r^{-2})$.
5. The radiation forces can result in quite interesting effects in electrodynamics, for example Rutherford scattering in the presence of an electromagnetic wave[31].

6. The secondary radiation forces can play a significant role in free-electron lasers or FELs (many interesting data may be found in the Internet by just typing "free electron lasers" and using Google or other search engines). At present, only primary radiation forces are taken into account in the FEL theory.
7. A lot of potential and many new applications may be developed here. This requires further research.

Bibliography

1. E. P. Mednikov, *Acoustic Coagulation and Precipitation of Aerosols* (Consultants Bureau, NY, 1965).
2. L. D. Rozenberg, (Ed.) *Physical Principles of Ultrasonic Technology* (Plenum, NY, 1973).
3. Yu. A. Kobelev, L. A. Ostrovskii and A. M. Sutin, *Pisma Zh. Eksp. Teor. Fiz.* **30**, 423 (1979).
4. T. G. Leighton, *The Acoustic Bubble* (Academic, London, 1994).
5. C. E. Brennen, *Cavitation and Bubble Dynamics* (Oxford U.P., London, 1995).
6. A. J. Walton, and G. T. Reynolds, *Adv. Phys.* **33**, 595 (1984).
7. L. A. Crum, *Physics Today* **47**, 22 (1994).
8. R. V. Jones, and B. Leslie, *Proc. R. Soc. Lond. A* **360**, 347 (1978).
9. A. Ashkin, *Science* **210**, 1081 (1980).
10. V. F. K. Bjerknes, *Fields of Force* (Columbia U. P., NY, 1906).
11. W. Konig, *Ann. Phys.* **42**, 353 (1891); **42**, 549 (1891).
12. A. G. Avertisyan, V. S. Arakelyan, O. V. Bagdasaryan and A. K. Dudoyan, *Sov. Phys. Acoust.* **31**, 227 (1985).
13. I. Akhtov, U. Parlitz and W. Lauterborn, *J. Acoust. Soc. Am.* **96**, 3627 (1994).
14. L. V. King, *Proc. R. Soc. Lond. A* **147**, 212 (1934).
15. L. D. Landau and E. M. Lifshitz, *Fluid Mechanics* (Pergamon, NY, 1959).
16. V. A. Krasilnikov, and V. V. Krylov, *Introduction to Physical Acoustics* (Nauka, Moscow, 1984).
17. K. Yosioka, Y. Kawasima and H. Hirano, *Acoustica* **5**, 173 (1955).
18. L. A. Crum and A. I. Eller, *J. Acoust. Soc. Am.* **48**, 181 (1970).
19. A. A. Doinikov, *J. Fluid Mech.* **267**, 1 (1994).
20. A. A. Doinikov, *Proc. R. Soc. Lond. A* **447**, 447 (1994).
21. A. A. Doinikov, *J. Acoust. Soc. Am.* **101**, 713 (1997).
22. A. A. Doinikov, *J. Acoust. Soc. Am.* **101**, 722 (1997).
23. A. A. Doinikov, *J. Acoust. Soc. Am.* **101**, 731 (1997).
24. A. A. Doinikov, *J. Acoust. Soc. Am.* **103**, 143 (1998).
25. S. T. Zavtrak, *J. Phys. A: Math. Gen.* **23**, 1493 (1990).
26. B. E. Nemtsov, *Pisma v ZTF* **9**, 858 (1983).
27. A. A. Doinikov and S. T. Zavtrak, *J. Acoust. Soc. Am.* **99**, 3849 (1996).
28. A. A. Doinikov and S. T. Zavtrak, *Ultrasonics* **34**, 807 (1996).

29. A. A. Doinikov and S. T. Zavtrak, *Phys. Fluids* **7**, 1923 (1995).
30. A. A. Doinikov, and S. T. Zavtrak, *J. Acoust. Soc. Am.* **102**, 1424 (1997).
31. S. T. Zavtrak and L. I. Komarov, *J. Phys. A: Math. Gen.* **23**, 5547 (1990).
32. L. D. Landau and E. M. Lifshitz, *Theory of Field* (Nauka, Moscow, 1973).
33. V. B. Berestetsky, E. M. Lifshitz and L. P. Pitaevsky, *Relativistic Quantum Theory* (Nauka, Moscow, 1968).
34. Q. Su, and J. H. Eberly, *Phys. Rev. Lett.* **64**, 862 (1990).

MASTER EQUATION FOR STRUCTURED RESERVOIRS

D. S. Mogilevtsev† and S. Ya. Kilin‡
Institute of Physics, Belarus National Academy of Sciences, F.Skarina Ave. 68, Minsk 220072 Belarus,
†E-mail: mogilev@dragon.bas-net.by
‡E-mail: kilin@dragon.bas-net.by

In this work a master equation for the interaction with structured reservoirs is derived. The equation is applicable when the "freezing" of atomic population decay and the existence of atom-photon bound state are possible. This equation may be used for arbitrarily strong excitation of the reservoir. It is shown that for a wide range of reservoir spectral densities this master equation can be reduced to the Markovian form. With the help of the obtained equation, spontaneous emission and resonance fluorescence near the band edge are considered. We demonstrate that the stationary state is strongly dependent on the initial state of decaying atom-field system. For resonance fluorescence, we confirm the prediction of positive stationary atomic inversion.

1. Introduction

One of the most striking effects of a quantum system interaction with the structured reservoir is "freezing" of spontaneous decay, when the process terminates after a finite period of time. Such a phenomenon is possible when the spectral density of the interaction Hamiltonian reaches zero in, at least, one point (general conditions of decay "freezing" are given in Ref.[1]). A particular and much recently discussed case is an interaction of atoms and molecules with radiation inside photonic crystals, i.e. systems with the gap in the spectrum of photonic states density. However, up to day, an adequate description of essentially non-Markovian interaction with structured bosonic reservoirs remains a major theoretical task. An adequate treatment of the reservoir memory effects becomes necessary, and standard Markovian master equations are no longer valid (for the recent review, see[2]). When the problem involves consideration of only a few photons in the reservoir, a number of approaches developed around the Schroedinger equation for the truncated wave function were suggested[3,4]. Consideration of "few-photon" problems led to prediction of such phenomena as "freezing" of atomic population decay and existence of the atom-photon bound state[3], transparency near band-edge[5] and single-atom switching[6]. Simulation of atomic decay in realistic photonic crystals confirmed predictions of very large Lamb shift closely connected to atom-photon bound state[7].

Going beyond a "few-photon" limitation appears to be rather complicated and to this day unresolved task. Exact treatment of multi-photon problems involves a calculation of the the multi-dimensional photon distribution functions, which is rather difficult. A number of ways to circumvent a problem were suggested. Reduction of non-Markovian problem to the Markovian one by introducing quasi-modes[8–10] is likely to be one of the most effective ways for investigation of the atomic interaction with radiation in structured reservoirs. The main idea of this method is to replace an ensemble of the reservoir modes interacting with the atom with few quasi-modes damping due to the interaction with Markovian reservoirs. The method is capable to provide a possibility to go beyond a "few-photon" restriction[10]. However, the major restriction imposed on the method in its current form is that the density of states is to be continuous and of rather specific form. An opportunity to apply it to reservoirs with gaps is limited to the densities reaching zero value in a finite number of points (such as, for example, a Fano-profile density[11]).

One should also mention a direct estimation of the density matrix of a two-level atom for solving the problem of resonance fluorescence into a structured reservoir[12]. Unfortunately, a numerical extensiveness of this technique has not still allowed to apply it to even most common examples of densities used for modelling. Much simpler are perturbative methods such as the one applied in the work[6]. It appeared, that even in frames of Born approximation it is possible to capture the "freezing" of decay. However, the method fails for some densities (for example, for the isotropic one). Another perturbative technique tailored to the conditions when the "freezing" of decay is possible[13], stems from the method of collective operators[1]. This method is based on a diagonalization of the atom-reservoir interaction Hamiltonian, and on appearance of a large parameter corresponding to a separation of one eigenvalue of the Hamiltonian from the rest; this makes all but one collective operators to be close to bosonic ones. This perturbation techniques still poses a limitation on the reservoir excitation.

Present work proposes a way to solve the "beyond a few photons" problem by unifying advantages of the quasi-modes and the collective operator methods. We demonstrate that, for the cases when the "freezing" of decay takes place in absence of interaction with external systems, the collective operator approach provides a possibility to reduce the problem of interaction with the structured reservoir to a simple master equation which, for a wide range of reservoir spectra, can be reduced to the Markovian form.

The outline of the papers is as follows: in the next Section we define the range of problems which our consideration is aimed at, and describe the method of collective operators. Section 3 is devoted to a description of the algebraic procedure transforming the initial atom-reservoir interaction Hamiltonian into the one describing an atom interacting with two field modes, of which only one is coupled to the reservoir. Section 4 is about the approximation allowing to obtain in Section 5 a master equation for the problem involving an interaction of an atom with reservoirs near band-gap edge, where the spontaneous decay of atomic population can be "frozen". The obtained master equation describes an interaction of the two-level atom with two field modes; only one is damped. For a wide range of densities, this equation can be reduced to the Markovian one, and, for some densities, the second mode can be adiabatically eliminated. This procedure is described in Section 6. Examples of spontaneous decay and resonance fluorescence are considered in Section 7; it is demonstrated that the equation is able to provide a correct stationary atomic population. For resonance fluorescence a prediction of a stationary positive atomic inversion is confirmed.

2. The problem and the collective operator approach

In this work we will obtain the master equation for problems described by the interaction Hamiltonian

$$H_{gen} = H_{out} + H_{res}, \tag{1}$$

where H_{out} describes interaction of a two-level atom with systems not included in the reservoir, and H_{res} is a standard Hamiltonian describing interaction of the same atom with an ensemble of reservoir modes. In the frame rotating at the frequency of atomic transition (here we use the system of units where $\hbar = 1$):

$$H_{res} = \sum_{j=1} \Delta_j b^\dagger b + \sum_{j=1} g_j(\sigma^+ b_j + b_j^\dagger \sigma^-), \tag{2}$$

where b_j and $b_j^\dagger$ are annihilation and creation operators for reservoir modes; $\sigma^\pm = |\pm\rangle\langle\mp|$; vectors $|\pm\rangle$ correspond to the lower and upper states of the atom; g_j are interaction constants; Δ_j are detunings of reservoir mode frequencies from the atomic transition frequency.

A starting point of our approach is a representation of H_{res} in terms of collective operators[13]. Introducing with the help of the unitary matrix U the collective operators

$$C_j = U_{j0}\sigma^- + \sum_{k=1} U_{jk} b_k,$$

the Hamiltonian (2) can be diagonalized[1,14]:

$$H_{res} = \sum_{j=0} \lambda_j C_j^\dagger C_j. \tag{3}$$

The elements of the transformation matrix are given by the following formula

$$U_{j0} = \Big(1 + \sum_{k=1} g_k^2/(\lambda_j - \Delta_k)^2\Big)^{-1/2}, \quad U_{jk} = U_{j0} g_k/(\lambda_j - \Delta_k). \tag{4}$$

Eigenvalues are provided by

$$\lambda_j = \sum_{k=1} g_k^2/(\lambda_j - \Delta_k). \tag{5}$$

It is useful to note that, apart from usual unitary relations for the coefficients U_{jk} defined by the equation (4), and the eigenvalues defined by the equation (5) the following relation holds

$$\sum_{j=0} \lambda_j U_{j0}^2 = 0. \tag{6}$$

The collective operators C_j satisfy the following commutation relations:

$$[C_j, C_k] = [C_j^\dagger, C_k^\dagger] = 0, \qquad [C_j, C_k^\dagger] = \delta_{jk} - 2U_{0j}U_{0k}\sigma^\dagger\sigma^-. \tag{7}$$

Taking into account relations (7), the Hamiltonian (3) leads to the equations of motion

$$i\frac{d}{dt}C_j - \lambda_j C_j + [H_{out}, C_j] = -2U_{0j} \sum_{k,l,n=0} \lambda_n U_{k0} U_{l0} U_{n0} C_k^\dagger C_l C_n. \tag{8}$$

For the case of spontaneous emission, when $H_{out} = 0$, and the initial state of the system belongs to the single-photon subspace, the action of the right-hand side of every equation (8) on the wave-function belonging to that subspace is zero. Thus for the single-photon problem

$$C_j(t) = C_j(0)\exp\left(-i\lambda_j t\right). \tag{9}$$

In particular, for the spontaneous emission of the completely excited atom into the initially empty reservoir, for the upper-state atomic population, one has

$$\langle \sigma^\dagger(t)\sigma^-(t)\rangle = \sum_{j,k=0} \langle C_j^\dagger(t) C_k(t)\rangle = |\sum_{j=0} U_{0j}^2 \exp(-i\lambda_j t)|^2. \tag{10}$$

One has to emphasize that, for the spontaneous emission, the collective operator method is completely equivalent to the traditional Schroedinger equation approach. However, it opens more possibilities even without any approximation. For example, it allows one to find an analytic solution for the decay problem with two photons in the reservoir[13].

3. The method and the procedure

The key point of our approach is a observation that, when the decay can be "frozen" and the atom-filed bound state created, it means that in the stationary state the atom remain entangled with the part of the reservoir. So, one should expect that there are a set of modal operators corresponding to this part, for which a non-Markovian effects of interaction with the atom are the most significant.

To find possible candidates on the role of such operators we use the collective operator method[13]. Collective operators C_j include both the atomic and field operators. The equations of motion (8) for them are quite complicated (some ways to deal with the system (8) were suggested in works[13]). However, by separating the atomic and the field parts in the collective operators, a significant simplification can be achieved. Representing collective operators as

$$C_j = U_{0j}\sigma^- + B_j, \quad B_j = \sum_{k=1} U_{jk} b_k, \tag{11}$$

the Hamiltonian (3) can be recast in the following form

$$\begin{aligned} H_{res} = \lambda_0 U_{00}(\sigma^\dagger B_0 + B_0^\dagger \sigma^-) + \lambda_0 B_0^\dagger B_0 + \\ + A^\dagger \sigma^- + \sigma^\dagger A + \sum_{j=1} \lambda_j B_j^\dagger B_j, \end{aligned} \tag{12}$$

where $A = A_r - \lambda_0 U_{00} B_0$. Here we do not specify which operator to chose from the set of B_j; any of them in place of B_0 in the equation (12) will be an appropriate choice for the procedure described below. However, in the next Section it will be seen that the choice of B_0 corresponding to the specific eigenvalue λ_0 is deliberate.

The operators B_0, $B_{j>0}$ and A are not mutually commuting. One needs to introduce mutually commuting collective modal operators. It can be done by the following linear transformation:

$$A \rightarrow \acute{A} = A - \frac{[B_0, A^\dagger]}{[B_0, B_0^\dagger]} B_0,$$

$$B_{j>0} \rightarrow \acute{B}_{j>0} = B_{j>0} - \frac{[B_0, B_{j>0}^\dagger]}{[B_0, B_0^\dagger]} B_0 - \frac{[\acute{A}, B_{j>0}^\dagger]}{[\acute{A}, \acute{A}^\dagger]} \acute{A}. \tag{13}$$

From the unitarity of U_{jk} and (4) one has

$$\begin{aligned}
\nu_0 =&[B_0, B_0^\dagger] = 1 - U_{00}^2,\\
\nu_A =&[\acute{A}, \acute{A}^\dagger] = \sum_{l=1} g_l^2 - \lambda_0^2 U_{00}^2 \nu_0^{-1},\\
\nu_{0j} =&[B_0, B_j^\dagger] = -U_{00} U_{j0},\\
\nu_{Aj} =&[\acute{A}, B_j^\dagger] = U_{j0}(\lambda_j + \lambda_0 U_{00}^2 \nu_0^{-1}),\\
\nu_{0A} =&[B_0, A^\dagger] = \lambda_0 U_{00}^3,\\
\nu_{jk} =&[\acute{B}_j, \acute{B}_k{}^\dagger] = \delta_{jk} - U_{0j} U_{0k} \nu_0^{-1} - \nu_{Aj} \nu_{Ak} \nu_A^{-1}.
\end{aligned} \tag{14}$$

where $j, k > 0$. Now normalizing with help of (14) the operators B_0, $\acute{A}$ and $\acute{B}_{j>0}$, we finally introduce

$$\bar{B}_0 = \acute{B}_0 \sqrt{\nu_0^{-1}}; \quad \bar{B}_j = \acute{B}_j \sqrt{\nu_{jj}^{-1}}; \quad \bar{A} = \acute{A} \sqrt{\nu_A^{-1}}. \tag{15}$$

Using operators $\bar{B}$, $\bar{A}$ and $\bar{B}_{j>0}$, H_{res} can be cast in the following form:

$$\begin{aligned}
H_{res} =& -\lambda_0 C^\dagger C + \lambda_0(\sigma^\dagger \sigma^- + \bar{B}^\dagger \bar{B}) +\\
&+\omega_A \bar{A}^\dagger \bar{A} + G(C^\dagger \bar{A} + \bar{A}^\dagger C) +\\
&+\sum_{j=1} \bar{\lambda}_j \bar{B}_j{}^\dagger \bar{B}_j + \sum_{j=1} \bar{g}_j (\bar{A}^\dagger \bar{B}_j + \bar{B}_j{}^\dagger \bar{A}),
\end{aligned} \tag{16}$$

where the new collective operator

$$C = \sigma^- - \bar{B} \frac{U_{00}}{\sqrt{1 - U_{00}^2}},$$

and, for $j > 0$,

$$\omega_A = \sum_{l=1} \lambda_l \frac{\nu_{Al}^2}{\nu_A}, \quad \bar{g}_j = \lambda_j \nu_{Aj} \sqrt{\frac{\nu_{jj}}{\nu_A}}, \quad \bar{\lambda}_j = \lambda_j \nu_{jj}, \quad G = \sqrt{\nu_A}. \tag{17}$$

The procedure of orthogonalization can be continued similarly to the one described above. Generally, it is possible introducing a set of independent bosonic operators

$$P_l = W_{0l}\bar{A} + \sum_{j=1} W_{lj}\bar{B}_j,$$

transform the Hamiltonian (16) to the following one

$$H_{res} = -\lambda_0 C^\dagger C + \lambda_0(\sigma^\dagger\sigma^- + \bar{B}^\dagger\bar{B}) + \\ +G\sum_{j=1} W_{j0}(C^\dagger P_j + P_j^\dagger C) + \sum_{j=1} \bar{\mu}_j P_j^\dagger P_j, \tag{18}$$

We will not discuss here the details of the procedure leading to the Eq. (18). In this paper we limit our consideration with the orthogonalization resulting in the Hamiltonian (16).

It should be stressed that Hamiltonians (2), (3) and Hamiltonians (16), (18) are completely equivalent and can be transformed one to another by linear transformations over the set of bosonic operators. Until now we have made no assumptions about U_{0j} and implemented no approximations.

4. Dominant coefficient approximation

Now let us consider the Hamiltonian (16) for the case when decay of the atom interacting with the reservoir can be "frozen". It is known that for the atomic transition frequency near the gap edge (one should note that for certain kinds of densities, for example, for the isotropic one, the frequency could even be outside the gap), one of the eigenvalues, λ_0, is separated from others[1,13]. It leads (as follows from the equation (4)) to the dominance of the corresponding matrix element $|U_{00}| >> |U_{j0}|$ for all $j > 0$ and, as consequence, demonstrates the formal mechanism underlaying the "freezing" of spontaneous decay. For the spontaneous emission of the initially excited atom into the initially empty reservoir, for the upper-state atomic population from the equation (10) it follows, that only the dominant term in the right side of the equation (10) survives, and the "frozen" stationary amplitude is

$$\langle\sigma^+(\infty)\sigma^-(\infty)\rangle = |U_{00}^4|. \tag{19}$$

Here and further in our discussion we assume that the atomic transition frequency is close to the high-frequency edge of the gap. We illustrate the separation of the eigenvalue λ_0 and "freezing" of decay with the example of the step-like one density of states extensively discussed in theoretical works. It is discretized as it was suggested and implemented in works[4,13], and the matrix U was found.

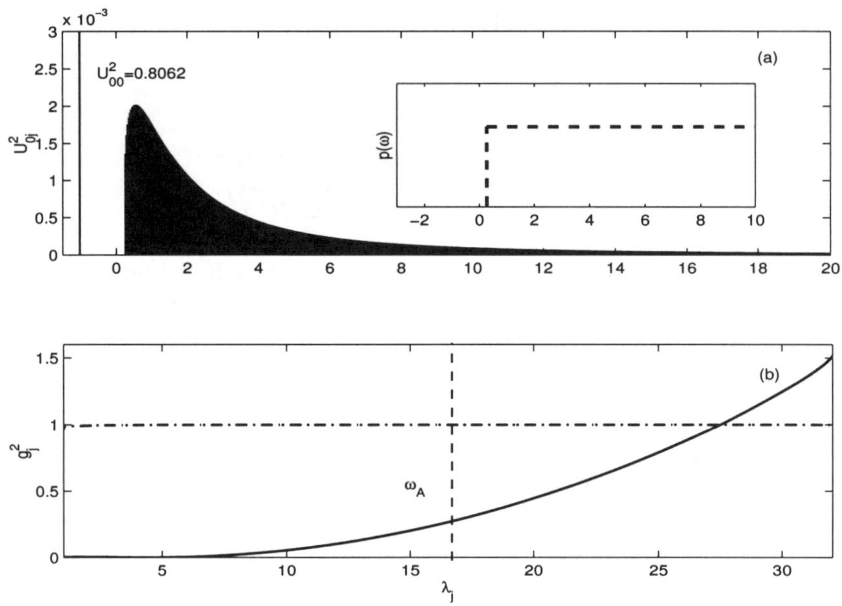

Figure 1. A distribution of matrix coefficients U_{0j}^2 for the step-like density of states (a) and distribution of the coefficients $\bar{g}_j^2$(b). Vertical dotted line denotes the frequency ω_A position; dash-dotted line shows the distribution of ν_{jj}. The inset in (a) shows the step density adopted for simulation. The distance between the atomic transition frequency and the the gap edge is $0.25D$, the frequency interval used for simulation is defined by $\omega_h - \omega_g = 32D$; $N = 1000$.

In Figure 1(a), a distribution of U_{0j} is shown; a dominance of U_{00} is demonstrated. It is assumed that all constants of interaction g_k are equal

$$g_k = g = \sqrt{\frac{1}{N}\int\limits_{\omega_g}^{\omega_h} d\omega p(\omega)} = \sqrt{\frac{D}{\pi N}(\omega_h - \omega_g)},$$

where $p(\omega)$ is the projected local density of states[15]. The parameter D reflects a respective strength of interaction between the atom and modes of the structured reservoir; ω_g is the frequency of the band edge; ω_h is the frequency of the upper border of the region chosen for the discretization (it is assumed to be arbitrarily large but finite); N is the number of discretization points. The set of frequencies for the reservoir is supposed to be equidistant. A few hundred discretization points are enough to achieve good convergence of the upper-state atomic population for the values of parameters used in our example. For our choice of parameters the eigenfrequency $\lambda_0 = -1.037D$, whereas $\omega_g - \omega_0 = 0.25D$; so the eigenfrequency corresponding to the collective operator C_0 is pushed far into the gap.

It is useful to note that the discretization procedure described in works[4,13] allows, in some cases, to find a simple approximation for the coefficients U_{0j} and eigenvalues λ_j far from the gap edge. Indeed, the discretization procedure implies that the distances between the adjacent points are inversely proportional to value of projected local density of states $\omega_{j+1} - \omega_j \sim p(\omega_j)^{-1}$. For non-decreasing monotonous density of states $p(\omega)$ (such, for example, as step-like or anisotropic ones) far from the gap edge, the eigenvalues λ_j tend to be very close to the corresponding frequencies ω_j. So from equations (4,5) it follows that

$$\lambda_j \approx \Delta_j + \frac{g_j^2}{\Delta_j}, \qquad \lambda_j^2 U_{0j}^2 \approx g_j^2. \tag{20}$$

As follows from this formulae, far from the gap edge the coefficients U_{0j} decrease in inverse proportion to the frequency. This leads to the conclusion that the steeper the density of states increases, the smaller frequency interval near the edge can be taken to provide good correspondence between the solution obtained as the result of discretization and the exact one for the continuous spectrum.

5. Master equation

Obviously, the Hamiltonian (16) is not much simpler to deal with than Hamiltonians (2), (3), because the operators $\bar{B}_j$ are not mutually independent. However, when the "freezing" of decay is possible, due to the relation $|U_{00}| >> |U_{0j}|$ the operators $\bar{B}_j$ (15) are close to the mutual independence. For the discussed example of the reservoir with step-like density of states, one has, for $j \neq k$: $\max|[\bar{B}_j, \bar{B}_k^\dagger]| < 10^{-2}$. For the same reason, far from the gap edge, one has $|\bar{g}_j| << \bar{\lambda}_j$. Assuming that the operators $\bar{B}_{j>0}$ do commute, and taking to the interaction picture with respect to

$$H_0 = \lambda_0(\sigma^\dagger\sigma^- + \bar{B}^\dagger\bar{B}) + \omega_A\bar{A}^\dagger\bar{A} + \sum_{j=1}\bar{\lambda}_j\bar{B}_j{}^\dagger\bar{B}_j \tag{21}$$

the following interaction Hamiltonian can be obtained

$$V = -\lambda_0 C^\dagger C + G(C^\dagger \bar{A} \exp(-i(\omega_A - \lambda_0)t) + h.c.) + \\ + \sum_{j=1} \bar{g}_j (\bar{A}^\dagger \bar{B}_j \exp\{-i(\bar{\lambda}_j - \omega_A)t\} + h.c.). \tag{22}$$

Now if we assume that the initial state of the reservoir modes (and, consequently, of all $\bar{B}_j$ modes) are vacuum, then taking $\bar{g}_j$ for small parameters, and applying Born approximation one obtains from the Hamiltonian (22) the following master equation for the density matrix averaged over the vacuum of the modes $\bar{B}_j$:

$$\frac{d\rho}{dt} = -i[\bar{H}_{out} - \lambda_0 C^\dagger C + \\ +G(C^\dagger \bar{A} \exp(-i(\omega_A - \lambda_0)t) + h.c.), \rho] + \\ + \int_0^t d\tau \Big(2Re(K(t-\tau))\bar{A}\rho(\tau)\bar{A}^\dagger - \\ -K(t-\tau)\bar{A}^\dagger \bar{A}\rho(\tau) - K^*(t-\tau)\rho(\tau)\bar{A}^\dagger \bar{A} \Big), \tag{23}$$

where $\bar{H}_{out}$ is the Hamiltonian H_{out} affected by the transformation to the interaction picture; the reservoir correlation function is

$$K(t) = \sum_{j=1} \bar{g}_j^2 \exp\{-i(\bar{\lambda}_j - \omega_A)t\}. \tag{24}$$

It is to be stressed that, when deducing the equation (24) no assumptions about excitation of reservoir modes were made. The equation (24) can be used for arbitrary H_{out}.

It is interesting to note that the direct application of Born approximation to the Hamiltonian (2) done in the work[6], where the original interaction constants g_j were chosen to play the role of small parameters, failed to capture the "freezing" for isotropic density of states when the atomic transition frequency was close to the gap edge but outside the gap. Moreover, recently it was demonstrated[16], that the master equation with the reservoir memory effects introduced in the way similar to those resulted from the Born approximation, may occasionally lead to non-physical features, for example, to non-positivity of the diagonal elements of the density matrix. So it seems useful to point out that the equation (24) is always able to capture the "freezing" of atomic population decay if the system described by the Hamiltonian (2) allows for it. Indeed, it is not hard to notice that the master equation (24) for $H_{out} = 0$ has a stationary solution

$$\rho_{st} = \sum_{\alpha,\beta} s_{\alpha\beta}|\alpha\rangle\langle\beta|, \tag{25}$$

where $\alpha, \beta = \psi, v$. The state $|v\rangle = |-\rangle|0\rangle_B|0\rangle_A$; the wave-vector $|\psi\rangle$ satisfies the following equations $C|\psi\rangle = \bar{A}|\psi\rangle = 0$, which gives

$$|\psi\rangle = (U_{00}|+\rangle|0\rangle_B + \sqrt{1 - U_{00}^2}|-\rangle|1\rangle_B)|0\rangle_A, \tag{26}$$

where $|0,1\rangle_{A,B}$ are Fock states for modes $\bar{A}$ and $\bar{B}$. If initially the atom is excited, and the reservoir is in the vacuum state, then this initial states can be represented as

$$|+\rangle|0\rangle_A|0\rangle_B = U_{00}|\psi\rangle + \sqrt{1 - U_{00}^2}|\chi\rangle,$$

where the wave-vector $|\chi\rangle$ is orthogonal to $|\psi\rangle$. Thus one may conclude that

$$Tr(\sigma^+\sigma^-\rho_{st}) = |U_{00}|^4, \tag{27}$$

which is exactly the value of "frozen" atomic population given by the equation (19).

6. Reduction to the Markovian form

Here we will not consider the exact solution of the equation (24) concentrating our attention on the situation when this equation can be reduced to the Markovian form. An important feature of the master equation (24) is that the frequency of the damped mode $\bar{A}$ may appear to be far from the gap edge and deep inside the continuum of the reservoir modes. As follows from the definition of ω_A (17), for a wide range of reservoir spectra (for example, for all three most commonly considered isotropic, step-like and anisotropic model densities; generally, it obviously true for any non-decreasing $p(\omega > \omega_g)$) for the atomic transition frequency sufficiently close to the gap edge, the frequency ω_A can satisfy $(\omega_A \gg |\omega_0 - \omega_g|)$; the distribution of $\bar{g}_j^2$ may be smooth near ω_A (an example for the step-like density is shown in Figure 1(b), and then the correlation function (24) is sharply peaked near $t = \tau$. These facts naturally lead to the idea of applying Markovian approximation to the equation (24) for such densities. One has also to note that in the discussed case a possibility to reduce the master equation (24) to the markovian one corresponds to the possibility to have a smooth spectrum of the coefficients W_{0j} in the Hamiltonian (18). However, here we are again leaving a more detailed discussion to be done elsewhere; as it will be seen below, our approach allows to obtain approximate solutions close to the exact ones both for the initial stage of evolution and in the long time limit; the stationary solution obtained with help of the approach coincide with exact ones.

Thus for the density $p(\omega)$ and the atomic transition frequency satisfying the conditions discussed above one can obtain the following Markovian master equation

$$\begin{aligned}\frac{d\rho}{dt} \approx -i[\bar{H}_{out} - \lambda_0 C^\dagger C + \delta(t)\bar{A}^\dagger \bar{A} + \\ +G(C^\dagger \bar{A}\exp(-i\omega_A t) + h.c.), \rho] + \\ +\gamma(t)\Big(2\bar{A}\rho\bar{A}^\dagger - \bar{A}^\dagger\bar{A}\rho - \rho\bar{A}^\dagger\bar{A}\Big),\end{aligned} \tag{28}$$

where the time-dependent relaxation rate $\gamma(t)$ and the frequency shift $\delta(t)$ are given by the real and imaginary parts of the integrated correlated function (24):

$$\gamma(t) + i\delta(t) = \int_0^t K(\tau)d\tau. \tag{29}$$

For $t \to +\infty$ both $\gamma(t)$ and $\delta(t)$ tend to the constant values. One has to note that for examples considered here the transition time can be comparable with the typical time-scale of the atomic evolution. So, further in our investigation it is assumed that we are interested in the system evolution on times far exceeding the transition time. Then the following estimation for the stationary values can be made:

$$\gamma = \pi\bar{p}(\omega_A), \quad \delta = P\int d\bar{\lambda}\frac{\bar{p}(\bar{\lambda})}{\bar{\lambda} - \omega_A}, \tag{30}$$

where $\bar{p}(\bar{\lambda})$ is a projected local density of states for the reservoir described by the set of $\bar{B}_j$.

If conditions leading to the approximations (20) are satisfied, then, far from the gap edge, the coefficients ν_j tend to unity (as it is illustrated for step-like density in Figure 1(b)), and from equations (17), (20) one has $\bar{g}_j^2 \approx \Delta_j^2 g_j^2/\nu_A$. Thus from formulae (30) it follows

$$\begin{aligned}\gamma = \pi p(\omega_A + \omega_0)\frac{\omega_A^2}{\nu_A}, \\ \delta = \frac{1}{\nu_A}P\int_{\omega_g}^{\omega_h} d\omega\frac{(\omega - \omega_0)^2 p(\omega)}{\omega - \omega_A - \omega_0},\end{aligned} \tag{31}$$

and the frequency ω_A is in fact the averaged over the reservoir spectrum detuning from the atomic transition frequency

$$\omega_A \approx \frac{1}{\nu_A} \int_{\omega_g}^{\omega_h} d\omega(\omega - \omega_0)p(\omega).$$

For example, for step-like density of states $\omega_A \approx (\omega_h + \omega_g - 2\omega_0)/2$. In fact, the quantity ν_A is a normalization constant: $\nu_A \approx \int_{\omega_g}^{\omega_h} d\omega p(\omega)$.

The master equation (28) can be simplified further if γ is sufficiently large to allow for adiabatic elimination of the mode A. For example, for the discussed here step-like density $\gamma \sim \omega_A$ and increases with increasing ω_h. It is worth noting that a similar phenomenon occurs also for isotropic and anisotropic densities. It is useful to note that γ might be large in comparison with λ_0 not only for extended spectra (when ν_A tends to infinity for $\omega_h \to \infty$), but also for limited spectra as, for example, the Fano-profile spectrum considered in the work[11] where, for such spectrum, the Markovian master equation similar to the one discussed here was obtained.

Adiabatically eliminating the mode A, one obtains the final Markovian equation of our consideration

$$\frac{ds}{dt} = -i[\bar{H}_{out} - (\lambda_0 + \bar{\delta})C^\dagger C, s] + \bar{\gamma}\Big(2CsC^\dagger - C^\dagger Cs - sC^\dagger C\Big),$$

where s is the density matrix ρ traced over the mode A; and

$$\bar{\gamma} = \frac{\nu_A \gamma}{\gamma^2 + (\omega_A + \delta)^2}, \quad \bar{\delta} = \frac{\nu_A(\omega_A + \delta)}{\gamma^2 + (\omega_A + \delta)^2}.$$

For step-like density of states, in the limit of large ω_h, one has $\bar{\gamma} \approx 4D/(4 + \pi^2)$ and $\bar{\delta} \approx 2\bar{\gamma}/\pi$; so the decay rate seems to be independent of λ_0 and on the position of atomic transition frequency. However, as it is to be seen from the next Section, due to the properties of the operator C, the dynamics appears to be strongly dependent on the parameter U_{00}.

7. Examples

In this Section, we consider some problems to which the master equation (32) can be applied. First of all, it is the traditional problem of a spontaneous decay of a single-photon initial state (for example, the initially fully excited atom's decay into the vacuum of the reservoir). We compare the obtained analytical solution with the exact one, obtained via the method of collective operators described in Section 2. Also we consider the multi-photon problem of decay of N-photon states of the correlated atom and the collective mode B_0. Physically, it would correspond to rather specifically arranged initial excitation of the reservoir. However, it allows us to illustrate a number of novel and unexpected features of the atom-field bound state dynamics. Notwithstanding the Markovianity of the master equation (32), the stationary state appears to be strongly dependent on the initial state. Finally, we look at the probing ground of the "beyond a few photons" problem of interaction with structured reservoirs, namely on the interaction of the atom with a classical driving field. We show, that to achieve higher values of the stationary atomic upper state population, one should pump the atom off-resonantly.

7.1. *Decay of a single-photon initial state*

To demonstrate the validity of our method let us start with the spontaneous emission of an initially excited atom into the vacuum of the reservoir ($H_{out} = 0$). It is easy to see that in this case (similarly to the discussed equation (24)) these is a stationary solution (25); where the vector

$$|\phi\rangle = U_{00}|+\rangle|0\rangle_B + \sqrt{1 - U_{00}^2}|-\rangle|1\rangle_B \tag{32}$$

satisfies $C|\phi\rangle = 0$. If the atom is initially excited, then the equation (32) for the stationary value of atomic population gives $\langle\sigma^+\sigma^-\rangle_{st} = |U_{00}|^4$, which is exactly the asymptotical "frozen" value of the atomic population given by the equation (10). It's interesting to notice that the vector

$$|\chi\rangle = \sqrt{1 - U_{00}^2}|+\rangle|0\rangle_B - U_{00}|-\rangle|1\rangle_B$$

is an eigenvector of the operator C^+C:

$$C^+C|\chi\rangle = (1 - U_{00}^2)^{-1}|\chi\rangle.$$

Thus in one-photon subspace one has $C^+C = (1 - U_{00}^2)^{-1}|\chi\rangle\langle\chi|$, and, naturally, from the master equation (32) it follows

$$\frac{d}{dt}\langle C^+(t)C(t)\rangle = -2G\langle C^+(t)C(t)\rangle$$

with the solution

$$\langle C^+(t)C(t)\rangle = \langle C^+(0)C(0)\rangle \exp\left(-2Gt\right), \tag{33}$$

where $G = \bar{\gamma}/(1 - U_{00}^2)$.

So instead of the atomic population the "population" of the collective atom-field object is subjected to exponential decay; and this decay is faster for the larger detuning of the atomic transition frequency from the gap edge. However, the time required to reach the stationary state simultaneously decreases, and for $U_{00} \to 1$ and $G \to \infty$ the atom tends to remain in it's initial state, since $|\phi\rangle \to |+\rangle|0\rangle_B$.

The atomic population behaves in a quite different way. Introducing the density matrix elements

$$s_{rf}(t) = \langle r|s(t)|f\rangle, \quad r, f = \chi, \phi$$

in the basis of vectors $|\chi, \phi\rangle$, from the master equation (32) one obtains $s_{\phi\phi}(t) = s_{\phi\phi}(0)$, and

$$\frac{d}{dt}s_{\chi\phi}(t) = -\frac{i(\lambda_0 + \bar{\delta}) + \bar{\gamma}}{1 - U_{00}^2}s_{\chi\phi}(t). \tag{34}$$

From equations (33,34) it follows for the atomic population

$$\langle\sigma^+(t)\sigma^-(t)\rangle = U_{00}^2 s_{\phi\phi}(0) + (1 - U_{00}^2)s_{\chi\chi}(0)\exp(-2Gt) + \\ +2U_{00}\sqrt{1 - U_{00}^2}\exp(-Gt)Re\left\{s_{\phi\chi}(0)\exp\left(-i\frac{\lambda_0 + \bar{\delta}}{1 - U_{00}^2}t\right)\right\}. \tag{35}$$

For the initially excited atom from the equation (35) it follows that

$$\langle\sigma^+(t)\sigma^-(t)\rangle = U_{00}^4 + (1 - U_{00}^2)^2\exp(-2Gt) + \\ +2U_{00}^2(1 - U_{00}^2)\exp(-Gt)\cos\left(\frac{\lambda_0 + \bar{\delta}}{1 - U_{00}^2}t\right). \tag{36}$$

Looking on this equation it may seem that (similarly to the equation (33)) it predicts an acceleration of the population decay when U_{00} approaches unity. However, it is not the case. For the initial stage of dynamics from the equation (36) one has simply

$$\langle\sigma^+(t)\sigma^-(t)\rangle \approx 1 - 2\bar{\gamma}t,$$

i.e. short-time evolution does not depend on the atomic detuning from the gap-edge. As it was discussed in the work[17], exact solutions for the atomic population decay near the gap edge reveal the same feature, and the rate of population decay on the initial stage depends on the $p(\omega)$ rather than on $\omega_0 - \omega_g$.

In Figure 2 we've illustrated solutions for the atomic population given by the equation (36) for different ω_0; for comparison exact solutions (10) provided by the method of collective operators are also shown. It can be seen that the Markovian master equation (32) provides a good agreement with the exact solution when $t \gg 1/\bar{\gamma}$, as it should be expected from the solution of the master equation.

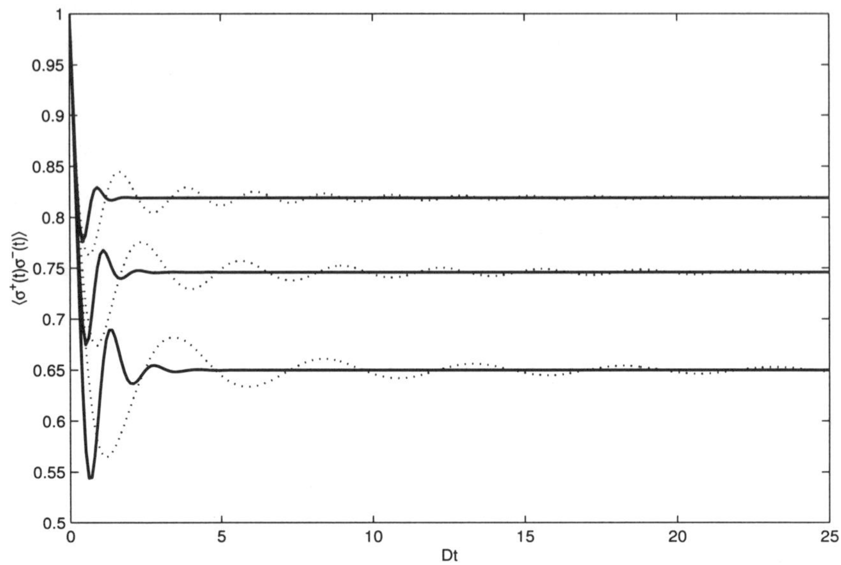

Figure 2. An illustration of atomic population dynamics described by the equation (32) for the spontaneous decay into the vacuum of the reservoir (solid lines); for comparison exact solutions given by the equation (10) are plotted (dotted lines). Curves correspond in descending order to $\omega_g - \omega_0 = 2D, 1.2D, 0.25D$. Other parameters are as for Figure 1.

7.2. *Decay of a multi-photon initial state*

Here we consider a decay of the multi-photon initial states of the following form

$$|\psi(0)\rangle = \cos(\alpha)|+\rangle|N-1\rangle + \sin(\alpha)|-\rangle|N\rangle. \tag{37}$$

This state belongs to the subspace with precisely N photons. Damping described by the master equation (32) induces transitions to the subspaces with lower number of photons. Generally, $s(t)$ can be represented as a sum of conditional density matrices

$$s(t) = \sum_{n=0} \rho_n(t),$$

where the matrix ρ_n belongs to the subspace with precisely $N-n$ photons. From (32) one gets the following equation for these conditional matrices

$$\frac{d\rho_n}{dt} = R^{(N-n)}\rho_n + J^{(N-n)}\rho_{n-1}. \tag{38}$$

Here $R^(k)$ and $J^{(k)}$ correspond to parts of the super-operators R and J (defined by $Js = \bar{\gamma} CsC^\dagger$ and $Rs = (-\bar{\gamma} + i(\bar{\delta} + \lambda_0))sC^\dagger C + h.c.$) acting on a subspace with k photons (see Appendix A). The equation (38) allows for easy deducing of an asymptotic solution. Using Laplace transform, from (38) one obtains

$$\rho_n(z) = \frac{1}{z - R^{(N-n)}}(-\rho_n(0) + J^{(N-n)}\rho_{n-1}(z)). \tag{39}$$

The asymptotic solution of (38) can be found, if one takes into account the following: the first, that $\rho_n(0) = 0$ for $n \neq 0$; the second, that for the Laplace transform one has for the function $f(t)$ and its Laplace transform $f(t \to +\infty) = \lim_{z\to 0}\{zf(z)\}$[18]. Then it can be seen from (39), that the asymptotic state of the system is a mixture of ρ^N and ρ^{N-1}; thus similarly to (25) it is possible to write

$$s_{st} = s_{\psi\psi}|\psi\rangle\langle\psi| + (1 - s_{\psi\psi})|v\rangle\langle v|, \tag{40}$$

where (Appendix A)

$$s_{\phi\phi} = \langle\phi|\Big\{\prod_{k=1}^{N-1} J^{(k)}\Big(R^{(k+1)}\Big)^{-1}\Big\}s(0)|\phi\rangle; \tag{41}$$

the product in (41) is assumed to be k-ordered.

In Figure 3(a) we show examples of atomic upper-state spontaneous decay dynamics for initial multi-photon state (37). For simulation of the master equation (32) the standard quantum Monte-Carlo method was used[19]. In Figure 3(b) results of (41) simulation are shown. One can see that the atomic population tends to the constant N-independent value for fixed α in the initial state (37). However, the asymptotic atomic population is extremely sensitive to the weights of component states in the initial superposition. In dependence on the phase β the population can change from nearly zero to U_{00}^2 even for fixed N (Fig. 4).

Such curious phenomenon can be explained by a very specific character of decay described by (32). Indeed, each act of photon emitting and jump to the subspace with lower number of photons (i.e. $s \to CsC^+$) creates an entanglement between the atom and the mode B_0 even if there were unentangled before the jump. The continuous evolution between jumps (governed by the operator C^+C) also entangles the atom and the mode. Obviously, results of such processes depend on the initial state notwithstanding the number of jumps and time interval required to reach

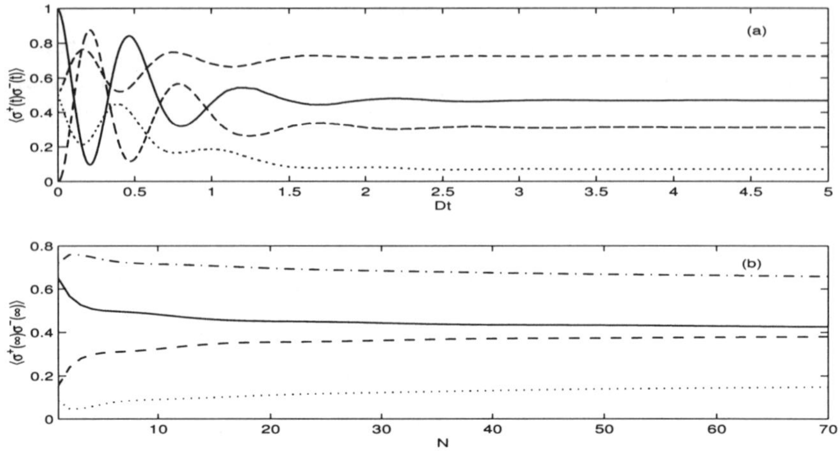

Figure 3. An illustration of atomic upper state population dynamics provided by numerical simulation of the equation (32) for spontaneous decay of a multi-photon state (37). In (a) examples of time-dependent dynamics for $N = 10$ are given for $\alpha = 0$ (solid line); $\alpha = \pi/2$ (dashed line); $\alpha = \pi/4$ (dash-dotted line); $\alpha = -\pi/4$ (dotted line). The summation was done over 7000 trajectories. In (b) asymptotic solutions for the upper-state atomic population provided by (41) are given in dependence on N. The values of α and corresponding lines are the same as in the previous figure. Other parameters are as for Figure 1.

the stationary state; the non-unitary evolution described by (32) does preserve phase information of the initial state.

The atom cannot hold more then one photon bounded in atom-photon bound state. However, this state "remembers" anything that was initially put in. The stationary state (40) of the atom+collective mode B_0 remains entangled is the atom-field entangled one with the concurrence[20] $c(s_{st}) = s_{\phi\phi}\sqrt{(x-1)^2+4x^3}/(x^2+1)$.

7.3. *Resonance fluorescence*

Finally, we apply our master equation (32) to the problem which became a testing ground for the approaches attempting to go "beyond a few photons". We consider a classically driven two-level atom the gap edge. We do not present here a detailed discussion of the problem limiting ourselves by demonstrating that our approach is able to capture features predicted in earlier work[6], namely, a possibility of positive stationary population inversion even for a modest intensities of driving field.

We consider both the cases when the driving frequency lies inside and

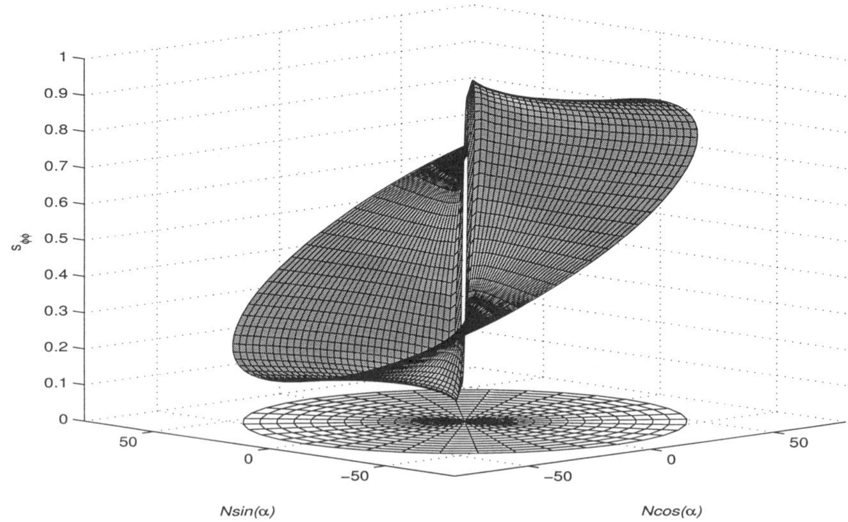

Figure 4. Illustration of $s_{\psi\psi}$ dependence on the respective phase α and number of photons N of the states in the initial superposition (37) as given by (41). Other parameters are as for Figure 1.

outside the gap in the continuum of reservoir states. In the first case the atom can be driven through the narrow band of allowed states created inside the gap, for example, by appropriate defect in the photonic crystal lattice[21]. The second case the problem is equivalent to one with the initial excitation of one of the reservoir modes with the coherent state of the amplitude Ω. The equation (32) can be directly applied to the problem of resonance fluorescence near the gap edge without any limitations on the driving intensity. If the atom is driven with classical field of the frequency $\omega_0 + \Delta_\Omega$, then

$$H_{out} = \Omega \exp\left(-i(\omega_0 + \Delta_\Omega)t\right)\sigma^+ + h.c.,$$

and in the basis rotating with the atomic transition frequency the equation (32) takes the following form

$$\frac{ds}{dt} = -i[(\Omega \exp\left(i(\lambda_0 - \Delta_\Omega)t\right)\sigma^+ + h.c.) - (\lambda_0 + \bar{\delta})C^\dagger C, s] + \\ +\bar{\gamma}\Big(2CsC^\dagger - C^\dagger Cs - sC^\dagger C\Big). \quad (42)$$

In Figures 5, 6 results of numeric simulation of (42) are given. It can be seen that comparatively weak excitation appears to be able to

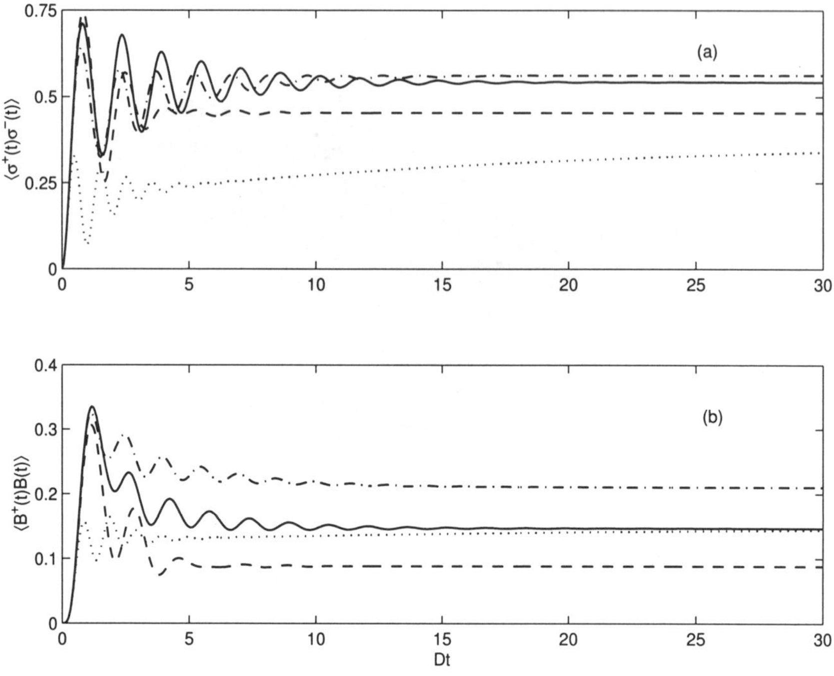

Figure 5. Atomic upper-state population (a) and average number of photons in the mode B_0 (b) provided by the numerical simulation of (42) are given for $\Delta_\Omega = -D, 0, 1.25D, 4D$ (dashed, solid, dash-dotted, dotted curves); for all curves $\Omega = 2D$. Other parameters are as for Figure 1.

produce a noticeable positive inversion. Additionally, to reach maximal positive inversion the atom should be driven off-resonantly (which was also pointed to in the work[6]). The equation (42) provides clues for qualitative explanation of these features. First of all, it is worth mentioning that, essentially, the master equation (42) describes a driven damped Jaynes-Cummings system, for which a possibility of a positive stationary inversion was predicted long ago[22]. The atomic driving produces a non-zero stationary average number of photons in the mode B_0. An increase in stationary atomic population is accompanied by increase in the average number of photons. As was discussed in the previous sub-Chapter, the system of atom+collective field is excited as a whole, and always behaves as an entangled one. Since the average number of the mode B_0 photons is less than unity for the amplitudes Ω considered here, one may conclude

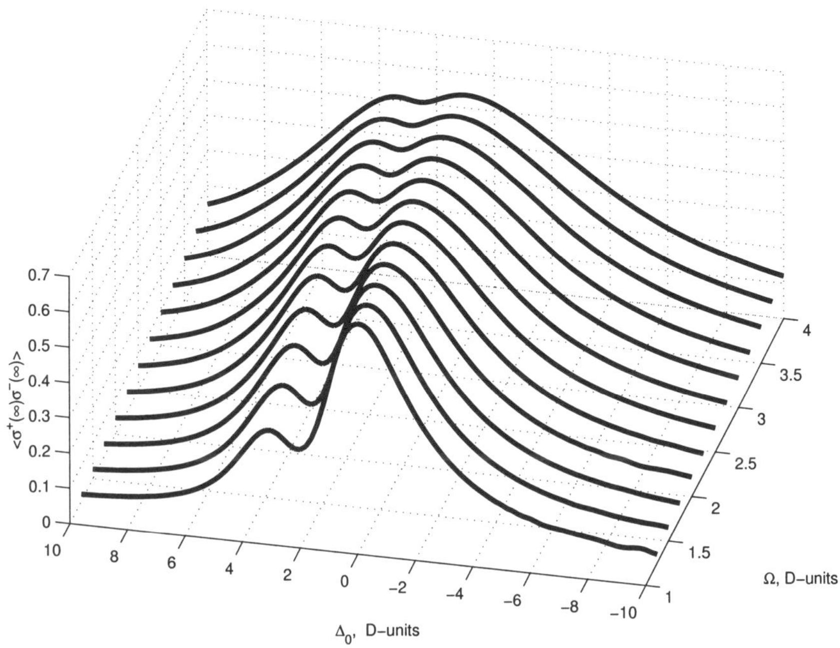

Figure 6. Dependence of stationary atomic upper-state population on the driving field detuning Δ_Ω and amplitude Ω as provided by the numerical simulation of (42). Other parameters are as for Figure 1.

that the system dynamics is limited mostly by one-photon sub-space. The driving removes the degeneracy between the lowest energy levels of the Jaynes-Cummings system, and shifts one of the resulting levels into the continuum. That explains positions of major maxima shown in Fig. 6. The driving field also splits the upper level (detuned from the frequency of the atomic transition by $-(\lambda_0 + \bar{\delta})/(1 - U_{00}^2)$), which is in the continuum near the position of the second maximum of the atomic upper-level stationary population. When the amplitude of the driving increases, the splitting increases also. Thus the major maximum shifts inside the continuum, both maxima becomes closer and the distribution of $\langle \sigma^+(\infty, \Delta_\Omega)\sigma^-(\infty, \Delta_\Omega\rangle)$ becomes wider.

It is useful to stress a difference in the nature of coupling strength Ω for in and out of the gap driving. If the atom is driven through the narrow band of allowed states inside the gap, than one should expect a strong coupling between the atom and the field[21]). As it was pointed in the beginning of this sub-Section, out of the gap driving is achieved by coherent excitation of one of the reservoir modes. So if the j-th mode is excited, then one has $\Omega = g_j\mu$, where μ stands for the amplitude of the coherent state of the mode. Obviously, the same Ω corresponds to much stronger driving field for the out of the gap driving.

8. Conclusions

To conclude, we have obtained a general master equation for treatment of atom-field interaction in structured reservoirs suitable for cases when effects of reservoir feedback are most important, namely, when the "freezing" of decay and existence of atom-photon bound states are possible. This was done by applying the collective operator method to the problem, and then by introducing collective field modes. In the derivation no suggestions about possible reservoir excitation were implemented. The method allows in a simple and illustrative way to go beyond a "few photon" restriction which was a major difficulty in the field so far. For wide range of densities of reservoir spectrum the master equation can be reduced to the Markovian form. We have illustrated our discussion with examples of spontaneous emission into the structured reservoir and resonance fluorescence near the gap edge. Consideration of spontaneous emission allow to determine time range when the obtained master equation allows for good agreement with the exact solution. As it is natural to expect from the Markovian equation, it works well for times exceeding the relaxation rate which is defined by the properties of the reservoir, and depends especially on the form of the projected local density of states in the nearest vicinity of the gap edge. We have also shown that for spontaneous decay of a multi-photon state of the atom and the collective mode coupled to it the stationary state of such a system is strongly dependent on it's initial state notwithstanding the markovian character of the evolution. For the resonance fluorescence the method allows to confirms the earlier prediction of stationary atomic inversion near the gap edge, and points to the way to explain this phenomena: the system is essentially a driven damped Jaynes-Cummings system.

The method can straightforwardly generalized for wide class of problems involving interaction with structured reservoirs, for example, for multi-level systems and systems of atom interacting with the reservoir field near gap-edge.

Acknowledgments

The authors gratefully acknowledge financial support from BRFFI.

Appendix A. Asymptotic solution for decay of multi-photon states

Here we derive the asymptotic solution of (39). Each matrix ρ_n can be represented as four-component vector

$$\bar{\rho}_n = [\rho_{qq}^{(n)}, \rho_{qr}^{(n)}, \rho_{rq}^{(n)}, \rho_{rr}^{(n)}]^T,$$

with elements $\rho_{qr}^{(n)} = \langle q_n|\rho_n(t)|r_n\rangle$, where $|q_n, r_n\rangle = |+\rangle|N-n-1\rangle_B$ or $|-\rangle|N-n\rangle_B$ (for $n = N$ it should be taken $|q, r\rangle = |-\rangle|0\rangle_B$).

In the basis of states $|q_n, r_n\rangle$ the super-operators $J^{(k)}$ and $R^{(k)}$ are the following 4×4 matrices:

$$R^{(k)} = \begin{pmatrix} 2\bar{\gamma}x^2k & -\tilde{\gamma}^*x\sqrt{k} & -\tilde{\gamma}x\sqrt{k} & 0 \\ -\tilde{\gamma}^*x\sqrt{k} & \tilde{\gamma}x^2k + \tilde{\gamma}^*y & 0 & -\tilde{\gamma}x\sqrt{k} \\ -\tilde{\gamma}x\sqrt{k} & 0 & \tilde{\gamma}^*x^2k + \tilde{\gamma}y & -\tilde{\gamma}^*x\sqrt{k} \\ 0 & -\tilde{\gamma}x\sqrt{k} & -\tilde{\gamma}^*x\sqrt{k} & 2\bar{\gamma}y \end{pmatrix}$$

where $y = 1 + x^2(k-1)$ and $\tilde{\gamma} = \bar{\gamma} + i\bar{\delta}_A$;

$$J^{(k)} = 2\bar{\gamma} \begin{pmatrix} x^2(k+1) & -x\sqrt{k+1} & -x\sqrt{k+1} & 1 \\ 0 & x^2\sqrt{k(k+1)} & 0 & -x\sqrt{k} \\ 0 & 0 & x^2\sqrt{k(k+1)} & -x\sqrt{k} \\ 0 & 0 & 0 & x^2k \end{pmatrix}.$$

From all matrices $R^{(k)}$ only one $R^{(1)}$ is degenerate (as it can be seen from the sub-Chapter 7.1, it is because of C^+C having the one-photon eigenstate with zero eigenvalue). So from (39) and properties of the Laplace transform it follows, that

$$\bar{\rho}_{N-1} = \lim_{z\to 0} \frac{z}{z + R^{(1)}} J^{(1)} \frac{1}{R^{(2)}} J^{(2)} \cdots \frac{1}{R^{(N)}} \bar{\rho}_0,$$

or, introducing the matrix transform $R_{diag}^{(1)} = U^{-1}R^{(1)}U$, one obtains

$$\bar{\rho}_{N-1} = UP^{(1)}U^{-1}J^{(1)} \frac{1}{R^{(2)}} J^{(2)} \cdots \frac{1}{R^{(N)}} \bar{\rho}_0,$$

where the matrix $P^{(1)}$ is the diagonal one with elements $P_{11}^{(1)} = 1$, $P_{jj}^{(1)} = 0$, $j = 2, 3, 4$.

The transformation matrix U is the one defining the transformation of $\bar{\rho}_{N-1}$ from the basis of states is diagonal in the basis of states $|q_{N-1}\rangle$,

$|r_{N-1}\rangle$ to the one of the states $|\phi\rangle$, $|\chi\rangle$. Taking this into account one finally obtains the formulae (41).

Bibliography

1. S. Ya. Kilin and D. S. Mogilevtsev, *Laser Physics* **2**, 153 (1992); S. Kilin and D. Mogilevtsev, *Opt. Spectr.* **74**, 583 (1993).
2. P. Lambropopoulos, G. M. Nikolopoulos, T. R. Nielsen, and S. Bay, *Rep. Prog. Phys.* **63**, 455 (2000).
3. V. P. Bykov, *Sov. J. Quantum Electron.* **4**, 863 (1975); E. Yablonovitch, *Phys. Rev. Lett.* **58**, 2059 (1987); S. John, *Phys. Rev. Lett.* **58**, 2486 (1987); S. John and J. Wang, *Phys. Rev. Lett.* **64**, 2418 (1990); Sajeev John and Tran Quang, *Phys. Rev.* A**50**, 1764 (1994); S. John and T. Quang, *Phys. Rev. Lett.* **74**, 3419 (1995).
4. Georgios M. Nikolopoulos, Soren Bay, and P. Lambropoulos, *Phys.Rev.* A**60**, 5079 (1999); Georgios M. Nikolopoulos and P. Lambropoulos, *J. Opt. B: Quantum Semiclass. Opt.* **3**, 115 (2001).
5. E. Paspalakis, N. J. Kylstra, and P. L. Knight, *Phys. Rev.* A**60**, R33 (1999).
6. Marian Florescu and Sajeev John, *Phys. Rev.* A**64**, 033801 (2001).
7. Xue-Hua Wang, Ben-Yuan Gu, Rongzhou Wang, and Hong-Qi Xu, *Phys. Rev. Lett.* **91**, 113904 (2003).
8. A. Imamoglu, *Phys. Rev.* A**50**, 3650 (1994).
9. B. M. Garraway, *Phys. Rev.* A**55**, 2290 (1997); A**55**, 4636 (1997).
10. B. J. Dalton, S. M. Barnett, and B. M. Garraway, *Phys. Rev.* A**64**, 053813 (2001); B. J. Dalton and B. M. Garraway, *Phys. Rev.* A**68**, 033809 (2003).
11. Soren Bay, P. Lambropoulos, and Klaus Molmer, *Phys. Rev.* A**57**, 3065 (1998).
12. M. W. Jack and J. J. Hope, *Phys. Rev.* A**63**, 043803 (2001).
13. D. Mogilevtsev and S. Kilin, *Opt. Spect.* **93**, 406 (2002); *Phys. Rev.* A**67**, 023815 (2003).
14. Similar formal diagonalization can be found, for example, in Rodney Loudon "The quantum theory of light", Oxford University Press, 1978.
15. Kurt Busch, Nipun Vats, Sajeev John, and Barry C. Saunders, *Phys. Rev.* A**62** 4251 (2000).
16. S. M. Barnett and S. Stenholm, *Phys. Rev.* A**64**, 033808 (2001).
17. D. Mogilevtsev and S. Kilin, "Kick-back" model for interaction with structured photonic reservoirs", to be published in Laser Physics (2004).
18. C. J. Savant "Fundamentals of the Laplace transformation", (McGraw-Hill, New-York, 1962).
19. K. Molmer, Y. Castin and J. Dalibard, *J. Opt. Soc. Am.* **B**10, 524 (1993).
20. Williams K. Wooters, *Phys. Rev. Lett.* **8**0, 2245 (1998).
21. J. D. Joannapoalos, R. D. Meade and J. N. Winn, "Photonic crystals: Molding the Flow of Light", Princeton, NJ:Princeton University Press, 1995.
22. C. M. Savage, *Phys. Rev. Lett.* **6**0, 1828 (1988); M. Lindberg and C. M. Savage, *Phys. Rev.* **A**38, 5182 (1988).

PARAMETRIC X-RADIATION: FROM THE THEORETICAL PREDICTION TO THE FIRST OBSERVATION AND APPLICATIONS

Vladimir Baryshevsky

Institute for Nuclear Research,
18 Bobruiskaya Str., 220050, Minsk, Belarus
E-mail: bar@inp.by

Ilya Feranchuk

Belarusian State University,
4 Fr. Skorina Ave., 220050, Minsk, Belarus
E-mail: fer@open.by

The current state of theoretical and experimental investigations of the parametric X-ray radiation (PXR) from relativistic charged particles in a crystal is briefly discussed. Various mechanisms of X-radiation in the crystal are analyzed from the common point of view, based on quantum electrodynamics in a medium. It is shown that PXR, diffraction radiation (DR) and coherent bremsstrahlung (CB) result in different contributions to the amplitude of the radiation process but because of their interference they can't be considered separately in the radiation intensity. The concept of high resolution parametric X-rays (HRPXR) is introduced and the universal radiation distribution is also considered. Possible applications of HRPXR are discussed as well.

1. What is parametric X-radiation?

It is well known nowadays that characteristics of electromagnetic processes in a medium and in vacuum are qualitatively different and even new radiation mechanisms can appear due to the coherent interaction of charged particles with atoms of the medium. But the first phenomenon of such type was observed in the experiment quite accidentally[1] and was rather surprising for most of theoreticians. It was Vavilov-Cherenkov optical radiation produced by a relativistic charge particle moving with a constant velocity in a medium. At present a clear theoretical explanation of this phenomenon is well known[2] and its applications take an important place for the detecting equipment in high-energy physics[3] . Later the series of the analogous phenomena were investigated both theoretically and experimentally in a lot of papers (see, for example,[4] and citation therein).

Let us remind that the Vavilov-Cherenkov radiation (VCR) is appeared when the charged particle moves in the medium with the velocity $\vec{v}$ exceeding the value of the phase velocity $c_{ph} = c/n$ of the light in the same medium. As the light velocity c in the vacuum has the most possible value for any movement, then the condition $v > c_{ph}$ can be fulfilled only for the media with the refraction index $n > 1$. If one takes into account the space and time dispersion $n(\vec{k}, \omega)$ of the real refraction index, the intervals for the permitted frequency ω and wave vector $\vec{k}$ for the VCR photons are defined by the inequality

$$1 - \frac{v}{c} \cos\theta n(\vec{k}, \omega) \leq 0, \tag{1}$$

with θ as the angle between $\vec{v}$ and $\vec{k}$.

As it was mentioned above, the investigations of VCR were referred to the optical and more soft range of the frequency of emitted photons. The reason was in the fact that in accordance with the very general consideration[5] the refraction index of any homogeneous medium for the X-ray wavelengths is less than unity: $n(\vec{k}, \omega) \simeq 1 - \frac{\omega_0^2}{\omega^2}; \quad \omega > \omega_0$, where ω_0 is the medium plasma frequency which belongs to the optical range. Therefore for a long time the Cherenkov radiation was seemed to be forbidden for the X-ray wavelengths.

But the physical picture can be qualitatively changed if the radiation process happens in a crystal characterized by the periodical space distribution of the electron density. It was well known that this circumstance leads to the diffraction of the external X-rays in a crystal[6] but for a long time it was not taken into account for the electromagnetic radiation produced from the charge particles inside the crystal. The essential transformation of the radiation process in a crystal was firstly considered theoretically in our work[7] for the general case of the dynamical diffraction[6] and almost at the same time in the papers[8,9] in the framework of the kinematical approximation which can be applied only for the thin crystals. Accordingly to[7] if the wave vector of the radiated photon satisfies to the condition of Bragg diffraction

$$(\vec{k} + \vec{h})^2 \approx k^2; \quad 2\vec{k}\vec{h} + \vec{h}^2 \approx 0, \tag{2}$$

with one of the reciprocal lattice vector of the crystal, the effective refraction index can become more than unity. Therefore the Cherenkov radiation channel happens to be opened for that $\vec{k}, \omega$ if the both conditions (1) and (2) are fulfilled together. This new mechanism was firstly called by us as the parametric X-radiation[10] because it was possible due to the new parameter of the crystal (lattice period) in comparison with a homogeneous media. This phenomena can be also considered as the X-ray analog of the well known optical effect called as parametric conversion of the one photon to the coherent superposition of two photons with another wave vectors.

For a long time this theoretical prediction was not confirmed experimentally in spite of its fundamental description[11,12] and clear analysis of its qualitative features[13,14]. But the special experiments were initiated by us and fulfilled due to the collaboration between Belarusian and Tomsk Universities for the first experimental observation of PXR[15,16]. At present there are several strong experimental groups in the world which investigate various characteristics of PXR (see, for example[17–19] and citation therein) and the qualitative features of this phenomenon is discussed in the special sections of several International conferences (for example[20]). It was remarkable, that all details of the PXR spectral and angular distributions, theoretically predicted in our papers, were confirmed experimentally with rather high accuracy[21].

But in the very last decade a series of problems connected with the X-ray production from the relativistic charge particles (for definiteness, electrons) in a crystal were discussed by many authors. These problems can be divided by the following groups: 1) the role of the dynamical diffraction effects in these processes; 2) what are the contributions of the various radiation mechanisms to the total intensity and 3) which applications of the X-ray radiation from the relativistic electrons in a crystal could be most essential?

This interest is conditioned by some qualitatively advanced experimental works has been appeared recently (for example,[22]) which can open the essentially new side in the PXR investigations and applications. The two crystal scheme of the detector was used for very high energy resolution of the emitted photon spectrum. In some sense this is analogous to the essential extension of the applications of the conventional diffractometry due to high resolution X-ray diffraction (HRXRD) which is now the very important part of the modern semiconductor and nanostructure technology[23]. Therefore it becomes actual to discuss in more detail the general approach for the description of the dynamical effects in the formation of X-ray spectra from the charged particles in crystals. The fine structure of such spectra can be considered as the high resolution parametric X-radiation (HRPXR) and it should be represented in some universal form as it was made for the kinematical PXR[14]. It will help to analyze the experimental data and consider the possible applications of HRPXR.

2. Quantum electrodynamics in a crystal

We'll calculate the cross-sections of various electromagnetic processes in a crystal on the basis of the special representation of the quantum electrodynamics (QED)[12]. The system of units is $\hbar = c = 1$, and the fine structure constant $e^2 \simeq 1/137$. Let us also select the ranges of the electron energy E and the photon frequency ω which are the most essential for the considered problem $E \geq 50MeV$; $\omega \sim 2-20keV$. Accordingly to[13], the electron energy in this range is optimal for the X-ray production in a crystal.

It is clear that the recoil effects are not important in the radiation processes with such ratio of the photon and electron energies and the classical formulas are also applicable. But we prefer to use the QED approach because: i) actually the quantum calculations are not more complicated than classical but remain correct for any energies, and ii) it permits one to make more clear the classification of various radiation processes as it will be shown below.

The quantum description of the electromagnetic processes in a crystal is based on the following representation of the Hamiltonian of the system

$$\hat{H} = \hat{H}_{e,c} + \hat{H}_{\gamma,c} + \hat{H}_{e,\gamma}, \tag{3}$$

which is differed from the vacuum QED[24], because the Hamiltonians for the electron and photon subsystems include both the operators $\hat{H}_e, \hat{H}_\gamma$ of the free fields and the coherent potentials of their interaction with a crystal. Therefore, the second quantization of these fields is based not on the vacuum plane waves but on the one-particle eigenfunctions of the operators $\hat{H}_{e,c}$, $\hat{H}_{\gamma,c}$.

Particularly, the wave functions $\Psi^{\pm}_{E}(\vec{r})$ for the electron states with the energy E and other quantum numbers ν are defined as the solutions of the Dirac equation

$$[(\vec{\alpha}\vec{\hat{p}}) + \beta m + U(\vec{r}) - E]\Psi^{\pm}_{\nu E}(\vec{r}) = 0. \tag{4}$$

Here α_i, β are the Dirac's matrices; $\vec{\hat{p}} = -i\vec{\nabla}$ is the momentum operator; index $(\pm)$ corresponds to the wave functions including asymptotically outgoing (ingoing) spherical waves which are used for the electrons in the initial (final) states[11,12]. Potential $U(\vec{r})$ describes the coherent interaction of the electron with the periodical field of a crystal and can be expressed through the amplitudes of its elastic scattering at the atom of the crystal unit cell[23]:

$$U(\vec{r}) = \sum_{\vec{h}} U_{\vec{h}} e^{i\vec{h}\vec{r}} H(z)H(L-z);$$

$$U_{\vec{h}} = -\frac{4\pi e^2}{h^2\Omega}\sum_j [Z_j - F_j(\vec{h})]\, e^{i\vec{h}\vec{R}_j} e^{-W_j(\vec{h})}, \tag{5}$$

where $\vec{h}$ is the reciprocal lattice vector; $\vec{R}_j$ is the coordinate of the atom with index j in the cell; $Z_j, F_j(\vec{h})$ are its nucleus charge and atomic scattering factor correspondingly; $e^{-W_j(\vec{h})}$ is the Debye-Waller factor and Ω is the volume of the unit cell.

It is important to note that the full set of the states for QED in the media should be constructed for the finite thickness L of the crystal which we took into account by means of the Heaviside functions $H(z)$ and $H(L-z)$ in the formula (3).

In its turn, the stationary states of the electromagnetic field with the frequency ω and quantum numbers μ in the crystal are defined by the vector potentials $\vec{A}^{\pm}_{\mu\omega}(\vec{r})$ which are the solutions of the Maxwell equations with the periodical permittivity of the crystal[6]:

$$\epsilon_{ij}(\vec{r},\omega) = \delta_{ij} + \chi_{ij}(\vec{r},\omega)H(z)H(L-z). \tag{6}$$

The susceptibility $\chi_{ij}(\vec{r},\omega)$ is well known for the X-ray wavelengths[26]:

$$\chi_{ij}(\vec{r},\omega) = \delta_{ij}\sum_{\vec{h}} \chi_{\vec{h}} e^{i\vec{h}\vec{r}};$$

$$\chi_{\vec{h}} = -\frac{4\pi e^2}{m\omega^2\Omega}\sum_j [F_j(\vec{h}) + f'(\omega) + if''(\omega)] e^{i\vec{h}\vec{R}_j} e^{-W_j(\vec{h})}, \tag{7}$$

with the f' and f'' as the anomalous dispersion corrections.

The Hamiltonian $\hat{H}_{e,\gamma}$ for the interaction between the electron and electromagnetic fields in a crystal has the same form as in the vacuum QED[24]

$$\hat{H}_{e,\gamma} = e\int d\vec{r}\hat{\Psi}^*(\vec{r})(\vec{\alpha}\hat{\vec{A}}(\vec{r}))\hat{\Psi}(\vec{r}), \tag{8}$$

but with the field operators which are represented as the series in the annihilation and creation operators corresponding to the states

$\Psi^{\pm}_{\nu E}(\vec{r}), \vec{A}^{\pm}_{\mu\omega}(\vec{r})$ instead of the plane waves[12]:

$$\hat{\Psi}(\vec{r}) = \sum_{\nu}[b_\nu \Psi^{-}_{\nu E}(\vec{r}) + b^{+}_{\nu}\Psi^{+}_{\nu E}]; \hat{\vec{A}}(\vec{r})) = \sum_{\mu}[a_\mu \vec{A}^{-}_{\mu\omega}(\vec{r}) + a^{+}_{\mu}\vec{A}^{+}_{\mu\omega}]. \quad (9)$$

Analogously to the vacuum QED, the amplitude of any electromagnetic process in the crystal, homogenous media or in some external field can be calculated by means of the perturbation theory on the operator $\hat{H}_{e,\gamma}$. It leads to the same form of the diagrammatic representation for these amplitudes but with some change of the physical sense of the lines on the diagrams: instead of thin lines corresponding to the plane waves in the vacuum QED, one should use the thick lines which define the wave functions of the particle or the photon taking into account in some non-perturbative way their interaction with the media. Each diagram uses the same vertex as for the vacuum QED. If one not considers the many-photon processes, the Feynman diagrams of the first order are most essential and actually it is enough in order to calculate the cross-section of any known electromagnetic process. Some examples of the diagrams are shown in the Fig. 1. In particular, the diagram a) corresponds to the vacuum QED and this process, evidently, is forbidden because of the conservation laws of the momentum and energy[24]. But for all other diagrams the conservation law for the momentum doesn't fulfil because of the external potential and they will describe the amplitudes of the real physical processes: b) defines the Cherenkov radiation in the homogeneous media with the constant refraction index n; c) corresponds to the synchrotron radiation from the electron in the magnetic field $\vec{H}$; d) can be used for calculation of the PXR amplitude; e) describes the amplitude as the coherent bremsstrahlung as the radiation from the channeled particles and f) corresponds to the most general case of the radiation in a crystal.

The rules for identification of the diagram with the analytical expression for the amplitude of the concrete process are the same as in vacuum QED except the momentum conservation. As for example, the analytical expression for the amplitude corresponding to the diagram Fig. 1 f) is the following[12]

$$M_{fi} = \delta(E - E_1 - \omega)T_{fi}; \quad T_{fi} = \int d\vec{r}(\Psi^{-}_{\nu_1 E_1})^*(\vec{\alpha}\vec{A}^{-}_{\mu\omega})^*(\Psi^{+}_{\nu E}). \quad (10)$$

The considered sketch for the QED in media is sufficient to make clear the technique of the calculations for the further sections.

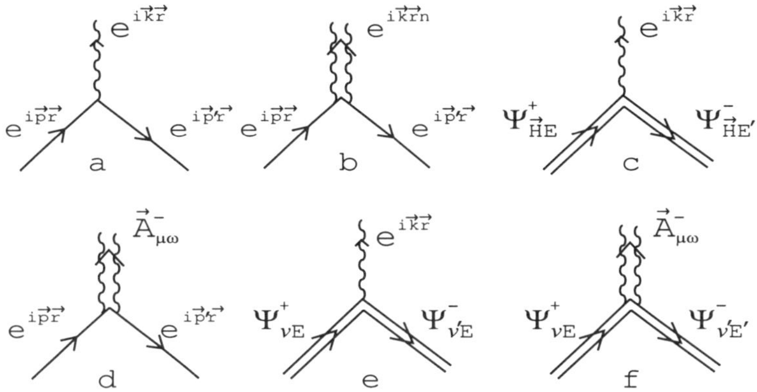

Figure 1. Feynmann diagrams for the quantum electrodynamics in media

3. Analytical expressions for the wave function and vector potential

Certainly, the equation (8) is not the full solution of the considered problem because one should find the appropriate expression for the electron and photon eigenstates. In general case one should use some approximation for the solution of the corresponding equations but it is important that these equations are homogeneous in the considered quantum approach[11,12] as distinct from the classical calculation[8,9].

Let us find firstly the electron wave function as the solution of the equation (2). Equation (2) can be "squared" in order to exclude the most of the spinor operators[24] and with an accuracy of $(U/E)^2$ it leads to

$$[-\Delta + m^2 - E^2 + 2EU(\vec{r}) - \ i(\vec{\alpha}\vec{\nabla})U(\vec{r})]\Psi_E(\vec{r}) = 0. \tag{11}$$

In the first stage we should find the stationary states in the infinite crystal and can use for this calculation the perturbation theory (PT) because of the condition (9)

$$\Psi_E(\vec{r}) \equiv \Psi_0 + \sum_{\vec{h}\neq 0} \Psi_{\vec{h}} \approx e^{i\vec{p}\vec{r}}[1 + \sum_{\vec{h}\neq 0} \hat{B}_{\vec{h}} e^{i\vec{h}\vec{r}}]\chi_E;$$

$$\hat{B}_{\vec{h}} = -\frac{2E + (\vec{\alpha}\vec{h})}{2(\vec{p}\vec{h}) + h^2} U_{\vec{h}}; p_z = \sqrt{E^2 - m^2 - p_\perp^2}, \tag{12}$$

with χ_E as the bispinor corresponding to the free electron[24].

It is well known for this problem that PT is unapplicable for the special geometry when the particle momentum is perpendicular to some set of the reciprocal lattice vector $(\vec{p}\vec{h}) \approx 0$. This case corresponds to the electron channeling and the nonperturbative method should be used for the solution of the equation (2) taking into account the zone spectrum for the transversal movement[11]. But the special case of the channeling is not essential for our consideration and we will suppose further that $(\vec{p}\vec{h}) \neq 0$ and the approximation (11) can be simplified because of the condition $2(\vec{p}\vec{h}) \gg h^2$.

Next step is to use the continuity conditions for the wave function on the crystal boundaries $z = 0$, $z = L$. The calculation leads to the following results

$$\Psi_E^{(\pm)} \approx e^{i\vec{p}\vec{r}}[1 + \varphi_E^{(\pm)}(\vec{r})]\chi_E;$$

$$\varphi_E^{(+)} = \sum_{\vec{h}\neq 0} \hat{B}_{\vec{h}} e^{i\vec{h}\vec{r}}\{C_h(z)H(z)H(L-z) + C_h(L)H(z-L)\};$$

$$C_h(z) = 1 - \exp[i(\tilde{p}_h - p_z - h_z)z]; \quad \tilde{p}_h = \sqrt{E^2 - m^2 - (\vec{p}_\perp - \vec{h}_\perp)^2};$$

$$\varphi_E^{(-)} = \sum_{\vec{h}\neq 0} \hat{B}_{-\vec{h}} e^{-i\vec{h}\vec{r}}\{C_{-h}(z-L)H(z)H(L-z) + C_{-h}(-L)H(-z)\} \quad (13)$$

It should be noted that the non-perturbed plane wave includes to the functions $\Psi_E^{(\pm)}$ from the left and right side of the crystal correspondingly.

Analogous problem for the vector potential $\vec{A}^{\pm}_{\mu\omega}(\vec{r})$ should be solved taking into account the diffraction of the emitted photon and the standard boundary conditions for the field on the interfaces of the crystal. In the case of the function $\vec{A}^{+}_{\mu\omega}(\vec{r})$ it is equivalent to the conventional X-ray diffraction problem[23]. The function $\vec{A}^{-}_{\mu\omega}(\vec{r})$ with another asymptotic can be calculated by means of the relation[12] $\vec{A}^{-}_{\vec{k}\omega}(\vec{r}) = [\vec{A}^{+}_{-\vec{k}\omega}(\vec{r})]^*$, which is the analog of the well known "reciprocity theorem". So, we can refer to the results of the dynamical diffraction theory[23] without discussion of the details. As for example and in order to introduce the necessary notations, we write here the expression only for $\vec{A}^{+}_{\mu\omega}(\vec{r})$ calculated in the two-beam approximation and for the Bragg boundary conditions[23], corresponding to the experiment[2]. Figure 2 shows the distribution of the primary $\vec{A}_{\vec{k}}$ and diffraction $\vec{A}_{\vec{k}}$ wave fields and their normalized analytical representations are:

$$\vec{A}^{(+)s}_{\vec{k}\omega}(\vec{r}) \equiv \vec{A}_{\vec{k}} + \vec{A}_{\vec{h}} = \sqrt{\frac{2\pi}{\omega}} e^{i\vec{k}\vec{r}}\{(\vec{e}_s + \vec{e}_{sh}e^{i\vec{h}\vec{r}}D_{sh}(L))H(-z) + (\vec{e}_s D_{s0}(L-z) + \vec{e}_{sh}e^{i\vec{h}\vec{r}}D_{sh}(L-z))H(z)H(L-z) + \vec{e}_s D_{s0}(L) \times H(z-L)\};$$

$$D_{s0}(z) = -\sum_{\mu=1}^{2} \gamma^0_{\mu s} \exp[i\omega\epsilon_{\mu s}\frac{L}{\gamma_0}];\ D_{sh}(z) = \beta\sum_{\mu=1}^{2} \gamma^h_{\mu s} \exp[i\omega\epsilon_{\mu s}\frac{L}{\gamma_0}]. \quad (14)$$

We use here the standard notations from the dynamical diffraction theory[23]: $\vec{e}_s, \vec{e}_{sh}$; $s = 1, 2$ are the polarization vectors for the primary and diffraction waves; $\gamma_0 = (\vec{k}\vec{N})/k$ with $\vec{N}$ as the unit vector of the normal to the crystal surface; $\beta = \gamma_0/\gamma_h$; $\gamma_h = (\vec{k}_h\vec{N})/k_h$; $\vec{k}_h = \vec{k} + \vec{h}$ (Fig. 2). The solutions of the dispersion equation which defines the refraction of the waves in a crystal are expressed by means of the components of the susceptibility (5) and has the following form[23]:

$$\epsilon_{\mu s} = \frac{1}{4}\{-\alpha_B + \chi_0(\beta+1) \pm \sqrt{[-\alpha_B + \chi_0(\beta-1)]^2 + 4\beta C_s^2\chi_h\chi_{-h}}\};$$

$$C_s = \cos 2\theta_B;\ \sin\theta_B = \frac{h}{2k};\ \alpha_B = \frac{2\vec{k}\vec{h} + h^2}{k^2}. \quad (15)$$

The dimensionless value α_B is very important for the diffraction theory, in our case it characterizes the deviation of the wave vector of the emitted photon from the exact Bragg condition corresponding to $\alpha_B = 0$. Additional parameters in Eq. (15) were appeared because of the boundary condition

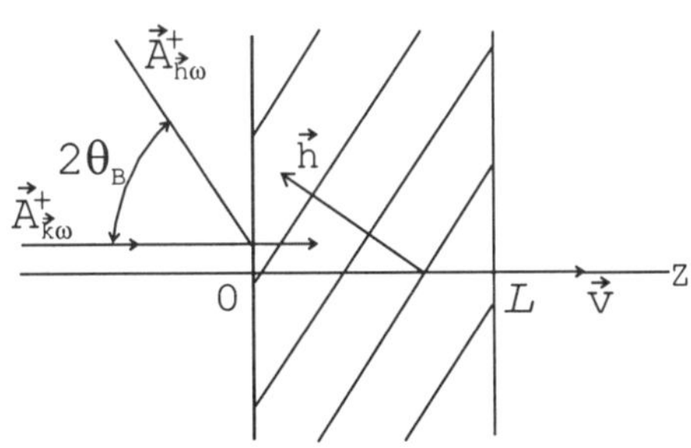

Figure 2. Directions of the wave propagations for the diffraction in Bragg geometry

and in the considered Bragg case they are defined as follows:

$$\gamma^{0}_{1(2)s} = \frac{2\epsilon_{2(1)s} - \chi_0}{2\epsilon_{(2(1)s} - \chi_0) - 2(\epsilon_{1(2)s} - \chi_0)\exp[i\omega(\epsilon_{2(1)s} - \epsilon_{1(2)s})L/\gamma_0]};$$

$$\gamma^{h}_{1(2)s} = \frac{-\beta C_s \chi_h}{2\epsilon_{(2(1)s} - \chi_0) - 2(\epsilon_{1(2)s} - \chi_0)\exp[i\omega(\epsilon_{2(1)s} - \epsilon_{1(2)s})L/\gamma_0]}. \quad (16)$$

The formulas (11)–(17) for eigenfunctions look a little bit inconveniently because of the different representations inside and outside of the crystal. But we stress that all these parts of the functions should be taken into account in calculating of the matrix element (8) in order to describe all essential radiation processes (see below Sec. 4).

4. Intensity of the PXR+CBS radiation

The number of photons emitted within the spectral interval $d\omega$ and with the angular spread $d\Omega$ from one electron in the crystal is expressed through the matrix element (8) by means of the standard rules of QED[24]

$$\frac{\partial^2 N}{\partial\omega\partial\Omega} = \omega^2 \int |T_{fi}|^2 \delta(E - E_1 - \omega)\frac{d\vec{p}_1}{(2\pi)^6}, \quad (17)$$

where the integration over the $E_1, \vec{p}_1$ as the energy and momentum of the electron in the final state is supposed.

As it was qualitatively illustrated in our papers[12,14] the angular distribution from the relativistic electron is separated actually to the narrow peak with the angular width $\Delta\theta \sim m/E$ along the electron velocity and to the set of non-overlapping peaks (reflexes) with the analogous angular widths along the directions corresponding to the various reciprocal lattice vectors $\vec{h}$ and each of these reflexes can be analyzed independently. Fig. 3 shows those three diagrams which give the contribution to one of such side reflex. Fig. 3 a) corresponds to the diffraction of the emitted photon that is to the amplitude of PXR. b) and c) describe the amplitudes of CBS conditioned by the diffraction of the electron in the initial or the final states correspondingly. This representation makes more clear the physical difference between the processes: in both cases there is quite a big momentum transfer when producing the photon with the wave vector $(\vec{k}+\vec{h})$, but in the PXR case the recoil is received by the crystal during the photon diffraction and in the CBS process it is compensated by the change of the electron momentum due to the diffraction by the crystallographic planes with the reciprocal lattice vector $(-\vec{h})$. In is interesting to note that the electron diffraction itself can not be resolved for the considered relativistic energy E (all of them are in the very small cone along the electron velocity) but one can see its secondary effect as the CBS peaks. It becomes possible because the small scattering angle for electron leads to the large angle for the emitted photons due to the condition $E \gg m$.

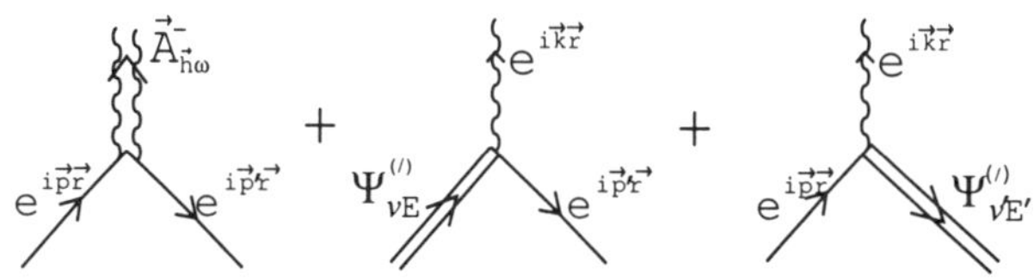

Figure 3. Diagrams for the (CBS + PXR) radiation

The wave functions corresponding on the lines of the diagrams from Fig. 3 for the considered Bragg geometry are defined by Eqs. (11), (15). Not very complicated calculation of the matrix elements with the average over the initial electron spin states and the summation over its final ones[24] lead to the following expression (the intensities of the photons with different polarization $s = 1, 2$ can be considered separately):

$$\frac{\partial^2 N_{hs}}{\partial\omega\partial\Omega} = \frac{e^2\omega}{4\pi^2}(\vec{e}_{hs}\vec{v}_h)^2|M_{PXR} + M_{CBS}|^2;$$

$$M_{PXR} = \sum_{\mu=1}^{2}\gamma_{\mu s}^{h}\left(\frac{1}{q_{0h}} - \frac{1}{q_{\mu hs}}\right)[\exp(iq_{\mu hs}L/\gamma_1) - 1];$$

$$M_{CBS} = -\frac{U_{-\vec{h}}}{(\vec{v}\vec{h})}\frac{[\exp(iq_{0h}L/\gamma_1) - 1]}{q_{0h}};$$

$$\vec{v}_h = (\vec{v} + \frac{\vec{h}}{\omega}); \; q_{0h} = [\omega - (\vec{v}\vec{k}_{-h})]/v; \; \vec{k}_{-h} = \vec{k} - \vec{h};$$

$$q_{\mu hs} = [\omega - (\vec{v}\vec{k}_{-h\mu s})]/v; \; \vec{k}_{-h\mu s} = \vec{k} - \vec{h} + \omega\epsilon_{\mu s}\vec{N}/\gamma_0. \qquad (18)$$

Only the leading terms in a power series of the parameters U/E and h/E are taken into account in M_{CBS}. The values $L_{0h} = 1/q_{0h}$ and $L_{\mu hs} = 1/q_{\mu hs}$ are known as the radiation coherent lengths for the vacuum and crystal correspondingly. The PXR matrix element actually includes the coherent superposition of both coherent lengths. It was suggested in some papers (for example,[22]) to consider the contribution, which is proportional to L_{0h} and conditioned by the crystal boundary, as a special radiation mechanism (diffraction transition radiation). But it seems to us that such separation is not very rigorous in the radiation intensity and we consider further M_{PXR}

as the single amplitude of the process conditioned by the diffraction of the emitted photons in a crystal.

One can see from the formula (19) that the total radiation intensity is defined by the interference of the PXR and CBS amplitudes. This phenomenon was firstly discussed theoretically in the paper[25] and in more details together with the experimental results was considered in[26]. It was also shown in our paper[27] that the PXR and CBS interference is the most essential for the nonrelativistic electrons when the angular distribution for both radiation mechanisms are undistinguishable. In the case of ultrarelativistic electrons considered in this paper the relative contributions of the both intensities can be estimated by the following way. The amplitude M_{PXR} is the most essential in the spectral $(\Delta\omega/\omega)$ and angular $(\Delta\theta)$ intervals where the wave vector of the emitted photon satisfies to the Bragg condition. In this case the parameters in the formulas (16), (17) are estimated as $\alpha_B \simeq \Delta\omega/\omega \simeq |\chi_{\vec{h}}|$, $|\gamma^h_{\mu s}| \simeq 1$.

In this narrow range the PXR contribution to the angular-spectral distribution has an order of magnitude

$$(\frac{\partial^2 N}{\partial\omega\partial\Omega})_{PXR} \approx \frac{e^2}{4\pi}(\vec{e}_{hs}\vec{v}_h)^2\omega L^2_{0h}. \tag{19}$$

At the same time its contribution to the integral intensity of the reflex is defined as follows[14]:

$$N_{PXR} \approx \frac{e^2}{4\pi}\frac{|\chi_h|^2}{\sin^2\theta_B}\omega_B L_a[1-\exp(-L/L_a], \tag{20}$$

where L_a is the crystal absorption length for the X-rays with the frequency $\omega_B = h/\sin\theta_B$, corresponding to the Bragg frequency in the PXR reflex[14].

The analogous estimations for the CBS contribution leads to

$$(\frac{\partial^2 N}{\partial\omega\partial\Omega})_{CBS} \approx \frac{e^2}{4\pi}(\vec{e}_{hs}\vec{v}_h)^2\omega L^2_{0h}|\frac{U_{\vec{h}}}{h}|^2; \quad |\frac{U_{\vec{h}}}{h}| \leq 10^{-2};$$

$$N_{CBS} \approx \frac{e^2}{4\pi}\frac{m^2}{E^2}|\frac{U_{\vec{h}}}{h}|^2\omega L_a[1-\exp(-L/L_a]. \tag{21}$$

It means that the maximal spectral intensity of PXR is a several orders larger than the analogous one for CBS, but CBS can give noticeable contribution to the integral intensity of the reflex. If one takes into account Eqs. (3) and (5) for the components of the potential and susceptibility the ratio of the CBS and PXR intensities in the same reflex can be calculated

$$\xi = \frac{N_{CBS}}{N_{PXR}} \approx [\frac{Z - F(\vec{h})}{F(\vec{h})}]^2(\frac{m^2}{Eh})^2\frac{1}{16\sin^4\theta_B}. \tag{22}$$

In other words, the CBS contribution should be taken into account in the following electron energy range

$$E \leq |\frac{Z - F(\vec{h})}{F(\vec{h})}|(\frac{m^2}{4h\sin^2\theta_B}). \tag{23}$$

This estimation is in the qualitativ agreement with the results of the paper[26]. It also shows that in the considered in this work energy range ($E \geq 50MeV$) this radiation mechanism is not very important except the case of the high-order harmonics when the atomic scattering factor $F(\vec{h})$ can be exponentially small. Fig. 4 shows the contribution of CBS to the integral intensity of the reflex calculated by numerical integration in the formula (19) together and without M_{CBS} for the experimental conditions corresponding to[22]. The intensity of the reflex in the Bragg backward geometry was calculated as the function of the angle ψ between the electron velocity $\vec{v}$ and the crystal surface normal $\vec{N}$ in the vicinity of $\psi = 0$. One can see that the effect of CBS is almost negligible for the electron energy $E = 85MeV$ but becomes important for $E = 30MeV$.

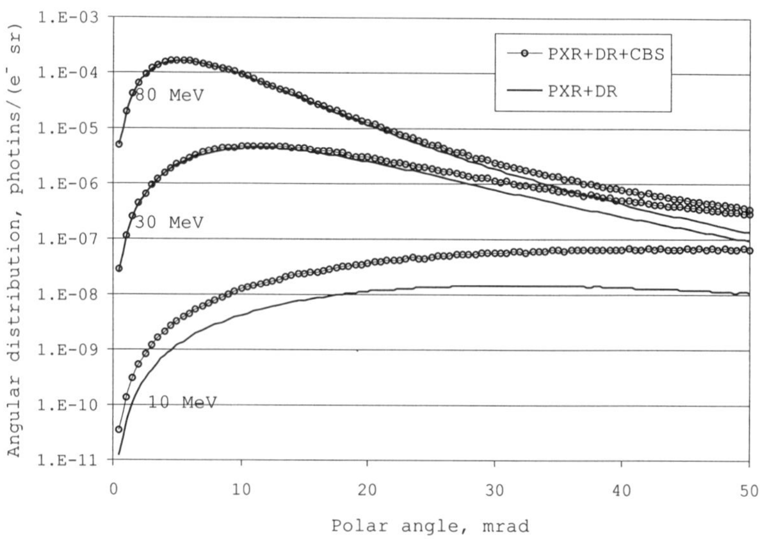

Figure 4. Contribution of CBS to the radiation intensity in dependence on the electron energy

5. Analysis of the dynamical diffraction effects in the case of HRPXR

In this section we will analyze the general formula (19) in the narrow angular-spectral range where the dynamical effects could be appeared. As it was mentioned above the CBS matrix element can be omitted in this case. Let us also introduce for the PXR reflex ($\vec{h}$) the special coordinate system where the frequency ω and the angles θ, φ of the emitted photons are counted relatively to the vector $\vec{k}_B$ when the Bragg condition ($\alpha_B = 0$) is satisfied exactly. In these variables the photon wave vector $\vec{k}$ with an accuracy of $|\chi_{\vec{h}}| \preceq 10^{-5}$ can be written as[14]:

$$\vec{k} = \vec{k}_B + \delta\vec{k}; \; \delta\vec{k} = \delta\omega\vec{n}_B + \omega_B\theta(\vec{n}_x\cos\varphi + \vec{n}_y\sin\varphi);$$
$$\delta\omega = \omega - \omega_B; \; \vec{k}_B = \omega_B\vec{v} + \vec{h}; \; \omega_B = \frac{h}{2v|\sin\theta_B|};$$
$$\vec{n}_B = \frac{\vec{k}_B}{k_B}; \; \theta_B = \frac{1}{2}\widehat{\vec{v}\vec{n}_B}, \tag{24}$$

and the unit vectors $\vec{n}_x \parallel [\vec{v}\vec{h}]$, $\vec{n}_y \parallel [\vec{n}_x\vec{n}_B]$ are in the plane perpendicular to $\vec{n}_B$.

One can express the most essential values in Eq. (19) by means of these variables with the same accuracy:

$$\alpha_B = -2\frac{\delta\omega}{\omega_B}\sin\theta_B - 2\theta\sin\varphi\sin 2\theta_B; \; \gamma_0 = (\vec{v}\vec{N}); \; \gamma_1 = (\vec{n}_B\vec{N});$$
$$q_{0h} = \frac{1}{2}\omega_B[\theta^2 + \theta_{ph}^2]; \; q_{\mu hs} = \frac{1}{2}\omega_B[\theta^2 + \theta_{ph}^2 - 2\epsilon_{\mu s}];$$
$$(\vec{e}_{hs})\vec{v}_h)^2 = \theta^2\nu_s; \; \nu_{1(2)} = \sin^2\varphi(\cos^2\varphi); \quad \theta_{ph}^2 = \frac{m^2}{E^2} + \theta_{sc}^2 + \theta_M^2. \tag{25}$$

Analogously to the model[14] we have introduced here the value θ_{ph}^2 in order to take into account qualitatively the influence of the electron multiple scattering (θ_{sc}^2) and the crystal mosaicity (θ_M^2) on the formation of the PXR reflex.

Equation (19) can be used for any crystal thickness L. Particularly, if $L < L_{ext} \simeq (\omega_B\chi_0)^{-1}$ (L_{ext} is the extinction length), these formula takes the same form as in the kinematical diffraction theory which was analyzed earlier in our paper[14]. Therefore in this paper we consider the opposite case $L > L_{ext}$ when the dynamical effects is supposed to be essential for HRPXR analogously to the conventional high resolution X-ray diffraction (HXRXD)[23]. Moreover, in order to make more clear the specific features of the HRPXR peaks let us consider the case of the thick crystal with $L > L_{abs} \simeq (2\omega_B \Im\chi_0)^{-1}$ (L_{abs} is the absorption length) when the oscillation structure of the diffraction peaks is not appeared[23].

With this supposition the only root with $\Im\epsilon_{\mu s} > 0$ of the equation (16) is essential and Eq. (19) is essentially simplified

$$\frac{\partial^2 N_{hs}}{\partial\omega\partial\Omega} = \frac{e^2}{\omega_B \pi^2} \left|\frac{\beta C_s \nu_s \chi_{\vec{h}}}{2\epsilon_{1(2)} - \chi_0}\right|^2 \left|\frac{1}{\theta^2 + \theta_{ph}^2} - \frac{1}{\theta^2 + \theta_{ph}^2 - 2\epsilon_{1(2)s}}\right|^2, \quad (26)$$

where the index 1(2) for various ω, θ should be chosen in dependence on the sign of $\Im\epsilon_{\mu s}$. We also stress that the analytical integration on the azimuth angle φ can not be made in general case because the value α_B depends on this variable.

This formula shows that there are two essentially different angular and spectral scales for characterization the structure of the PXR reflex. One of them (low resolution scale — LRS) is defined by the angular dependence of the radiation coherent length:

$$(\delta\theta)^2 \preceq |\chi_{\vec{h}}|; \; (\delta\theta)_{LRS} \preceq \sqrt{|\chi_{\vec{h}}|} \sim 10^{-2} \; - \; 10^{-3}. \quad (27)$$

In the scope of LRS one can integrate Eq. (27) over one of the variable (θ or *omega*) and find the universal forms of distributions have been considered earlier in[14] and have been investigated in many experimental. Actually these distributions are not dependent on the dynamical effects and their amplitudes are defined by the value L_{abs}. Analogous picture takes place in the conventional low resolution diffraction when the intensity of the reflex is also proportional to L_{abs}[23].

But there is also high resolution scale (HRS) in Eq. (27) which is defined by the Bragg condition for the emitted photons, that is

$$|\alpha_B| \preceq |\chi_{\vec{h}}|; \; (\delta\theta)_{HRS} \simeq \frac{\delta\omega}{\omega_B} \preceq |\chi_{\vec{h}}| \sim 10^{-5} \; - \; 10^{-6}. \quad (28)$$

Certainly, the experimental investigation of such thin structure of the PXR reflex should be connected with the two-crystal detector technique which is widely used in modern HRXRD but was firstly applied recently[22] for HRPXR analysis.

Therefore it is actual to consider the most essential features of the distribution (27). It proves that analogously to kinematical PXR theory[14] and the conventional dynamical diffraction theory[23] this distribution can be represented in some universal form. Let us introduce new variables having the scale of the order of unity:

$$\eta_s = \frac{-\beta\alpha_B + \chi_0(\beta - 1)}{\varkappa_s}; \; x_s = \frac{\theta^2 + \theta_{ph}^2}{\varkappa_s}; \; \varkappa_s = 2C_s\sqrt{\beta\chi_{\vec{h}}\chi_{-\vec{h}}}. \quad (29)$$

In this variables the distribution of the PXR photons can be represented as

$$\frac{\partial^3 N_{hs}}{\partial \eta_s \partial x_s \partial \varphi} = \frac{e^2 \nu_s^2}{4\pi^2 \beta \sin\theta_B} I(\eta_s, x_s), \tag{30}$$

with the a little bit different scales but the same universal function $I(\eta, x)$ for both polarizations:

$$I = \frac{x - u_{ph}}{x^2 |\eta + sign(\eta)\sqrt{\eta^2 - 1}|^2} |1 - \frac{x}{x - \eta - sign(\eta)\sqrt{\eta^2 - 1} - u_0}|^2;$$

$$u_{ph} = \frac{\theta_{ph}^2}{\varkappa_s};\ u_0 = \frac{\chi_0}{\varkappa_s};\ x > u_{ph};\ -\infty < \eta < \infty. \tag{31}$$

Here the function $sign(\eta)$ is (+1) for $\eta > 0$ and (-1) for $\eta < 0$. Analogously to HRXRD the scans of the distribution on the dimensionless variables for HRPXR can be realized by various experimental scans both on angles and frequency.

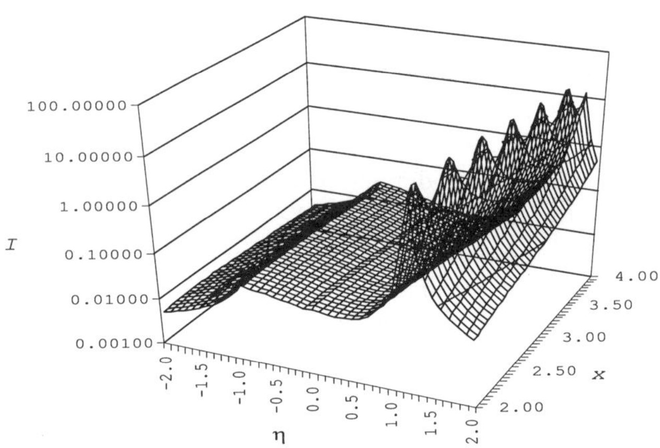

Figure 5. Universal function for the HRPXR spectral-angular distribution (u_{ph} = 1.5; $u_0 = 1.0 - i\ 0.1$)

Figure 5 shows the characteristic form of the universal function $I(\eta, x)$. One can see that it includes the well known Darvin step[23] conditioned by the second factor in (32) and more narrow Cherenkov peak due to the last term in (32). The ratio of amplitudes of these peaks depend on the values x, u_{ph}, u_0.

It is easy to estimate by means of Eq. (32) the maximal dimensionless spectral intensity for PXR:

$$(\frac{\partial^3 N_{hs}}{\partial \eta_s \partial x_s \partial \varphi})_{max} \simeq \frac{e^2 \nu_s^2}{4\pi^2 \beta \sin\theta_B} Q^2;$$

$$Q = \frac{\Re \chi_0}{\Im(\chi_0 - \varkappa_s)} \sim 10 \ - \ 100, \tag{32}$$

which is achieved in the very narrow spectral range $\frac{\delta\omega}{\omega} \simeq \Im\chi_0 \sim 10^{-6} - 10^{-7}$.

6. Conclusions

In the paper we have investigated in details the role of the dynamical effects for high resolution parametric X-radiation. The general formulae for description of this phenomena and other radiation mechanisms in the same wavelength range are obtained. These expressions are also represented in a simplified analytical form which can make more clear the qualitative features of the considered effects and find the optimal conditions for their observation.

High spectral intensity of HRPXR could open new directions for application of such radiation for the crystal and nanostructure analysis. In particular, the estimation for the spectral intensity of the synchrotron radiation (SR) from one electron, analogous to (33), leads to the value $e^2/2\pi^{13}$, which can be essentially less than for HRPXR. Certainly, in the most cases the integral number of SR photons from the storage ring with a great electron current is much larger than the quantity of PXR quanta from linear accelerator with a quite small average current. But the applications of HRPXR as the X-ray source for structure analysis could be of interest for the spectral-sensitive experiments where the spectral density of the radiation in the narrow spectral range is necessary. Some of possible applications in this field were considered in the papers[28–31].

Bibliography

1. P. A. Cherenkov, *Doklady Akademy of Science USSR* **2**, 451 (1934); S. I. Vavilov, *Doklady Akademy of Science USSR* **2**, 457 (1934).
2. I. E. Tamm and I. M. Frank, *Doklady Akademy of Science USSR* **14**, 107 (1937).
3. J. V. Jelley, *Cherenkov Radiation and its Applications* (Pergamon Press, London), (1958).
4. V. L. Ginzburg, *Theoretical Physics and Astrophysics* (Nauka, Moscow), (1975).

5. I. D. Feranchuk, L. Gurskii, L. I. Komarov et al., *Acta Cryst. A* **58**, 370 (2002).
6. R. W. James, *The Optical Principles of the Diffraction of X-rays.* (London:G.Bell and Sons Ltd)(1960).
7. V. G. Baryshevsky and I. D. Feranchuk, *Sov.Phys.JETP* **34**, 502 (1972).
8. M. L. Ter-Mikaelyan, *High Energy Electromagnetic Processes in Condensed Media* (Wiley, NY)(1972).
9. G. M. Garibyan and C. Yang, *Sov.Phys.JETP* **34**, 495 (1972).
10. V. G. Baryshevsky and I. D. Feranchuk, *Phys. Lett. A* **34**, 502 (1976).
11. V. G. Baryshevsky, *Channeling, Radiation and Reactions in Crystals under the High Energies*, (Belarusian University, Minsk) (1982).
12. V. G. Baryshevsky and I. D. Feranchuk, *J. de Phys. (Paris)* **44**, 913 (1983) .
13. V. G. Baryshevsky and I. D. Feranchuk, *Nucl. Instr. Meth.* **228**, 490 (1985).
14. I. D. Feranchuk and A. V. Ivashin, *J. de Phys. (Paris)* **46**, 1981 (1985).
15. Yu. N. Adishchev, V. G. Baryshevsky, S. A. Vorob'ev et al., *Pis'ma Zh. Eksp. Teor. Fiz.* **41** 295 (1985).
16. V. Baryshevsky, V. Danilov, I. Feranchuk et al., *Phys. Lett. A* **110**, 477 (1985).
17. K. Brenzinger, C. Herberg, B. Limburg, et al., *Zh. Phys. A* **358**, 107 (1997).
18. J. Freudenberger, H. Genz, V. Morokhovskyi et al., *Phys. Rev. Lett.* **84**, 270 (2000).
19. T. Akimoto, M. Tamura, J. Ikeda et al., *Nucl. Instr. and Methods A* **459**, 78 (2001).
20. 5th Int. Symp. on Radiation from Relativistic Electrons in Periodic Structures, *Topical Issue of Nucl. Instr. Meth. B* **201**, 1 (2003) .
21. A. V. Schagin, *Radiation Physics and Chemistry* **61**, 283 (2001).
22. H. Backe, C. Ay, N. Clawiter et al., *Proc. Int. Symp. on Channeling - Bent Crystals - Radiation Processes*, (Frankfurt am Main), 41 (2003).
23. A. Authier, S. Lagomarsino, B. Tanner (Ed), *X-ray and Neutron Dynamical Diffraction, NATO ASI Seies*, (Plenum Press, New York) (1996).
24. A. Akhezer, V. Berestetzky, *Quantum Electrodynamics*, (Nauka, Moscow) (1969).
25. S. Blazhevich, G. Bochek, V. Gavrikov et al., *Phys. Lett. A* **195**, 210 (1994).
26. V. Morokhovsky, J.Freudenberger, H. Genz et al., *Phys. Rev. B* **61**, 3347 (2000).
27. I. Feranchuk, A. Ulyanenkov, J. Harada and J. Spence, *Phys. Rev. E* **62**, 4225 (2000).
28. I. D. Feranchuk, A. V. Ivashin and I. V. Polikarpov, *J. of Phys D: Appl. Phys.* **21**, 831 (1988).
29. V. G. Baryshevsky and I. D. Feranchuk, *Phys. Lett. A* **76**, 452 (1980).
30. V. G. Baryshevsky and I. D. Feranchuk, *Phys. Lett. A* **102**, 141 (1984).
31. I. D. Feranchuk and A. V. Ivashin, *Krystallographiya* **34**, 39 (1989).